ECHINODERM PALEOBIOLOGY

Life of the Past

James O. Farlow, editor

ECHINODERM PALEOBIOLOGY

Edited by William I. Ausich and Gary D. Webster

Indiana University Press

Bloomington & Indianapolis

This book is a publication of
Indiana University Press
601 North Morton Street
Bloomington, IN 47404-3797 USA

http://iupress.indiana.edu

Telephone orders: 800-842-6796
Fax orders: 812-855-7931
Orders by e-mail: iuporder@indiana.edu

Library of Congress Cataloging-in-Publication Data

Echinoderm paleobiology / edited by William I. Ausich and Gary D. Webster.
    p. cm. — (Life of the past)
  Includes bibliographical references and index.
  ISBN-13: 978-0-253-35128-9 (cloth)
  1. Echinodermata, Fossil. 2. Paleobiology. I. Ausich, William
I. (William Irl), date– II. Webster, G. D. (Gary D.), date–
  QE781.E26 2008
  563'.9—dc22
                                                    2007046932

1 2 3 4 5 13 12 11 10 09 08

CONTENTS

ECHINODERM PALEOBIOLOGY

INTRODUCTION

William I. Ausich and Gary D. Webster

To the dismay of paleontology undergraduate students, echinoderms possess a bewildering number of polygonal plates, all with names and oriented with A through E or AB through DE designations. This undergraduate burden is the delight of an echinoderm paleontologist. Because of this array of characters, echinoderms provide a wealth of data for phylogenetic and evolutionary paleoecologic studies. This situation is analogous in many ways to fossil vertebrates, and, like vertebrates, echinoderms are unique in two other important aspects. First, they are taphonomically volatile, most beginning to disarticulate within a few days after death. Complete fossil echinoderms must have been buried alive. Thus, bedding surfaces with complete fossil echinoderms have a minimum of time-averaging biases and a maximum of ecologic time information. These occurrences are snapshots of ancient seafloor communities. Second--and again, unlike almost all other invertebrates--the feeding apparatus of fossil pelmatozoans, arms and brachioles, are routinely preserved on complete echinoderm fossils. Thus, inferences about ancient feeding ecology can be made with confidence.

Collectively, these attributes of the echinoderm fossil record provide ideal and nearly unique data from which we may expect high-resolution answers to paleobiologic and evolutionary questions. Echinoderms are ideal to ask the sophisticated questions now being addressed with geohistorical data. A recent National Research Council (2005) report convincingly argued that a national research agenda is emerging to develop a more complete understanding of global environmental and biodiversity changes. In recent years, intense study of global change issues on short timescales (several thousand years) has been underway; however such short time frames do not allow for a full understanding of the natural range of variability in the climate system and in the biosphere. This report concludes,

> Only geohistorical data—the organic remains, biogeochemical signals, and associated sediments of the geological record—can provide a time perspective sufficiently long to establish the full range of natural variability of complex biological systems, and to discriminate natural perturbations in such systems from those included or magnified by humans. Such data are crucial for acquiring the necessary long-term states, both like and unlike those of the present day, and provide the empirical framework needed to discover the general principles of biosphere behavior necessary to predict future change and inform policy managers about the global environment. (National Research Council, 2005, p. 11)

Echinoderms, especially pelmatozoan echinoderms, were dominant faunal elements in many shallow marine settings during the Paleozoic. In fact crinoids help to define the Paleozoic fauna (Sepkoski, 1981). They were dominant both in terms of the abundance and diversity in many settings, and they were dominant organisms forming the complex, epifaunally tiered communities of the Paleozoic. Paleozoic marine ecosystems cannot be understood as needed without a full understanding of echinoderms. Similarly, echinoids may be dominant elements in post-Paleozoic communities (Chapters 6, 18) and some Paleozoic communities (Chapter 4). Echinoids can play an important role on reefs and can be dominant bioturbators. Further, asteroids are commonly keystone predators.

Most echinoderm studies during the twentieth century and earlier were devoted to the description of new crinoid faunas. This was essential to build a taxonomic, assemblage, and geographic database. Continuing to build this database remains a high priority. However, during the 1970s, the next phase of echinoderm study gradually began. By using established data, collecting new and different information, applying new techniques, and asking biologically relevant questions, the paleobiologic study of echinoderms was underway, and this field continues to expand rapidly.

Principal among those who began and encouraged the paleobiologic study of crinoids was N. Gary Lane (1930–2006), coauthor of Chapter 17. The authors dedicate this volume to our friend, mentor, and colleague, whose untimely death prevented him from seeing the publication of his chapter and this volume. Many aspects of modern echinoderm paleobiology began in the laboratory of Gary Lane, both at UCLA and at Indiana University. Many examples of his breakthrough works are cited throughout Echinoderm Paleobiology, and notable among these were field sampling techniques, feeding biology of Mississippian crinoids, community paleoecology, and internal anatomy of Paleozoic crinoids. Those who study fossil echinoderms as well as those who work on the paleobiology of fossil assemblages are indebted to the early work of Gary Lane and to his enthusiasm and encouragement of students and colleagues.

Echinoderm Paleobiology highlights modern study of fossil echinoderms, conveying both the data encoded in echinoderm fossils and many approaches that can be adapted to other fossil groups. Echinoderm Paleobiology is organized into five parts, including echinoderm paleoecology, functional morphology, and paleoecology; evolutionary paleoecology; morphology for refined phylogenetic studies; innovative new applications of data encoded in echinoderms; and development of new and improved crinoid data sets.

References

National Research Council 2005. The Geological Record of Ecological Dynamics: Understanding the Biotic Effects of Future Environmental Change. National Academies Press, Washington, D.C., 200 p.

Sepkoski, J. J., JR. 1981. A factor analysis description of the Phanerozoic marine fossil record. Paleobiology, 7:36–53.

PART 1. FUNCTIONAL MORPHOLOGY, PALEOECOLOGY, AND TAPHONOMY

INTRODUCTION TO PART 1

William I. Ausich and Gary D. Webster

The foundation of the paleobiology paradigm is a fundamental understanding of fossils as once living organisms. Relatively little marine ecology was known about some echinoderm groups before the paleobiology movement, so in many ways, the study of the biology and paleobiology of certain echinoderms advanced together. Understanding the ecology of living echinoderms helped our understanding of ancient organisms at the same time that issues concerning the function and paleoecology of fossils helped to frame questions that could be addressed among living echinoderms (e.g., Lane, 1963, 1973; Macurda and Meyer, 1974; Meyer, 1979; Ausich, 1980; Kammer and Ausich, 1987; Baumiller et al., 1991). Although these previous studies have provided some answers, they raised new questions that were in turn addressed in subsequent studies. This productive interplay between fossil and ancient echinoderms has been an integral and exciting aspect of echinoderm paleobiology. Echinoderm paleontologists continue to learn new information about the paleobiology of fossil echinoderms and in the process learn that what was once regarded as a truth may require revision.

Baumiller, Gahn, Hess, and Messing (Chapter 1) and Nebelsick (Chapter 6) consider behavior of living forms and directly apply this to understanding the preservation and postures of fossil echinoderms. It is generally assumed that pelmatozoans require attachment to the substratum in order to be a functional adult, although exceptions are known including attachment to floating objects, floats that can reattach, prehensile columns, and crinoids that lack a column as an adult. As argued by Guensburg and Sprinkle (1992), attachment was not only a necessary aspect for individuals, but in many ways, pelmatozoans and other epifaunal suspension feeders first became abundant during the Early Ordovician when widespread hardground habitats were common. The paleoecology of echinoderms was undoubtedly driven by many factors, but Brett, Deline, and McLaughlin (Chapter 2) demonstrate that this remained an important distributional factor among Late Ordovician crinoids.

Chapters 3 and 5 tackle two vexing problems in crinoid paleoecology: microcrinoids and *Uintacrinus*. Washed residues of Paleozoic marine sediments contain microfossils and abundant fragmentary macrofossils, but many also contain microcrinoids. Defined as crinoids with a thecal height of less than 2 mm, microcrinoids include a few juveniles of larger, "normal" crinoids, but microcrinoids also include an array of miniaturized adults that offer numerous paleobiologic, paleoecologic, and evolutionary questions. Sevastopulo (Chapter 3) examines how these miniature crinoids lived. *Uintacrinus* is a stemless Cretaceous crinoid, long considered pelagic or nektic. Webber, Meyer, and Milsom (Chapter 5) provide further evidence for the hypothesis that this crinoid lived on the seafloor and that its surprising morphology is not for a pelagic way of life.

Echinoids are considered in Chapters 4 and 5. Most Paleozoic echinoids are constructed in a way that makes them susceptible to rapid disarticulation after death, whether that disarticulation is real or a taphonomic artifact. Because whole echinoids are rare in Paleozoic rocks, it is commonly assumed that they were not a significant element of Paleozoic communities. Schneider (Chapter 4) discusses Pennsylvanian Lagerstätten where echinoids dominated. There is little question that echinoids played a significant role in many post-Paleozoic communities. Nebelsick (Chapter 6) takes advantage of the solid construction of living clypeasteroids to establish a preservational gradient that can be used to understand fossils of these echinoids, regardless of the state of preservation.

References

Ausich, W. I. 1980. A model for niche differentiation in Lower Mississippian crinoid communities. Journal of Paleontology, 54:273–288.

Baumiller, T. K., M. Labarbara, and J. D. Woodley. 1991. Ecology and functional morphology of the isocrinid *Cenocrinus asterius* (Linneaus) (Echinodermata: Crinoidea). Bulletin of Marine Science, 48:731–738.

Guensburg, T. E., and J. Sprinkle. 1992, Rise of echinoderms in the Paleozoic evolutionary fauna: signficance of paleoenvironmental controls. Geology, 20:407–410.

Kammer, T. W., and W. I. Ausich. 1987. Aerosol suspension feeding and current velocities: distributional controls for late Osagean crinoids. Paleobiology, 13:379–395.

Lane, N. G. 1963. The Berkeley crinoid collection from Crawfordsville, Indiana. Journal of Paleontology, 37:1001–1008.

Lane, N. G. 1973. Paleontology and paleoecology of the Crawfordsville fossil site (Upper Osagian: Indiana). University of California Publications in Geological Sciences, 99, 141 p.

Macurda, D. B. Jr., and D. L. Meyer. 1974. Feeding posture of modern stalked crinoids. Nature, 247:394–396.

Meyer, D. L. 1979. Length and spacing of the tube feet in crinoids (Echinodermata) and their role in suspension feeding. Marine Biology, 51:361–369.

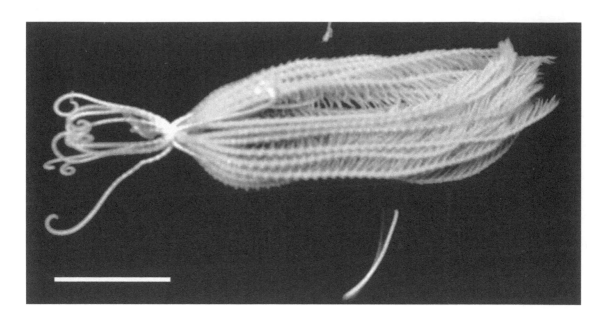

Figure 1.1. *Posture of a comatulid crinoid, **Stylometra spinifera** (Carpenter, 1881), after tumbling with sediment and water for 76 minutes. Scale = 20 mm.*

TAPHONOMY AS AN INDICATOR OF BEHAVIOR AMONG FOSSIL CRINOIDS

1

Tomasz K. Baumiller, Forest J. Gahn, Hans Hess, and Charles G. Messing

Taphonomic processes lead to the loss of biological information as tissues degrade and skeletal elements become broken, abraded, and, ultimately, chemically modified. Information loss is not random but rather is a function of an organism's ecology, morphology, and behavior. Thus, taphonomists can use predictable patterns of preservation to ameliorate the negative effects of biotic degradation.

Introduction

The multiplated skeletons of echinoderms are not particularly resistant to postmortem processes and disarticulate easily (Meyer and Meyer, 1986). Nevertheless, a sufficiently large number of specimens have escaped these taphonomic filters to leave a rich record of complete, or nearly complete, specimens (Brett and Baird, 1986; Meyer et al., 1989; Gahn and Baumiller, 2004). For benthic crinoid fossils, a high degree of articulation is typically associated with rapid burial and little transport (Simms, 1999a). Can the mode of preservation of well-articulated crinoids provide any insights into their behavior? Obviously, one way in which it can is if an organism is killed and buried instantly--what might be referred to as the Pompeii effect. In this instance, the organism might be preserved in a posture characterizing its normal behavior--that is, behavior under typical environmental conditions. However, even when rapid burial occurs, it is likely that the organism has had some time to respond to the rapidly changing conditions, and in that case, it might be buried in its trauma posture. Although the trauma response does not represent normal behavior, it nevertheless provides insights into the organism's functional abilities--or, perhaps, functional limits. The hypothesis we will test, therefore, is that the postures of well-preserved fossil crinoids should correspond to postures that characterized either their normal live behaviors or their trauma behaviors. We will do this by characterizing the normal and trauma behaviors, and by examining the mode of preservation in fossil comatulids and isocrinids. Finally, we will examine the mode of preservation of some stalked Paleozoic taxa and attempt to interpret their behaviors in light of these data.

COMATULID BEHAVIOR. Rapid burial events are attributable to large sediment loads and may be associated with increases in flow velocities; therefore, it is worth exploring the response of comatulids to such changing conditions. The most detailed in situ observations of comatulids to increas-

Behavior and Taphonomy

ing fluid velocities have been made by Meyer and his collaborators (Meyer, 1973, 1997; Meyer and Macurda, 1980; Meyer et al., 1984). These observations revealed that comatulids generally respond to increasing flow rates either by modifying and sometimes deflecting their filter, or by crawling to find a protective crevice.

For comatulids experiencing high flow velocities, the arms become "compressed into a cylinder oriented downcurrent" (Meyer and Macurda, 1980, p. 79), and in this posture, the crinoid is reminiscent of a shaving brush (Seilacher, personal commun., 2003). Observations of comatulid postures during burial events have not been made in situ, but experiments with specimens tumbled in sediment-laden water, meant to simulate burial events, revealed that comatulids compress their arms tightly into a cylinder and maintain the shaving brush posture after hours of tumbling (Fig. 1.1; Baumiller, 2003).

In addition to the shaving brush response, deteriorating conditions, such as increasing current velocities or sediment load, might also induce comatulids to crawl. During crawling, comatulids maintain the "oral surface uppermost" (Clark, 1915, p. 13), with their arms arranged radially around the central disk. Clark noted that even when comatulids are dropped mouth down in the water column, or when placed mouth down on the bottom, they always righted themselves, resulting in the usual mouth-up orientation.

Given that the living position of comatulids and stalkless crinoids is generally mouth up and that under extreme conditions they may assume a shaving brush posture, it is reasonable to expect that those would be the postures of well-preserved fossil comatulids--that is, those with arms and calyx largely intact. Thus, the most common mode of preservation should be the shaving brush, with the arms compressed into a cylinder and the oral-aboral axis parallel to the substrate. Occasionally, the rapid burial of comatulids in normal living or crawling positions should lead to a mouth-up posture with the oral-aboral axis perpendicular to the substrate and the arms arranged radially around the disk--the starburst-up mode of preservation. The fact that the mouth-down behavior has not been observed in extant comatulids suggests that the starburst-down posture should be rare in this group.

COMATULID TAPHONOMY. As a test of the above prediction, we examined the orientation of the stalkless crinoid, *Paracomatula helvetica* Hess, 1951, on several slabs from Hottwil, Switzerland (Klingnau Formation, Bajocian). None of the 42 specimens are preserved starburst down, but 11 are starburst up (Fig. 1.2); 23 are characterized by a shaving brush mode of preservation. An additional eight are either on their side or mouth up but neither starburst nor shaving brush; instead, their arms have an irregular arrangement neither radiating out nor compressed into a cylinder (Table 1.1).

The above observations confirm the predictions that mode of preservation is not independent of posture in live stalkless crinoids. Of course, this concept is not new; for example, Taylor (1983) reported that of several hundred specimens of *Ailsacrinus abbreviatus* Taylor, 1983, an essentially stemless Jurassic millericrinid, ~50% were preserved mouth up; this pat-

Figure 1.2. *Lower surface of a slab with two specimens of **Paracomatula helvetica** Hess, 1951, from Hottwil (Jurassic). The specimens are preserved starburst up and mouth up (away from the viewer), with the arms arranged radially around the central disk. Scale = 10 mm.*

tern was consistent with the usual living position of this crinoid and rapid in situ burial. Below, we apply the taphonomic approach to determine whether it can offer insights into the crawling abilities of stalked crinoids.

ISOCRINID BEHAVIOR. Knowledge of stalked crinoid behavior has grown greatly through the use of research submersibles. For instance, such research has shown that isocrinids predominantly live under low to moderate flow velocities (Breimer and Lane, 1978). In the normal feeding posture, the isocrinid stalk is subvertical with the proximal portion bent sharply downstream such that the oral-aboral axis of the crown orients parallel to the current direction with the oral surface directed downstream (Fig. 1.3.1). The extended arms are slightly recurved into the current, forming a concavo-convex filter, which has been named the parabolic filtration fan posture (Macurda and Meyer, 1974).

Isocrinids have also been observed under slack current conditions, where they maintain a so-called wilted flower posture with the arms flexed aborally, while the oral-aboral axis of the crinoid is vertical (Macurda and Meyer, 1974; Messing, 1985, Baumiller et al. 1991). Much less is known about their response to high current velocities, which make it difficult for submersibles to operate, and researchers have had limited opportunity to study deep-water crinoids under such conditions. However, available data allow us to generalize about isocrinids' response to high current velocities. The data consist of photos taken by a time-lapse camera deployed at 420

Taxon	Age	Locality	Starburst Up	Starburst Down	?	Shaving Brush
Paracomatula helvetica	Jur	Hottwil, Switzerland	11	0	8	23

Abbreviation.—?, not representative of any category.

Table 1.1. *Mode of preservation of fossil comatulids.*

m for ~6 weeks' recording of isocrinids' responses to velocities ranging from less than 100 mm/s to well above 500 mm/s (Messing et al., personal commun.), and manipulation of isocrinids in situ with a submersible slurp gun (Fig. 1.4). In the latter case, a slurp gun was used in reverse to generate a high-velocity stream directed at the isocrinid *Neocrinus decorus* Wyville-Thomson, 1864, and its response was recorded with a video camera. Both the time-lapse photographs and the slurp gun video revealed that as current velocity was increased, the isocrinids changed their posture from the parabolic filtration fan posture to a shaving brush posture with the stalk and crown deflected downstream and the long axes of the arms aligned parallel to the current with their distal tips pointing downstream (Figs 1.3., 1.2, 1.4). More importantly, the slurp gun experiment revealed that the change in posture can occur rapidly, within seconds of the onset of high flow velocities. The isocrinid posture is similar to the high flow velocity posture described for comatulids by Meyer and Macurda (1980).

Figure 1.3. *Schematic diagram showing the postural response of a stalked crinoid with well-developed muscular arm articulations to increasing current. 1, Parabolic filtration fan posture (PFFP). 2, Arms compressed parallel to oral-aboral axis. 3, Shaving brush posture. Length of arrows are proportional to current velocity.*

The above behaviors describe postures of isocrinids attached to the substrate by the distal end of their stalk with the crown elevated above the bottom. However, it is now an established fact that isocrinids are also capable of crawling. Direct evidence of isocrinid crawling became available through in situ observations (Messing et al., 1988) and laboratory flow-tank studies (Baumiller et al., 1991). These studies revealed that isocrinids could relocate by crawling with their arms and dragging the stalk along the substrate. Recently, data obtained from submersibles showed that they could move much faster (30–40 mm/s; Baumiller and Messing, 2005, personal commun.) than had been previously recognized (0.1 mm/s; Birenheide and Motokawa, 1994).

Two types of crawling behavior have been described: the slow mode (Baumiller et al., 1991; Birenheide and Motokawa, 1994) and the fast mode

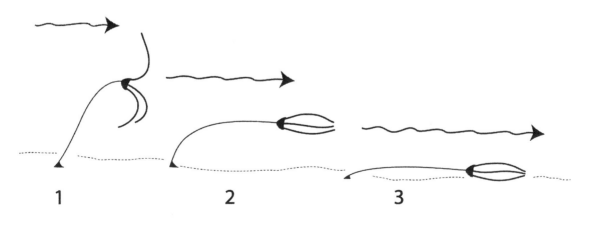

1   **2**   **3**

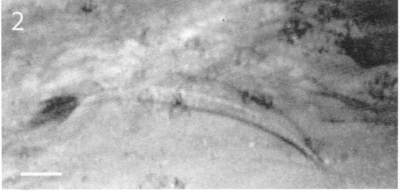

Figure 1.4. *Sequence of video frames from a slurp gun experiment. 1, Initial response of the isocrinid,* **Neocrinus decorus**, *to an increase in sediment load and current velocity. 2, Posture of* **N. decorus** *after approximately 30 seconds from onset of high sediment/high current flow. Scale = 100 mm.*

(Baumiller and Messing, 2005, personal commun.) mode. In the slow mode (Fig. 1.5.1), the stalk is dragged behind the crown along the bottom; its proximal portion is bent sharply such that the oral-aboral axis of the crown is subvertical; and the arms are arranged radially and are only slightly flexed aborally such that their long axes are roughly parallel to the substrate. This is analogous to the mouth-up crawling posture of comatulids. In the fast mode (Fig. 1.5.2), the stalk is also dragged behind the crown along the substrate, but the proximal portion remains nearly straight, so the stalk and the oral-aboral axis of the crown are horizontal. In this posture, only a portion of the strongly aborally flexed arms are in contact with the substrate; others face away from it. Finally, tumbling experiments with isocrinids have revealed that, just like comatulids, their posture when tumbled with sediment is the shaving brush, with the arms straight and compressed into a cylinder or a cone (Fig. 1.6; Baumiller, 2003).

ISOCRINID TAPHONOMY. The observed behaviors of isocrinids lead to predictions about their mode of preservation. The most probable burial posture of a stalked crinoid is the shaving brush, as it represents the behavioral response of the crinoid to high flow velocities and to tumbling with sediment (Figs 1.3.3, 1.6). Alternatively, burial during crawling would lead to a star-burst-up mode of preservation (Fig. 1.5.1). To test these predictions, we examined the mode of preservation of the Jurassic isocrinid *Chariocrinus an-*

Figure 1.5. *Schematic diagram of isocrinid crawling postures. 1, Fast mode. 2, Slow mode.*

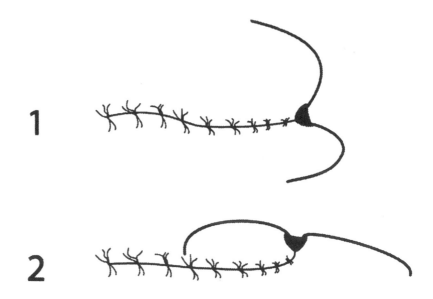

Figure 1.6. *Posture of a stalked crinoid, the isocrinid **Neocrinus decorus**, while tumbling in a Plexiglas tumbler. Scale = 100 mm.*

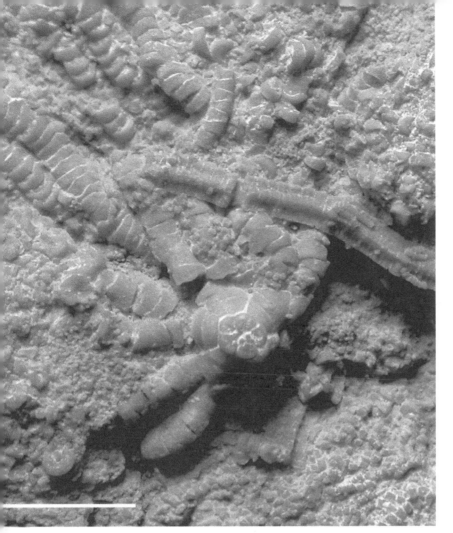

Figure 1.7. *Lower surface of a slab with specimens of the isocrinid* **Chariocrinus andreae** *from Reigoldswil (Jurassic). The specimen is preserved starburst up and mouth up (away from the viewer), with the arms arranged around the central disk. Scale = 10 mm.*

dreae Desor, 1845, from several localities (Table 1.2). Although the shaving brush mode of preservation was most common (112 of 143 well-preserved specimens), 19 were starburst up (Fig. 1.7) and 12 were partially splayed on their side, but with the arms not compressed into a cylinder. No specimens of *C. andreae* were preserved mouth down; such a mode of preservation is expected to be rare, given that a mouth-down posture has not been observed among extant isocrinids. *Chariocrinus andreae* shares most morphological features with extant isocrinids, suggesting that it was capable of crawling; the preservation data seem to support its crawling abilities.

Implications for Other Stalked Crinoids

The comatulid and isocrinid examples confirm that the postures of well-preserved fossils are not independent of their mode of life or their trauma behaviors. We have demonstrated that well-preserved fossils of these taxa are most commonly represented by shaving brush postures, which indicate the trauma response. The starburst-up postures are also common. For comatulids, starburst-up postures may represent instantaneous burial in live position or while crawling, but for isocrinids, only the latter interpretation is feasible.

Taxon	Age	Locality	Starburst Up	Starburst Down	?	Shaving Brush
Chariocrinus andreae	Jur	Glattweg, Switzerland	1	0	1	13
Chariocrinus andreae	Jur	Munzachberg, Switzerland	13	0	1	9
Chariocrinus andreae	Jur	Lausen, Switzerland	4	0	5	46
Chariocrinus andreae	Jur	Reigoldswil, Switzerland	1	0	3	11
Chariocrinus andreae	Jur	Munzachberg, Switzerland	13	0	1	9

Abbreviation.—?, not representative of any category

Table 1.2. *Mode of preservation of fossil isocrinids*

We have examined only a single isocrinid species for which starburst orientations can be determined, and we predict that for all isocrinids and other stalked crawlers, the shaving brush and starburst up postures would be the most common when preserved. On the other hand, stalked crinoids that were incapable of crawling generally should not be preserved starburst up because the mouth-up postures, such as the wilted flower posture, do not represent behaviors likely to be preserved. To test this prediction, we examined Paleozoic crinoids that lacked morphological characteristics of crawlers (Table 1.3)--that is, the well-developed muscular arm articulations that characterize post-Paleozoic articulates and Paleozoic advanced cladids (Simms and Sevastopulo, 1993; Simms, 1999b). Absence of muscles and especially of a fulcral ridge on the articular facet imply limited relative movement of brachials and arm flexibility, a major constraint on locomotion. Also, Paleozoic crinoids in our data lack those features associated with the ability to reattach to the substrate, a characteristic necessary for eleutherozoic life.

Examination of postures in our sample revealed that the shaving brush posture was the most common (Fig. 1.8), and the starburst-up posture was rare. Surprisingly, a starburst-down mode of preservation proved to be relatively common among these taxa, in contrast to its absence among comatulids and isocrinids (Fig. 1.9).

The taphonomy of the stalked Paleozoic taxa suggests that although they may have shared some behavioral traits with the isocrinids, they differed in some important respects. For one, their behavioral repertoire did not include mouth-up crawling. Also, although the common shaving brush postures indicate that they may have had a trauma response similar to the comatulids and isocrinids, their frequent starburst-down mode of preservation implies that they may not have been able to compress their arms as quickly into a cylinder or that they lived in environments where some burial events were more rapid. Both scenarios are conjectural because we have no modern analogs for mouth-down behavior or trauma response, but the former is consistent with what is known about the differences in the functional morphology of comatulids and isocrinids relative to the Paleo-

Table 1.3. *(opposite) Mode of preservation of fossil Paleozoic stalked crinoids.*

Taxon	Age	Locality	Starburst Up	Starburst Down	?	Shaving Brush
Arthroacanatha carpenteri Hinde, 1885	Dev	Michigan, USA	0	13	0	34
Gennaeocrinus variabilis Kesling and Smith, 1962	Dev	Michigan, USA	0	14	0	70
Several (Franzén, 1983)	Sil	Gotland, Sweden	0	0	2	242
Glyptocrinus decadactylus Hall, 1847	Ord	Maysville, Kentucky USA	2	33	0	227

Abbreviation.— ?, not representative of any category

zoic taxa we examined. Comatulids and isocrinids are able to respond rapidly to deteriorating conditions (current velocity, sediment load) by changing their posture from a variety of filtration fan arrays to a shaving brush. An equally rapid response is unlikely to have been possible for crinoids lacking muscular arm articulations, such as the Paleozoic crinoids we examined. In part, this is because, even though crinoid ligament may be contractile (Birenheide and Motokawa, 1996), its contraction speed is much slower than that of muscle (Motokawa et al., 2004). Also, the lack of a fulcral ridge would eliminate the lever effect. Under this scenario, the Paleozoic crinoids we examined may have had limited arm mobility and

Figure 1.8. *Upper surface of the Nar slab (Silurian from Gotland described by Franzén, 1983) with specimens of camerate and flexible crinoids in shaving brush posture. Note stalk alignment. Scale = 200 mm.*

Figure 1.9. *Upper surface of a slab from the Bell Shale, Rockport, Alpena County, Michigan (Devonian), with specimens of the camerate **Gennaeocrinus variabilis** in the mouth-down (starburst-down) posture. Scale = 100 mm.*

could have modified their postures, but unlike the isocrinids, they could do this only slowly. During rapid burial events, the drag on a permanently attached crinoid in a feeding posture, such as the parabolic filtration fan posture, would force the crown downstream and toward the substrate. For isocrinids, the arms would be compressed tightly into a cylinder. Alternatively, with rapid enough burial, Paleozoic taxa with more limited arm mobility could have the crown forced mouth down into the substrate, resulting in a starburst-down posture (Fig. 1.10). Under this scenario, the ratio of starburst-down to shaving brush modes of preservation is a function of the rate at which the arms can flex orally; a high ratio indicates a slow rate of flexure (Paleozoic taxa of this study), a low ratio indicates a fast rate (isocrinids and comatulids), and intermediate ratios indicate intermediate rates of flexure (advanced cladids?).

Whereas the above scenario ascribes taphonomic differences between the Paleozoic taxa and the isocrinids and comatulids to differences in functional morphology, a similar pattern could result from differences in

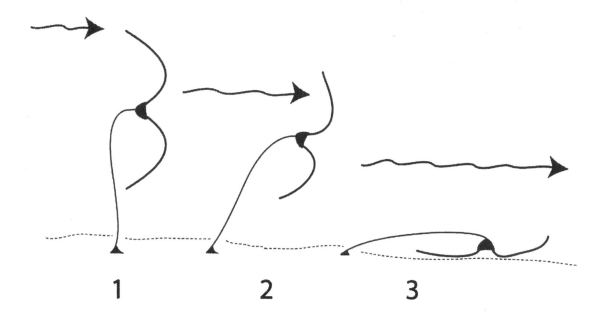

1   2   3

the environments they inhabited--specifically, differences in the rapidity of burial events. For example, if burial events were much more rapid for the Paleozoic taxa we examined than for the isocrinids and comatulids, the common starburst-down mode of preservation among the former would result. Because we have no paleoenvironmental data that would allow us to test for these differences, at present, we consider this scenario more ad hoc than the scenario based on functional morphology.

Reconstructing the behavior of extinct organisms, although a difficult task, is essential for paleontologists interested in ecological questions as well as for those interested in assessing the importance of ecology in evolution. A variety of approaches can provide insights into behavior and function of extinct organisms, including those based on comparison of homologous or analogous structures among fossil and living organisms (or fossils and mechanical analogues), biomechanics, ichnology, and taphonomy (Plotnick and Baumiller, 2000). Undoubtedly, such approaches will also prove to be the most important guide for detecting crawling abilities in stalked crinoids (Baumiller and Messing, 2005). Several features that are necessary for crawling have been identified: a detachable stalk; some mode of reattachment of the stalk to the substrate, such as flexible cirri; and arms that are both flexible and strong enough to generate the power and recovery strokes to pull the animal along the bottom. Such features characterize the post-Paleozoic clade that includes holocrinids, comatulids, isocrinids, and perhaps some of the Late Paleozoic "advanced cladids" but excludes the cyrtocrinids, bourgueticrinines, and hyocrinids (Simms and Sevastopulo, 1993; Simms, 1999b). A more detailed analysis of the functional morphology of extant crawlers and of the distribution of crawling characters in Paleozoic fossils will be needed to ascertain this claim. Here, we have em-

Conclusions

Figure 1.10. *Schematic diagram showing the hypothesized postural response of a stalked crinoid with ligamentary (nonmuscular) arm articulations to increasing current. Length of arrows are proportional to current velocity.*

phasized the taphonomic approach as a way of reconstructing the behavior of fossil crinoids, with special emphasis on crawling in stalked crinoids.

In our approach, the basic premise is that live postures and trauma behaviors should affect the mode of preservation of fossils. On the basis of studies of live crinoids, we predicted that two modes of preservation should be common: a shaving brush posture representing the behavioral response to high current velocities and other environmental disturbances, and a starburst-up posture representing the mouth-up live orientation during feeding in comatulids and crawling in comatulids and isocrinids. We tested this prediction by comparing postures of living comatulids, a group that lacks a stalk and generally lives in a mouth-up orientation, to the mode of preservation of representative fossils. We found that the Jurassic paracomatulid *Paracomatula helvetica* was predominantly represented by shaving brush and starburst-up postures. A similar pattern emerged when we examined the Jurassic isocrinid *Chariocrinus andreae*. Broad generalities are difficult to draw from only two examples; nevertheless the phylogenetic position and morphological traits of these two taxa argue strongly for a behavioral similarity with extant comatulids and isocrinids, making the congruence between behavioral interpretations of crawling based on taphonomy and morphology compelling.

A further prediction of the taphonomic approach is that the mode of preservation for noncrawling stalked crinoids should rarely include the starburst-up posture, because it is difficult to imagine a mouth-up live or trauma posture for these taxa. Our analysis of Paleozoic crinoids that lack the morphological features typically associated with crawling confirmed that prediction; we found only two specimens in the starburst-up posture. However, the starburst-down posture was relatively common, suggesting that high flow velocities may produce such a response in crinoids that combine a permanent mode of attachment of the stalk with nonmuscular arms with limited flexibility and mobility. If the latter is correct, one might predict that relative mobility of arms could be reflected in the ratios of starburst-down to shaving brush postures; high ratios characterize taxa with limited arm mobility, intermediate ratios those with some mobility, and low ratios those with highly flexible, muscular arms.

The taphonomic method we have outlined for reconstructing crawling abilities in fossil stalked crinoids must be viewed as preliminary, and it requires further tests. Furthermore, the taphonomic method is best used to complement other approaches for reconstructing behavior or function; the most compelling interpretations are those supported by a variety of methods.

Acknowledgments

We thank the submersible and ship crews from the Harbor Branch Oceanographic Institution, through whom the field portion of this research was carried out. B. Deline, University of Cincinnati, provided orientation data for *Glyptocrinus decadactylus*. W. I. Ausich, S. Donovan, and D. L. Meyer provided comments that greatly improved the chapter. This work was partly funded under NSF grants EAR-9004232 (CGM), EAR-9104892 (TKB), EAR-9218467 (CGM), EAR-9304789 (TKB), and EAR-9628215 (CGM).

References

Baumiller, T. K. 2003. Experimental and biostratinomic disarticulation of crinoids: taphonomic implications, p. 243–248. *In* J.-P. Feral and B. David (eds.), Echinoderm Research 2001, A. A. Balkema, Rotterdam.

Baumiller, T. K., and C. G. Messing. 2005. Crawling in stalked crinoids: in situ observations, functional morphology, and implications for Paleozoic taxa. Geological Society of America Abstracts with Programs, 37(5):65.

Baumiller, T. K., M. Labarbera, and J. D. Woodley. 1991. Ecology and functional morphology of the isocrinoid *Cenocrinus asterius* (Linnaeus) (Echinodermata: Crinoidea): in situ and laboratory experiments and observations. Bulletin of Marine Science, 48:731–748.

Birenheide, R., and T. Motokawa. 1994. Morphological basis and mechanics of arm movement in the stalked crinoids *Metacrinus rotundus* (Echinodermata, Crinoidea). Marine Biology, 121:273–283.

Birenheide, R., and T. Motokawa. 1996. Contractile connective tissue in crinoids. Biological Bulletin, 191:1–4.

Breimer, A., and N. G. Lane. 1978. Paleoecology, p. T316–T347. *In* R. C. Moore and C. Teichert (eds.), Treatise on Invertebrate Paleontology. Pt. T. Echinodermata 2. Geological Society of America and University of Kansas Press, Lawrence.

Brett, C. E., and G. C. Baird. 1986. Comparative taphonomy: a key to paleoenvironmental interpretation based on fossil preservation. Palaios, 1:207–227.

Carpenter, P. H. 1881. Report on the results of dredging under the supervision of Alexander Agassiz, in the Gulf of Mexico (1877–1878), the Caribbean Sea (1878–1879), and the East Coast of the United States (1880) by the U.S.C.S. Str. "Blake," Lieut.-Commander C. D. Sigsbee, U.S.N., and Commander J. R. Bartlett, U.S.N. Commanding. XVI. Preliminary Report on the Comatulae. Harvard University Museum of Comparative Zoology Bulletin, 9:151–170.

Clark, II. L. 1915. The comatulids of Torres Strait: with special reference to their habits and reactions. Papers from the Department of Marine Biology, Carnegie Institute, Washington, 8:97–125.

Desor, E. 1845. Résumé de les etudes sur les crinoides fossilies de la Suisse. Bulletin de la Societe Neuchateloise des Sciences Naturelles, 1:211–222.

Franzén, C. 1983. Ecology and taxonomy of Silurian crinoids from Gotland. Acta Universitatis Upsaliensis, 665:1–31.

Gahn, F. J., and T. K. Baumiller. 2004. A bootstrap analysis for comparative taphonomy applied to Early Mississippian (Kinderhookian) crinoids from the Wassonville Cycle of Iowa. Palaios, 19:17–38.

Hall, J. 1847. Palaeontology of New York, volume 1, Containing descriptions of the organic remains of the lower division of the New-York system (equivalent of the Lower Silurian rocks of Europe). Natural History of New York, Albany, New York, 6, 338 p.

Hess, H. 1951. Ein neuer Crinoide aus dem mittleren Dogger der Nordschweiz (*Paracomatula helvetica* n. gen. n. sp.). Eclogae Geologicae Helvetiae, 43:208–216.

Hinde, G. J. 1885. Description of a new species of crinoids with articulating spines. Annals and Magazine of Natural History, series 5, 5:157–173.

Kesling, R. V., and R. N. Smith. 1962. *Gennaeocrinus variabilis*, a new species of crinoid from the Middle Devonian Bell Shale of Michigan. University of Michigan Contributions from Museum of Paleontology, 17:173–194.

Macurda, D. B., JR., and D. L. Meyer. 1974. Feeding posture of modern stalked crinoids. Nature, 247:394–396.

Messing, C. G. 1985. Submersible observations of deep-water crinoid assemblages in the tropical western Atlantic Ocean, p. 185–193. *In* B. F. Keegan and B. D. S. O'Connor (eds.), Proceedings of the Fifth International Echinoderm Conference, Galway. A. A. Balkema, Rotterdam.

Messing, C. G., M. C. Rosesmyth, S. R. Mailer, and J. E. Miller. 1988. Relocation movement in a stalked crinoid (Echinodermata). Bulletin of Marine Science, 42:480–487.

Meyer, D. L. 1973. Feeding behavior and ecology of shallow water unstalked crinoids (Echinodermata) in the Carribbean Sea. Marine Biology, 22:105–130.

Meyer, D. L. 1997. Reef crinoids as current meters: feeding responses to variable flow. Proceedings of the 8th International Coral Reef Symposium, 2:1127–1130.

Meyer, D. L., and D. B. Macurda Jr. 1980. Ecology and distribution of the shallow-water crinoids (Echinodermata) of Palau and Guam (Western Pacific). Micronesica, 16:59–99.

Meyer, D. L., and K. B. Meyer. 1986. Biostratinomy of Recent crinoids (Echinodermata) at Lizard Island, Great Barrier Reef, Australia. Palaios, 1:294–302.

Meyer, D. L., W. I. Ausich, and R. E. Terry. 1989. Comparative taphonomy of echinoderms in carbonate facies: Fort Payne Formation (Lower Mississippian) of Kentucky and Tennessee. Palaios, 4:533–552.

Meyer, D. L., C. A. Lahaye, N. D. Holland, A. C. Arenson, and J. R. Strickler. 1984. Time-lapse cinematography of feather stars (Echinodermata: Crinoidea) on the Great Barrier Reef, Australia: demonstrations of posture changes, locomotion, spawning and possible predation by fish. Marine Biology, 78:179–184.

Motokawa, T., S. Osamu, and R. Birenheide. 2004. Contraction and stiffness changes in collagenous arm ligaments of the stalked crinoid *Metacrinus rotundus* (Echinodermata). Biological Bulletin, 206:4–12.

Plotnick, R., and T. K. Baumiller. 2000. Invention by evolution: functional analysis in paleobiology. Paleobiology, 26:305–321.

Simms, M. J. 1999a. Middle Jurassic of southern England, p. 197–202. *In* H. Hess, W. I. Ausich, C. E. Brett and M. J. Simms (eds.), Fossil Crinoids. Cambridge University Press, Cambridge.

Simms, M. J. 1999b. Systematics, phylogeny and evolutionary history, p. 31–40. *In* H. Hess, W. I. Ausich, C. E. Brett and M. J. Simms (eds.), Fossil Crinoids. Cambridge University Press, Cambridge.

Simms, M. J., and G. D. Sevastopulo. 1993. The origin of articulate crinoids. Palaeontology, 36:91–109.

Taylor, P. D. 1983. *Ailsacrinus* gen. nov., an aberrant millericrinid from the Middle Jurassic of Britain. Bulletin British Museum Natural History (Geology), 37:37–77.

Wyville-Thomson, C. 1864. Sea lilies. Intellectual Observer, 6:1–11.

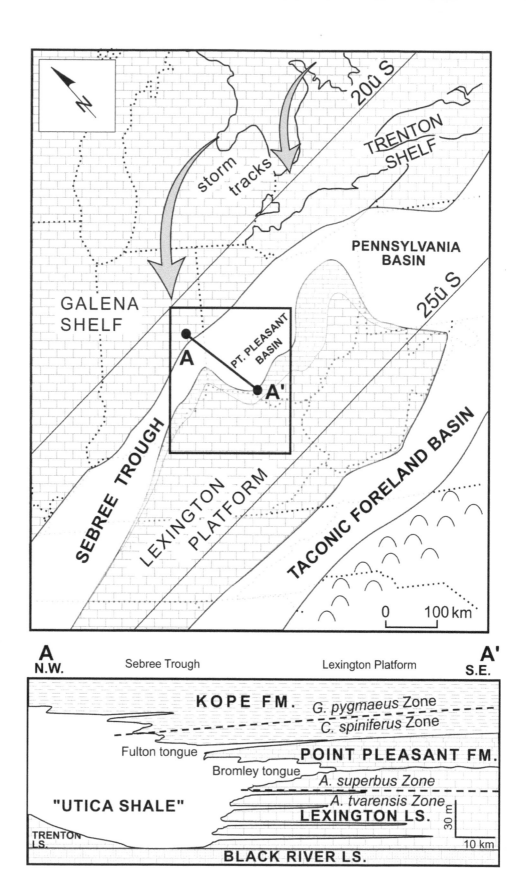

ATTACHMENT, FACIES DISTRIBUTION, AND LIFE HISTORY STRATEGIES IN CRINOIDS FROM THE UPPER ORDOVICIAN OF KENTUCKY

2

Carlton E. Brett, Bradley L. Deline, and Patrick I. McLaughlin

A major challenge of paleobiology is relating functional morphology of extinct organisms to patterns of distributional ecology. Relationships to substrate may be particularly important for organisms such as stalked echinoderms, which are permanently attached. Crinoids also afford excellent opportunities to reconstruct paleoautecology because they preserve most morphological aspects in their endoskeletons. Thus, feeding behavior, column height, and attachment strategies may be reconstructed in some detail. However, because of the propensity of crinoid's multiclement skeletons toward disarticulation, the degree to which this morphology can be reconstructed for any given species depends on extraordinary preservation of articulated skeletons in rapidly buried seafloors or obrution Lagerstätte deposits (Brett and Seilacher, 1991; Taylor and Brett, 1996; Ausich et al., 1999a). Once identified, however, even disarticulated remains may be useful for determining the presence and relative abundance of crinoids in different facies (Brett, 1985; Meyer et al., 2002).

The dependency of crinoid studies on Lagerstätten is both an advantage and a disadvantage. Assemblages with articulated crinoids are rare and scattered. However, those that are known record snapshots of populations at geological instants, more akin to the census samples of modern ecologists than to time-averaged ensembles typical of fossil assemblages. The structure of these populations, together with patterns of distribution of crinoids, may be analyzed to determine aspects of the life history of the species.

Studies of modern stalked crinoids have provided important insights into modes of life of ancient counterparts (Meyer, 1973; Rasmussen, 1977; Hess, 1999; Baumiller and Messing, this volume). However, extant isocrinids differ in significant ways from early Paleozoic counterparts. Isocrinids have muscular articulations in their arms that afford considerable flexibility and even locomotion (Ausich and Baumiller, 1996); conversely, Ordovician crinoids lacked muscular articulations and had little or no capacity for locomotion or for adjusting their position. Modern isocrinid stalked crinoids possess whorled flexible cirri for strong attachment in strong currents. Isocrinids shut down feeding in the face of currents in excess of ~50

Introduction

Figure 2.1. *1, Location map for studied sections in the tristate area of Kentucky, Ohio, and Indiana. 2, Paleogeographic reconstruction of Cincinnati area showing shelf-to-ramp profile during Edenian (early Cincinnatian).*

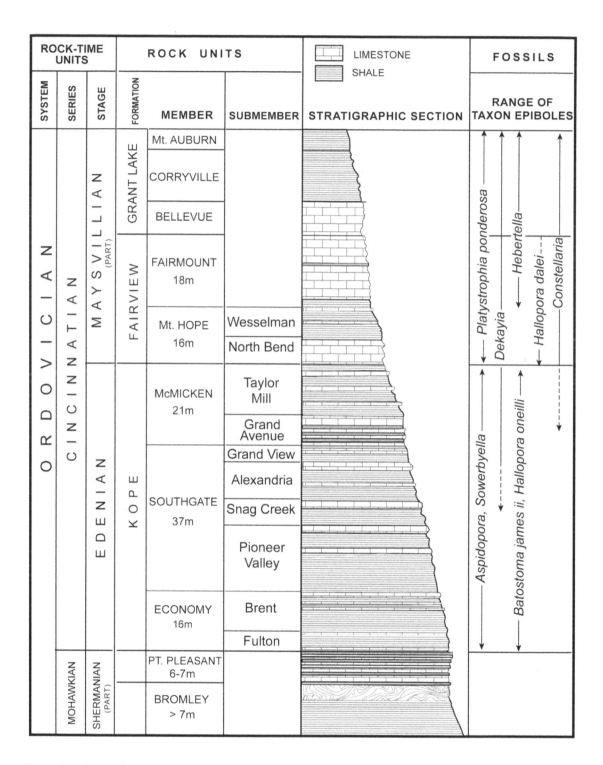

Figure 2.2. Generalized stratigraphic column for the Lexington and lower Cincinnatian of the northern Kentucky, Ohio, and Indiana tristate area of the Cincinnati Arch.

cm/s (Messing, personal commun., 2005) but maintain their position. Most Ordovician crinoids lacked such firm rooting and almost none had cirri. This presents a number of challenges to their interpretation.

Moreover, living isocrinids are confined to deep sea bathyal to abyssal environments typically deeper than 600 m; unlike comatulids, stalked crinoids are absent from shallow shelf settings and apparently have been since the Cretaceous, when stalked crinoids shifted to offshore environments, possibly in response to increasing predation intensity in the Mesozoic marine revolution (Meyer and Macurda, 1977; Meyer and Ausich, 1983; Vermeij, 1987). In contrast, crinoids constitute a significant element of most Ordovician benthic communities from outer shelf or deep ramp muddy bottom facies to high-energy shallow shoals and reefs.

In this chapter, we explore the relationship between distributions of Late Ordovician crinoids and their modes of attachment, feeding strategies, and life history strategies. We also will focus on several spectacular new occurrences of crinoids and their implications for crinoid paleoecology. New evidence from these deposits raises further questions about the modes of life, life history strategies, and living sites of the crinoids. Insight into these issues requires a combination of observation of complete skeletal material, including modes of attachment, column length, population structure, and a comparative taphonomic approach that suggests distinct modes of life and death in various crinoids.

Study Area Materials and Methods

Upper Ordovician strata exposed in the Cincinnati Arch in northern Kentucky and adjacent regions (Fig. 2.1) contain abundant and superbly preserved crinoid material in a range of facies recording varied environments from deep dysoxic settings below all but the deepest storm waves to shallow shoal settings. The stratigraphy and correlations of these rocks have been studied in detail for more than a century (Ulrich and Bassler, 1914; Cressman, 1973; Weir et al., 1984; Hay, 1981, 1998; see summary in Davis and Cuffey, 1998). Recent studies have emphasized sequence stratigraphy and depositional environments (Holland, 1993; Holland and Patzkowsky, 1998, 2004; Pope and Read, 1997a, 1977b; Brett and Algeo, 2001; Brett et al., 2003, 2004; McLaughlin et al., 2004). Particular emphasis has been directed recently toward interpretation of fossil associations and gradients, characterized quantitatively, and cyclicity (Holland et al., 2001; Holland and Patzkowsky, 2004; Miller et al., 2001; Meyer et al., 2002). These studies provide a temporal and paleoenvironmental framework for the present study of crinoid distributional ecology.

Crinoids are abundant in portions of the Upper Ordovician (Chatfieldian) Lexington Limestone and the Cincinnatian Eden, Maysville, and Richmond groups (Fig. 2.2; Ausich, 1999). This study deals with crinoids of the upper Lexington Formation (Stamping Ground–Strodes Creek, Devils Hollow, and Point Pleasant members) and the overlying Kope, Fairview, and Grant Lake formations (Edenian–Maysvillian). Material used in this study comprises both completely articulated specimens in several obrution deposits and abundant disarticulated material, particularly hold-

Species	Cemented Discoidal	"Lichenocrinid"	Distal Coils	Radiating Cirri	Unknown	Approximate Column Length (cm)
Disparids						
Anomalocrinus incurvus	X	—	—	—	—	15–20
Cincinnaticrinus pentagonus	—	X	—	—	—	10–15
Cincinnaticrinus varibrachialus	—	X	—	—	—	10–15
Dystactocrinus constrictus	—	—	—	—	X	?
Ectenocrinus simplex	—	X	—	—	—	10–15
Iocrinus subcrassus	—	—	X	—	—	15–20
Ohiocrinus brauni	—	—	—	—	X	?
Tenuicrinus longibasalis	—	—	—	—	X	?
Monobathrid camerates						
Canistrocrinus typus	—	—	—	—	X	?
Compsocrinus miamiensis	—	—	—	—	X	?
Glyptocrinus decadactylus	—	—	X	—	—	5–40
Glyptocrinus fornshelli	—	—	—	—	X	?
Pycnocrinus dyeri	—	—	X	—	—	5–40
Xenocrinus baeri	—	—	X	—	—	?
Diplobathrid camerates						
Gaurocrinus nealli	—	—	X	—	—	?
Pararchaeocrinus regulosus	—	—	—	—	X	?
Ptychocrinus parvus	—	—	—	—	X	?
Rhaphanocrinus sculptus	—	—	X	—	—	?
Cladids						
Cupulocrinus polydactylus	X	—	—	—	—	?
Dendrocrinus caduceus	—	—	—	—	X	?
Merocrinus curtus	X	—	—	—	—	50–70
Plicodendrocrinus casei	—	—	—	X	—	?
Quinquecaudex cincinnatiensis	—	—	—	—	X	?

fasts. We have made use of the extensive collections of the Cincinnati Museum Center and collections presently housed at the University of Cincinnati. In particular, we incorporate data from three major accumulations of articulated crinoids from near the base and the top of the Stamping Ground Member near Swallowfield and Georgetown, Kentucky, respectively, and the Fairview–Miamitown interval near Maysville, Kentucky (see Brett and Algeo, 2001, and Taha McLaughlin, 2005, for details on localities in Kentucky), in addition to hardground collections from the Point Pleasant Member of the Lexington Limestone and the overlying Kope Formation at various localities in Hamilton County, Ohio, and northern Kentucky.

Crinoid calyx heights (base of calyx to top of radial plates) and column diameters were measured to the nearest tenth of a millimeter with digital calipers. Column lengths for complete crinoids were measured to the nearest millimeter with string to conform to curvature.

Anomalocrinus holdfast populations were measured on hardgrounds and bryozoan substrates. To examine population structure of different cohorts, holdfasts were assigned to one of five categories on the basis of degree of wear, as follows: 1) perfectly preserved with columnal facet sharply defined, wrinkled exterior stereom intact, and/or portions of distal pentameres intact; 2) column facet slightly worn, no columnals; external stereom in place but slightly worn; 3) column facet worn, internal canals barely exposed; 4) column facet worn, internal canals extensively exposed; 5) holdfast heavily degraded with column facet worn to near obscurity; commonly only marked by remnant basal disk. In addition, color and overgrowth relationships with other holdfasts and bryozoans were recorded. Descriptive statistics (mean, variance) were computed with Excel. Other statistics were calculated with PAST (Hammer et al., 2001).

Table 2.1. *Maximum column lengths and holdfast types of common Late Ordovician crinoid species.*

Attachment and Column Morphologies in Upper Ordovician Crinoids

Late Ordovician crinoids from the Cincinnati Arch region exhibit varying stem lengths and distinct modes of attachment that may relate to their facies distributions and life history strategies. These include cemented discoidal holdfasts, multiplated "lichenocrinid" holdfasts, and distal coils. Unlike most living crinoids, none of these crinoids possessed flexible cirri, pseudocirri, or other appendages (Table 2.1).

SIMPLE DISPARID CRINOIDS. Simple disparids crinoids had long, robust columns in proportion to their small crowns. In addition, *Ectenocrinus simplex* Hall, 1847, and *Cincinnaticrinus varibrachialis* Warn and Strimple, 1977, possessed small crowns with low density and open filtration fans (Fig. 2.3; Brower, 2005).

Cincinnaticrinus Warn and Strimple, 1977, was attached by multiplated holdfasts (formerly termed *Lichenocrinus*; hereafter referred to as lichenocrinus-type holdfasts), which were typically cemented to shells or reworked concretionary cobbles (Warn, 1973; Ausich, 1999; Brower, 2005, personal commun., 2005). Presumably these attachments are underrepresented relative to other types of cemented holdfasts because they were subject to disarticulation after decay. The column of *Cincinnaticrinus* and

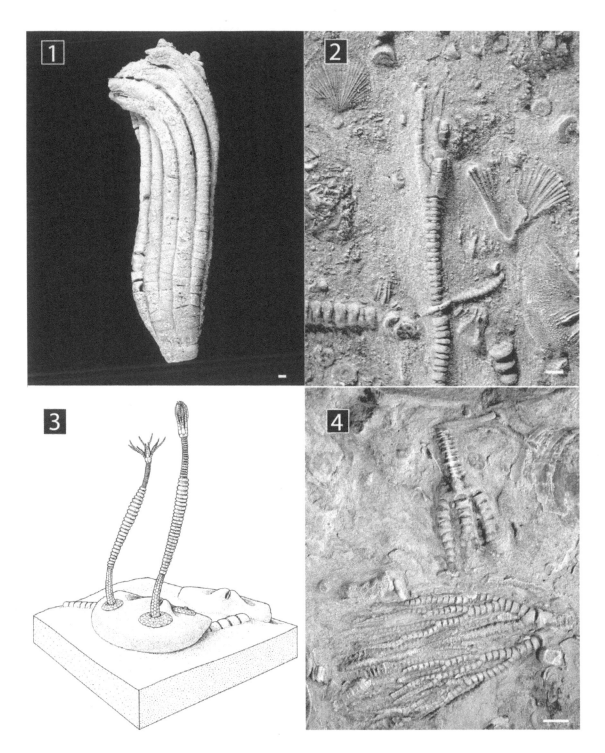

Figure 2.3. Common disparid crinoids of the late Mohawkian and Cincinnatian. *1, **Ectenocrinus simplex** crown and column, Kope Formation, CMC 11595. Scale bar = 1 mm. 2, **Cincinnaticrinus** crown and column, CMC 11600. Scale bar = 1 mm. 3, Reconstruction of **Cincinnaticrinus** sp. with attached lichenocrinus-type holdfast. Reprinted with permission from Ausich (1999). 4, **Iocrinus subcrasus** Meek and Worthen, 1865, Corryville Formation, Stonelick Creek, Stonelick, Ohio, CMC 44363. Scale bar = 5 mm.*

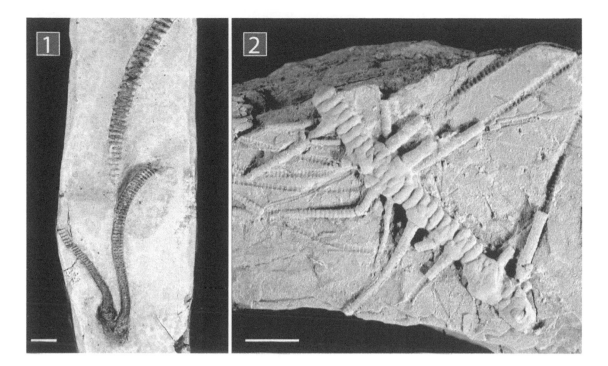

Ectenocrinus Miller, 1889, underwent a dramatic change upward from meric into holomeric pentangular to circular columnals (Fig. 2.4A).

The holdfast of the common disparid crinoid *Ectenocrinus* has rarely been identified but is apparently similar to that of *Cincinnaticrinus*, except that the associated dististele is trimeric rather than pentameric, a finding based on complete specimens of *E. simplex* from the Galena Group in Minnesota and the Trenton Group (Rust Member) in the Walcott-Rust quarry in New York State (Brower, 1992). Undoubtedly, many lichenocrinid holdfasts that have been attributed to *Cincinnaticrinus* actually belonged to *Ectenocrinus*.

Survey of *Cincinnaticrinus* holdfasts in collections of the Cincinnati Museum Center shows that a majority were attached to brachiopods, primarily *Rafinesquina* Hall and Clarke, 1892; a few were on bryozoans, and a few were on hardgrounds or cobbles, trilobite fragments, other crinoids, bivalved mollusks, or coral (Table 2.2). Additionally, juveniles of *Cincinnaticrinus* had a propensity for settling on columns of conspecific adults. Brower (2005) illustrated an example of lichenocrinus-type holdfasts on columns of *C. varibrachialis* from the Mohawkian Rust Limestone (Trenton Group) in New York State. Similarly, Figure 2.4C shows a column of *C. varibrachialis* from the Edenian Kope Formation in Ohio with several small lichenocrinus-like holdfasts and dististeles attached that presumably belonged to this species. These observations demonstrate gregarious settlement of juveniles in the vicinity of established adults; whether the settling larvae were genuinely selecting for adults or merely utilizing the limited substrates available in their environments is unclear.

Both *Cincinnaticrinus* and *Ectenocrinus* frequently occur in large, monospecific clusters, with masses of current aligned columns (crinoid

Figure 2.4. *Lichenocrinus-type holdfasts. 1, Specimen of the brachiopod* **Rafinesquina** *showing attached lichenocrinus-type holdfasts of* **Cincinnaticrinus** *with intact pentameric dististeles, Kope Formation, USNM 86647. 2, Column of* **Cincinnaticrinus** *with several minute lichenocrinus-type holdfasts, attached, presumably belonging to other individuals of* **Cincinnaticrinus***, Kope Formation, Southgate Member, Alexandria submember; Sycamore Creek, Indian Hill, Ohio, CMC 50683. Scale bars = 5 mm.*

Table 2.2. *Attachments of lichenocrinus type to various substrates.*

Substrate	n
Brachiopod	107
Bryozoan	23
Crinoid columnals	20
Reworked concretions	7
Bivalved mollusks	5
Trilobite fragments	4
Coral	2
Total	168

Note.—All lichenocrinus-type holdfasts from the Cincinnati Museum collections of the Cincinnatian series of Ohio and Kentucky were surveyed.

logjams) in mudstones (Fig. 2.5; *Ectenocrinus* logjam). The similar size of the columns suggests that they represent cohorts of similar-aged individuals (Fig. 2.6). One sample from the middle Kope Formation with cross-sectional area of about 200 cm^2 yielded a total of 842 parallel-oriented columns (average density ~5 columns/cm^2) and 30 associated crowns of *Ectenocrinus*. Calyx heights were approximately normally distributed and ranged from 2.1 mm to nearly 7 mm (mean = 4.3; SD = 0.85 mm); complete column lengths could not be determined because the logjam is broken, but maximum column lengths were 29.2 cm.

DISPARID IOCRINUS. The disparid *Iocrinus* Hall, 1866, possessed a slender pentagonal column approximately 15–20 cm long. The column is heteromorphic with one level of nodals, and sections of the stem are typically curved. This crinoid has no evidence of a primary cemented holdfast. Rather, numerous specimens have looped distal coils, in some cases attached to other crinoid columns. In distal coils, the columnals are cuneiform or wedge-shaped, tapering toward the inner part of the coil (Fig. 2.7). Measurements of calyx heights from a single population have a nor-

Figure 2.5. *Bundles (logjams) of parallel, aggregated crinoid columns. 1, 2, Longitudinal and cross-sectional views of **Ectenocrinus simplex**.* Note intact crowns. Kope Formation, Economy Member, Southgate, Kentucky, CMC 38442. Scale bar = 5 mm.

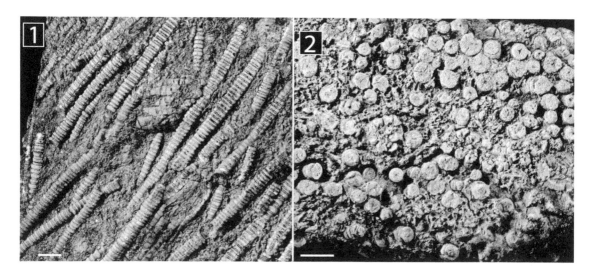

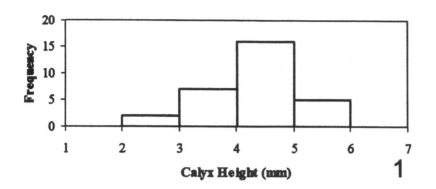

N=30, M=4.31, SD=0.85

Calyx Height (mm)

1

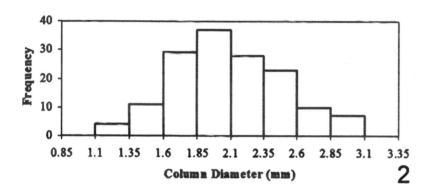

N=149, M=2.09, SD=0.41

Column Diameter (mm)

2

Figure 2.6. *Histograms of 1, columnal diameters and 2, calyx heights from an **Ectenocrinus** logjam population, lower Kope Formation, Economy Member, Southgate, Kentucky. Population size (N), mean (M), and standard deviation (SD) given above each histogram, CMC 38442.*

mally distributed, unimodal size frequency distribution with a relatively low variance (Fig. 2.8).

CEMENTED SOLID HOLDFASTS IN *MEROCRINUS*. The cladid crinoid *Merocrinus* Walcott, 1883, was similar to the disparids in possessing a long column (up to 80 cm in an incomplete column; Fig. 2.9). However, the column was strictly homeomorphic throughout, composed of thin, circular, holomeric columnals, with a small, round lumen. There are no nodals, and the column is nearly cylindrical with no tendency toward tapering. The profile of the column is nearly straight-sided, and isolated pluricolumnals are almost entirely straight or slightly curved.

The holdfast is poorly known; surprisingly few specimens possess the column intact. However, a single specimen from the lower Kope, Fulton submember, which is the only source for *Merocrinus curtus* Ulrich, 1879, has a small, solid, discoidal holdfast cemented to a bryozoan with an intact distal column resembling that of *Merocrinus* (Fig. 2.9). This single specimen suggests permanent, cemented attachment in *M. curtus.* Surprisingly, few such holdfasts have been noted in association with *Merocrinus*

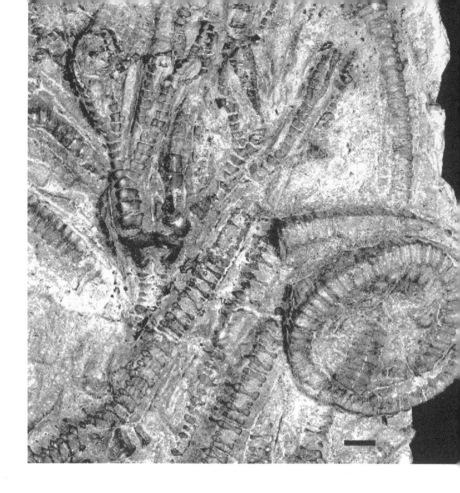

Figure 2.7. *Distal column of **Iocrinus subcrassus** showing coiled holdfast with cuneiform columnals, Corryville Formation, USNM 530470. Scale bar = 3 mm.*

Figure 2.8. *Size frequency distribution of a population of **Iocrinus subcrassus** from Corryville Formation, Stonelick Creek, Stonelick, Ohio, CMC 44363.*

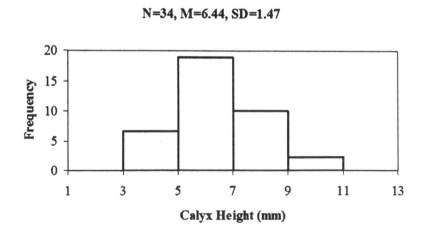

N=34, M=6.44, SD=1.47

columns; possibly some have been identified as *Anomalocrinus* Meek and Worthen, 1865.

CAMERATE CRINOIDS: DISTAL COILS. Brower and Kile (1994) illustrated a specimen of *Pycnocrinus gerki* Kolata, 1986, with a small digitate holdfast attached to a brachiopod, whereas numerous specimens of both *Pycnocrinus* Miller, 1883, and *Glyptocrinus* Hall, 1847, from the Upper Ordovician exhibit coiled, distally tapered columns that may or may not have a minute cemented plate at the base of the column.

A recently discovered obrution deposit of *Glyptocrinus decadactylus* Hall, 1847, from the upper Fairview Formation (Miamitown Shale equivalent) at Maysville, Kentucky, provides important insights into the paleoautecology of these crinoids. A slab now displayed at the Cincinnati Museum Center (CMC 50668) has an extraordinary density of these crinoids, with nearly 230 individuals per square meter (Fig. 2.10, 2.11). Measurement of

Figure 2.9. ***Merocrinus curtus***. *1, Probable holdfast and distal column, Kope Formation, Fulton submember, Kentucky, CMC 6432. 2, Parallel-aligned **Merocrinus curtus** columns; note single crown. Kope Formation, Fulton beds, Bradford, Kentucky, CMC 6431. Scale bar = 5 mm.*

Figure 2.10. *Overview of large slab of* **Glyptocrinus decadactylus** *showing numerous articulated crowns and intact columns some with distal coils. Upper Fairview Formation (Miamitown Shale equivalent), U.S. Rte. 62–68, Maysville, Kentucky, CMC 50668. Scale bar = 1 cm. Photo by Brenda Hunda.*

more than 100 well-preserved calyxes from this slab shows a normally distributed population with a mean calyx height of 11.3 mm (SD = 2.4 mm; Fig. 2.12). More than 10% of the glyptocrinids from the Maysville occurrence have complete columns, and these have a range of lengths from 4 to more than 25 cm (Table 2.3). Column length has a moderately strong correlation with calyx height, although with considerable scatter (r^2 = 0.44, Pearson's correlation: p < 0.001; Fig. 2.13).

All of the Maysville specimens with complete columns have coils at the tapered distal end of the column (Fig. 2.11). Columnals are distinctly cuneiform, with the widest diameters on the outsides of the curves. Three specimens are preserved coiled around other columns or bryozoans, but most show open loops of rather consistent diameter, about 8 to 12 mm (N = 50; mean = 88 mm; SD = 2.3 mm).

Figure 2.11. *Detail of distally coiled holdfasts of* **G. decadactylus** *from the Maysville slab shown in Figure 2.10, CMC 50668.*

Figure 2.12. *Histogram of calyx heights of* **G. decadactylus** *from the Maysville slab shown in Figure 2.10.*

N=103, M=11.3 mm, SD= 2.43 mm

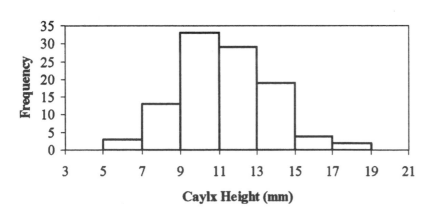

Caylx Height (mm)

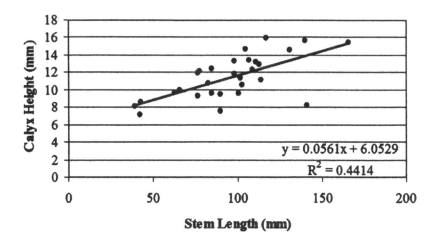

Figure 2.13. *Scatterplot of column length versus calyx height for 28 complete specimens of* **G. decadactylus**, *upper Fairview, Maysville, Kentucky.*

A newly discovered extraordinary crinoid bed in the upper part of the Stamping Ground Member of the Lexington Limestone at Georgetown, Kentucky, features masses of nonaligned, articulated *Glyptocrinus* buried in a cross-bedded skeletal grainstone (calcarenite) throughout a thickness of about 15 to 20 cm (Fig. 2.14). Stacked obrution deposits occur on the tops of multiple foreset beds, each approximately 5 cm thick and mutually separated by thin mud partings. Excellent preservation indicates that the crinoids were preserved near their living site. A slab measuring 13 by 80 cm yielded densities of 33 to more than 100 individuals per square meter. This occurrence must represent multiple stacked obrution deposits, almost certainly associated with a single catastrophic event. The size-frequency distribution of the crinoids on each bedding plane is similar to that of the Maysville occurrence (Fig. 2.15).

CEMENTED HOLDFASTS: ANOMALOCRINUS. The large disparid crinoid *Anomalocrinus* provides an important contrast to the taxa just discussed. Well-preserved specimens indicate that this crinoid possessed a large crown up to 85 mm in height with strongly ramulate/pinnulate arms, a rarity for a disparid crinoid (Fig. 2.16). *Anomalocrinus* was solidly attached by means of a robust cemented discoidal holdfast (Podolithus-type holdfast; Sardeson, 1908). These structures have elaborate ramifying canals, possibly utilized in providing nutrition to secretory tissues, and a coating of wrinkled stereom (Fig. 2.17).

Anomalocrinus occurs in a range of facies, representing moderately deep- to shallow-water environments but always associated with hard substrates (McLaughlin and Brett, 2007). The Point Pleasant member (informal designation) contains hardgrounds at multiple stratigraphic levels that are heavily encrusted by *Anomalocrinus*. Many of these surfaces are developed on truncated, trough cross-bedded, skeletal grainstones, and in some instances, holdfasts with portions of intact distal column are buried in a similar cross-bedded skeletal matrix. *Anomalocrinus* holdfasts of similar size also encrust reworked concretionary cobbles and bryozoans in mudstone matrix of the Kope Formation (Wilson, 1985), suggesting a distinctly lower energy setting.

Well-preserved holdfasts occur in densities of up to several dozen individuals on single 10 cm² cobbles, commonly positioned close to one an-

Calyx Height (mm)	Stem Length (cm)	Distal Coil Width (mm)
10.6	10.3	7.67
11.37	10.2	7.8
12.1	7.8	8.63
9.55	10.1	9.17
11.81	9.8	9.19
8.23	14.1	11.71
13.3	9.8	15.67
13.25	11.1	11.8
12.95	11.3	9.05
11.97	11.2	—
10.76	8.3	9.07
8.13	3.9	5.92
15.63	14	14.92
8.64	4.3	6.8
12.33	10.9	9.6
13.42	10.7	9.24
7.58	9	10.7
12.41	8.5	6.95
7.01	29	21.26
14.59	13.1	8.02
7.15	4.2	9.58
11.2	11.4	8.43
15.25	7.5	—
9.71	6.3	6.63
15.9	11.7	8.91
12.2	6.3	—
15.45	16.6	9.38
7.85	12.5	—
9.32	7.7	6.63
11.95	7.7	6.72
9.98	6.6	5.96
9.51	9	9.84
9.62	8.5	7.9
9.32	8.2	—
14.72	10.5	9.86

Table 2.3. *Calyx heights, column lengths, and the width of the distal coil for **Glyptocrinus decadactylus** from the obrution deposit illustrated in figure 2.10, upper Fairview Formation, Maysville, Kentucky.*

Note. Measurements from slab (Cincinnati Museum 50668); individual specimens are not labeled.

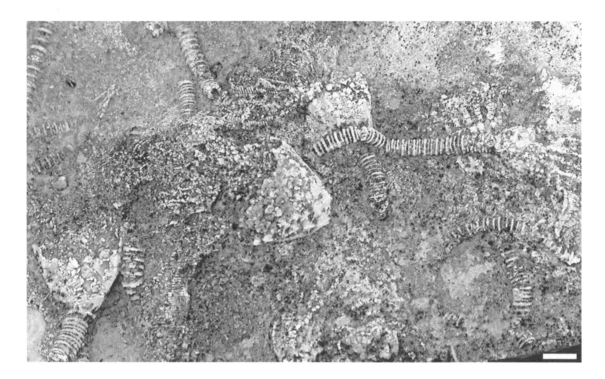

Figure 2.14. *Detail of the slab of* **Glyptocrinus** *sp., Strodes Creek Member of the Lexington Limestone, Georgetown, Kentucky, CMC 50875.*

other. *Anomalocrinus* occurs primarily on hardgrounds and reworked concretions; it is rare on small skeletal substrates such as brachiopods (McLaughlinand Brett, 2007). A particularly spectacular specimen features a moundlike bryozoan colony approximately 30 cm in maximum diameter with more than 200 *Anomalocrinus* holdfasts encrusting and partially embedded in the zoarium (Fig. 2.18).

Size frequency plots of *Anomalocrinus* holdfasts indicate that two or more cohorts of crinoids were present on certain substrates (Fig. 2.19). Moreover, in some assemblages, holdfasts exhibit varying states of corrosion from pristine, with the external wrinkled stereomic covering intact and even with pentameric columnals articulated; to slightly corroded, with internal canals just visible; to highly corroded basal plates. A total of 180

Figure 2.15. *Size-frequency histograms for* **Glyptocrinus** *populations on two different bedding planes in the occurrence from the Strodes Creek Member of the Lexington Limestone, Georgetown, Kentucky.*

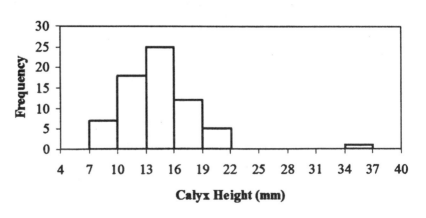

N=68, M=14.36, SD=3.91

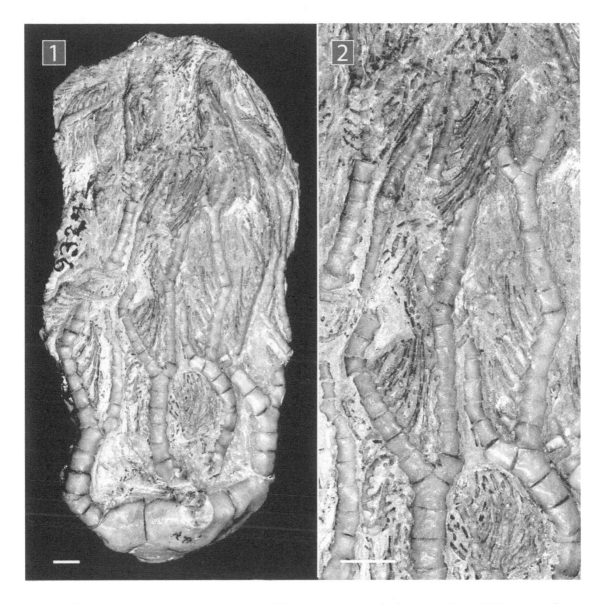

measurable specimens were assigned to one of five categories on the basis of overgrowth relationships and preservational quality. Maximum outer diameters of the holdfasts and diameters of column facets were measured. Both measurements were similar in all five taphonomic categories (Table 2.4; overall average external diameter [ED] = 15.88, SD = 5.72; mean columnal facet diameter [CFD]: 7.1 mm, SD = 2.01). A slight but nonsignificant increase in mean size was evident in the more altered categories (Table 2.4, e.g., for the perfectly preserved category 1: N = 30; mean ED = 6.46, SD = 1.52; mean CFD = 6.25 mm, SD = 1.43 mm; category 5: N = 21 mean ED = 7.41 mm, SD = 2.01 mm; CFD = 7.4 mm, SD = 2.21 mm). This may reflect the fact that older populations contained more gerontic individuals and/or a taphonomic bias against small and more fragile holdfasts. Both pristine and corroded populations from the same surface have similar size-frequency distributions, with low juvenile mortality. Most one-sided *t* tests

Figure 2.16. *Anomalo-crinus* crown showing densely ramulate (pinnu-late?) arms. 1, 2, Two views of the crown, USNM 93242. Scale bars = 5 mm.

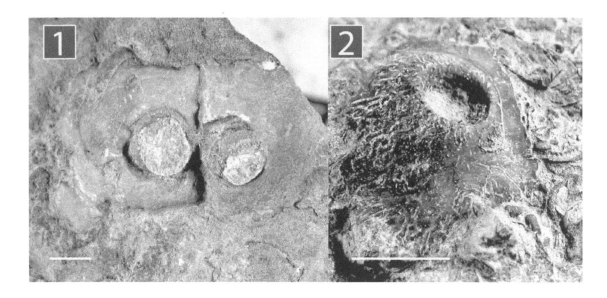

Figure 2.17. *Holdfasts of* **Anomalocrinus***. 1, Pristine specimens with pink, wrinkled stereom in a doublet of holdfasts grown closely adjacent to one another, hardground from lower Clays Ferry (Kope) Formation, Clays Ferry, Kentucky, CMC 50876. 2, Moderately corroded holdfast (category 4), Kope Formation, CMC 50877. Scale bars = 5 mm.*

failed to reject the hypothesis of equivalent means among the different categories, although an *F* test indicates significant differences in the variances of the populations (Table 2.4).

Discussion: Paleoautecology and Distribution of Late Ordovician Crinoids

WEAKLY ATTACHED DEEPER-WATER OPPORTUNISTIC CRINOIDS. The smaller disparid crinoids *Cincinnaticrinus* and *Ectenocrinus* were weakly attached to small substrates. Although Warn and Strimple (1977, p. 43–53) argued that the lichenocrinus-type holdfasts were only temporary and that adult *Cincinnaticrinus* became detached and free-living, the occurrence of relatively large, intact pentameric columns attached to these holdfasts indicates that the crinoids retained the holdfast through life (also see Guensburg, 1984; Brower, 2005, p. 153; Fig. 2.4C). Brower (2005) further argued that such crinoids with a bulk density only slightly greater than seawater and a slender profile, offering little drag, would not have had a problem with support or detachment under low-energy conditions. Disparids of this sort, lacking cirri on the distal column, would have had little or no ability to reattach and would have been at significant disadvantage if detached.

The retention of a highly sutured pentameric distal column afforded some flexibility at the attachment, whereas the remainder of the column was robust and relatively inflexible. This morphology may have permitted flexibility at the holdfast in the face of minor currents. Attachment to brachiopod shells or bryozoans also provided some stability, evidently sufficient for growth under some environmental conditions.

Cincinnaticrinus and *Ectenocrinus* typify the deepest-water facies of the Cincinnatian (Miller et al., 2001; Meyer et al., 2002); and indeed, their remains are abundant in truly basinal facies in the Ohio subsurface (Kirchner, 2005). It is probable that these taxa were confined to areas below average or frequent storm wave base. However, such crinoids were evidently subject to major disruption during large storms. The common logjams of

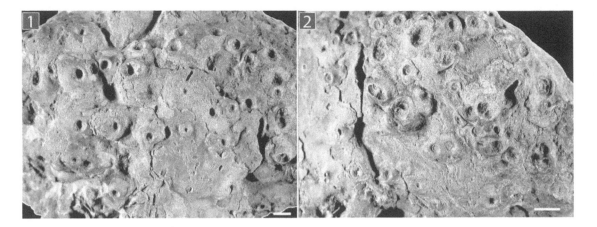

both *Cincinnaticrinus* and *Ectenocrinus* represent dislodgment, local current transport, and subsequent interference and pileup, as columns became lodged. The alignment of distal columns articulated to lichenocrinus-type holdfasts on brachiopods (Fig. 2.5) indicates that the storm currents oriented these specimens while still anchored and before breakage of the proximal columns. Individual horizons with local logjams have been traced across multiple localities for more than 10 km in some instances (CEB, personal observation). Thus, these crinoids show evidence of a weedy, opportunistic mode of life, with rapid establishment of enormous populations over large tracts of seafloor. Evidently, storm disturbance was sufficiently rare that crinoids were not adapted to withstand strong currents and were highly vulnerable to deep storm waves and/or gradient currents.

The long, cylindrical, and nontapering column of *Merocrinus* would have had little flexibility and served as a rigid support rod. Its considerable length may have been adaptive in elevating the crown above dysoxic, low-energy seafloors. Limited evidence indicates that *Merocrinus* was permanently attached to small skeletal substrates, such as ramose bryozoans, by a small holdfast. This mode of life seemingly would have confined *Merocrinus* to relatively quiet water; and indeed, the crinoid is known exclusively from deeper-water facies of the Fulton submember in the Kope Formation, where it occurs together with intact remains of the olenid trilobite *Triarthrus* Green, 1832, which has been argued to represent a possible chemosymbiotic, dysoxic adapted trilobite (Fortey, 1999). The deeper Trenton facies of New York State have also yielded *Merocrinus*. However, logjams of these crinoids indicate that they were occasionally detached and aligned by storm-generated currents.

Iocrinus co-occurs with other small, slender disparids, although it is somewhat more common in slightly shallower facies, as indicated by faunal gradient analysis (Holland et al., 2001; Meyer et al., 2002). The distinctly cuneiform nature of the distal columnals in *Iocrinus* indicates that these were permanent attachments, not flexible, prehensile structures, like sea horse tails. These coils were not formed in response to storm dislodgment. Evidently these columnals conformed to the curvature of the column,

Figure 2.18. *Two views of tightly clustered **Anomalocrinus** holdfasts on a single large bryozoan colony from the middle Kope Formation, upper Economy Member, Pioneer Valley submember of Brett and Algeo (2001); exposures behind Corporex/PMI Food Equipment Group Co. off Rt. 17, near Covington, Boone Co., Kentucky, CMC 50878.*

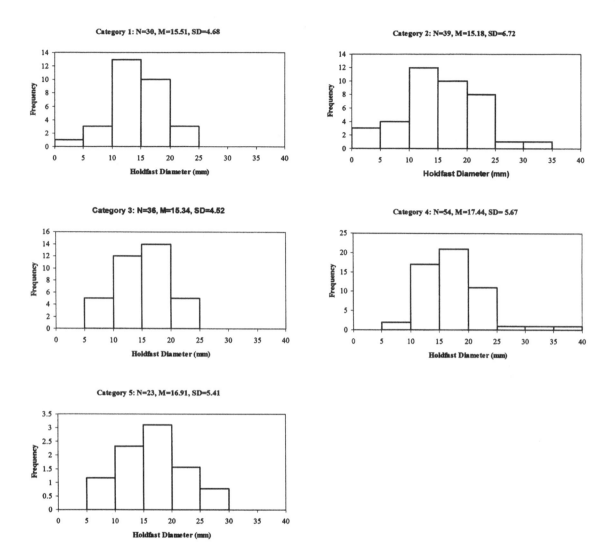

Figure 2.19. Size-frequency histograms of overall holdfast diameters for each of five preservational categories of **Anomalocrinus** holdfasts on bryozoan colony shown in Figure 3.18. Population size (N), mean (M), and standard deviation (SD) are listed above each histogram.

proving long-term growth in this position. At present, it is not certain whether the distal column was initially cemented and then broke free and had a period of initial flexibility (Kelly, 1978).

The occurrence of masses of similar-sized individuals suggests that dense monospecific populations representing, perhaps, single spatfalls were common (Fig. 2.8). Again, high-density populations of these crinoids have been found, and their rather consistent size indicates that they may reflect single cohorts that were disrupted and buried in masses. A weedy, opportunistic mode of life is again suggested.

GLYPTOCRINID CAMERATES: WEAKLY ATTACHED SHALLOW-WATER OPPORTUNISTS. Crinoids in shallower-water carbonate deposits from the Late Ordovician of the Cincinnati region tend to be dominated by a few camerate species with densely pinnulate arms, primarily the monobathrids *Glyptocrinus* and *Pycnocrinus* (Meyer et al., 2002). According to aerosol filtration theory, the high-density filtration fans of these crinoids are adapted to high current velocities to force water through the filter (Ausich,

Characteristic	Stem Diameter (mm)				
Preservation category	1	2	3	4	5
N	30	36	35	57	21
Mean	6.46	6.72	6.99	7.65	7.41
Standard Deviation	1.52	2.67	1.87	1.63	2.01

Characteristic	Holdfast Diameter (mm)				
Preservation category	1	2	3	4	5
N	30	39	36	54	23
Mean	14.50	15.18	15.34	17.44	16.91
Standard Deviation	4.68	6.72	4.53	6.67	5.41

Table 2.4. *Mean and standard deviation of* **Anomalocrinus** *column and holdfast diameters on bryozoan colony for the different preservation categories, Kope Formation, Covington, Kentucky.*

1980, 1999; Ausich et al., 1999b). This inference appears to be corroborated by the common occurrence of glyptocrinids in shallow-water environments. The distinctive calyx plates and columnals of glyptocrinid camerates form a major constituent of many bioclastic limestones, reflecting shallow, shoal-type settings of skeletal sediments. This might be construed to indicate that many crinoidal shoal deposits represent transported debris (Aigner, 1985). However, new discoveries of well-preserved assemblages of glyptocrinids in calcarenite deposits suggest that the crinoids lived in or near these high-energy settings, but herein lies a paradox. In order for the filtration fan process to be effective, the crinoids would need to have been firmly tethered in place; otherwise they would have drifted with currents instead of allowing currents to pass through the filtration mesh. Yet as noted above, these crinoids did not possess strongly cemented or radicular holdfasts.

Again, the wedge shape of distal columnals of glyptocrinids indicates that these were permanent attachments. If the distal tip of this column remained cemented by a small plate, as suggested by the observation of Brower and Kile (1994), the development of the coil would have involved a gradual spiraling of the entire crinoid animal around its attachment substrate, an upright support. In fact, a number of specimens do show evidence of having coiled around erect substrates, including bryozoans and other crinoid columns. However, most show a relatively consistent small-diameter open loop, supporting the notion that most individuals were attached to nonpreservable objects such as sponges or seaweeds.

The *Glyptocrinus* pocket from the Lexington Limestone at Georgetown raises a further intriguing paradox because the unstable shifting skeletal sand would seemingly inhibit colonization of the seafloor by echinoderms. One resolution must be that even these high-energy grainstone deposits were only reworked infrequently. Indeed, the excellent preservation of these crinoids in such grainstones provides evidence that they lived nearby. These facies were probably deposited below normal wave base, and/or local crinoid patches were in some way sheltered from normal wave action and most minor storms. These high-density patches appear to represent, at most, a few coexisting cohorts of crinoids that colonized the seafloor in large masses despite their vulnerability to mass mortality and burial. They persisted by rapid, opportunistic colonization of the seafloor between disruptive events. Dense patches of crinoids may also have survived in microtopographic lows, e.g., troughs of skeletal sand dunes.

It is notable that the columns of these camerates were relatively short; the highest are generally less than 30 cm high. The crown would have only occupied an intermediate tier height, If these columns became partially recumbent during growth, as is evident for some camerates, they may never have extended high into the water column. Concentrations in low-lying areas among bryozoan thickets or between megaripples would have been relatively sheltered from stronger currents or waves that could have dislodged them.

Further, we suggest that high densities of these crinoids may have aided in stabilization. The closely adjacent individuals evidently formed a tangled mass of mutually intertwined columns (Fig. 2.11) that, en masse, may have served to anchor the individual crinoids against dislodgment except under extraordinary current velocities.

ANOMALOCRINUS: AN ATTACHED EQUILIBRIUM SPECIES. *Anomalocrinus* presents a far different case from the glyptocrinids. Functionally, these large-crowned ramulate/pinnulate crinoids probably belonged to a similar feeding guild as uniserial camerate crinoids, in contrast to other disparids, although the somewhat more open filtration mesh of the arms may have permitted *Anomalocrinus* to occupy lower-energy conditions. Thus, unlike camerates, *Anomalocrinus* appears to occupy a broad range of depth-related facies, although constrained by large, stable substrates, especially hardgrounds, reworked concretions, and large bryozoan mounds. *Anomalocrinus* holdfasts rarely occur on small shells or small bryozoans. Presumably this crinoid was adapted for settlement on large or massive hard surfaces, frequently occupied by earlier generations of encrusters, including other *Anomalocrinus*, because of its substantially larger size and larger drag increased its probability of dislodgement. The pinnulate fans of these crinoids would have produced substantial drag in high-energy settings. Nonetheless, such crinoids were evidently capable of surviving on wave-swept hardgrounds.

This genus appears to have proliferated during certain sedimentation phases, primarily those associated with sediment-starved transgressive regimes, in which large areas of cemented hard substrate were available (McLaughlin and Brett, in press). During alternate times, *Anomalocrinus* became rare and perhaps survived only in isolated populations on large shells, massive bryozoans, and other skeletons.

There is strong evidence that many generations of *Anomalocrinus* successively occupied particular tracts of hard substrate. Most hardgrounds feature holdfasts in a wide array of preservational states, indicating multiple generations. In some cases, hardground cobbles and reworked concretions have holdfasts on all sides, indicating repeated colonization even after reworking. A heavily holdfast-encrusted bryozoan sphere from the Kope Formation is particularly instructive because it records multiple cross-cutting relationships as well as varied modes of preservation of the holdfasts. In many cases, highly corroded holdfasts, assigned to types 4 or 5, were overgrown by bryozoans and/or other holdfasts of types 3 to 4, and these in turn were encrusted by later generations of encrusters, including type 1 holdfasts. This implies that early generations of *Anomalocrinus* grew large holdfasts, died, were disarticulated, and then were corroded and mineralized before encrustation. Thus, the time required for a single step in this cross-cutting series was not only that of the lifespan of the crinoid (probably a decade or more, based on modern large crinoids; Messing, personal commun.) but also time for disarticulation and corrosion of the holdfast. Cross-cutting relationships indicate at least five generations of *Anomalocrinus* on this substrate. Moreover, an unknown but probably large number of generations may be embedded in the bryzoan colony, as indicated by outlines of holdfasts that are completely enclosed within the bryozoan mass. Obviously, larval settlement in these crinoids was gregarious, as indicated by the common occurrence of doublets or closely adjacent mutually overgrown holdfasts; thus, these were multigenerational.

Such evidence of time-averaging indicates that multiple generations of these crinoids occupied some tracts of hard seafloor for prolonged periods of time--obviously decades, and perhaps centuries. Moreover, the fact that mean population sizes and basic distributions are similar for different preservational cohorts provides strong indication of both low juvenile mortality and relatively stable population structure through time.

It is also notable that no logjams of *Anomalocrinus* have been observed to date, despite evidence for relatively high-density populations from holdfasts. Conversely, the lower ends of the columns were frequently partially buried upright. This evidence suggests that these crinoids were able to withstand many storm disturbance events and remain intact even with partial burial. Presumably a majority of these crinoids survived mass mortalities to die of natural causes and disintegrate without burial. Collectively, this evidence indicates that *Anomalocrinus* behaved as an equilibrium (K-selected) species specialized to stable hard substrates. It maintained stable populations in given areas over many generations.

COMPARISON WITH OTHER LATE ORDOVICIAN CRINOID ASSOCIATIONS. The Mohawkian–Cincinnatian crinoid associations discussed herein form part of a spectrum of Late Ordovician echinoderm associations. Only a few studies provide detailed data on holdfasts for entire assemblages, the most notable being the studies of Lewis (1982), Guensburg (1984), and Brower (1994, 1995; personal commun., 2005). Frest et al. (1999) presented a more limited observation of holdfasts. These represent slightly older (Turinian), diverse echinoderm assemblages, but they present interesting similarities and contrasts to the present study.

Tables 2.5 and 2.6 compares the pelmatozoan faunas of several well-known echinoderm assemblages from the Upper Ordovician Turinian, Chatfieldian (Kirkfieldian, Shermanian), and early Cincinnatian (Edenian, Maysvillian) in eastern North America and summarizes information on holdfast types. It is notable that most Late Ordovician pelmatozoans had small, solid holdfast disks. In particular, most rhombiferans appear to have had tiny cemented distal tips and rested freely on the substrate. Most camerates, except *Reteocrinus* Billings, 1859, appear to have lacked ce-

			Stem Diameter		
	1	2	3	4	5
1	X	0.789	0.276	0.721	0.097
2		X	0.560	0.031	0.294
3			X	0.079	0.458
4				X	0.598
5					X

			Holdfast Diameter		
	1	2	3	4	5
1	X	0.451	0.289	0.008	0.057
2		X	0.906	0.084	0.300
3			X	0.067	0.235
4				X	0.705
5					X

Table 2.5. *T-tests for significance of differences in means and standard deviations of column diameters and maximum diameters of **Anomalocrinus** holdfasts of five distinct preservation categories, Kope Formation, Covington, Kentucky.*

mented holdfasts and possessed distally coiled columns; *Iocrinus* is a rare example of a noncamerate that utilized distal coiling. Small, gracile disparids were attached, at least as juveniles, by lichenocrinus-type holdfasts. Robust, well-cemented holdfasts, including discoidal and rare radices, were typical of large calyx cladids (e.g., *Carabocrinus* Billings, 1857) and somewhat analogous disparids *Anomalocrinus* and *Cleiocrinus* Billings, 1857, typically found in association with large skeletal substrates and hardgrounds.

As shown in Tables 2.5 and 2.6, the Cincinnatian Kope and Fairview assemblages stand out as low-diversity crinoid-dominated associations. In contrast to most other Late Ordovician assemblages, the Cincinnatian assemblages generally lack blastozoans and robust cladids, such as *Carabocrinus*. This is clearly not simply a matter of loss of these taxa in the later Ordovician, because assemblages of equivalent age in the upper Mississippi Valley show the persistence of abundant and diverse pelmatozoan associations, including blastozoans, with many genera common to the diverse Turinian and Chatfieldian echinoderm associations.

The Cincinnatian crinoid-dominated assemblages are most similar to those of the Rust Formation in the Trenton Falls area of New York in that both contain an abundance of gracile disparid (*Cincinnaticrinus, Ectenocrinus, Iocrinus*) and cladid (*Merocrinus, Dendrocrinus* Hall, 1852) crinoids and glyptocrinid camerates, and few blastozoans or robust cladids. The Trenton occurrence represents a relatively deep shelf setting typified by episodic pulses of fine carbonate and terrigenous mud sedimentation. We interpret the Cincinnatian Rust-type associations as stressed, opportunistic crinoid populations. It is notable that similar associations are known from thick siliciclastic successions in the Martinsburg and Snake Hill formations of the Appalachian foreland basin (Bassler and Moodey, 1943; CEB, personal observation). Most members of this association (e.g., cincinnaticri-

Stem Diameter				
	1	2	3	4	5
1	X	0.003	0.279	0.002	0.066
2		X	0.039	0.001	0.380
3			X	0.372	0.366
4				X	0.076
5					X

Holdfast Diameter				
	1	2	3	4	5
1	X	0.151	0.425	0.620	0.830
2		X	0.020	0.250	0.280
3			X	0.159	0.338
4				X	0.832
5					X

nids, glyptocrinids) are also present in low abundance in more diverse pelmatozoan associations, as would be predicted of opportunistic species. A combination of low energy, high turbidity, and perhaps oxygen stress may have excluded more specialized blastozoans and large crinoids. The latter, in particular, may have persisted because of its ability to colonize large skeletal substrates and temporary hardgrounds and may have been an equilibrium species with life strategies more akin to those of *Anomalocrinus*, which stands out as an exception to the typical opportunistic assemblages of the Cincinnatian. Surprisingly, *Anomalocrinus* seems to be of minimal importance in most typical Late Ordovician carbonate shelf assemblages, appearing only rarely in the Lebanon Limestone assemblages (Tables 2.5 and 2.6).

Given the eurytopic, opportunistic nature of type Cincinnatian crinoids, one might anticipate that these organisms would have had extinction resistance; and indeed, there is evidence that these associations persisted for much of the Late Ordovician. However, none of the major genera typical of these associations survived the Hirnantian extinctions near the end of the Ordovician. Conversely, some specialized and rare members of the high-diversity assemblages (e.g., acolocrinids, glyptocystitids, protaxocrinids) did persist. Why hardy and broadly adapted organisms such as the cincinnaticrinids and *Ectenocrinus* failed to survive the fluctuations of the Late Ordovician remains a major mystery.

Table 2.6. *F-tests for significance of differences in means and standard deviations of column diameters and maximum diameters of* **Anomalocrinus** *holdfasts of five distinct preservation categories, Kope Formation, Covington, Kentucky.*

Summary

The results of this study suggest the following generalizations about Late Ordovician crinoids:

1. Different modes of attachment were associated with different life history strategies and environmental distribution patterns.

2. Small, slender, long-stemmed disparids were weakly anchored by lichenocrinus-type or small discoidal holdfasts and were confined to deeper-water, low-energy sites.
3. High density clumps of weakly attached camerate crinoids in shallow water may have had a behavioral strategy that favored anchorage of interconnected masses of semirecumbent columns.
4. Despite frequent mass mortalities, loosely attached Ordovician crinoids presumably survived because of rapid proliferation as r-selected generalists. Disruptive events were sufficiently rare that they did not provide a significant selective factor for firm attachments.
5. Firmly attached holdfasts were associated with the less common large crinoid *Anomalocrinus*, which was a hard substrate specialist. These crinoids show moderate population sizes, persistence in given areas over prolonged time spans, and broad facies distribution; and they rarely, if ever, occur in logjams.

Acknowledgments

We acknowledge the assistance of several people who made this study possible. Several members of the Cincinnati Dry-Dredgers (amateur paleontology association) contributed material used in this study. R. Fine excavated and donated the large crinoid encrusted bryozoan mass from the Kope Formation; the students of the 2005 Field Geology of the Cincinnati Region course recovered the large glyptocrinid slab from the Stamping Ground Member at Georgetown; the Cincinnati Museum slab of glyptocrinids from Maysville was excavated by D. Cooper from a horizon originally discovered by F. Gahn and meticulously cleaned by S. Vergiles. B. Hanke, F. Gahn, and D. Meyer provided assistance with the photographs used in the study. B. Hanke and F. Gahn facilitated access to specimens in the Cincinnati Museum and U.S. National Museum Springer Collection, respectively. D. Meyer also made specimens of *Anomalocrinus*, *Merocrinus*, and *Ectenocrinus* available for study. S. Taha McLaughlin provided editorial assistance. W. I. Ausich and T. E. Guensburg provided reviews that improved the chapter.

References

Aigner, T. 1985. Storm Depositional Systems: Dynamic Stratigraphy in Modern and Ancient Shallow Marine Sequences: Lecture Notes in the Earth Sciences 3. Springer-Verlag, Berlin, 174 p.

Ausich, W. I. 1980. A model for niche differentiation in Lower Mississippian crinoid communities. Journal of Paleontology, 54:273–288.

Ausich, W. I. 1999. Upper Ordovician of the Cincinnati, Ohio area, USA, p. 75–80. *In* H. Hess, W. I. Ausich, C. E. Brett, and M. Simms (eds.), Fossil Crinoids. Cambridge University Press, Cambridge.

Ausich, W. I., and T. K. Baumiller. 1996. Crinoid stalk flexibility: theoretical predictions and fossil stalk postures. Lethaia, 29:47–59.

Ausich, W. I., C. E. Brett, and H. Hess. 1999a. Taphonomy, p. 50–54. *In* H. Hess, W. I. Ausich, C. E. Brett, and M. Simms (eds.), Fossil Crinoids. Cambridge University Press, Cambridge.

Ausich, W. I., C. E. Brett, H. Hess, and M. J. Simms. 1999b. Crinoid form and function, p. 2–30. *In* H. Hess, W. I. Ausich, C. E. Brett, and M. Simms (eds.), Fossil Crinoids. Cambridge University Press, Cambridge.

Bassler, R. S., and M. W. Moodey. 1943. Bibliographic and faunal index of Paleozoic Pelmatozoan echinoderms. Geological Society of America Special Papers, 45, 734 p.

Billings, E. 1857. New species of fossils from Silurian rocks of Canada. Canada Geological Survey Report of Progress 1853–1856, 1856:247–345.

Billings, E. 1859. On the Crinoideae of the Lower Silurian rocks of Canada. Canadian Organic Remains, Decade 4. Canada Geological Survey, 72 p.

Brett, C. E. 1985. Pelmatozoan echinoderms on Silurian bioherms in western New York and Ontario. Journal of Paleontology, 59:820–838.

Brett, C. E., and T. J. Algeo. 2001. Stratigraphy of the Upper Ordovician Kope Formation in its type area (northern Kentucky), including a revised nomenclature, p. 47–64. *In* T. J. Algeo and C. E. Brett (eds.), Sequence, Cycle, and Event Stratigraphy of Upper Ordovician and Silurian Strata of the Cincinnati Arch Region. Kentucky Geological Survey Guidebook 1, Kentucky Geological Survey, Lexington.

Brett, C. E., and A. Seilacher. 1991. Fossil Lagerstätten: a taphonomic consequence of event sedimentation, p. 283–297. *In* G. Einsele, W. Ricken, and A. Seilacher (eds.), Cycles and Events in Stratigraphy. Springer-Verlag, New York.

Brett, C. E., T. J. Algeo, and P. I. McLaughlin. 2003. Use of event beds and sedimentary cycles in high-resolution stratigraphic correlation of lithologically repetitive successions, p. 315–350. *In* P. J. Harries, (ed.), High-Resolution Approaches in Stratigraphic Paleontology. Kluwer Academic Publishers, The Netherlands.

Brett, C. E., P. I. McLaughlin, S. R. Cornell, and G. C. Baird. 2004. Comparative sequence stratigraphy of two classic Upper Ordovician successions, Trenton shelf (New York–Ontario) and Lexington Platform (Kentucky–Ohio): implications for eustasy and local tectonism in eastern Laurentia. Palaeogeography, Palaeoclimatology, Palaeoecology, 210:295–329.

Brower, J. C. 1992. Hybocrinid and disparid crinoids from the Middle Ordovician Galena Group (Dunleith Formation) of northern Iowa and southern Minnesota. Journal of Paleontology, 66:973–993.

Brower, J. C. 1994. Camerate crinoids from the Middle Ordovician (Galena Group, Dunleith Formation) of northern Iowa and southern Minnesota. Journal of Paleontology, 68:570–599.

Brower, J. C. 1995. Eoparisocrinid crinoids from the Middle Ordovician (Galena Group, Dunleith Formation) of northern Iowa and southern Minnesota. Journal of Paleontology, 69:351–366.

Brower, J. C. 2005. The paleobiology and ontogeny of *Cincinnaticrinus variabrachialus* (Warn and Strimple, 1997) from the Middle Ordovician (Shermanian) Walcott-Rust Quarry of New York. Journal of Paleontology, 79:152–174.

Brower, J. C., and K. M. Kile. 1994. Paleoautecology and ontogeny of *Cupulocrinus levorsoni* (Kolata), a Middle Ordovician crinoid from the Guttenberg Formation of Wisconsin, p. 25–44. *In* E. Landing (ed.), Studies in Stratigraphy and Paleontology in Honor of Donald W. Fisher. New York State Museum Bulletin, 431.

Cressman, E. R. 1973. Lithostratigraphy and depositional environments of the Lexington Limestone (Ordovician). U.S. Geological Survey Professional Paper, 768, 61 p.

Davis, R. A., and R. J. Cuffey. 1998. Sampling the layer cake that isn't: the stratigraphy and paleontology of the "type Cincinnatian." Ohio Division of Geological Survey Guidebook, 13, 194 p.

Frest, T. J., C. E. Brett, and B. D. Witzke. 1999. Caradocian–Gedinnian echinoderm associations of Central and Eastern North America, p. 638–783. *In* A. J. Boucot and J. D. Lawson (eds.), Paleocommunities--A Case Study from the Silurian and Lower Devonian. Cambridge University Press, Cambridge.

Fortey, R. 1999. Olenid trilobites as chemoautotrophic symbionts. Acta Universitatis Carolinae, Geologica, 43(1–2):355–356.

Green, J. 1832. A Monograph on the Trilobites of North America. Philadelphia, Pennsylvania, 93 p.

Guensburg, T. E. 1984. Echinodermata of the Middle Ordovician Lebanon Limestone, central Tennessee. Bulletins of American Paleontology, 86(319), 100 p.

Hall, J. 1847. Palaeontology of New York, volume 1, Containing descriptions of the organic remains of the lower division of the New-York system (equivalent of the Lower Silurian rocks of Europe). Natural History of New York, Albany, New York, 6, 338 p.

Hall, J. 1852. Palaeontology of New York, volume 2, Containing descriptions of the organic remains of the lower middle division of the New-York system. Natural History of New York, New York, D. Appleton & Co., 6, 362 p.

Hall, J. 1866. Descriptions of new species of Crinoidea and other fossils from the Lower Silurian strata of the age of the Hudson-River Group and Trenton Limestone. Albany, New York, 17 p.

Hall, J., and J. M. Clarke. 1892. An introduction to the study of the genera of Palaeozoic Brachiopoda. New York Geological Survey, 8(1), 367 p.

Hammer, Ø., D. A. T. Harper, and P. D. Ryan. 2001. PAST: Paleontological Statistics Software Package for Education and Data Analysis. Palaeontologia Electronica, 4(1).

Hay, H. B. 1981. Lithofacies and formations of the Cincinnatian Series (Upper Ordovician), southeastern Indiana and southwestern Ohio. Unpublished Ph.D. dissertation, Miami University, Oxford, Ohio, 236 p.

Hay, H. B. 1998. Paleogeography and paleoenvironments, Fairview through Whitewater formations (Upper Ordovician, southeastern Indiana and southwestern Ohio), p. 120–134. *In* R. J. Cuffey and R. A. Davis (eds.), Sampling the Layer Cake that Isn't: The Stratigraphy and Paleontology of the Type-Cincinnatian. Ohio Division of Geological Survey Guidebook, 13.

Hess, H. 1999. Tertiary, p. 237–244. *In* H. Hess, Ausich, W. I., C. E. Brett, and M. Simms (eds.), Fossil Crinoids. Cambridge University Press, Cambridge.

Holland, S. M. 1993. Sequence stratigraphy of a carbonate-clastic ramp: the Cincinnatian Series (Upper Ordovician) in its type area. Geological Society of America Bulletin, 105:306–322.

Holland, S. M., A. I. Miller, and D. L. Meyer. 2001. Sequence stratigraphy of the Kope–Fairview Interval (Upper Ordovician, Cincinnati, Ohio area), p. 93–102. *In* T. J. Algeo and C. E. Brett (eds.), Sequence, Cycle, and Event Stratigraphy of Upper Ordovician and Silurian Strata of the Cincinnati Arch Region. Kentucky Geological Survey Guidebook, 1.

Holland, S. M., and M. E. Patzkowsky. 1998. Sequence stratigraphy and relative sea level history of the Middle and Upper Ordovician of the Nashville Dome, Tennessee. Journal of Sedimentary Research, 68:684–699.

Holland, S. M., and M. E. Patzkowsky. 2004. Ecosystem structure and stability: Middle Upper Ordovician of central Kentucky, USA. Palaios, 19:316–331.

Kelly, S. M. 1978. Functional morphology and evolution of *Iocrinus*, an Ordovician disparid inadunate crinoid. Unpublished M.A. thesis, Indiana University, Bloomington, 80 p.

Kirchner, B. 2005. An integrated analysis of lateral trends in the faunal and sedimentologic character of meter-scale limestone-mudrock cycles in the Kope

Formation of the Cincinnati region. Unpublished Ph.D. dissertation, University of Cincinnati, Ohio.

Kolata, D. R. 1986. Crinoids of the Champlainian (Middle Ordovician) Guttenberg Formation–Upper Mississippi Valley region. Journal of Paleontology, 60:711–718.

Lewis, R. D. 1982. Holdfasts, p. 57–64. *In* J. Sprinkle (ed.), Echinoderm Faunas from the Bromide Formation (Middle Ordovician) of Oklahoma. University of Kansas Paleontological Contributions Monograph, 1.

McLaughlin, P. I., and C. E. Brett. 2007. Signatures of sea level rise in a mixed carbonate-siliciclastic foreland basin succession: case study from the Upper Ordovician of the Cincinnati Arch. Palaios, 22:245–267.

McLaughlin, P. I., C. E. Brett, S. L. Taha McLaughlin, and S. R. Cornell. 2004. High-resolution sequence stratigraphy of a mixed carbonate-siliciclastic, cratonic ramp (Upper Ordovician; Kentucky-Ohio, USA): insights into the relative influence of eustasy and tectonics through analysis of facies gradients. Palaeogeography, Palaeoclimatology, Palaeoecology, 210:267–294.

Meek, F. B., and A. H. Worthen. 1865. Descriptions of new species of Crinoidea, etc. from the Paleozoic rocks of Illinois and some of the adjoining states. Proceedings of the Academy of Natural Sciences of Philadelphia, 17:143–155.

Meyer, D. L. 1973. Feeding behavior and ecology of shallow-water unstalked crinoids (Echinodermata) in the Caribbean Sea. Marine Biology, 22:105–129.

Meyer, D. L., and W. I. Ausich. 1983. Biotic interactions among Recent and among fossil crinoids, p. 377–427. *In* P. L. McCall (ed.), Biotic Interactions in Recent and Fossil Benthic Communities. Plenum Press, New York.

Meyer, D. L., and B. MacUrda Jr. 1977. Adaptive radiation of the comatulid crinoids. Paleobiology, 3:74–82.

Meyer, D. L., A. I. Miller, S. M. Holland, and B. F. Datillo. 2002. Crinoid distribution and feeding morphology through a depositional sequence Kope and Fairview formations, Upper Ordovician, Cincinnati Arch Region. Journal of Paleontology, 76:725–732.

Miller, A. I., S. M. Holland, D. L. Meyer, and B. F. Dattilo. 2001. The use of faunal gradient analysis for intraregional correlation and assessment of changes in sea-floor topography in the type Cincinnatian. Journal of Geology, 109:603–613.

Miller, S. A. 1883. *Glyptocrinus* redefined and restricted, *Gaurocrinus*, *Pycnocrinus* and *Compsocrinus* established, and two new species described. Journal of the Cincinnati Society of Natural History, 6:217–234.

Miller, S. A. 1889. North American Geology and Paleontology. Western Methodist Book Concern, Cincinnati, Ohio, 664 p.

Pope, M. C., and J. F. Read. 1997a. High-resolution surface and subsurface sequence stratigraphy of the Middle to Late Ordovician (late Mohawkian–Cincinnatian) foreland basin rocks, Kentucky and Virginia. American Association of Petroleum Geologists Bulletin, 81:1866–1893.

Pope, M. C., and J. F. Read. 1997b. High-resolution stratigraphy of the Lexington Limestone (late Middle Ordovician), Kentucky, U.S.A.: a cool-water carbonate-clastic ramp in a tectonically active foreland basin, p. 410–429. *In* N. James (ed.), Cool-Water Carbonates. SEPM (Society for Sedimentary Geology) Special Publication, 56.

Rasmussen, H. W. 1977. Function and attachment of the stem in Isocrinidae and Pentacrinitidae: review and interpretation. Lethaia, 10:51–57.

Sardeson, F. W. 1908. Discoid crinoidal roots and *Camarocrinus*. Journal of Geology, 16:239–254.

Taylor, W. L., and C. E. Brett. 1996. Taphonomy and paleoecology of echinoderm Lagerstätten from the Silurian (Wenlockian) Rochester Shale. Palaios, 11:118–140.

Taha McLaughlin, S. L. 2005. Sequence stratigraphy and faunal patterns of the middle Lexington Limestone (Upper Ordovician) in central Kentucky. Unpublished M.S. thesis, University of Cincinnati, Ohio, 143 p.

Ulrich, E. O. 1879. Descriptions of new genera and species of fossils from the Lower Silurian about Cincinnati. Journal of the Cincinnati Society of Natural History, 2:8–30.

Ulrich, E. O., and R. S. Bassler. 1914. Report on the Stratigraphy of the Cincinnati, Ohio Quadrangle. U.S. Geological Survey Open-File Report, Washington, D.C., 122 p.

Vermeij, G. 1987. Evolution and Escalation: An Ecological History of Life. Princeton University Press, Princeton, New Jersey, 527 p.

Walcott, C. D. 1883. Descriptions of new species of fossils from the Trenton Group of New York. New York State Museum of Natural History Annual Report, 35:207–214.

Warn, J. M. 1973. The Ordovician crinoid *Heterocrinus*, with reference to brachial variability in *H. tenuis*. Journal of Paleontology, 47:10–18.

Warn, J. M., and H. L. Strimple. 1977. The disparid inadunate superfamilies Homocrinacea and Cincinnaticrinacea (Echinodermata: Crinoidea), Ordovician–Silurian. Bulletins of American Paleontology, 72:1–138.

Weir, G. W., W. L. Peterson, and W. C. Swadley. 1984. Lithostratigraphy of Upper Ordovician strata exposed in Kentucky. U.S. Geological Survey Professional Paper, 1151-E, 121 p.

Wilson, M. A. 1985. Disturbance and ecologic succession in an Upper Ordovician cobble-dwelling hardground fauna. Science, 228:575–577.

Genus	Family/Superfamily	Stratigraphic Range
Allagecrinus Carpenter and Etheridge. 1881	Allagecrinidae	Devonian–Pennsylvanian
Allocatillocrinus Wanner, 1937	Allagecrinidae	Mississippian
Desmacriocrinus Strimple, 1966	Allagecrinidae	Mississippian
Kallimorphocrinus Weller, 1930	Allagecrinidae	Pennsylvanian
Litocrinus Lane and Sevastopulo, 1982b	Allagecrinidae	Mississippian–Permian
Thaminocrinus Strimple and Watkins, 1969	Allagecrinidae	Mississippian
Trophocrinus Kirk, 1930	Allagecrinidae	Mississippian
New genus A	Allagecrinidae	Mississippian
New genus B	Allagecrinidae	Mississippian
Acaraiocrinus Wanner, 1924*	Codiacrinacea	Mississippian–Permian
Clistocrinus Kirk, 1937*	Codiacrinacea	Mississippian–Pennsylvanian
Coenocystis Girty, 1908*	Codiacrinacea	Pennsylvanian–Permian
Cranocrinus Wanner, 1929	Codiacrinacea	Mississippian–Permian
Dichostreblocrinus Weller, 1930*	Codiacrinacea	Mississippian–Permian
Hypocrinus Beyrich, 1862	Codiacrinacea	Mississippian–Permian
Lageniocrinus de Koninck and Le Hon, 1854*	Codiacrinacea	Mississippian–Permian
Lampadosocrinus Strimple and Koenig, 1956*	Codiacrinacea	Mississippian–Pennsylvanian
Monobrachiocrinus Wanner, 1916	Codiacrinacea	Mississippian–Permian
Neolageniocrinus Arendt, 1970*	Codiacrinacea	Mississippian–Permian
Okacrinus Arendt, 2002*	Codiacrinacea	Mississippian
Genus C	Codiacrinacea	Mississippian
Amphipsalidocrinus Weller, 1930	Amphipsalidocrinidae	Silurian–Permian

Note.—*Streptostomocrinus Yakovlev*, 1927, is regarded as a junior synonym of *Acaraiocrinus* Wanner, 1924 (cf. Arendt, 2002). Extensions to published ranges are based on unpublished information.

* Abrachiate taxa.

Table 3.1. *Genera of microcrinoids that occur in the Carboniferous, their families, and their total stratigraphic ranges.*

PALEOBIOLOGY OF CARBONIFEROUS MICROCRINOIDS

3

George D. Sevastopulo

Microcrinoids are micromorph crinoids, defined arbitrarily as having an upper limit thecal height of 2 mm. Large samples of microcrinoids contain individuals at various stages of growth, with the largest individuals clearly not the juveniles of macrocrinoids in the same fauna. The smallest microcrinoid had a maximum dimension of the theca at maturity of less than 1 mm. Disparid, cladid, articulate, and possibly camerate crinoids had micromorph representatives. Their stratigraphic range is from the Silurian (or possibly Ordovician) to Cretaceous. Many Mesozoic microcrinoids were pelagic, whereas all Paleozoic forms were benthic.

Miniaturization of crinoids occurred in different clades as a result of different evolutionary and developmental mechanisms. It led to particular adaptations in feeding and reproduction. This chapter explores the paleobiological aspects of small size in Carboniferous microcrinoids. Terminology follows Ubaghs (1978). Specimens are held in the collections of the Illinois Geological Survey (catalog numbers prefixed ISGS), the Department of Geology, University of Iowa, Iowa City (catalog numbers prefixed SUI), and the Geological Museum, Trinity College, Dublin (catalog numbers prefixed TCD).

Crinoid genera that include Carboniferous micromorph representatives are listed in Table 3.1. With the exception of *Amphipsalidocrinus* Weller, 1930, which is tentatively classified as a camerate crinoid (Lane et al., 1985), they are assigned to the disparid family Allagecrinidae Carpenter and Etheridge, 1881, and the cladid superfamily Codiacrinacea Bather, 1890. The classification of cyathocrinine, cladid crinoids in Moore et al. (1978) is followed here rather than that of Arendt (1970), pending revision for the forthcoming edition of the *Treatise on Invertebrate Paleontology*, part T. Illustrations of the microcrinoid taxa listed are in Moore et al. (1978), with the exception of the following: *Litocrinus* (Lane and Sevastopulo, 1981, 1982b); new genus A, which will be proposed to accommodate allagecrinids such as *"Kallimorphocrinus" extensus* Wright, 1952, that have multifaceted radials in the A, B, and D rays but lack an anal notch; new genus B, which is discussed below; *Okacrinus* (Arendt, 2002); and genus C, which is discussed below.

Introduction

Locality	Paris Landing	Carrière Lemay	Feltrim Cover	Slieve-more Sligo	Floraville, Illinois	Weller Falls
Age	Tn	Tn	Tn	Vis	Vis	Mosc
Allagecrinus	—	—	—	1	—	—
Allocatillocrinus	—	—	—	—	1	—
Kallimorphocrinus	—	—	—	—	—	1
Litocrinus	1	—	2	3	—	—
New genus A	—	1	1	—	—	—
Cranocrinus	—	—	—	1	—	—
Dichostreblocrinus	—	—	—	—	—	1
Hypocrinus	—	—	—	1	—	—
Lampadosocrinus	1	—	1	—	—	—
Monobrachiocrinus	—	—	—	1	—	—
Neolageniocrinus	1	—	1	1	—	1
Genus C	—	—	—	1	—	—
Amphipsalidocrinus	1	—	1	1	—	1
Total number of species	4	1	6	10	1	4

Note.—See text for locality information and discussion.

Abbreviations.—Tn, Tournaisian; Vis, Viséan; Mosc, Moscovian.

Abundance, Species Richness, and Paleoenvironments of Carboniferous Microcrinoid Faunas

Table 3.2. *Number of species of microcrinoids in selected faunas of Carboniferous age.*

Microcrinoids were common constituents of Carboniferous benthic assemblages, but information about the environments they inhabited is incomplete, partly because of collection bias and partly because of the generally serendipitous discovery of microcrinoid faunas. Almost all faunas have been recovered from mudstone or shale because these rocks are amenable to chemical and physical disaggregation. However, many of the shales that yield microcrinoids are partings within or are interbedded with limestone. Silicified thecae, such as those described by Lane and Sevastopulo (1982a, 1982b) and Weller (1930) from the Perth Limestone (Moscovian), Redwood Creek, Warwick County, Indiana (Weller Falls in Table 3.2), and rare calcitic thecae found in residues after the etching of limestone in acetic acid show that microcrinoids inhabited environments floored by carbonate as well as siliciclastic sediment.

Although no systematic study has been undertaken, anecdotal evidence suggests that at least half the Carboniferous faunas that contain ossicles of larger crinoids also contain microcrinoids. Typically, the number of individuals of the most common species of larger crinoid in a washed sample of shale (estimated from the number of disarticulated radials) is less than the number of thecae of the most common microcrinoid in the sample. The species richness of six selected microcrinoid faunas that have been carefully studied, and that span the range of diversity encountered thus far in the Carboniferous, is listed in Table 3.2.

The Paris Landing fauna from the base of the New Providence Shale Formation (Tournaisian), Paris Landing, Tennessee, was partially described by Lane and Sevastopulo (1981). Smith (1978) interpreted the paleoenvironment as a soft, clay-rich, relatively deep-water, dysoxic basin floor, where sedimentation was slow. The fauna associated with the microcrinoids contains a few larger crinoids, blastoids, holothuroids, small brachiopods, solitary rugose corals, and abundant ostracodes. The species richness (one species in each of four genera) is comparable to that of other North American Tournaisian faunas described by Peck (1936) and Strimple and Koenig (1956). In terms of numbers of individuals, the microcrinoid fauna is dominated by *Litocrinus punctatus* (Lane and Sevastopulo, 1981).

The Tournaisian fauna from Carrière Lemay (Table 3.2) at Vaulx-les-Tournai, Belgium (Hibo and Tourneur, 1989; Sevastopulo, personal observation), is interesting in that it consists of a single species of microcrinoid, an undescribed species of new genus A, even though the accompanying macrofauna contains a diverse assemblage of larger crinoids, identified from disarticulated plates. Other elements of the fauna include a blastoid, ophiuroids, echinoids, a holothuroid, several species of brachiopods, bryozoans, solitary rugose and tabulate corals, and ostracodes. The paleoenvironment is considered to have been one of fairly deep water on the outer part of a ramp (Gaillard et al., 1999).

The five genera and six species of microcrinoids from the late Tournaisian Feltrim Cover Mudstone (Table 3.2) occur in dark mudstone of deep-water origin that overlies a Waulsortian carbonate bank (Hudson et al., 1966; Sevastopulo, 1969). In terms of abundance, the microcrinoid fauna is dominated by an undescribed species of *Litocrinus*. The associated fauna includes more than 19 species of larger crinoids, as well as blastoids, echinoids, a cyclocystoid, ophiuroids, brachiopods, bryozoans, solitary rugose and tabulate corals, ostracodes, trilobites, and cephalopods.

The most species-rich microcrinoid fauna, with at least 10 species distributed among eight genera, is from the late Viséan Meenymore Formation, Slievemore, County Sligo, Ireland (Sevastopulo, 2002). The habitat of the microcrinoids is interpreted as having been a shallow-water, marine environment, floored by clay and carbonate mud, some of which had lithified on the sea floor to form a hard substrate. In terms of abundance, the microcrinoid fauna is dominated by an undescribed species of *Litocrinus*. The accompanying fauna contains several taxa of larger crinoids, blastoids, asteroids, ophiuroids, a cyclocystoid, echinoids, and a holothuroid, as well as brachiopods, cephalopods, rare corals, and bryozoans.

Only a single species of microcrinoid, *Allocatillocrinus carpenteri* (Wachsmuth, 1882), occurs in washings of shale from the famous Ridenhower (Paint Creek) Formation (late Viséan) locality, near Floraville, Illinois, which has yielded numerous specimens of the blastoid *Pentremites* Say, 1820, as well as a diverse fauna of larger crinoids, asteroids, brachiopods, bryozoans, and corals. The largest specimens of *Allocatillocrinus carpenteri* known from other localities have thecae slightly larger than 2 mm, but the modal size is smaller. The shale has been interpreted as having been deposited in a shallow-water, back-barrier setting by Langhorne and Read (2001).

A large fauna of silicified microcrinoids (four species and four genera) has been obtained from the Perth Limestone Member of the Staunton Formation, Redwood Creek, Warren County, Indiana (Lane and Sevastopulo, 1982a, 1982b; Weller, 1930), referred to as the Weller Falls fauna in Table 3.2. In terms of abundance, the microcrinoid fauna is dominated by *Kallimorphocrinus astrus* Weller, 1930. The Perth Limestone is either of latest Atokan or earliest Desmoinesian age (Rexroad et al., 1998), equivalent to the Moscovian Stage. Rexroad et al. (1998) concluded that the limestone was formed in a shallow, nearshore, marine environment, of generally low to moderate energy.

In summary, Carboniferous microcrinoids are common and occur in a range of shallow- to deep-water environments. Individual genera appear to span a large range of depths. Faunas may be diverse, with as many as 10 species and eight genera in a single fauna. Allagecrinids, such as *Litocrinus* and *Kallimorphocrinus* Weller, 1930, are generally the numerically dominant elements of a fauna.

Relation to the Substrate

Examples of microcrinoids with their columns attached and with holdfasts preserved are extremely rare. In some specimens of allagecrinids, a few proximal columnals are preserved attached to the aboral cup. They are low, with a distinctive articular surface (see Lane and Sevastopulo, 1981, pl. 1, figs. 5, 6; and Lane and Sevastopulo, 1982b, pl. 3, fig. 13). In some specimens, they form a tapering proxistele, approximately the same height as the aboral cup. More complete columns are known in *Allocatillocrinus carpenteri* (Wachsmuth, 1882; Springer, 1923; Burdick and Strimple, 1982, pl. 10, figs. 11, 17; Sevastopulo, personal observation). The holotype (Illinois Geological Survey specimen ISGS2440), in which the cup is 2.2 mm high, has a portion of stem 49.4 mm long preserved. A larger specimen in the collections of the University of Iowa (SUI 48933), with a cup 2.74 mm high, has 92.4 mm of column preserved. A small specimen (SUI 48940 B) has a cup 1.25 mm high; its stem, which is judged to be complete, is 23.7 mm long. Specimen SUI 32479 has a theca 0.8 mm high, and probably had five arms; 8.9 mm of stem is preserved attached to the cup. These data suggest that the crown of *Allocatillocrinus* Wanner, 1937, and by inference probably of other allagecrinids, was maintained at a relatively high elevation above the substrate. Some of the columns of the suite of specimens of *Allocatillocrinus* discussed above are slightly curved, particularly at the level of the proxistele, suggesting that the proximal part of the stem was fairly flexible.

Mapes et al. (1986) reported a large number of *Allagecrinus coronarius* Gutschick, 1968 (regarded here as a junior synonym of *Allocatillocrinus carpenteri* [Wachsmuth, 1882]), within the living chamber of a large orthoconic nautiloid in the Imo Formation (Late Mississippian) of Arkansas. The concentration of thecae, mostly juveniles, within the shell contrasted with the much lower concentration outside the shell. Mapes et al. (1986) suggested that the microcrinoids were a colony begun by larvae, which had settled on the shell and which grew and reproduced before the whole assemblage was killed by an influx of sediment.

In contrast to the situation in allagecrinids, the few columns of codiacrinaceans preserved are short and rigid. The most complete specimen from the Carboniferous is an immature individual of *Monobrachiocrinus* Wanner, 1916 (Sevastopulo, 2002, fig. 4h), in which the column and holdfast are 1.6 mm in length (Fig. 3.1.1). The holdfast was clearly cemented to a firm substrate. Several other specimens of *Monobrachiocrinus* in the same fauna have short, straight lengths of column attached to the aboral cup. Arendt (1970, pl. 10, fig. 3; pl. 12, fig. 7) illustrated thecae of *Neolageniocrinus schichanensis* Arendt, 1970, and *Hemistreptacron abrachiatum* Yakovlev, 1926, also with short straight lengths of column attached. The most extreme case of the reduction of the stem is in genus C from the late Viséan fauna at Slievemore, County Sligo. The crinoid (Figs. 3.1.2, 3.1.3) consists of a low cylindrical tube, which was cemented to the substrate at its lower end where the cylinder was partially closed. The upper end exhibits five facets, to each of which a thecal plate would have been attached. By analogy with *Pilidiocrinus heckeri* Arendt, 1970 (genus C may be synonymous with *Pilidiocrinus* Wanner, 1937), the tube consists of fused infrabasals, and the plates in the next circlet were basals.

In summary, the small amounts of data available suggest that allagecrinid and codiacrinacean microcrinoids had different strategies regarding the position of the crown relative to the substrate. Allagecrinids had relatively long, slightly flexible columns, as seems to have been the case for many disparid crinoids, whereas several codiacrinaceans had short, rigid columns, which were cemented to firm substrates, rather similar to representatives of the post-Paleozoic, articulate family Eugeniacrinitidae.

Figure 3.1. *Codiacrinacean microcrinoids with short stems from the Meenymore Formation (Mississippian; late Viséan), Slievemore, County Sligo, Ireland, ×40). 1, **Monobrachiocrinus** sp. with complete stem and holdfast (TCD 54335). 2, 3, Genus C, showing fused infrabasals that were cemented to a firm substrate. 2, lateral view of a specimen with infrabasal circlet higher on one side (TCD 54336). 3, Oblique lateral view of a cylindrical circlet showing the distal margins of the infrabasals (TCD 54337).*

Feeding

The arms of a small number of species of allagecrinids are known from well-preserved material. The brachials are uniserial and, with the exception of the proximal facet of the first brachial, have articular surfaces that are nearly flat, suggesting that there was little movement between individual brachials (Figs. 3.2.1, 3.2.2). The hypothesis that each arm was essentially rigid is supported by the occurrence in several taxa—for example, *Litocrinus punctatus* (Lane and Sevastopulo, 1981, text-fig.1) and *Kallimor-*

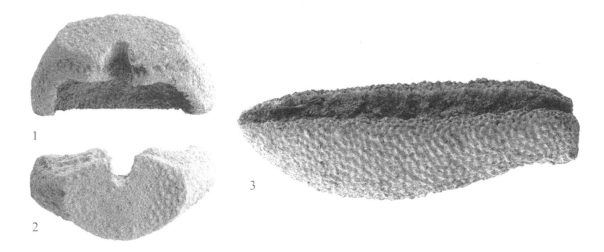

Figure 3.2. *Brachials of allagecrinid microcrinoids. 1, 2, First primibrachial of* **Litocrinus** *sp. nov. from the Meenymore Formation (Mississippian; late Viséan), Slievemore, County Sligo, Ireland (TCD 54338), ×60). 1, Adoral view with straight proximal articular surface at the base. 2, Nearly flat distal articular surface. 3, Oblique lateral view (×40) of second primibrachial of* **Kallimorphocrinus strimplei** *(Kirk, 1936) (TCD 54339) from the Dewey Limestone (Pennsylvanian), east of Dewey, NE1/4, NE1/4, sec. 27, T. 27 N, R. 13 E, Washington Co., Oklahoma, USA, showing scalloped margin of the ambulacral groove that accommodated cover plates.*

phocrinus astrus Weller, 1930 (Lane and Sevastopulo, 1982b, pl. 3, figs. 1, 2, 4)—of arms formed of only two or three brachials. When an allagecrinid was not feeding, the arms were held parallel to the polar axis of the theca with adjoining arms in lateral contact. In order to feed, the microcrinoid moved its arms outward. Each arm rotated about the fulcrum provided by the straight, narrow, proximal surface of the first brachial (Fig. 3.2.1) and the transverse ridge of the articular facet of the radial. It is not clear how far the arms could have opened, but the geometry of the radial articular facets suggests that it was less than 90°. When in the feeding position, each arm would have been held rigid by stiffening of the outer ligament, which was inserted in a deep pit in the radial, abaxial of the transverse ridge. Return to the resting position would have involved relaxation of the outer ligament and contraction of the inner ligaments, or of muscles, although there is no evidence in the architecture of the stereom for the latter.

Lane and Sevastopulo (1981) speculated that *Litocrinus punctatus* fed using cilia rather than podia. This now seems unlikely for two reasons. First, a nerve cord emerged from the theca through a canal immediately adaxial of the transverse ridge of the radial articular facet and passed through the first brachial to lie on the floor of the adoral groove of more distal brachials. It seems most likely that it served to innervate podia. Second, the adoral grooves of the brachials were almost certainly roofed with covering plates as tentatively inferred for *Litocrinus punctatus* by Lane and Sevastopulo (1981, p. 24). Their interpretation is supported by the morphology of brachials of *Kallimorphocrinus strimplei* (Kirk, 1936), which clearly have scalloped margins of the adoral grooves with which cover plates would have articulated (Fig. 3.2.3). The most probable way in which the cover plates would have been opened was by extension of the podia beneath. In small allagecrinids the adoral groove of the arms was narrow; for example, it was only 40 μm wide in small specimens of *Litocrinus punctatus* (Lane and Sevastopulo, 1981, p. 26). This suggests that only small particles of food were ingested and that allagecrinids exploited a narrower range of food resources than larger crinoids.

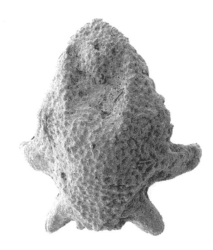

1   2

Food particles would have been transported down the adoral grooves toward the theca, but in order for them to enter the theca, the oral plates would have had to open. Lane and Sevastopulo (1981, p. 24) set out the arguments supporting the hypothesis that the orals were capable of opening outward, probably as a result of the extension of underlying oral podia. The most compelling of these are the lack of a canal into the interior of the theca through which food could pass (Fig. 3.3.1) and the intricate architecture of the margins of the orals, which ensured a close fit when the orals closed (Figs. 3.3.1, 3.3.2).

The suite of specimens of *Allocatillocrinus carpenteri*, which was referred to above in connection with the growth of the column, also provides information about the growth of the arms. As in other allagecrinids, the smallest specimens were armless and presumably fed using either podia around the mouth when the orals opened or absorbed dissolved nutrients from seawater. Arms developed in a prescribed order with the largest specimen studied having 29 arms. For five specimens, in which the number of arms ranges from five to 25, confident estimates can be made of the length of the arms. A log/log plot of the volume of the theca (calculated by treating it as a frustum) against the length of the food-gathering system (the product of the number of arms and their length) fits an allometric equation (Fig. 3.4) with an exponent of approximately 0.68, a result within the range reported for several Ordovician camerate crinoids by Brower (1974). Comparison with the ontogeny of the food gathering system of other disparid crinoids described by Brower (2005) is less straightforward because he computed the relationship between total crown volume and length of arms. As expected for a filter-feeding organism, the arms (the food-gathering system) grew more rapidly than the theca, which housed the bulk of the soft tissue.

An anal tube, consisting of a uniserial column of armlike plates and a corresponding large opening into the theca, is present in some genera of allagecrinids (for example, *Allagecrinus* Carpenter and Etheridge, 1881) but not in others (for example, *Litocrinus* Lane and Sevastopulo, 1982b), which have no anal vent. There is no obvious correlation between the

Figure 3.3. *Relation of oral and radial circlets in allagecrinids. 1, 2,* **Litocrinus** *sp. nov. from the Meenymore Formation Mississippian (late Viséan), Slievemore, County Sligo, Ireland. 1, Oblique lateral C-ray view of theca (TCD 54340) showing the absence of a food canal from the radial facet into the interior, ×60. Note the interoral suture. 2, External view (×75) of CD oral (TCD 54341).*

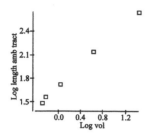

Figure 3.4. *Log-log plot of thecal volume against length of the ambulacral tract for five specimens of **Allocatillocrinus carpenteri**, Renault Formation (Chesterian), near Walters Creek, about 6.4 km east of Waterloo, SE1/4, SW1/4, SW1/4, sec. 22, T. 2 S, R. 19 W, Paderborn Quadrangle, Monroe Country, Illinois.*

presence and absence of the anal tube and other attributes of allagecrinid taxa. It seems unlikely that taxa without an anal vent had a blind gut and more probable that the feces were expelled when the orals were open.

Despite commonly reproduced reconstructions of *Monobrachiocrinus* (Wanner, 1920) showing a single short arm lined with tube feet, there is little direct evidence of the arms of codiacrinaceans. Wanner (1929, pl. 2, fig. 13) illustrated a specimen of *Tenagocrinus sulcatus* Wanner, 1916, with arms shorter than the height of the tegmen. It remains to be proven whether all codiacrinaceans had short arms, but clearly the reduction of the number of arms to one in *Monobrachiocrinus* and none in genera such as *Neolageniocrinus* Arendt, 1970, indicates that reduction of length of the ambulacral tract was common. In abrachiate Carboniferous microcrinoid taxa (Table 3.1), feeding must have taken place with the orals open, presumably by the activities of podia around the mouth. That the orals of codiacrinaceans did open is shown by the intricate ridge and groove structure on the margin of the orals (Figs. 3.5.1, 3.5.2), which served to guide them when they were closing so that they fitted tightly together. The straight fulcrum-like contact between the orals and underlying plates, and the flat triangular areas on the proximal parts of the orals and the corresponding distal parts of the underlying plates in several taxa served to limit the outward movement of the orals (Fig. 3.5.3). In taxa such as *Neolageniocrinus* that lack an anal vent, feces would have been dispersed when the orals were open.

In summary, the feeding strategies of adult allagecrinid and codiacrinacean microcrinoids was in all likelihood far different, with allagecrinids probably having proportionately much longer ambulacral tracts than codiacrinaceans

Predation

Arm regeneration in fossil crinoids has been taken as a proxy for nonlethal predation (Baumiller and Gahn, 2005). Among the suite of specimens of *Allocatillocrinus carpenteri*, referred to above, there are 12 specimens with complete or partially preserved arms; of these, one (SUI 32476) shows regeneration of the arms from a level just above their base. Also, Lane and Sevastopulo (1982b, pl. 3, fig. 7) illustrated a brachial of *Kallimorphocrinus astrus* in which the tip was regenerating. The identity of predators of microcrinoids is not known, but they are likely to have been taxa different from those that preyed on larger crinoids.

Reproduction

Before discussing the reproductive biology of Carboniferous microcrinoids, it is useful to summarize the current state of knowledge of reproduction in Recent crinoids and the small amount written about fossil crinoids. The summary below relies heavily on Holland's (1991) excellent review, which contains references to earlier literature.

Knowledge of the reproduction of Recent crinoids is biased. Relatively complete information is available for several comatulids, but little is known about stalked forms. It appears that crinoids are almost exclusively gonochoric. For some comatulids at least, the sex ratio is 1:1. The gonads are

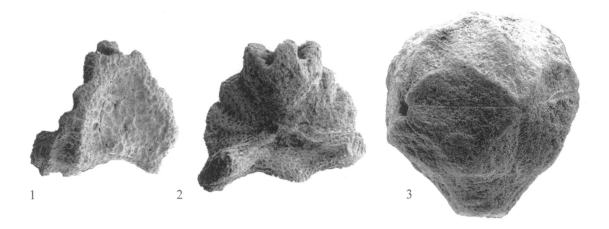

external to the theca, being housed in the proximal pinnules (and, less commonly, in the arms), except in the genus *Cyathidium*, where they are within the theca (Heinzeller and Fechter, 1995). The eggs of comatulids are typically shed and fertilized externally. In a few taxa, the eggs either attach to the exterior of the mother, enter brood pouches on the pinnules, or are fertilized within the oviducts. The frequency of spawning ranges from once to many times a year. The fecundity of females of nonbrooding species has been estimated to range over at least two orders of magnitude, with females of the least fecund species investigated producing 9000 eggs per year. The eggs are generally denser than seawater and fall to the bottom after being shed, but in some taxa, they are lighter and float. The fertilized eggs develop generally to pelagic lecithotrophic, doliolaria larvae, which spend a short time (typically several days) in the plankton before settling. After settling, the larva metamorphoses to a juvenile crinoid, which initially has no mouth but relatively soon develops a mouth, podia, and arms. In those taxa in which the larvae are brooded, development may proceed past metamorphosis before the juveniles leave the mother.

The early ontogeny of stalked crinoids is known for a single species. Nakano et al. (2003) described the development of *Metacrinus* from eggs, 350 μm in diameter, to auricularia and then to lecithotrophic doliolaria larvae.

There is little direct evidence regarding reproduction in Paleozoic crinoids. Most authorities have concluded that they shared the primitive condition for echinoderms of internal gonads (Lane, 1984), although they do not possess external gonopores, in contrast to many blastozoans. Because few Recent crinoids exhibit sexual dimorphism of the arms or pinnules, diagnostic indicators of extrathecal gonads in fossils are likely to be rare. However, the large ratio of the size of the theca to the size of the crown in camerates and in many cladids and flexibles is consistent with the hypothesis that their gonads were internal. Most disparid crinoids have relatively small thecae and few have enlarged anal sacs, which Lane (1984) suggested might house the gonads in some cladids. However, an undescribed taxon of allagecrinid microcrinoid (new genus B in Table 3.1) has a theca that bulges out in the DE interrradius (Figs. 3.6.1–3.6.3). Unlike the case of *Trophocrinus* Kirk, 1930,

Figure 3.5. *Feeding in codiacrinacean microcrinoids. 1, 2, Inner and outer views (×20) of an oral of* **Cranocrinus** *sp. nov. (TCD 54342), Mississippian (late Viséan), Slievemore, County Sligo, Ireland. 3, C-ray view (×30) of* **Lageniocrinus** *sp. nov. (TCD 54343) showing the straight hinge between the orals and basals. Roque Redonde Formation (Mississippian; late Viséan), Roque Redonde, near Valhain, France.*

Figure 3.6. *Genus B gen. nov.* from shales at the top of the Waulsortian Limestone, Carrickboy Quarry, County Longford, Ireland (×30). 1, Lateral view. CD interray view showing swollen DE interray (TCD 54344). 2, 3, Lateral view. CD interray and oral views showing swollen DE interray (TCD 54345).

discussed below, the bulge is formed by an outward extension of the thecal wall, and it is tempting to correlate it with the position of a single internal gonad. If it does accommodate a gonad, however, its location in the DE interray is not what would have been anticipated from the situation in blastozoans, where the gonopore occurs in the CD interray.

Irrespective of the weak evidence for the location of gonads in disparid microcrinoids, it is clear that in abrachiate codiacrinaceans, such as *Neolageniocrinus*, the gonad or gonads must have been internal. In view of the small thecal volume, this raises interesting questions about the size of the eggs and the reproductive strategy of such microcrinoids. The smallest eggs recorded for Recent comatulids are ~100 μm in diameter (Holland, 1991). Taking the internal volume of a large microcrinoid to be an estimated 10^{-3} mL, the maximum number of eggs of 100 μm diameter that could have been accommodated would have been less than 1700 if the theca had been completely filled by the gonad. Even this clearly extravagant figure suggests a low level of fecundity compared with Recent crinoids that broadcast their gametes.

The size of the eggs would have governed the number that could have been accommodated in a microcrinoid. In general, eggs that develop to planktotrophic larvae in Recent noncrinoid echinoderms are smaller than those developing to lecithotrophic larvae. Because lecithotrophy is considered to be a derived condition in echinoderms (Strathmann, 1978), some, if not all, Paleozoic crinoids may have had planktotrophic larvae and perhaps smaller eggs than those of Recent crinoids. The smallest eggs of any echinoderm with obligate planktotrophic larvae are ~50 μm in diameter (in the apodid holothurian *Synaptula reciprocans* Forskål, 1775; Mortensen, 1937). However, it is generally held that one of the outcomes of miniaturization in marine invertebrates is the production of lecithotrophic larvae and the brooding of larvae (Strathmann and Strathmann, 1982). That brooding occurred in some disparid microcrinoids is shown by the allagecrinid genus *Trophocrinus*. Six species of *Trophocrinus*, all from the Tournaisian of North America (Kirk, 1930; Peck, 1936; Strimple and Koenig, 1956), are distinguished from the otherwise similar genus *Litocrinus* by the possession of a hoodlike structure attached to two radials. Kirk (1930, p. 211) suggested that this structure was a brood pouch in *Trophocrinus tumidus* Kirk, 1930, and noted that an accompanying specimen was similar in every respect except for the presence of the pouch. Peck (1936, p. 286) confirmed the observations of dimorphism for *Trophocrinus exsertus* Peck, 1936, and *T. corpulentus* Peck, 1936. Strimple and Koenig (1956) concluded that the pouch was unlikely to be a brood pouch, partly because in *Trophocrinus bicornis* Strimple and Koenig, 1956, two pouches were present. They tentatively interpreted the structures as current deflectors used in food gathering. Data from a fauna from the Rockford Limestone of Indiana supports Kirk's original interpretation, and his and Peck's observation that *Trophocrinus* spp. occur as two dimorphs, with one, interpreted as the male, lacking the brood pouch (personal observation). If, as seems probable, the pouches were used for brooding, the embryos could have been up to several hundred microns in maximum dimension, if one embryo was present in a brood pouch at any one time.

In an early classification of crinoids, Wachsmuth and Springer (1886) referred allagecrinid and codiacrinacean microcrinoids to the Branch Larviformia of the Suborder Inadunata. It is interesting to inquire to what extent allagecrinids and codiacrinaceans exhibit larval morphology and to speculate on the evolutionary and developmental significance of miniaturization in these crinoids.

Discussion

The phylogeny of disparid crinoids is not clear, and the ancestry of allagecrinids is not known. Nevertheless, it is likely that they were derived from relatively small crinoids because in general, disparids have small thecae relative to those of most cladids and camerates. Thus, allagecrinid microcrinoids have mature thecae whose height is generally more than one-tenth of those of representative species of such typical disparid genera as *Cincinnaticrinus* Warn and Strimple, 1977, *Ectenocrinus* Miller, 1889, *Homocrinus* Hall, 1852, *Pisocrinus* de Koninck, 1858, and *Synbathocrinus* Phillips, 1836. The only features that might be considered as larval in the adults of allagecrinids are their small size and the retention of the oral circlet of five plates in some taxa. The orals were resorbed in other taxa (Lane and Sevastopulo, 1982; Sevastopulo, personal observation). In the case of *Trophocrinus*, it should be possible to establish when the crinoids were sexually mature. This has not been attempted rigorously, but anecdotal evidence suggests that specimens with a brood pouch had developed at least some arms and were close to the maximum size known for each species. If this can be extrapolated to other allagecrinids, it suggests that the relationship between somatic and gonadal development in allagecrinids was similar to that in other crinoids. Therefore, allagecrinids have not evolved by a mechanism that can be described as heterochronic, except inasmuch as they were probably smaller than their ancestors.

In contrast, many codiacrinaceans, such as *Monobrachiocrinus*, had fewer than five arms or, like *Neolageniocrinus*, never developed radial plates. Because large collections of these crinoids commonly include a range of sizes, the largest of which retain juvenile characteristics, it is clear that they are paedomorphic; development of the gonads had proceeded to maturity while somatic development was at a larval stage relative to other crinoids. Because many codiacrinaceans never progress past larval morphology, they provide examples of neoteny (McKinney, 1999).

Another aspect of the evolution of microcrinoids that is worthy of further study is the increase in maximum size of representatives of many clades through the Carboniferous and Permian. For example, many of the allagecrinids and codiacrinaceans illustrated by Wanner (1929) from the Permian of Timor are two to three times as large as mature individuals of species of the same genera from the Mississippian. In the case of several of the Permian codiacrinaceans, despite their larger size, somatic development still never progressed past the larval stage.

Acknowledgments

Both G. Lane and H. Strimple were partners with me in studies of Carboniferous microcrinoids; the material that they made available and the ideas that they shared with me are gratefully acknowledged. D. B. Macurda Jr. made available washed samples from the Floraville locality. M. Aretz made available the specimen of *Lageniocrinus* illustrated in Figure 3.5. Reviews by T. W. Kammer and G. D. Webster led to improvements. D. John, N. Leddy, and C. Reid of the Centre for Microscopy and Analysis, Trinity College, Dublin, helped with the scanning electron microscopy.

References

Arendt, Yu. A. 1970. Morskie lilii gipokrinidy [Crinoids Hypocrinidae]. Trudy Paleontologicheskogo Instituta, Akademiya Nauk SSSR, Number 128, 220 p.

Arendt, Yu. A. 2002. Early Carboniferous echinoderms of the Moscow Region. Paleontological Journal, 36(Supplement 2):S115–S184.

Bather, F. A. 1890. British fossil crinoids. II. The classification of the Inadunata. Annals and Magazine of Natural History, series 6, 5:310–334.

Baumiller, T. K., and F. J. Gahn. 2005. Testing predator-driven evolution with Paleozoic crinoid arm regeneration. Science, 305:1453–1455.

Beyrich, H. E., Von. 1862. Gebirgsarten und versteinerungen von Koepang auf Timor. Zeitschrift der Deutschen Geologischen Gesellschaft, 14:537.

Brower, J. C. 1974. Ontogeny of camerate crinoids. University of Kansas Paleontological Contributions, Paper 72, 53 p.

Brower, J. C. 2005. The paleobiology and ontogeny of *Cincinnaticrinus varibrachialus* Warn and Strimple, 1977 from the Middle Ordovician (Shermanian) Walcott-Rust Quarry of New York. Journal of Paleontology, 79:152–174.

Burdick, D. W., and H. L. Strimple. 1982. Genevievian and Chesterian crinoids of Alabama. Geological Survey of Alabama Bulletin, 121, 277 p.

Carpenter, P. H., and R. Etheridge Jr. 1881. Contributions to the study of the British Paleozoic crinoids.—No. 1. On *Allagecrinus*, the representative of the Carboniferous limestone series. Annals and Magazine of Natural History, 7:281–298.

De Koninck, L. G. 1858. Sur quelques Crinoides paleozoiques nouveaux de l'Angleterre et de l'Ecosse. Bulletin de la Academie Royal des Sciences, des Lettres et des Beaux-Arts de Belgique, series 2, 4:93–108.

De Koninck, L. G., and H. Le Hon. 1854. Recherches sur les crinoides du terrain carbonifère de la Belgique. Academie Royal de Belgique, Memoir 28, 215 p.

Forskål, P. 1775. Descriptiones—Animalium Avium, Amphibiorum, Piscium, Insectorum, Vermium; quæ in Itinere Orientali Observavit Petrus Forskål. Post mortem auctoris edidit Carsten Niebuhr. Copenhagen, Hauniæ, 164 p.

Gaillard, C., M. Hennebert, and D. Olivero. 1999. Lower Carboniferous *Zoophycos* from the Tournai area (Belgium): environmental and ethologic significance. Geobios, 32:513–524.

Girty, G. H. 1908. The Guadalupian fauna. U.S. Geological Survey Professional Paper 58. 512 p.

Gutschick, R. C. 1968. Late Mississippian (Chester) *Allagecrinus* (Crinoidea) from Illinois and Kentucky. Journal of Paleontology, 42:987–999.

Hall, J. 1852. Palaeontology of New York, volume 2, Containing descriptions of the organic remains of the lower middle division of the New-York system. Natural History of New York, 6. D. Appleton & Co.. New York, 6, 362 p.

Heinzeller, T., and H. Fechter. 1995. Microscopical anatomy of the Cyrtocrinid *Cyathidiium meteorensis* (sive foresti) (Echinodermata Crinoidea). Acta Zoologica, 76:25–34.

Hibo, D., and F. Tourneur. 1989. Fragments d'echinodermes aux environs du "Gras-Délit" (Ivorien) dans la carrièrre Lemay à Vaulx-les-Tournai. Miscellanea Geologica Liège, 9:2.

Holland, N. D. 1991. Echinodermata: Crinoidea, p. 247–299. *In* A. C. Giese, J. S. Pearse, and V. B. Pearse (eds.), Reproduction of Marine Invertebrates. Volume VI: Echinoderms and Lophophorates. Boxwood Press, Pacific Grove, California.

Hudson, R. G. S., M. J. Clarke, and G. D. Sevastopulo. 1966. The palaeoecology of a lower Viséan crinoid fauna from Feltrim, Co. Dublin. Scientific Proceedings of the Royal Dublin Society, 2:273–286.

Kirk, E. 1930. *Trophocrinus*, a new Carboniferous crinoid genus. Journal of the Washington Academy of Science, 20:210–212.

Kirk, E. 1936. A new *Allagecrinus* from Oklahoma. Journal of the Washington Academy of Sciences, 26:162–165.

Kirk, E. 1937. *Clistocrinus*, a new Carboniferous crinoid genus. Journal of the Washington Academy of Science, 27:105–111.

Lane, N. G. 1984. Predation and survival among inadunate crinoids. Paleobiology, 10:453–458.

Lane, N. G., and G. D. Sevastopulo. 1981. Functional morphology of a microcrinoid: *Kallimorphocrinus punctatus* n. sp. Journal of Paleontology, 55:13–28.

Lane, N. G., and G. D. Sevastopulo. 1982a. Microcrinoids from the Middle Pennsylvanian of Indiana. Journal of Paleontology, 56:103–115.

Lane, N. G., and G. D. Sevastopulo. 1982b. Growth and systematic revision of *Kallimorphocrinus astrus*, a Pennsylvanian microcrinoid. Journal of Paleontology, 56:244–259.

Lane, N. G., G. D. Sevastopulo, and H. L. Strimple. 1985. *Amphipsalidocrinus*; a monocyclic camerate microcrinoid. Journal of Paleontology, 59:79–84.

Langhorne, B. S., JR., and J. F. Read. 2001. Discrimination of local and global effects on Upper Mississippian stratigraphy, Illinois Basin, U.S.A. Journal of Sedimentary Research, 71:985–1002.

Mapes, R. H., N. G. Lane, and H. L. Strimple. 1986. A microcrinoid colony from a cephalopod body chamber (Chesterian; Arkansas). Journal of Paleontology, 60:400–404.

McKinney, M. L. 1999. Heterochrony: beyond words. Paleobiology, 25:149–153.

Moore, R. C., N. G. Lane, and H. L. Strimple. 1978. Order Cladida, p. T578–

T759. *In* R. C. Moore and C. Teichert (eds.), Treatise on Invertebrate Paleontology. Pt. T. Echinodermata 2. Geological Society of America and University of Kansas Press, Lawrence.

Miller, S. A. 1889. North American Geology and Paleontology. Cincinnati, Western Methodist Book Concern, 664 p.

Mortensen, D. T. 1937. Contributions to the study of the development and larval forms of echinoderms III. Kongelige Danske Videnskabernes. Selsklab Skrifter Kjobehavn. Ninth series, 7, 1:1–65.

Nakano, H., T. Hibino, T. Oji, Y. Hara, and S. Amemiya. 2003. Larval stages of a living sea lily (stalked crinoid echinoderm). Nature, 421:158–160.

Peck, R. E. 1936. Lower Mississippian microcrinoids from the Kinderhook and Osage groups of Missouri. Journal of Paleontology, 10:282–293.

Phillips, J. 1836. Illustrations of the Geology of Yorkshire, or a Description of the Strata and Organic Remains. Part 2, The Mountain Limestone Districts, 2nd edit. London, John Murray, 253 p.

Rexroad, C. B., L. M. Brown, J. Devera, and R. J. Suman. 1998. Conodont biostratigraphy and paleoecology of the Perth Limestone Member, Staunton Formation (Pennsylvanian) of the Illinois Basin, U.S.A., p. 247–259. *In* H. Szaniawski (ed.), Proceedings of the Sixth European Conodont Symposium (ECOS VI) Palaeontologia Polonica, 58.

Say, T. 1820. Observations on some species of zoophytes, shells, etc., principally fossil. American Journal of Science, 2:34–45.

Sevastopulo, G. D. 1969. On some Irish Lower Carboniferous crinoids. Unpublished Ph.D. dissertation, University of Dublin, 353 p.

Sevastopulo, G. D. 2002. Fossil "lilies of the ocean" and other echinoderms from Carboniferous rocks of Ireland. John Jackson Memorial Lecture 2002. Royal Dublin Society Occasional Papers in Science, 25:1–15.

Smith, D. P. 1978. Paleontology and paleoecology of the basal New Providence Shale (Osagian; Mississippian) at Paris Landing, Tennessee. Unpublished M.A. thesis, Indiana University, Bloomington, 172 p.

Springer, F. 1923. On the fossil crinoid family Catillocrinidae. Smithsonian Miscellaneous Collection, 76:1–41.

Strathmann, R. R. 1978. The evolution and loss of feeding larval stages in marine invertebrates. Evolution, 32:894–914.

Strathmann, R. R., and M. F. Strathmann. 1982. The relationship between adult size and brooding in marine invertebrates. American Naturalist, 119:91–101.

Strimple, H. L. 1966. Some notes concerning the Allagecrinidae. Oklahoma Geology Notes, 26:99–111.

Strimple, H. L., and J. W. Koenig. 1956. Mississippian microcrinoids from Oklahoma and New Mexico. Journal of Paleontology, 30:1225–1247.

Strimple, H. L., and W. T. Watkins. 1969. Carboniferous crinoids of Texas with stratigraphic implications. Palaeontographica Americana, 6:139–275.

Ubaghs, G. 1978. Skeletal morphology of fossil crinoids, p. T58–T216. *In* R. C. Moore and C. Teichert (eds.), Treatise on Invertebrate Paleontology. Pt. T. Echinodermata 2. Geological Society of America and University of Kansas Press, Lawrence.

Wachsmuth, C. 1882. Descriptions of two new species of crinoids from the Chester Limestone and Coal Measures of Illinois. Illinois State Museum of Natural History, Bulletin, 1:40–43.

Wachsmuth, C., and F. Springer. 1886. Revision of the Palaeocrinoidea: Part III. Discussion of the classification and relations of the brachiate crinoids, and conclusion of the generic descriptions. Proceedings of the Academy of Natural Sciences of Philadelphia, 1886:64–226.

Wanner, J. 1916. Die Permischen echinodermen von Timor, I. Teil. Palaontologie von Timor, 11:1–329.

Wanner, J. 1920. Über Armlose Krinoiden aus dem jüngeren Palaeozoikum. Verhandelingen Geologie Mijnbouw Genootoch, Geologic Series, 5:21–35.

Wanner, J. 1924. Die permischen Krinoiden von Timor. Jaarbook van net Mijnwezen Nederlandes Oost-Indie, Verhandlungen (1921), Gedeelte 3, 348 p.

Wanner, J. 1929. Neue Beitrage zur Kenntnis der Permischen Echinodermen von Timor. I. *Allagecrinus*. II. *Hypocrinites*. Dienst van den Mijnbouw in Nederlandsch-Indie, Wetenschappelijke Mededeelingen 11, 116 p.

Wanner, J. 1937. Neue beiträge zur kenntnis der Permischen echinodermen von Timor, VIII–XIII. Palaeontographica, Supplement 4, 212 p.

Warn, J. M., and H. L. Strimple. 1977. The disparid inadunate superfamilies Homocrinacea and Cincinnaticrinacea (Echinodermata. Crinoidea), Ordovician–Silurian, North America. Bulletins of American Paleontology, 72:1–13.

Weller, J. M. 1930. A group of larviform crinoids from Lower Pennsylvanian strata of the eastern Interior Basin. Illinois Geological Survey Report of Investigations, 21:1–43.

Wright, J. 1952. Crinoids from Thornton Burn, East Lothian. Geological Magazine, 89:320–327.

Yakovlev, N. N. 1926. A fauna of armless crinoids and crinoids with an incomplete number of arms in the Permo-Carboniferous of the Urals. Izvestiy Geologicheskogo Komiteta (Bulletin Committee Geology) Leningrad (1925), 44.2.

Yakovlev, N. N. 1927. Sur l'homologie dans la structure de la face ventrale du calice de Cystoidea et de Crinoidea. Doklady Academii Nauk SSSR (Comptes Rendus de l'Academie Sciences de l'URSS), 3:54–56.

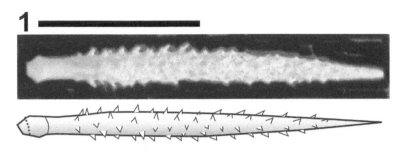

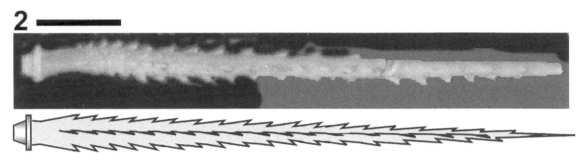

Figure 4.1. *Examples of ornate* **Archaeocidaris** *spines. Scale bars = 1 cm. 1, Round, solid spine with small spinules,* **Archaeocidaris** *sp. from the Colony Creek Shale, Pennsylvanian, north-central Texas. 2, Spine with triangular cross section and thorns along edges, causing a serrated appearance;* **A. brownwoodensis** *from a black shale in the Winchell Formation, Pennsylvanian, north-central Texas.*

THE IMPORTANCE OF ECHINOIDS IN LATE PALEOZOIC ECOSYSTEMS

4

Chris L. Schneider

Modern regular echinoids are conspicuous and important members of many marine communities. As scavengers, predators, and herbivores, these organisms are contributors in local trophic webs, and in some ecosystems, they are crucial in shaping the marine community.

In some marine communities, echinoids can have extreme effects on ecosystem structure and biodiversity. The negative effects on kelp forests by echinoids has been known for decades; echinoids are responsible for denudation of kelp forests off the coasts of western North America and Europe during increases in population size and density (Estes and Palmisano, 1974; Estes and Duggins, 1995; Estes et al., 1998; Sala et al., 1998). In the Caribbean, echinoids have positive effects on coral reef structure and biodiversity through herbivory of algae, which outcompete corals through overgrowth and shadowing (Hughes, 1994). Conversely, when echinoids are removed from the coral reef, such as occurred during the 1983–1984 decimation of *Diadema antillarum* Philippi, 1845, through disease, algae rapidly outcompeted corals, resulting in coral die-off and rapid decline in overall reef biodiversity (Lessios, 1988; Hughes, 1994).

Late Paleozoic echinoids may be perceived as being lesser members of marine communities in comparison to their more conspicuous sessile neighbors, such as brachiopods, bryozoans, and crinoids. This perception is partly the result of the structure of the echinoid test. All Paleozoic echinoid tests were loosely plated, with abutting or overlapping plates connected by soft tissue, rather than the interlocking plates of rigid-test modern echinoids. Modern echinoids have been shown to disarticulate within days to weeks after death (see review in Ausich, 2001). Because of the nature of their tests, Paleozoic echinoids may have disarticulated more rapidly into many small plates and spines (Kier, 1977) and are less likely to be included in community studies than better-preserved clades.

The typical late Paleozoic organisms—brachiopods, bryozoans, and crinoids—are much more commonly mentioned in paleoecological studies than echinoids. This may be one issue explaining the perceived lack of prominence of echinoids in late Paleozoic ecosystems. Although paleoecological research has provided much information about the biodiversity and dynamics of late Paleozoic ecology, many paleoecological studies focus on a portion of the ecosystem, such as brachiopods or crinoids; few investigate complete fossil assemblages. Paleozoic echinoid workers are few, and most attention on echinoids has focused on the description of new taxa or evolutionary patterns, rather than the autecology of echinoids or their role in Paleozoic ecosystems.

Introduction

The shallow marine habitat of late Paleozoic echinoids differed greatly from earlier in the Paleozoic. The record of carbonate rocks declined sharply during the Pennsylvanian as a result of extensive continental glaciation influences on eustacy (Walker et al., 2002; Tucker and Wright, 1990, and references therein). Gone were the complex coral and stromatoporoid reefs of the Devonian. Instead, muddy environments and algal buildups in the shallow marine environment were common, and these facies tracked rapid transgression and regression of Pennsylvanian and Permian shorelines. Ecosystems of the Pennsylvanian and Permian may have experienced a long interval of low evolutionary turnover (data in Bambach et al., 2004). However, evidence suggests ecological expansion during this interval, such as the radiation of some productid brachiopods during the Early Permian (Brunton and Lazarev, 1997). Further, a potential precursor to the Mesozoic marine revolution occurred with communities beginning to have Modern components because of the expansion of gastropods and bivalves in diversity and into new environments (Miller and Sepkoski, 1988; Sawyer and Leighton, 2005) and the increase in echinoid species diversity (Kier, 1965). Although crushing predation traces decreased during the Carboniferous (Alexander, 1986), predation pressure on marine benthos may have increased. First, predation scars usually represent repair. Lack of predation scars may indicate an increase in successful predation rather than a decrease in predators, because as predators become more effective, fewer prey organisms survive to be able to repair skeletal material. Second, many organisms evolved or expanded spinosity. For example, productid brachiopods radiated and became highly spinose, and archaeocidarid echinoids, which evolved large spines, radiated and dominated echinoid diversity.

Late Paleozoic echinoids were important to marine communities through contributions to biodiversity within ecosystems, their presence in most shallow marine ecosystems, their overall abundance in fossil assemblages, and their role in defining community structure and type. This chapter will review the known ecology of late Paleozoic echinoids after the Late Devonian biotic crisis, with particular emphasis on archaeocidarid paleoecology, and also present the results of a new study analyzing the presence of echinoids in Pennsylvanian marine communities from north-central Texas.

Late Paleozoic Echinoid Ecology

Modern regular echinoids are known from nearly all fully marine environments, from polar oceans to the tropics and estuaries to the deep ocean. All known Paleozoic echinoids were recovered from sediments on continental shelves or epeiric seas. No echinoids are known from abyssal and greater depths of Paleozoic oceans, leading to the suggestion that all Paleozoic echinoids were restricted to the shallow marine realm. Whether this lack of deep-sea echinoids is a result of the lack of preservation of deep ocean environments or truly the restriction of echinoids to the shallow ocean, interpretations of Paleozoic echinoid paleoecology is limited to that of continental shelf and epeiric sea environments.

Paleozoic echinoids were much less common than those in Mesozoic and Cenozoic shallow marine environments. Paleozoic echinoid richness is thought to have peaked during the Devonian (genera; Smith, 1984) or Mississippian (species; Kier, 1965). However, diversity may be underestimated for the late Paleozoic. First, there are indeed many fewer genera in the Pennsylvanian; however, one of these genera is *Archaeocidaris* M'Coy, 1849, the late Paleozoic echinoid genus with the greatest number of species. For instance, an early paper by Kier (1965) cited eight species for *Archaeocidaris*, but Jackson (1912) recognized 40 species. Since the publication by Jackson in 1912, at least 14 additional species of *Archaeocidaris* have been described, and more remain undescribed. Although most species are Mississippian in age, approximately half of the species described since Jackson (1912) are of Pennsylvanian and Permian age, and many of the undescribed specimens are from Pennsylvanian and Permian rocks.

Smith (1984) suggested that the Aristotle's lanterns of Paleozoic echinoids were not adapted for active predation because of weaker teeth and generalized lantern and instead were scavengers or herbivores. To date, no predation traces from echinoid teeth are known from any preserved Paleozoic skeletal material. Because predation has been much studied by various paleoecologists (see reviews in Kelley et al., 2003), it is possible that Smith is correct and archaeocidarids were incapable of active predation. However, recent evidence of what are likely preserved stomach contents in late Paleozoic echinoids presents the possibility of predation (Schneider, 2001). Sediment globs within the test of Pennsylvanian *Archaeocidaris* echinoids from Texas are distinctly different than surrounding sediment within and outside the tests and contain articulated and disarticulated crinoid cirrals (Schneider, 2001). Preliminary observation of the cirrals indicates no evidence of degradation, which would provide evidence of predation. Baumiller et al. (2001) considered the increased articulation of crinoid material in cidaroid gut contents, compared with the articulation rates of crinoid debris in surrounding sediments, to be evidence of predation rather than scavenging. However, their examination of echinoid gut contents was also supported by in situ observation of echinoid-crinoid interactions. Study of late Paleozoic predation by echinoids on crinoids is ongoing, but in any event, incorporation of cirrals into gut material of these echinoids suggests at least scavenging behavior by *Archaeocidaris*.

Aggregation behavior is common among many mobile echinoderm clades, and modern echinoids are no exception. Modern regular echinoids aggregate in the presence of food (Rodriguez and Farina, 2001), in the presence of predators (Hagen and Mann, 1994), under circumstances that cause a high density of individuals (Hagen and Mann, 1994), and during spawning (Lamare and Stewart, 1998; Young et al., 1992). Although no quantitative study has yet been undertaken regarding echinoid aggregation in the Paleozoic, there are several reports of abnormally high densities of late Paleozoic archaeocidarids in Lagerstätten. In all high-density echinoid occurrences, all specimens were well preserved in various states of articulation to slight disarticulation. In one Pennsylvanian locality near Brownwood, Texas, literally thousands of well-preserved articulated echinoids have been recovered by professional and amateur collectors since the 1970s (Schneider et al., 2005; Warme and Olson, 1971). These echinoids occurred in a shallow, possibly estuarine or baylike environment and were preserved by several instantaneous burial events. Echinoids occur in several closely spaced horizons at this locality, and all individuals bear little or no indication of transport other than overturning, which suggests repeated events of aggregation and burial. Because of the amount of terrigenous plant material in the surrounding sediments, and because of the similarity between this Carboniferous environment and the shallow estuarine environment of observed Modern echinoid aggregations during gamete release (Lamare and Stewart, 1998), these aggregation events are suggested to result from abundant food sources (plants) or seasonal breeding (Schneider, 2001). Other horizons of highly abundant echinoids are known from the Pennsylvanian of Kansas (Leighton and Kaplan, personal commun.) and a somewhat older locality in north-central Texas (D. Harlow, personal commun.) in the United States.

Like their modern counterparts, the size and ornament of *Archaeocidaris* spines are conspicuous and highly variable. Many *Archaeocidaris* spine lengths are equal to or surpass the diameter of the test (i.e., A. *brownwoodensis* Schneider et al., 2005; see others in Jackson, 1912), creating a halo of large and rather sharp spines that retain their sharpness through hundreds of millions of years of taphonomic and diagenetic processes. Most *Archaeocidaris* taxa contain spines with ornament. These spines commonly appear in two major morphological types (except those with smooth, straight spines): round, solid spines with a dense array of small, projecting, needlelike spinules, often arranged in a radial distribution around the spine; or solid spines with a triangular cross section that contain ordered rows of thorns that project at an angle from the spine shaft, creating serrated edges along the spine (Fig. 4.1). Like all echinoderms, the stereom of the spines was filled with living tissue, and the echinoid was capable of regeneration of damaged spines; regenerated spines are known from the Pennsylvanian Vanport Limestone, Allegheny Group, of Ohio (Hoare and Sturgeon, 1976). Considering the number of spines per echinoid (commonly more than 100 spines per adult individual), the spine size, the complexity of ornament, the ability to regenerate spines, and the mobility of spines, a high portion of the total echinoid metabolism was dedicated to spine growth, maintenance, and movement. Hence, large, ornate, energy-expensive spines were an important morphological feature for echinoids and would have required a metabolism fit to support such spines. Did these energy-expensive spines require a higher metabolism for *Archaeocidaris* compared with contemporaneous and earlier nonarchaeocidarid echinoids, and therefore jump-start the evolution of modern echinoids? Unfortunately, the answer to this question is difficult at best.

It has been suggested that large, ornate archaeocidarid spines may have been important in predator deterrence (Kier, 1965). Predation traces declined rapidly after the Mississippian, which could suggest a subsequent decline in predation pressure; however, the opposite has been interpreted to be the case (Signor and Brett, 1984). Instead of a decline in predation pressure, the decrease in predation traces in brachiopods has been interpreted to be an increase in predator efficiency—that is, the predator is more successful at taking prey, and fewer prey organisms are left alive to repair shell material (Alexander, 1986). Repair scars on crinoid anal spines persist into the Carboniferous (Baumiller and Gahn, 2004) and occur at high levels in gastropods when compared with brachiopods and bivalves (Sawyer and Leighton, 2005). In addition to these patterns, there is a Devonian through Carboniferous increase in spinose brachiopods and a radiation of productids (Signor and Brett, 1984; Leighton, 2003) plus an increase in ornament and spinosity in archaeocidarid echinoids (Kier, 1965), gastropods (Sawyer and Leighton, 2005), and crinoids (Meyer and Ausich, 1983; Signor and Brett, 1984). All of these morphological trends are possibly a response to increased predation pressure (Kier, 1965; Baumiller and Gahn, 2004; Sawyer and Leighton, 2005).

Like all modern cidaroid echinoids, the early cidaroids possessed spines that lacked an epithelium. Smith (1984) suggested that the lack of epithelium on *Archaeocidaris* spines would have been ideal for attracting epibionts. Many modern cidaroids, which also have bare and often ornate spines, host epizoans (Heterier et al., 2004; Nebelsick, Pawson, and Donovan, personal commun.). Recently, this phenomenon of encrusted spines on *Archaeocidaris* was proven for the late Paleozoic A. *brownwoodensis* (Schneider, 2003a). The ornate spines of this echinoid bore two taxa of bryozoans and the small, attached brachiopod *Crurithyris* George, 1931, in abundance and on a majority of specimens. Test diameters of A. *brownwoodensis* individuals were independent of epizoan abundance, suggesting little or no influence of epizoans on growth rate, a proxy for energy expended by the echinoid in moving heavily encrusted spines. Because of the lack of effect on echinoids and the benefits provided to the epizoans, substrate, and protection by the halo of ornate spines radiating from the echinoid, this relationship between the echinoids and its epizoans is determined to be commensal.

A preliminary study of Pennsylvanian through Early Permian echinoid spines from Texas and their epizoans suggested surprising interactions between echinoids and epizoans (Schneider, 2002). Five echinoid taxa were compared for ornament complexity and epizoan abundance and diversity. Consistently, echinoid spine complexity increased through time, as did epizoan abundance, diversity, and coverage (Fig. 4.2).

Like spines on many other organisms, spines on archaeocidarids probably served more than one purpose. In productid brachiopods, spines have been suggested to deter predation, act as a snowshoe mechanism to stabilize the brachiopod on soft substrate, orient the brachiopod in current, attach to substrates as juvenile brachiopods, deter epizoans, and attract potential algal encrusters (see review in Alexander, 2001). It is quite possible that spines in brachiopods could have been utilized for more than one of those functions. Similarly, spines on archaeocidarids may have also served a dual purpose. Spines in the earliest archaeocidarids are not as ornate or large as their descendants. It is possible that echinoid spines may have originally evolved for antipredatory defense. Later, as ornamentation intensified, encrusting organisms found acceptable substrate on bare archaeocidarid spines and by camouflaging echinoid individuals physically and/or

Figure 4.2. *Epibionts on Carboniferous and Permian echinoid spines from north-central Texas. 1, Percentage of spines in each sample that are encrusted with epibionts and total richness of epibionts in the sample. 2, Spine morphologies of echinoids from various localities and the lithologies in which specimens occurred in north-central Texas, in stratigraphic order from oldest (bottom) to youngest (top). Marble Falls Formation (limestone), articulated undescribed **Archaeocidaris** sp., Bashkirian age; specimen is a composite reconstruction of two spine segments; Winchell Formation (black shale), complete spine from the holotype of **A. brownwoodensis**, Kasimovian age; Thrifty Formation (shale), a composite reconstruction from three spine segments of a disarticulated **Archaeocidaris** sp., Ghezelian age; Colony Creek Shale, complete spine of a disarticulated **Archaeocidaris** sp., Ghezelian age; Wichita Formation (shale), complete spine from disarticulated **Archaeocidaris?** sp., Asselian age.*

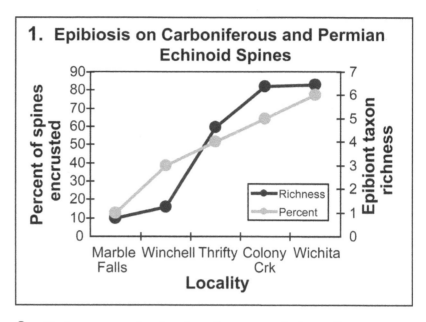

1. Epibiosis on Carboniferous and Permian Echinoid Spines

2. Spine morphologies from each locality:

Wichita fm.

Thrifty Fm.

Colony Creek Sh.

Winchell Fm.

Marble Falls Fm.

■■■ 1 cm

chemically, may have provided positive feedback for the evolution of increasingly ornate spines.

In a Bray-Curtis ordination of Pennsylvanian ecosystems from Texas and Kansas, echinoids were key taxa that determined community distribution in ordination space (Schneider, 2003a). Results from this study suggested that echinoids were common to ecosystems containing minimal three-dimensional community structure; in other words, echinoids were least common in three-dimensional dense bryozoan thickets and crinoid gardens that contained complex spatial and tiering structure. Instead, echinoids were most common in communities dominated by sediment-hugging organisms, such as brachiopods and large bivalves, and containing low abundances of erect organisms such as crinoids. Echinoids most often co-occurred with *Composita* Brown, 1849 (Pearson's correlation; $p < 0.01$), and several large productid brachiopods (Pearson's correlation; $p < 0.05$), organisms common to those low-lying, two-dimensional communities.

Echinoid spine substrates, especially in the soft-substrate environments of the late Paleozoic, were crucial to settling epibiont larvae where hard substrates were limited. Other organic substrates, such as large bivalve and brachiopod shells, also may have provided potential substrate, but not all surfaces were acceptable to epibiont attachment. Large, alate brachiopods with moderate ornament such as costae or plicae, and some smooth-shelled brachiopods were acceptable attachment substrates in Devonian and Carboniferous ecosystems (see reviews in Taylor and Wilson, 2003; Schneider and Webb, 2004). Those late Paleozoic communities containing echinoids commonly also contained abundant productid brachiopods, which were rarely encrusted (Schneider and Webb, 2004), and varying abundances of athyrid and large spiriferid brachiopods, which were rarely to frequently encrusted (Schneider and Webb, 2004). In late Paleozoic ecosystems, encrusting organisms such as bryozoans, spirorbids, corals, and small attaching brachiopods frequently were substantial contributors to richness and biovolume (Schneider, 2003b; Schneider and Webb, 2004). Therefore, archaeocidarids contributed to the tiering and trophic complexity in two-dimensional paleocommunities by providing surfaces for the settlement of diverse, attached filter feeders and predators.

Archaeocidarid, lepidocentrid, and echinocystoid echinoids are known from the Pennsylvanian Winchell Formation of north-central Texas (Schneider et al., 2005; Fig. 4.3). The most spectacular occurrence of these echinoids is in the series of seven Lagerstätten horizons in the aggregation described above, but numerous other remains of these echinoids, primarily spines, plates, demipyramids, and teeth, were recovered during a community paleoecological study within the formation. These samples provide an excellent opportunity to investigate the importance of echinoids in late Paleozoic marine communities, in that data encompassed the time interval of the known range of existence of *Archaeocidaris brownwoodensis*, which is conspicuous in many horizons throughout the locality. The other three echinoids known from the locality are much less common. They in-

Case Study from the Late Carboniferous of North-Central Texas

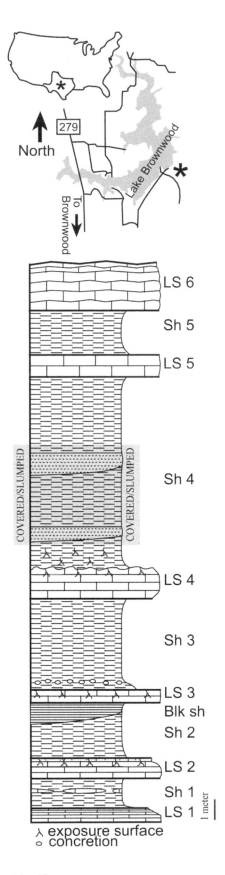

Figure 4.3. *Locality map and stratigraphic column for the Winchell Formation, Brownwood locality, north-central Texas. Section height totals 26.6 m. Unit names are informal. Abbreviations: Blk sh, black shale; LS, limestone; Sh, shale.*

clude *A. apheles* Schneider et al., 2005, a smooth-spined echinoid; *Elliptechinus kiwiaster* Schneider et al., 2005, an unusual elongate, upright lepidocentrid with numerous unornamented small spines; and an unnamed echinocystoid with expanded ambulacra of diverse plate morphologies relative to interambulacra.

METHODS. Samples of all macrofossil data were taken from carbonate and shale samples at a single locality near Brownwood, Texas, in the Winchell Formation. Carbonate samples were point counted on the outcrop surface, tabulating any macrofossil greater than approximately 1 mm in diameter that fell within the view of a hand lens at 2- or 5-cm intervals up to 200 points or the available extent of the outcrop. Shale samples were bulk collected and screened, or in the case of the unit containing the echinoid Lagerstätten, collected in large slabs and prepared along echinoid-bearing and other fossiliferous horizons. Point counting in thin section can be a proxy of biovolume (Watkins, 1996), and this method has also been applied to limestone surfaces in the field (Wilson, 1982; Schneider, 2003b). However, with high abundances of fossils, point counting also may be used for abundance proxies. Abundance on point-counted surfaces could then be compared with abundances recovered in shale samples. In all, 29 carbonate and 20 shale samples were collected.

The importance of echinoids in Winchell communities can be determined in three ways: 1) abundance in fossil samples, 2) rank abundance in fossil samples, and 3) diversity of echinoids in the locality. In addition to tabulating abundance and rank abundance of echinoids per sample, results were tested by comparing percent abundances and frequency of occurrence in shale and limestone samples by Mann-Whitney U and χ^2 tests to determine preference of echinoids for either environmental regime. In addition, abundance and rank abundance results were compared with that of crinoids to test for preservational influence on echinoids because of the similarity between mineralogy and structure of crinoid ossicles and echinoid plates and spines. For instance, if crinoids and echinoids were of low abundance in shales compared with limestones, then either both taxa preferred lime-mud environments, or preservation in terrigenous mud regimes was poor. Conversely, if crinoids were abundant in shales and echinoids were less common when compared with limestone samples, then preservation in shales would be deemed to be good, because crinoids and echinoids likely would have experienced the same taphonomic processes in shales.

In many modern communities, more than one echinoid taxon is present. The presence of more than one echinoid taxon in samples from the Brownwood locality may suggest a similar dynamic to echinoids in modern communities. For each sample, echinoids were tabulated by taxon to note diversity of echinoids within each lithologic unit.

RESULTS. Echinoids were significantly more abundant in limestone samples; they were less abundant in shale samples (Mann-Whitney U, $z = 5.045$; $p < 0.01$; Table 4.1). This is not attributed to poor preservation of echinoderms in shales compared with limestones because crinoids had a contrasting pattern of increased abundance in shales compared with limestones ($z = 2.698$; $p < 0.01$). In limestone point counts, echinoids were

Unit*	Height above Base of Limestone 1 (m)	Echinoid Material for Entire Sample of Macrofossils (%)
LS 6	26.6	16.7
	26.6	0
Sh 5	22.5	0
	22.5	0
LS 5	22.25	7
	22.1	24
Sh 4	18.5	1
LS 4	11.3	54.2
	11	28.6
	11	51.9
	10.8	50
Sh 3	6.8	0
	6.6	0
	6.6	0.4
LS 3	6.4	10.4
	6.4	10.6
	6.35	59.3
Blk Sh	5.3	7
	5.3	0
	5.3	0
	5.3	2.9
	5.3	7.4
	5.3	7.9
	5.3	23.3
Sh 2	3.15	1
	3.15	3.5
	3.15	0.3
	3.15	5.3
LS 2	3.0	59.4
	3.0	18.2
	2.8	22.8
	2.8	15.8
	2.5	27.3
	2.5	4.2
	2.5	1.7
	2.15	27
	2.15	7.1
Sh 1	1.5	3.5
	1.5	5.1
	1.5	11.7

Table 4.1. *Percentage of echinoid material per sample, as a proxy for biovolume, in stratigraphic order.*

Table 4.1. *cont.*

Unit*	Height above Base of Limestone 1 (m)	Echinoid Material for Entire Sample of Macrofossils (%)
LS 1	0.7	13.8
	0.7	14.1
	0.7	71.4
	0.65	18
	0.65	67.6
	0.65	58.9
	0.65	43.9
	0.5	58.3
	0.5	56.3

Note.—Cf. with Figure 4.3. Average echinoid material per sample, 20.0%; average echinoid material per limestone sample, 31.0%; average echinoid material per shale, 4.0%.

Abbreviations.—Blk Sh, black shale; LS, limestone; Sh, shale.

* Unit is lithologic unit at the Brownwood locality of the Winchell Formation, north-central Texas; refer to Figure 4.3.

commonly the highest or among the highest ranking taxon in abundance but did not have similar ranks in shales, except for the echinoid Lagerstätten in the black shale, where echinoids were the second most abundant taxon after the encrusting brachiopod *Crurithyris*.

Archaeocidaris brownwoodensis was recovered more frequently from limestones, occurring in 28 of the 29 limestone samples, but only 14 of the 20 shale samples ($\chi^2 = 31.0$, df = 1, $p < 0.01$), although they were commonly in low abundances in the shale samples. Of the 28 *A. brownwoodensis*-bearing carbonate samples, 15 samples contained at least one other echinoid taxon, whereas only one shale sample, that of the Lagerstätten in the black shale, contained other echinoid taxa (Table 4.2). Several samples contained two other echinoid taxa, but the most conspicuous co-occurrence of multiple taxa was in the black shale Lagerstätten, which contained three other echinoid taxa and was the only nonlimestone sample to record multiple echinoid taxa. An echinocystoid was the most common echinoid other than *A. brownwoodensis*, occurring in 11 samples, followed by A. *apheles* at four samples and *Elliptechinus kiwiaster* in two samples.

DISCUSSION OF LIMESTONE PREFERENCE. In carbonate facies at the Brownwood locality of the Winchell Formation, echinoids were among the dominant members of the marine paleocommunities. Echinoids commonly maintained high abundances and top rank in limestone bedding plane point counts, suggesting that echinoids were a major contributor to the calcified biovolume. Conversely, echinoid material is rare in fossil assemblages from siliciclastic environments. This pattern is not the result of poor preservation in siliciclastic regimes because crinoids, which have similar mineralogy and crystalline microstructure, are more common in shales than in limestones at the Brownwood locality, suggesting that the dissolution of echinoid skeletal material is not a bias. Likewise, the possibil-

Echinoid	LS 1	LS 2	Blk Sh	LS 3	LS 4	LS 5	Total
Archaeocidaris apheles	1	—	—	—	1	—	2
Elliptechinus kiwiaster	—	1	—	—	—	—	1
Echinocystoid	3	3	—	1	—	1	8
Archaeocidaris apheles and echinocystoid	1	—	—	—	1	—	2
Elliptechinus kiwiaster and echinocystoid	—	—	—	—	—	1	1
All three echinoid species	—	—	1	—	—	—	1
Total samples with multiple taxa	5	4	1	1	2	2	15
Total samples analyzed per unit	11	10	1	4	5	2	35

Note.—Only those units with co-occurring echinoid taxa are shown; all shales except the black shale contain only *A. brownwoodensis*, and Limestone 6 also contains only *A. brownwoodensis*. All echinoids except *A. brownwoodensis* are rare.

Abbreviations.—Blk Sh, black shale; LS, limestone.

Table 4.2. *Other echinoid species co-occurring with* **Archaeocidaris brownwoodensis** *in the Winchell Formation.*

ity of transport into carbonate regimes of the Brownwood locality is minimal; in some limestone units, echinoid plates and spines remain in close association, suggesting that disarticulation occurred, but plates and spines were not transported far from the place of death. This suggests that echinoids, during the time of deposition of the Brownwood locality, preferred carbonate regimes.

Three other localities within the Winchell Formation also contain echinoids, and in each, echinoids are most common in the limestone facies. At one locality near Perrin, Texas, north of Brownwood, *Archaeocidaris* spines and plates (although a different, undescribed species than either Brownwood *Archaeocidaris*) occur in the carbonate horizons of an alternating phylloid algal mound-siliciclastic accumulation. Closer to Brownwood, *A. brownwoodensis* spines occur in carbonate horizons of two localities. At each of these localities, carbonate sedimentation was greatly influenced by phylloid algae, and horizons containing echinoid spines were fairly diverse. Echinoids occurred in shale only in one horizon at the Perrin locality, but echinoids in this shale sample were rare.

Although this study suggests that *A. brownwoodensis* and contemporaneous echinoids in the Midland Basin preferred carbonate environments, late Paleozoic echinoids were not restricted to limestone facies. In one locality in north-central Texas, not far from the Brownwood locality but upsection from the Winchell Formation in the Colony Creek Shale, literally thousands of unidentified *Archaeocidaris* spines have been recovered along one horizon in a dense concentration. Echinoid plates are disproportionately rare in the spine horizon, as observed during my repeated visits to the locality. Other fossils associated with the spine accumulation are the bivalve *Orthomyalina* Newell, 1942, and the bryozoan *Rhombopora* Meek, 1872 (personal observation; Sprinkle, personal commun.). Other important echinoid localities in Texas contain abundant spines, and two have produced complete echinoid tests, although at much reduced abundances than that of the Colony Creek Shale (Schneider, 2002; Sprinkle, personal commun.)

Echinoids appear to be more prevalent in limestones than in shales. Most species in Jackson (1912), where lithology is recorded, are from limestones. In limestones, echinoid tests that have been rapidly buried remain intact even during exposure as a result of the hardness of limestone. In many shales, particularly muddy shales, exposure often results in disassociation of articulated plates and spines. In the Lagerstätten at Brownwood, most plates and spines disarticulate on exposure to air and water; only a few plates remain cemented together. In the Colony Creek Shale, at a different locality than that containing an overabundance of spines, nearly articulated echinoid tests were exposed on the weathered surface of the muddy shale, but after collection, echinoid plates and spines disassociate if they are not stabilized with a fixative (Sprinkle, personal commun.).

Most described species of archaeocidarid echinoids occur in limestones; few are from shales. Because of the low number of described articulated echinoids recovered from shales and the high number of species descriptions from limestones, it is possible that times of high echinoid diversity, such as the Mississippian, may be driven by the volume of carbonate rock.

Among later fossil and Modern echinoids, most regular species preferentially inhabit high-energy, hard substrates, such as reef, interreef, rocky littoral, and rocky sublittoral environments. Few shallow-water regular echinoids are known to exist on muddy or sandy substrates, whether carbonate or siliciclastic (e.g., Carter et al., 1989; Le Loeuff, 1993). Distribution of Modern regular echinoid skeletal material in marine sediments generally reflects their original environment of deposition—that of shallow-water hard substrate facies (Greenstein, 1991; Nebelsick, 1995, 1996). However, this high-energy, hard substrate preference in more recent regular echinoids is not present in their late Paleozoic ancestors. For instance, echinoids are known abundantly from all carbonates at the Brownwood, Texas, locality; four limestone units are micritic or argillaceous, including limestone unit 4, with closely associated clusters of plates and spines throughout the unit suggesting little disturbance of echinoid remains. The remaining two limestone units are carbonate sands, with fragmentary echinoid remains. Echinoid plates and spines are rare in most shales from Brownwood except for the echinoid Lagerstätten horizons, which occur in a laminated sandy black shale. None of the paleoenvironments at the Brownwood locality have been interpreted to be hard substrates, and two, limestone unit 4 and the black shale, were probably low-energy environments. Similarly, across the Pennsylvanian and Permian Eastern Shelf of the Midland Basin, hard substrates such as reefs and rocky environments were rare and potentially limited to algal buildups. Archaeocidarid and other echinoid remains are abundant in many limestones and shales, which originally were soft-substrate environments.

Smith (1984) reported that the highest generic diversity of Paleozoic echinoids occurred during the Devonian, but Kier (1965) and echinoid counts from Jackson (1912) indicated that the highest species diversity was during the Mississippian. These peaks in Paleozoic echinoid diversity co-incide with times of high global sea level and increased accumulation of carbonate rock (Tucker and Wright, 1990, and references therein). It is possible that the poor preservation of articulated specimens and/or acces-sibility of specimens in shales, combined with the simultaneous occur-rence of peak Paleozoic echinoid diversity and increased carbonate accu-mulation, may have influenced current understanding of Paleozoic echinoid diversity. Further work, including more investigation of shale fa-cies, detailed investigation into the firmness of echinoid-bearing substrates, archaeocidarid taphonomy, and description of known but undescribed spe-cies, will shed light on the problem of a carbonate bias for the late Paleo-zoic echinoid record.

ECHINOID DIVERSITY IN THE LATE PALEOZOIC. In all but one of the limestones and in the Lagerstätten at Brownwood, multiple echinoid taxa coexisted in marine communities. Although one taxon, *Archaeocidaris brownwoodensis*, was by far the most abundant echinoid, the presence of plates and spines of other echinoid taxa suggests that the coexistence of multiple echinoid taxa was not uncommon at this locality.

Multiple echinoid taxa are also known from other late Paleozoic lo-calities. Mississippian echinoid diversity is high, and one cause of this is the number of carbonate formations that contain multiple echinoids. More than 30 species are known from the limestones of the Burlington, Keokuk, and St. Louis formations of the upper Midwestern United States. Because of these echinoderm-rich limestones and the exceptional preservation in many localities in these limestones, it is evident that localities and rock units containing multiple echinoid taxa are important factors in determin-ing echinoid diversity through time.

In the northernmost locality of the Winchell Formation near Perrin, Texas, in the flank beds and core of a phylloid algal mound, two species of *Archaeocidaris* coexisted (personal observation). The dominant taxon was an unidentified species, represented by hundreds of spines and plates throughout the outcrop along limestone horizons and in one shale sample. The other taxon, A. *apheles*, is represented by two spine fragments in the shale sample. In this instance, because of the low abundance of the A. *apheles* spine fragments, the fragments may have been transported, but if transport occurred, it was likely from nearby localities within the Winchell Formation. No A. *brownwoodensis* spines were recovered from this phylloid algal mound environment, suggesting either a regional differentiation of *Archaeocidaris* along the depositional strike of the Winchell Formation or the environmental preferences of taxa. However, A. *brownwoodensis* is known from a spine and plate samples in the phylloid algal facies at Brown-wood and nearby Winchell localities. Therefore, the environmental prefer-ences of taxa were probably not the cause of differences in echinoid distri-bution between the Brownwood and Perrin localities.

In the Vanport Limestone of the Allegheny Group (Pennsylvanian) in Ohio, two distinct echinoid taxa, an archaeocidarid and a lepidesthid, both unidentified, were represented by hundreds of echinoid plates, spines, and lantern elements. In their study, Hoare and Sturgeon (1976) were able to use the sizes of disarticulated demipyramids to reconstruct the populations of each taxon in the limestone, and the results suggested co-occurrence of the taxa during deposition of the carbonate sediment, and a growth series of each taxon. Archaeocidarid plates, spines, and lantern elements outnumbered those of the lepidesthid, also suggesting the dominance of the archaeocidarid in the community.

The Argentine Limestone in the upper Farholm Formation in Alberta contains one abundant species of *Albertechinus* Stearn, 1956, and possibly the spines of another, undescribed echinoid (Stearn, 1956). *Albertechinus* tests are not associated with spines, and the spines at the locality have an unusual morphology for late Paleozoic echinoids, with an expanded flange on the distal end of the spine. As Stearn noted, the spines may be from *Albertechinus*, but he does not rule out the possibility that the spines are from a different, unknown species.

These studies suggest that the coexistence of multiple echinoid taxa was not uncommon in late Paleozoic communities. Although the number of taxa per community did not reach the richness of echinoid species in some Modern ecosystems (e.g., Bolanos et al., 2005; Duran-Gonzalez et al., 2005; Laguarda-Figueras et al., 2005), the presence of more than one species is in concert with the increasing prominence of echinoids in late Paleozoic communities and the onset of resource partitioning among echinoids. If echinoid jaws were not strong enough for active predation, then remaining available resources within a community would have been shared among all co-occurring echinoid species. At present, no strong evidence exists as to exact echinoid diet. However, if different resources are utilized or if the same resource is utilized differently by separate taxa, jaw apparatuses should be distinct between taxa.

In the study by Hoare and Sturgeon (1976) that looked at the populations of an archaeocidarid and a lepidocentrid, two characteristics of the demipyramids— the size of demipyramids and shape of the foramen magnum in each demipyramid—were examined to determine the distinctive of each population and the size distribution of individuals. For each characteristic, the two taxa plotted separately in morphospace, indicating that demipyramid morphologies differed between echinoid taxa. Might these two echinoids, the abundant archaeocidarid and the less abundant (although not rare) lepidocentrid, have evolved different jaw mechanisms in order to minimize interspecific competition? Currently, no study has examined this for echinoids, although competition in the fossil record has been studied for trilobites (Robison, 1975) and brachiopods (Hermoiyan et al., 2002). Given the number of occurrences of multiple taxa, a study to determine potential lantern morphological variation to limit competition for food resources will be possible in the future.

Conclusions

Although late Paleozoic echinoids are frequently overlooked in paleoecological studies, they contributed greatly to the diversity and complexity of shallow marine ecosystems. Paleozoic echinoids reached peak species diversity in the Mississippian, but this phenomenon may be artificial, given global sea-level decrease and the subsequent decline in carbonate rock record during the Pennsylvanian and Permian. Similarly, echinoids appear to have preferred carbonate facies, although they are known from many shale horizons in the Brownwood Lagerstätten described here. The apparent preference of echinoids for carbonate environments may be an artifact of poor preservation and inherent problems with collecting articulated specimens from muddy shales. Late Paleozoic echinoids were typical of structurally simple, two-dimensional communities that contained mostly individuals living at the sediment-water interface, as opposed to the three-dimensional bryozoan thickets and crinoid gardens characterized by complex tiering and spatial structure.

The spines of late Paleozoic echinoids are important, not only to the echinoid for defense against predators, but also to the biodiversity and structure in paleocommunities. Spines of archaeocidarids were frequently encrusted by epibionts, especially after the evolution of complex spine ornament. Particularly in the Pennsylvanian and Permian, spines were host to diverse and abundant epibionts; this phenomenon resulted in an increase in diversity and tiering complexity in the two-dimensional ecosystems in which some echinoid taxa lived.

Aggregation behavior, the dense concentration of individuals through migration of large numbers of echinoids, can be interpreted for several echinoid Lagerstätten, but the quantitative assessment of this behavior is lacking. Reasons for late Paleozoic echinoid aggregation cannot yet be determined. In the Late Carboniferous Winchell Formation of north-central Texas, predation pressure can be ruled out, but ample terrigenous plant material as a potential food source and breeding are potential causes.

In many late Paleozoic communities, more than one echinoid species is present, ranging from two to four taxa. Within these communities, one species, usually an archaeocidarid, is the most abundant taxon; other echinoids are commonly rare by comparison. It is possible that the ecological separation of various echinoid taxa allowed coexistence in communities in the late Paleozoic.

Although much is known about late Paleozoic echinoid paleoecology, much potential work remains. These and other paleoecological studies of late Paleozoic echinoids will contribute greatly to the understanding of the mixing of Modern and Paleozoic faunas and the resulting diversity patterns and organism interactions during the late Paleozoic precursor to the Mesozoic marine revolution.

Acknowledgments

This chapter would not have happened for many years without the existence of this volume in honor of G. Lane; hence, I would like to dedicate this paper to his memory, in honor of the great work he has accomplished and his inspiration to the younger generation of echinoderm paleontologists. I would like to thank my reviewers and editor, J. Nebelsick, C. D. Sumrall, and W. I. Ausich, for excellent comments, and L. Leighton for assistance with early drafts. I am grateful to D. Harlow, A. Molineux, D. Ryder, and J. Sprinkle for discussions, advice, specimens, and/or field assistance in earlier portions of this work, and S. K. Donovan and A. Smith for later discussions and advice. Funding for various parts of this research was granted by the Austin Paleontological Society, the Central Texas Paleontological Society, the Geological Society of America, and the Jackson School of Geosciences Foundation at the University of Texas. Last, I acknowledge my cat, Shale, for dubious assistance in typing the manuscript.

References

Alexander, R. R. 1986. Frequency of sublethal shell-breakage in articulate brachiopod assemblages through geologic time. Biostratigraphie du Paleozoique, 4:159–166.

Alexander, R. R. 2001. Functional morphology and biometrics of articulate brachiopod shells, p. 145–170. In S. J. Carlson and M. R. Sandy (eds.), Brachiopods Ancient and Modern: A Tribute to G. Arthur Cooper. Paleontological Society Papers, 7.

Ausich, W. I. 2001. Echinoderm taphonomy, p. 171–227. In M. Jangoux and J. M. Lawrence (eds.), Echinoderm Studies, Volume 6. A. A. Balkema, Rotterdam.

Bambach, R. K., A. H. Knoll, and S. C. Wang. 2004. Origination, extinction, and mass depletions of marine diversity. Paleobiology, 30:522–542.

Baumiller, T. K., and F. J. Gahn. 2004. Testing predator-driven evolution with Paleozoic crinoid arm regeneration. Science, 305:1453–1455.

Baumiller, T. K., R. Mooi, and C. G. Messing. 2001. Cidaroid-crinoid interactions as observed from a submersible, p. 3. In M. Barker (ed.), Echinoderms 2000. A. A. Balkema, Rotterdam.

Bolanos, N., A. Bourg, J. Gomez, and J. J. Alvarado. 2005. Diversidad y abundancia de equinodermos en la laguna arrecifal del Parque Nacional Cahuita, Caribe de Costa Rica. Revista de Biologia Tropical, 53:285–290.

Brown, T. 1849. Illustrations of the Fossil Conchology of Great Britain and Ireland: With the Description and Localities of All the Species. Smith, Elder, London, 273 p.

Brunton, C. H. C., and S. S. Lazarev. 1997. Evolution and classification of the Productellidae (Productida), upper Paleozoic brachiopods. Journal of Paleontology, 71:381–394.

Carter, B. D., T. H. Beishel, W. B. Branch, and C. M. Mashburn. 1989. Substrate preferences of late Eocene Priabonian–Jacksonian echinoids of the Eastern Gulf Coast. Journal of Paleontology 63:495–503.

Duran-Gonzalez, A., A. Laguarda-Figueras, F. A. Solis-Martin, B. E. Buitron Sanchez, C. G. Ahearn, and J. Torres Vega. 2005. Equinodermos (Echinoderms) de las agues mexicanos del Golfo de Mexico. Revista de Biologia Tropical, 53:53–68.

Estes, J. A., and J. F. Palmisano. 1974. Sea otters: their role in structuring nearshore communities. Science, 185:1058–1060.

Estes, J. A., and D. O. Duggins. 1995. Sea otters and kelp forests in Alaska: generality and variation in a community ecological paradigm. Ecological Monographs, 65:75–100.

Estes, J. A., M. T. Tinker, T. M. Williams, and D. F. Doak. 1998. Killer whale predation on sea otters linking oceanic and nearshore ecosystems. Science, 282:473–476.

George, T. N. 1931. *Amocoelia* Hall and certain similar British Spiriferidae. Geological Society of London Quarterly Journal, 87:30–61.

Greenstein, B. J. 1991. An integrated study of echinoid taphonomy: predictions for the fossil record of four echinoid families. Palaios 6:519–540.

Hagen, N. T., and K. H. Mann. 1994. Experimental analysis of factors influencing aggregating behavior of the green sea urchin *Strongyolcentrotus droebachensis* (Muller). Journal of Experimental Marine Biology and Ecology, 7:107–126.

Hermoiyan, C. S., L. R. Leighton, and P. Kaplan. 2002. Testing the role of competition in fossil communities using limiting similarity. Geology, 30:15–18.

Heterier, V., C. De Ridder, B. David, and T. Rigaud. 2004. Comparative biodiversity of ectosymbionts in two Antarctic cidaroid echinoids, *Ctenocidaris spinosa* and *Rhynchocidaris triplopora*, p. 201–205. *In* T. Heinzeller and J. Nebelsick (eds.), Echinoderms. Munich, Swets.

Hoare, R. D., and M. T. Sturgeon. 1976. Echinoid remains from the Pennsylvanian Vanport Limestone (Allegheny Group), Ohio. Journal of Paleontology, 50:13–24.

Hughes, T. P. 1994. Catastrophes, phase shifts, and large-scale degradation of a Caribbean coral reef. Science, 265:1547–1551.

Jackson, R. T. 1912. Phylogeny of the Echini, with a revision of Paleozoic species. Memoirs of the Boston Society of Natural History, 7, 491 p.

Kelley, P. H., M. Kowalewski, and T. A. Hansen. 2003. Predator-Prey Interactions in the Fossil Record. Topics in Geobiology, Volume 20. Kluwer Academic/Plenum Publishers, New York, 464 p.

Kier, P. M. 1965. Evolutionary trends in Paleozoic echinoids. Journal of Paleontology, 39:436–465.

Kier, P. M. 1977. The poor fossil record of the regular echinoid. Paleobiology, 3:168–174.

LaGuarda-Figueras, A., F. A. Solis-Martin, A. Duran-Gonzalez, C. G. Ahearn, B. E. Buitron Sanchez, and J. Torres-Vega. 2005. Equinodermos (Echinodermata) del Caribe Mexicano. Revista de Biologia Tropical, 53:109–122.

LaMare, M. D., and B. G. Stewart. 1998. Mass spawning by the sea urchin *Evechinus chloroticus* (Echinodermata: Echinoidea) in a New Zealand fiord. Marine Biology, 132:135–140.

Le Lceuff, P. 1994. The benthic fauna of the Guinea continental shelf (trawling grounds), compared to the fauna of the Cote-D'Ivoire. Revue d'Hydrobiologe Tropicale 26:229–252.

Leighton, L. R. 2003. Predation on brachiopods, p. 215–237. *In* P. H. Kelley, M Kowalewski, and T. A. Hansen (eds.), Predator-Prey Interactions in the Fossil Record, Topics in Geobiology, Volume 20, Kluwer/Plenum, New York.

Lessios, H. A. 1988. Mass mortality of *Diadema antillarum* in the Caribbean: what have we learned? Annual Review of Ecology and Systematics, 19:371–393.

M'Coy, F. 1849. On some new Paleozoic Echinodermata. Annual Magazine of Natural Historyc series 2, 3:244–254.

Meek, F. B. 1872. Report on the paleontology of eastern Nebraska, with some remarks on the Carboniferous of that district, p. 83–239. *In* F. V. Hayden. Final report of the U.S. Geological Survey of Nebraska and portions of the adjacent territories. House Executive Document, 19, Washington, D.C.

Meyer, D. L., and W. I. Ausich. 1983. Biotic interactions among Recent and among fossil crinoids, p. 377–427. *In* M. J. S. Tevesz and P. L. McCall (eds.), Biotic Interactions in Recent and Fossil Benthic Communities. Plenum Press, New York.

Miller, A. I., and J. J. Sepkoski Jr. 1988. Modeling bivalve diversification; the effect of interaction on a macroevolutionary system. Paleobiology, 14:364–369.

Nebelsick, J. H. 1995. Actuopaleontologic investigations on echinoids: the potential for taphonomic interpretation, p. 209–214. *In* R. Emson, A. Smith, and

A. Campbell (eds.), Echinoderm Research 1995. A. A. Balkema, Rodderdam.

Nebelsick, J. H. 1996. Biodiversity of shallow-water Red Sea echinoids: implications for the fossil record. Journal of the Marine Biology Association, United Kingdom, 76:185–194.

Newell, N. D. 1942. Late Paleozoic pelecypods, Mytilacea. Kansas State Geological Survey Report, 10, 115 p.

Philippi, R. A. 1845. Beschreibung einiger neuen Echindermen nebst kritischen Bemerkungen über einige weniger bekannte Arten. Archiv für Naturgeschichte 11:344–359.

Robison, R. A. 1975. Species diversity among agnostoid trilobites. Fossils and Strata, 4:219–226.

Rodriguez, S. R., and J. M. Farina. 2001. Effect of drift kelp on the spatial distribution pattern of the sea urchin *Tetrapygus niger*: a geostatistical approach. Journal of the Marine Biology Association, United Kingdom, 81:179–180.

Sala, E., C. F. Boudouresque, and M. Harmelin-Vivien. 1998. Fishing, trophic cascades, and the structure of algal assemblages: evaluation of an old but untested paradigm. Oikos, 83:15.

Sawyer, J. A., and L. R. Leighton. 2005. Predator prey dynamics in Pennsylvanian near-shore shales: effects of molluscan invasion into normal marine habitats. Geological Society of America Annual Meeting Abstracts with Programs, 37:13–14.

Schneider, C. L. 2001. Heaps of echinoids in a Pennsylvanian echinoderm lagerstatten: implications for fossilized behavior. Paleobios, 21:113.

Schneider, C. L. 2002. Hitchin' a ride: epibionts on *Archaeocidaris* echinoids. Geological Society of America Abstracts with Programs, 34(7):34.

Schneider, C. L. 2003a. Hitchhiking on Pennsylvanian echinoids: epibionts on *Archaeocidaris*. Palaios, 18:435–444.

Schneider, C. L. 2003b. Community paleoecology of the Pennsylvanian Winchell Formation, north-central Texas. Unpublished Ph.D. dissertation, University of Texas, Austin, 321 p.

Schneider, C. L., J. Sprinkle, and D. Ryder. 2005. Pennsylvanian (Late Carboniferous) echinoids from the Winchell Formation, north-central Texas, USA. Journal of Paleontology, 79:745–762.

Schneider, C. L., and A. Webb. 2004. Where have all the encrusters gone? Encrusting organisms on Devonian versus Mississippian brachiopods. Geological Society of America Abstracts with Programs, 36(7):111.

Signor, P. W., III, and C. E. Brett. 1984. The mid-Paleozoic precursor to the Mesozoic marine revolution. Paleobiology, 10:229–245.

Smith, A. 1984. Echinoid Palaeobiology. Allen and Unwin, London, 190 p.

Stearn, C. W. 1956. A new echinoid from the Upper Devonian of Alberta. Journal of Paleontology, 30:741–746.

Taylor, P. D. and M. A. Wilson. 2003. Palaeoecology and evolution of marine hard substrate communities. Earth-Science Reviews, 62:1–103.

Tucker, M. E., and V. P. Wright. 1990. Carbonate Sedimentology. Blackwell Science, Oxford, 482 p.

Walker, L. J., B. H. Wilkinson, and L. C. Ivany. 2002. Continental drift and Phanerozoic carbonate accumulation in shallow-shelf and deep-marine settings. Journal of Geology, 110:75–87.

Warme, J. E., and R. W. Olson. 1971. Stop 5: Lake Brownwood Spillway, p. 27–43. *In* R. F. Perkins (ed.), Trace Fossils: A Field Guide to Selected Localities in Pennsylvanian, Permian, Cretaceous, and Tertiary Rocks of Texas and Related Papers Society for Sedimentary Geology Field Trip, April 1–3,

1971. Louisiana State University School of Geoscience Miscellaneous Publication,71-1.

Watkins, R. 1996. Skeletal composition of Silurian benthic marine faunas. Palaios, 11:550–558.

Wilson, M. A. 1982. Origin of brachiopod-bryozoan assemblages in an Upper Carboniferous limestone: importance of physical and ecological controls. Lethaia, 15:263–273.

Young, C. M., P. A. Tyler, J. L. Cameron, and S. G. Rumrill. 1992. Seasonal breeding aggregations in low density populations of the bathyal echinoid *Stylocidaris lineata*. Marine Biology, 113:603–612.

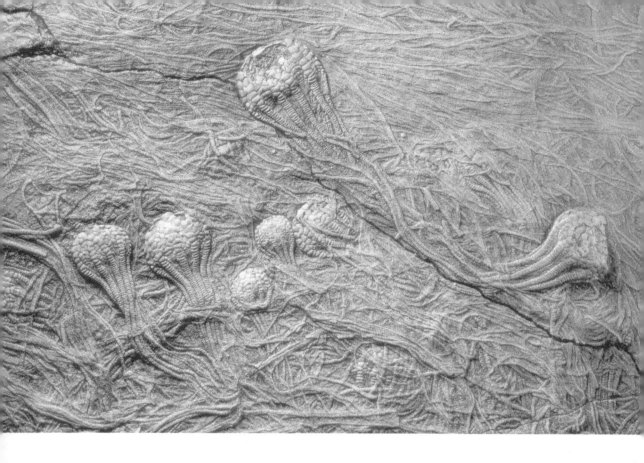

NEW OBSERVATIONS ON TAPHONOMY AND PALEOECOLOGY OF *UINTACRINUS SOCIALIS* GRINNELL (CRINOIDEA; UPPER CRETACEOUS)

5

Andrew J. Webber, David L. Meyer, and Clare V. Milsom

Introduction

With atypically long arms and a stemless, globular calyx, *Uintacrinus socialis* Grinnell, 1876, has fascinated and challenged paleontologists for more than a century. The basic structure of *Uintacrinus* Grinnell, 1876, is quite unlike any other crinoid, with arm lengths commonly exceeding 1 m—by far an extreme in crinoids. Although *Uintacrinus* occurs in the Upper Cretaceous of North America, Europe, and Australia, these crinoids are best known from the Niobrara Formation of Kansas, where they are exceptionally preserved as thin lenses of hundreds of individuals in a striking display of entangled arms and calyxes (Fig. 5.1; Hess, 1999). Large *Uintacrinus* aggregations from North America are true fossil Lagerstätten in the sense of both preservational quality and quantity of specimens, and they contain a wealth of paleobiological information (Seilacher, 1970). The lower surface of each *Uintacrinus* lens preserves up to hundreds of articulated crinoids, but the upper surface is covered with completely disarticulated skeletal ossicles. This mode of preservation is best explained by mass mortality followed by surficial disarticulation and microbial sealing, rather than by gradual accumulation (Meyer and Milsom, 2001). Thus, *Uintacrinus* aggregations represent life assemblages that provide a snapshot of *Uintacrinus* populations rather than a time exposure of successive populations. These lenses potentially can provide valuable data on the ecology of these Mesozoic crinoids, including life habit, population structure, and environmental conditions.

Although museums in North America and Europe display many spectacular *Uintacrinus* slabs that were collected in the late nineteenth and early twentieth centuries, precise locality and stratigraphic data for most of these slabs are usually lacking (Appendices 5.1 and 5.2). This limits the amount of paleobiological data that can be derived. Beginning in 1990, discoveries of large *Uintacrinus* aggregations in Kansas and Colorado prompted renewed study. The goal of this research is to examine these new findings and existing museum collections to improve understanding of the stratigraphic distribution, taphonomy, and life habits of *Uintacrinus*.

Specific objectives of this research are as follows. 1) Fieldwork based on recent findings in Kansas and Colorado has enabled, for the first time,

Figure 5.1. *Uintacrinus socialis* Grinnell specimens from a slab on display at Staatliches Museum für Naturkunde at Karlsruhe, Germany. This slab is from the Smoky Hill Member of the Niobrara Formation, Gove County, Kansas. Photo courtesy of S. Rietschel.

a determination of the stratigraphic occurrence of *Uintacrinus* with more precision and an examination of in situ lenses with respect to the enclosing strata. For this reason, reported collection localities have been revisited to examine the stratigraphic position, associated fauna, and preservation of *Uintacrinus* specimens in the field. 2) Given the taphonomic quality of *Uintacrinus* lenses, analyses of population structure represented by large aggregations of *Uintacrinus* have provided information on specimen size and age distribution. 3) Several studies have noted (Bassler, 1909; Struve, 1957) a peculiar radial pattern in which specimens lie with arms oriented toward the center of the aggregation like spokes. Consequently, the orientation of individual crinoids within large aggregations has been quantified to determine whether patterns of preferred orientation or alignment are present that might provide further information about *Uintacrinus* life habits and mass mortality.

Stratigraphic Occurrence

Uintacrinus has a narrow stratigraphic range, confined to the Santonian Stage of the Cretaceous Period, approximately 85 million years ago, but occurs almost worldwide in deposits in Germany, France, England, Italy, Russia, Australia, and North America (Bather, 1896; Springer, 1901; Miller et al., 1957; Rasmussen, 1961; Miller, 1968; Cobban, 1995). In Europe, *Uintacrinus* is a guide fossil for middle and upper Santonian strata, and the uppermost occurrences of *Uintacrinus anglicus* Rasmussen, 1961, and *Marsupites testudinarius* (Schlotheim, 1820) mark the Santonian–Campanian boundary (Bailey et al., 1983, 1984). However, Mitchell (1995) reported *Uintacrinus anglicus* occurrences from the lower Campanian. Most *Uintacrinus* specimens come from chalk deposits, but they also have been described in shale, sandstone, and marl (Milsom et al., 1994; Cobban, 1995). In North America, large slabs of *Uintacrinus* specimens are best known from deposits in the Smoky Hill Member of the Niobrara Formation in Kansas. According to Hattin (1982), *Uintacrinus* is restricted to the *Clioscaphites choteauensis* Biozone, a lower upper Santonian biostratigraphic range zone in North American Cretaceous deposits (Cobban, 1962). Miller (1968) also placed Kansas specimens of *Uintacrinus* in the *Clioscaphites choteauensis* Biozone, as well as the stratigraphically lower middle Santonian *Clioscaphites vermiformis* Biozone.

Well-preserved specimens also occur in equivalent strata of the Mancos Shale in western and southwestern Colorado (Cobban, 1995). Pike (1947) found specimens in the Upper Santonian *Desmoscaphites bassleri* Biostratigraphic Biozone, which is higher than the *Clioscaphites choteauensis* Biozone. Leckie et al. (1997) reported fragments from the *Scaphites leei* III Biozone, making it the youngest reported occurrence in North America. A well-preserved slab reposited in the Denver Museum of Natural History (DMNH 6000) was also discovered in the Mancos Shale, reportedly from the Lower Santonian *Clioscaphites saxitonianus* Biozone, making this the oldest specimen collected in North America (fossil locality report on file with BLM, courtesy of H. Armstrong). Other North American specimens are from the Mancos Shale in Utah and the Upper Santo-

Figure 5.2. *Field sites in the Smoky Hill Member of the Niobrara Chalk; Gove County, Kansas. Site K1, the Blue Knoll, location of the slab housed in the University of Wisconsin Geological Museum; NW1/4 SW1/4 Sec. 9, T14S, R33W. Site K4, the Martin Quarries; Sec. 27, T14S, R32W.*

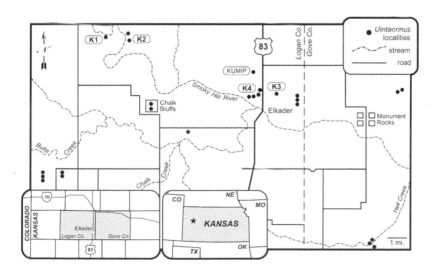

Figure 5.3. *Map showing the field localities of this study (K1–K4) and those of past **Uintacrinus** collections in Logan and Gove Counties, Kansas. Redrawn from Miller et al. (1957).*

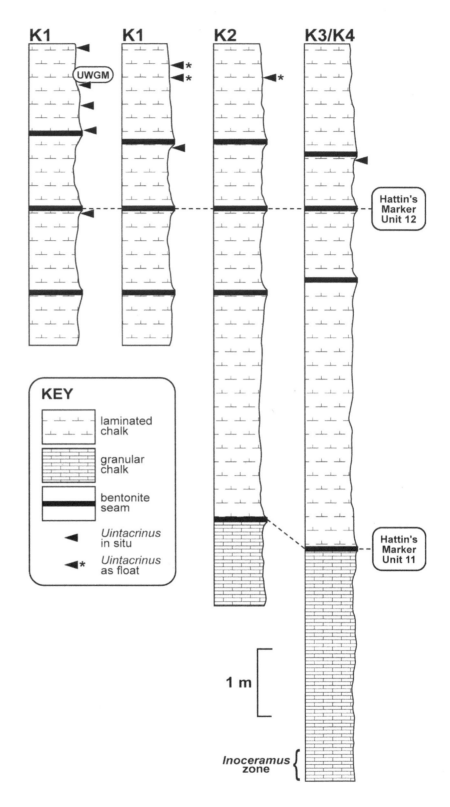

Figure 5.4. *Stratigraphic relation and correlation of **Uintacrinus** horizons in the Smoky Hill Member of the Niobrara Chalk, Gove and Logan Counties, Kansas, based on Hattin (1982). The localities K1–K4 can be found in Figure 5.3.*

KEY

laminated chalk

granular chalk

bentonite seam

Uintacrinus in situ

Uintacrinus as float

1 m

Inoceramus zone

nian *Desmoscaphites erdmanni* and *D. bassleri* Biozones of the Cody Shale in Wyoming and Montana (Collier, 1929; Thom et al., 1935; Keefer and Troyer, 1956, 1964). *Uintacrinus* also has been reported from Texas (R. Kaiser, personal commun., 1997), Vancouver (Whiteaves, 1904), and Alberta (Warren and Crockford, 1948).

Field localities in Kansas and Colorado noted in previous publications, reports, museum collections, and recent findings were revisited dur-

Figure 5.5. *Photographs of primary field site in the Smoky Hill Member of the Mancos Shale, Mesa County, Colorado, NE1/4 NW1/4 SW3 Sec. 26, T14S, R32W.*

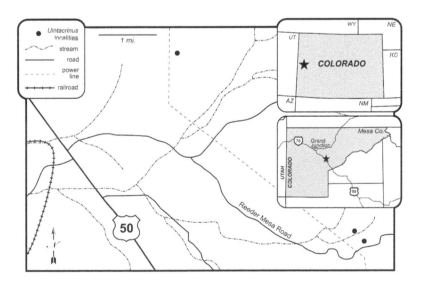

Figure 5.6. *Field sites in Mesa County, Colorado, southeast of Grand Junction.*

New Observations on Taphonomy and Paleoecology   97

ing the course of the present study. The stratigraphic position of each new specimen was recorded where possible, and each outcrop was correlated with others when necessary. In Kansas, outcrops were located in chalk beds of the Smoky Hill Member of the Niobrara Formation in Logan County and in western Gove County near the abandoned town of Elkader (Figs. 5.2 and 5.3). Occurrences of *Uintacrinus*, both as lenses of many individuals and as single specimens, are not restricted to a single layer of chalk but span an interval of approximately 2 m above marker unit 12 of Hattin (1982; Fig. 5.4). This marker unit is stratigraphically higher than the *Clioscaphites choteauensis* Biozone, which is restricted to marker units 9 and 10. Because no specimens were found stratigraphically lower than this biozone during the course of this study, it is difficult to confirm the reported lowest occurrence of *Uintacrinus* in Kansas. It is evident that *Uintacrinus* is not restricted to the *Clioscaphites choteauensis* Biozone in the Smoky Hill Member deposits in western Kansas. In Colorado, all localities revisited during the present study occurred in the Mancos Shale in Mesa County near Grand Junction (Figs. 5.5 and 5.6). All specimens were found within an approximately 12-m interval, although there is no detailed stratigraphy of the Mancos Shale in this area on which to log the precise position of *Uintacrinus*. Leckie et al. (1997) reported that *Uintacrinus* occurs in either the Smoky Hill Member or the stratigraphically lower Cortez Member of the Mancos Shale in Mesa Verde National Park, Montezuma County, Colorado.

Analysis of *Uintacrinus* Aggregations

Slabs containing large numbers of *Uintacrinus* have been collected for more than a century and are now housed in various museums and university collections (Appendices 5.1 and 5.2). Many of these collections and new field collections from the Smoky Hill Member of the Niobrara Formation in Kansas and the Mancos Shale in Colorado were analyzed for calyx size distribution, specimen orientation, stratinomic occurrence of preserved specimens, and associated fauna among *Uintacrinus* specimens.

CALYX SIZE MEASUREMENTS AND POPULATION STRUCTURE. The diameter of each individual specimen within a given aggregation was measured approximately perpendicular to the oral-aboral axis of the specimen. All measured specimens occurred on the lower surface of each slab, and only specimens with enough of the calyx preserved or exposed were considered for diameter measurement. One hindrance to taking direct diameter measurements is that calyxes of *Uintacrinus* are potentially distorted by severe compression. However, Harris (1974) noted that it is unlikely that the shape of any body fossil actually spreads laterally because the sediments and other surrounding matrix that are also being compressed would resist this horizontal extension. In this study, ping-pong balls and other hollow balls subjected to compression did not spread laterally to any appreciable degree when surrounded by other material. The models took two basic shapes after compression. In one, the top sides collapsed inward and the bottom sides retained their original shape, so the final form resembled a cup. In the other, this shape was flattened further into a saucer shape, yet

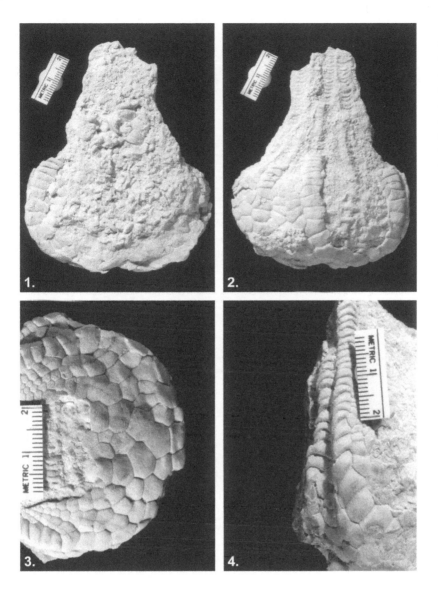

Figure 5.7. *Effects of compression on **Uinta-crinus** calyxes. 1, The upper surface is flattened and disarticulated. 2, The lower surface is well preserved. 3, Several plates along the edge overlap. 4, The edges of many calyxes show accordion-type folding.*

still without horizontal extension. Likewise, the lateral spread of an organism resulting from a compression is resisted by the lateral spread of the surrounding material.

In this manner, it is possible that the lower side of *Uintacrinus* acted like the lower side of the ping-pong balls in Harris (1974) and nearly retained its original shape, whereas the upper side collapsed inward. In this case, the increase in variation of measured diameters as a result of compression may be only minor, and the measured diameters of individuals are close to their original diameter. However, crinoids have contents that are soft and fluid, and their outer surfaces are composed of articulated plates rather than a uniform material. It would be expected that a crinoid would behave more complexly in compression than a ping-pong ball—that is, plates may undergo folding and overlapping under one another. Still, lateral extension would not be expected.

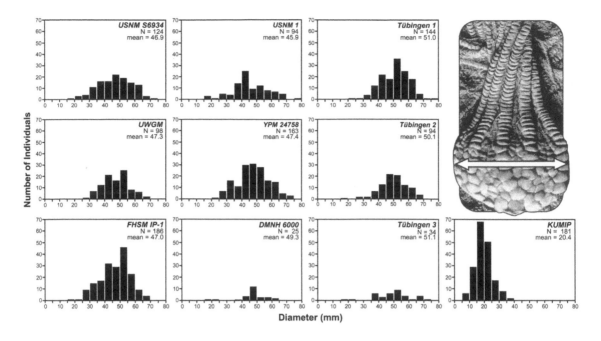

Figure 5.8, *Calyx diameters of 10* **Uintacrinus** *slabs. Each histogram is labeled with the name of the slab (Appendices 5.1 and 5.2), the number of measurable specimens, and mean calyx diameter. Also figured is a* **Uintacrinus** *calyx showing how diameter is measured.*

Several calyxes of *Uintacrinus* that were free from the slab matrix were examined to ascertain their compressed shape. In the case of *Uintacrinus*, the top side is disarticulated and shows evidence of collapse (Fig. 5.7.1). In addition, the well-preserved lower side is not hemispherical, as would be expected, but somewhat flatter, much like the saucer shape resulting from compression in Harris's study (Fig. 5.7.2). On close examination, slight overlapping of calyx plates occurs on the articulated side (Fig. 5.7.3), as well as accordion-like folding along the calyx margin (Fig. 5.7.4). This modification of shape accounts for the departure of the articulated sides of the calyxes from a hemispherical shape, but the measured diameter of each calyx nevertheless is similar to its original diameter.

Figure 5.9. *Cumulative frequency of* **Uintacrinus** *calyx diameters.*

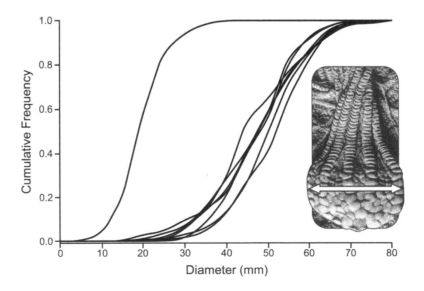

Ten slabs were analyzed for calyx size distributions (Appendices 5.1 and 5.2). All slabs have a unimodal bell-shaped distribution without obvious multiple peaks (Fig. 5.8). *Uintacrinus* calyxes on these slabs have a mean diameter of 46 to 47 mm, with the exception of three slabs housed at Geologisches-Paläontologisches Institut at Tübingen University (Tübingen 1–3), which have means ranging from 50 to 51 mm, and a slab at the University of Kansas (KUMIP), which has a mean diameter of 20.4 mm. Calyx diameters from each slab also have been plotted on a size versus cumulative frequency diagram (Fig. 5.9). This has been done to test for a normal distribution and to determine whether there were any groupings in the distribution of calyx sizes that would either be obliterated or artificially enhanced by a histogram with discrete size intervals. Each curve represents the population on one slab. The Tübingen 3 slab has been left out because it contains a much lower number of measurable specimens. All curves in this graph are S shaped, which indicates a normal distribution (Sokal and Rohlf, 1981). In addition, the curves have no size-group clustering, and all fall within close proximity to one another, with the exception of the KUMIP slab.

The analysis of calyx size distributions has revealed unimodal, bell-shaped curves for each *Uintacrinus* population. Moreover, the distribution of calyx sizes among populations is highly comparable, with one exception. The KUMIP slab is the only large assemblage with significantly smaller individuals; however, a small piece located in the Sternberg Museum at Fort Hays State University also contains small crinoids. The locality information on this piece suggests that it was collected from the same site as the KUMIP specimen. Another piece containing exclusively small individuals, which was figured in Springer (1901), was collected from this same locality and is now housed in the Harvard Museum of Comparative Zoology.

CALYX ORIENTATION. This study addressed preferred alignment in *Uintacrinus* specimens by quantifying the degree to which calyxes are aligned on 12 slabs from museum collections (Figs. 5.10 and 5.11). The orientation of individuals was measured according to the direction of the oral side of those specimens lying with their oral-aboral axis in the plane of bedding. Individuals with similar orientations would have oral sides facing nearly the same direction. Each slab was divided into quadrants to enable recognition of a radial pattern, as has been observed from several museum specimens. Slabs with a complete or partial radial pattern would have a significant number of calyxes in a given quadrant oriented approximately toward the center of the slab. Some of the slabs that do not have radial orientation patterns nevertheless had significant directional trends, which were also measured. The significance (at $\alpha = 95\%$) of directional trends from each quadrant was assessed by Rayleigh's test as described in Davis (2002). With this test, a mean orientation is calculated and can be plotted as a resultant vector (Figs. 5.10 and 5.11). The length of this resultant vector reflects the amount of spread in the data set, with tighter clusters having longer resultant vectors. Because sample size also plays a part in evaluating the significance of patterns, the resultant vector was standardized by the number of samples.

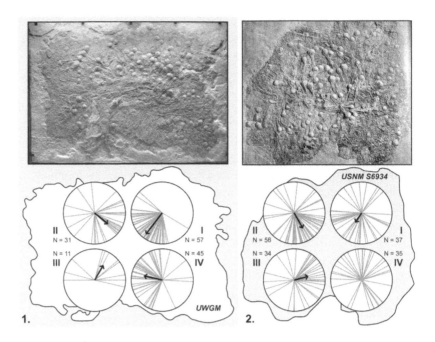

Figure 5.10, *Orientations of **Uintacrinus** calyxes on two slabs from the Smoky Hill Member of the Niobrara Chalk, Kansas. Each diagram is from one of four quadrants (labeled I–IV) on each slab. Each gray line indicates the orientation of a single calyx. Also shown with arrows are sample standardized resultant vectors indicating statistically significant mean orientations of calyxes from a given quadrant. Standardized resultant lengths have values from 0 to 1, with maximum values at the edge of each circle. Longer resultant vectors signify tighter clustering of orientations. If no resultant vector is present, then directional trends are not significant. 1, Slab now on display at the University of Wisconsin Geology Museum. 2, National Museum of Natural History, Smithsonian Institution, USNM S 6934.*

Eight of the 12 slabs analyzed for orientation patterns were collected from deposits in the Smoky Hill Member of the Niobrara Formation in Kansas. One slab now on exhibit in the Denver Museum of Natural History (DMNH 6000) was collected from the Smoky Hill Member of the Mancos Shale in Colorado near Grand Junction. The three slabs on display at the Tübingen University (Tübingen 1, 2, and 3) are from the Niobrara Formation in Kansas but lack specific locality data. Of these slabs, the strongest radial pattern is on the slab housed at the University of Wisconsin Geology Museum (UWGM), in which all quadrants have a significant directional trend toward the center of the slab (Fig. 5.10.1). On this slab, the oral side of most of the individuals faces a central point, and the arms of each crinoid extend inward in a spokelike pattern. Calyxes from a slab located at the U.S. Natural History Museum (USNM S6934) appear to be aligned in a radial pattern (Fig. 5.10.2). Analysis of this slab bears this out, although only three quadrants have statistically significant trends toward the center.

Many of the remaining slabs have weak or partial radial alignment of calyxes (Fig. 5.11). A slab from the Sternberg Museum at Fort Hays State University (FHSM IP-1) has calyxes from each quadrant aligned nearly toward the center of the slab. A small specimen on display at the Fick Fossil Museum (FFM, not figured), which was at one time attached to the FHSM IP-1 slab, has a strong parallel alignment. It is unknown, however, exactly how this smaller piece fits on its larger counterpart. Slabs from Staatliches Museum für Naturkunde Karlsruhe (SMNK) and one from the Senckenberg Museum of Natural History (SMF) have this pattern in only two quadrants.

On the remaining seven slabs, five have oriented calyxes in at least one quadrant. On the Tübingen 1 and KUMIP slabs, the orientation of calyxes

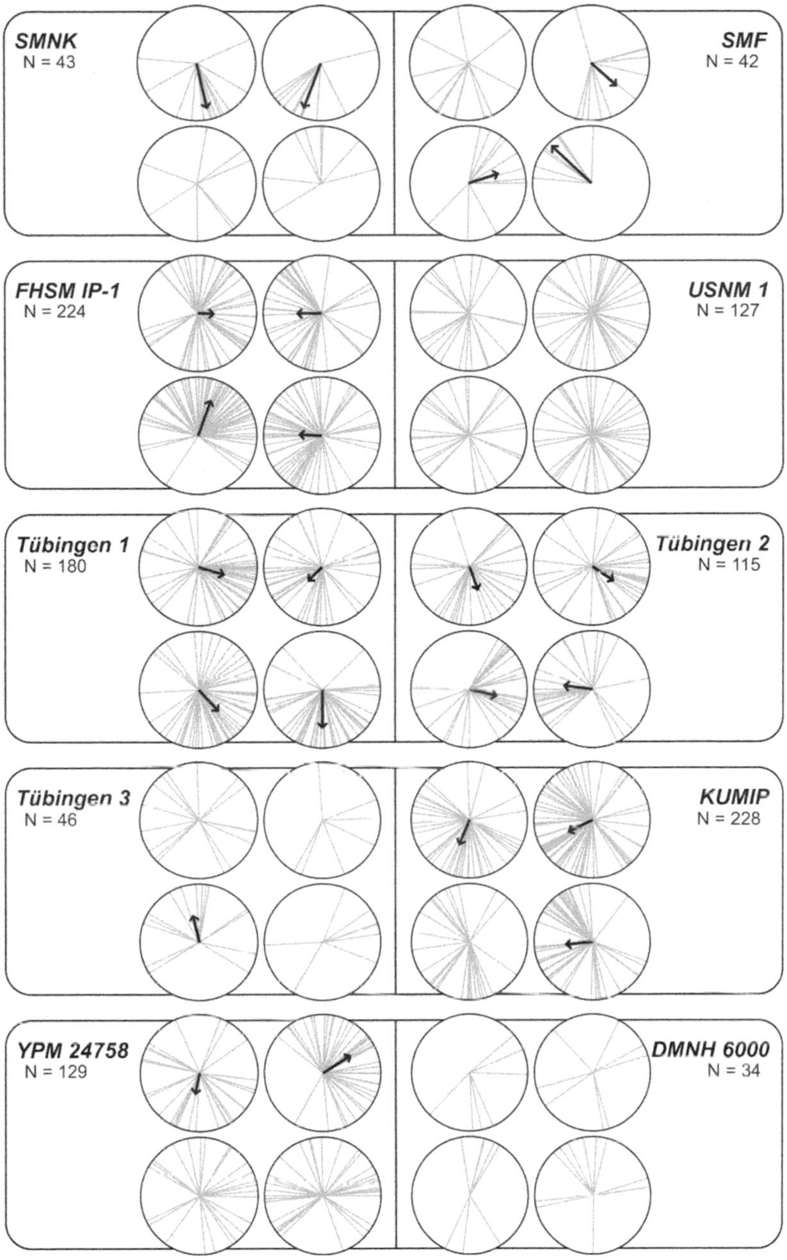

Figure 5.11. *Orientation data from 10* **Uintacrinus** *aggregations.*

is nearly uniform across the entire slab. On other slabs, many calyxes are clustered into similarly oriented bundles, but these bundles do not necessarily reveal any preferred directional alignment when compared with one another (slabs Tübingen 2, Tübingen 3, and YPM 24758 from the Yale Peabody Museum). Two slabs (USNM 1 and DMNH 6000) have no significant pattern at all.

One small aggregation collected during the present study from the Mancos Shale of Colorado (CMC 50687) has some alignment in a south-southwestern direction. For this case and others, it is not always possible to know whether the slab is a complete aggregation or a part of a larger cluster.

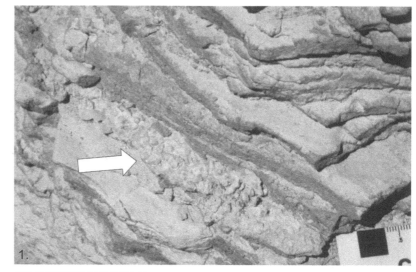

Figure 5.12. *Occurrence of **Uintacrinus** found during fieldwork in the Smoky Hill Member of the Niobrara Chalk. 1, In situ lens of **Uintacrinus** specimens. 2, Articulated parts of two single calyxes preserved in situ. Arrow points to one calyx, and the tip of the hammer pick points to the other.*

Figure 5.13. *Occurrence of baculite fossils with **Uintacrinus** lenses. 1, Specimen on the YPM 24758, first reported by Beecher (1900). Photo by W. K. Sacco, courtesy of Yale Peabody Museum. 2, One of several baculites found on the KUMIP slab. 3, 4, Baculites on the top and bottom of a small slab at the Sternberg Museum.*

Partial clusters would not necessarily have a pattern common to the entire aggregation. Indeed, many slabs in museums were cut from larger lenses, in particular those from Springer's (1901) locality 1, in which the original lens extended for some 15 m along the outcrop and for some 6 m into the outcrop.

STRATINOMIC OCCURRENCE OF UINTACRINUS IN THE FIELD. In Kansas and Colorado, *Uintacrinus* typically is preserved in thin beds that have undergone extreme compaction and consist exclusively of crinoidal material. Springer (1901) noted that the top sides of *Uintacrinus* slabs and fragments consisted of disarticulated plates, whereas articulated calyxes and arms comprised the lower side. Aggregations of *Uintacrinus* found in situ during this study in Kansas and Colorado confirmed this mode of preservation (Fig. 5.12.1). In situ specimens also occur as delicate parts of individuals, typically calyxes, or as small, fragmented lenses (Fig. 5.12.2). Float material, on the other hand, occurs as fragments of larger crinoid beds.

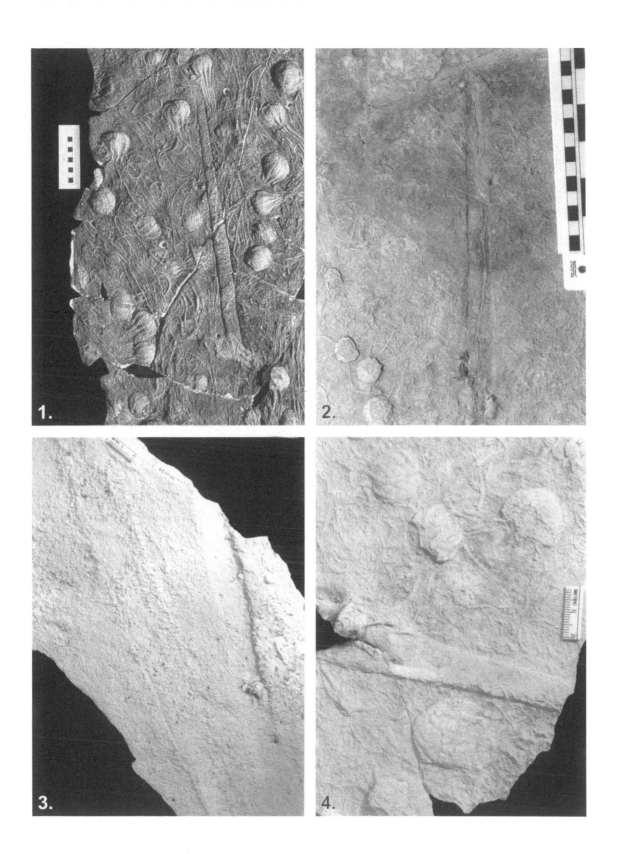

OTHER FAUNA FOUND ON SLABS. Beecher (1900) noted the presence of two subparallel grooves on the slab now residing at the Yale Peabody Museum (Fig. 5.13.1). He described these as possibly belonging to a baculite associated with the lens of *Uintacrinus*. Since then, no other baculites have been reported on any other slab. During the course of this study, however, baculite impressions were found on both sides of several pieces of *Uintacrinus* material. The KUMIP slab has three readily apparent baculite molds on its lower side (the upper side is coated in plaster), and perhaps more than that have been obscured (Fig. 5.13.2). The small fragment from the Sternberg Museum that most likely came from the same area as the KUMIP specimen has three baculite molds on the top surface aligned side by side and one on the bottom (Fig. 5.13.3, 5.13.4). The AMNH slab contains a mold of a baculite encrusted with small oysters on the lower side. In all of these, the original aragonitic material of the baculite has been dissolved, leaving only a mold of its shell, and on most, many small specimens of the oyster *Pseudoperna* Logan, 1899, cover the original location of the baculite. One small aggregation in the Mancos Shale of Colorado (CMC 50688) has a baculite on the lower surface, preserved as a thin dark brown remnant of the shell. This is the only baculite occurrence with *Uintacrinus* reported from the Mancos Shale.

Interpretations

UINTACRINUS MODE OF LIFE: BENTHIC OR PELAGIC? The unique morphology and occurrence of *Uintacrinus* have fueled a lengthy debate as to whether this crinoid lived a benthic existence like other crinoids or a pelagic lifestyle, either swimming actively or drifting passively by means of a flotation mechanism. The pelagic model was favored in two of the early major studies on *Uintacrinus* (Bather, 1896; Springer, 1901) and became widely adopted. Evidence favoring this lifestyle included the lack of any stalk, cirri, or other attachment structure; a globular calyx of thin plates; and a worldwide distribution over a restricted stratigraphic interval. *Uintacrinus* is an ideal index fossil for the Santonian Stage of the Upper Cretaceous, and its interpretation as a pelagic form is consistent with the characteristics commonly associated with other taxa having high biostratigraphic utility. Seilacher and Hauff (2004) cited additional morphological factors favoring a pelagic mode of life: incorporation of the proximal arms and pinnules into the thin-plated, balloon-like calyx that could have housed a gas- or oil-filled swim bladder (Breimer and Lane, 1978). However, because the arms branch only once and lack flexibility near the calyx, they may be poorly suited as a filtration fan in either a pelagic or benthic mode. Instead, Seilacher and Hauff proposed that *Uintacrinus* was a hemipelagic drifter, maintaining slight buoyancy close to the bottom, dragging the long arms along the substratum as a passive deposit feeder.

Other workers advocated a benthic mode of life for *Uintacrinus*, beginning with Jaekel (1918) and Dacqué (1921), and followed by Struve (1957), Milsom et al. (1994), Meyer and Maples (1997), Hess (1999), and Meyer and Milsom (2001). Milsom et al. (1994) presented functional morphological arguments supporting a benthic mode of life. First, the calyx plates and brachials do not show evidence of lightening, and the calyx volume corresponds to the exceptional length of the arms. Second, the lack of flexibility in the proximal arms (also cited by Seilacher and Hauff, 2004) indicated poor swimming ability. Muscle fossae in the thin brachial articulations are weakly developed. If the calyx possessed a special buoyancy mechanism, orientation would have been mouth downward. Most *Uintacrinus* specimens are oriented on their sides, and calyxes occur on the basal surface of thin crinoidal lenses. Only rarely are oral surfaces preserved downward. Furthermore, there is no direct evidence that the globular calyx housed a unique buoyancy mechanism.

ESTABLISHMENT OF UINTACRINUS AGGREGATIONS ON THE SEAFLOOR. Although we cannot exclude the possibility of a pelagic mode of life for *Uintacrinus*, we prefer a benthic lifestyle as the most parsimonious explanation of the evidence. A global distribution does not necessitate a free-swimming life habit if there was a planktotrophic larval stage in the development of the organism, which would facilitate a wide distribution in a benthic organism (Milsom et al., 1994). Crinoids are known to have a simple pear-shaped larva (doliolaria) with ciliary bands that enable a free-swimming lifestyle (Breimer, 1978). Even though this larval form has only weak swimming ability, sufficient residence time in the planktonic stage may allow wide dispersal with the assistance of currents. A modern comatulid crinoid species, *Tropiometra carinata* (Lamarck, 1816), has a wide distribution, occurring from the Indian Ocean to the Caribbean (Mortensen, 1920; Clark, 1947). *Uintacrinus* may have had similar larval dispersion as a benthic organism because it is most closely related to comatulid crinoids. On the basis of the close phylogentic relationship with comatulids, it is reasonable to postulate that *Uintacrinus* also had a stalk in its larval stage, which, like other comatulids, was eventually lost. The larvae of modern crinoids attach to some sort of firm substrate, even including other crinoids (Breimer, 1978). The larva eventually detaches, the stem is shed during further development, and the maturing crinoid has some degree of mobility and may attach to a wider variety of substrates with cirri. The larvae of *Uintacrinus* also may have needed to attach to a firm substrate on the seafloor. The baculites associated with several lenses of *Uintacrinus* populations might have facilitated the attachment of larvae by providing a suitable substrate.

Dense aggregations of living comatulid crinoids or ophiuroids (Warner, 1979) are possible modern analogs to *Uintacrinus*. Some comasterid comatulids, to which *Uintacrinus* is most closely related (Milsom et al., 1994), also lack cirri as attachment structures, but nevertheless are benthic, utilizing the arms and pinnules for support (Messing et al., 2006). Hess (1999) also regarded *Uintacrinus* as benthic and likened their dense aggregations to beds of eelgrass. Although *Uintacrinus* lacked a means of attachment to the substratum, the large, globular calyx itself, filled with soft tissue and coelomic fluid, may have provided a means of implantation on or into the soft substratum—a kind of iceberg strategy (Hess, 1999). Alternatively, the soft substrates may have been replaced by relatively firmer substrates during brief periods of time in which *Uintacrinus* could populate the seafloor. Evidence for this may be present in layers containing abundant oysters (*Pseudoperna*) that occur in the sediments surrounding those in which *Uintacrinus* were preserved. In other locations in the Niobrara Formation and the Mancos Shale, *Pseudoperna* is an epifaunal filter feeder attached to baculite shells and *Inoceramus* Sowerby, 1814, valves. This would indicate that these oysters were opportunistic, establishing themselves whenever possible. Their existence with the absence of any other firm substrate in laminated sediments suggests that the seafloor was at least somewhat more cohesive during brief periods to permit cementation.

The passive suspension feeding mode of living crinoids, either stalked or unstalked, relies on fixation to the substratum and deployment of a filtration fan to intercept food particles from passing currents (Meyer, 1982). The fine-grained sediments of the Cretaceous Interior Sea do not indicate strong current energy, but the evidence of some current alignment of *Uintacrinus* presented here shows that the environment was by no means stagnant. The low-diversity biofacies, in which *Uintacrinus* is associated only with inoceramid clams and oysters, is indicative of fluctuating, low-oxygen conditions. The benthic boundary layer may have been thick enough to create inhospitable conditions close to the sediment-water interface for passive suspension feeders (Bottjer and Ausich, 1986). In this setting, we suggest that the unique morphology of *Uintacrinus*, with stout, inflexible brachials at the base of the arms, enabled the support of the extreme arm length upward. Lacking a conventional stalk, *Uintacrinus* was able to tolerate this environment because it had longer arms than any other known crinoid, up to 1.25 m in length. This length enabled the crinoids to feed in the conventional passive suspension feeding manner above any boundary layer conditions of stagnation or low oxygenation.

POPULATION STRUCTURE. An analysis of *Uintacrinus* assemblages has determined that calyx sizes are normally distributed and unimodal. This may indicate single-age cohort groupings within populations or size sorting by current modification or transportation. Taphonomic evidence suggests that an enclosing microbial mat grew around remains shortly after death (Meyer and Milsom, 2001). The growth of this mat would hinder the postmortem dispersal of remains by currents, as indicated by the presence of disarticulated ossicles on the tops of slabs that would likely be transported by currents of even minimal energy. With this in mind, the population with a smaller, unimodal calyx size distribution (locality 2 of Springer, 1901) probably represents either a juvenile population or a population with stunted growth. The close geographic proximity of the KUMIP slab to others and the same depositional characteristics suggest that environmental conditions must not have been too different from one population to the next, and therefore, a stunted population is not likely. For this reason, this aggregation is interpreted as a juvenile population consisting of a single, yet smaller, age cohort. It should be noted that Springer regarded the small individuals of this locality (locality 2) to represent juveniles (*fide* Springer notebook for 1898, Springer Library, Natural History Museum, Smithsonian Institution).

DEATH OF UINTACRINUS ASSEMBLAGES. Once established, *Uintacrinus* populations, consisting of single cohorts, matured to a maximum size until dying as a result of some mass mortality event. Suggested mechanisms for mass mortality include intense storm activity, rapidly increased sedimentation rates, or anoxia events (Hess, 1999). It is also possible that *Uintacrinus* was killed by water toxicity induced by volcanic ash input. Meyer and Milsom (2001) proposed that *Uintacrinus* mortality came from a sudden drop in oxygen content rather than from sediment obrution, and the death assemblage was exposed for a short time on the seafloor (see Savrda and Bottjer, 1989, for a discussion of fluctuations in seafloor oxygenation for the Niobrara Formation of Colorado).

Studies of living crinoids show that soft tissue decay and skeletal disarticulation occur rapidly after death unless the crinoid is rapidly buried (Meyer, 1971; Liddell, 1975). *Uintacrinus* aggregations are known for their exquisite preservation of articulated calyxes and arms, but this high degree of articulation is restricted to the lower surfaces of the crinoidal lenses, as reported by Springer (1901) and confirmed by new discoveries of in situ lenses in both the Niobrara Formation of Kansas and the Mancos Shale of Colorado (Meyer and Milsom, 2001). Furthermore, soft tissue of the tegmen is sometimes preserved in Niobrara material. If *Uintacrinus* aggregations lived high in the water column, it is unlikely that dead individuals would remain highly articulated after sinking tens of meters to the seafloor. Decay and disarticulation occurred on the exposed surface but were retarded by the formation of a microbial mat that enveloped the surface of the aggregation (Meyer and Milsom, 2001). The lower surface of the crinoids retained high articulation against the soft calcareous ooze or mud, while the microbial mat stabilized the upper surface, preventing dispersal of disarticulated skeletal material. Although the contrasting lower/upper surface preservational style occurs in pseudoplanktonic crinoids like *Seirocrinus* Gislén, 1924, that sunk into an anoxic basin, it is also known in typical benthic, attached crinoids (Meyer and Milsom, 2001).

CURRENT MODIFICATION OF UINTACRINUS REMAINS. There are many instances of preferred orientation of crinoid remains in the fossil record in which arms and stems are oriented parallel or perpendicular to moving currents (Hess et al., 1999). The nearly parallel orientation of many *Uintacrinus* calyxes suggests current modification either during or after the death of individuals. The occurrence of clusters of calyxes having different preferred orientations across single slabs may indicate that currents moving in multiple directions aligned individuals as they sat exposed on the seafloor. It is also possible that eddy currents swirled calyxes into multiple orientations, which would explain the radial alignment of crowns with arms directed toward the center of the aggregation. This may have been produced as eddies swirled the crinoids into a spokelike pattern, leaving entangled arms near the center of aggregations and pulling the calyxes outward. Varying current energy during certain times may have played an important role at or just above the seafloor, carrying gametes some distance away from a population, allowing for filter feeding, and perhaps orienting individuals during or shortly after death.

Conclusions

1. *Uintacrinus* aggregations are snapshots of life assemblages presenting information about population structure and taphonomy.

2. By assessing new field collections and revisiting older sites, it has been determined that *Uintacrinus* is not restricted to a single biostratigraphic interval. Specimens are from the *Clioscaphites saxtonianus* Biozone up through the *Desmoscaphites bassleri* Biozone in both the Smoky Hill Member of the Niobrara Chalk in Kansas and the Smoky Hill Member of the Mancos Shale in Colorado.

3. An analysis of calyx size from 10 *Uintacrinus* slabs reveals normal distributions with no size clusters. All aggregations are clustered around similar means, with one exception, which has smaller individuals. For this reason, and on the basis of taphonomic evidence, each population of *Uintacrinus* is interpreted to represent a single age cohort, and therefore one spawning or recruitment event. The slab with smaller calyxes most likely represents a juvenile population that has not reached maturity, as the others have.

4. There are significant directional, sometimes radial, trends in the orientation of calyxes on many slabs. This pattern may be the result of current modification during or shortly after the death of *Uintacrinus* populations. These trends suggest the presence of bottom currents that may have been associated with these mass mortality events.

Acknoweldgments

Research grants to AW from the American Museum of Natural History, the Caster Fund, University of Cincinnati, and the Paleontological Society, and to DM from the National Geographic Society (grant 5656-96) supported this research. For assistance in the field, we thank C. Bonner, S. Barbour, R. Krause, J. LaRock, and P. Lask. We thank H. Armstrong and B. Fowler, Bureau of Land Management, for assistance with access to field localities in Colorado. For access to museum specimens, we thank N. Landman, American Museum of Natural History, New York; F. Collier, Museum of Comparative Zoology, Harvard University; A. Hart, Natural History Museum, University of Kansas; S. Rietschel Staatliches Museum für Naturkunde, Karlsruhe, Germany; T. Baumiller, Museum of Paleontology, University of Michigan; E. Schindler, Senckenberg Museum of Natural History, Frankfurt, Germany; J. Thompson, Smithsonian Institution; G. Liggett, Sternberg Museum, Fort Hays State University; A. Seilacher, University of Tübingen, Germany; K. Westphal, Geological Museum, University of Wisconsin; and R. White, Yale Peabody Museum. For assistance with image processing, we thank J. Yocis, Photographic Services, University of Cincinnati.

References

Bailey, H. W., A. S. Gale, R. N. Mortimore, A. Swiecicki, and J. Wood. 1983. The Coniacian-Maastrichtian stages of the United Kingdom, with particular reference to southern England. Newsletters on Stratigraphy, 12:29–42.

Bailey, H. W., A. S. Gale, R. N. Mortimore, A. Swiecicki, and C. J. Wood. 1984. Biostratigraphical criteria for the recognition of the Coniacian to Maastrichtian stage boundaries in the Chalk of north-west Europe, with particular reference to southern England. Bulletin of the Geological Society of Denmark, 33:31–39.

Bassler, R. S. 1909. Some noteworthy accessions to the Division of Invertebrate Paleontology in the National Museum. Smithsonian Miscellaneous Collections, 52:267–269.

Bather, F. A. 1896. On *Uintacrinus*; a morphological study. Proceedings of the Zoological Society of London, 1895:974–1003.

Beecher, C. E. 1900. On a large slab of *Uintacrinus* from Kansas. American Journal of Science, 159:267–268.

Bottjer, D. J., and W. I. Ausich. 1986. Phanerozoic development of tiering in soft substrata suspension-feeding communities. Paleobiology 12:400–420.

Breimer, A. 1978. General morphology recent crinoids, p. T9–T58. *In* R. C. Moore and C. Teichert (eds.), Treatise on Invertebrate Paleontology. Pt. T. Echinodermata 2. Geological Society of America and University of Kansas Press, Lawrence.

Breimer, A., and N. G. Lane. 1978. Paleoecology, p. T316–T347. *In* R. C. Moore and C. Teichert (eds.), Treatise on Invertebrate Paleontology, pt. T. Echinodermata 2. Geological Society of America and University of Kansas Press, Lawrence.

Clark, A. H. 1947. The comatulids. Volume 1, Part 4B. A monograph of the existing crinoids. Smithsonian Institute U.S. Museum Bulletin, 82, 473 p.

Cobban, W. A. 1962. Late Cretaceous *Desmoscaphites* range zone in the Western Interior region. U.S. Geological Survey Professional Paper, 0450-D:140–144.

Cobban, W. A. 1995. Occurrences of the free-swimming Upper Cretaceous crinoids *Uintacrinus* and *Marsupites* in the Western Interior of the United States. U.S. Geological Survey Bulletin, 2113-C:1–6.

Collier, A. J. 1929. The Kevin-Sunburst oil field and other possibilities of oil and gas in the Sweetgrass arch, Montana. U.S. Geological Survey Bulletin, 812-B:57–189.

Dacqué, E. 1921. Vergleichende biologische Vormenkunde der fossilen niederen Tiere. Bornträger, Berlin, 777 p.

Davis, J. C. 2002. Statistics and Data Analysis in Geology. John Wiley and Sons, New York, 646 p.

Gislén, T. 1924. Echinoderm studies. Zoologiska Bidrag från Uppsala, 9, 330 p.

Grinnell, G. B. 1876. On a new crinoid from the Cretaceous formation of the West. American Journal of Science, 3:81–83.

Harris, T. M. 1974. *Williamsoniella lignieri*: its pollen and the compression of spherical pollen grains. Palaeontology, 17:125–148.

Hattin, D. E. 1982. Stratigraphy and depositional environment of Smoky Hill Chalk Member, Niobrara Chalk, Upper Cretaceous of the type area, western Kansas. Kansas Geological Survey Bulletin, 225, 108 p.

Hess, H. 1999. *Uintacrinus* beds of the Upper Cretaceous Niobrara Formation, Kansas, USA, p. 225–232. *In* H. Hess, W. I. Ausich, C. E. Brett, and M. J. Simms, Fossil Crinoids. Cambridge University Press, Cambridge.

Hess, H., W. I. Ausich, C. E. Brett, and M. J. Simms. 1999. Fossil Crinoids. Cambridge University Press, 275 p.

Hovey, E. O. 1902. A remarkable slab of fossil crinoids. American Museum Journal, 2(2):11–14.

Jaekel, O. 1918. Phylogenie und System der Pelmatozoen. Paläontologische Zeitschrift, 3:1–128.

Keefer, W. R., and M. L. Troyer. 1956. Stratigraphy of the Upper Cretaceous and

Lower Tertiary rocks of the Shotgun Butte area, Fremont County, Wyoming. U.S. Geological Survey Oil and Gas Investigations, Chart OC-56.

Keefer, W. R., and M. L. Troyer. 1964. Geology of the Shotgun Butte area, Wyoming. U.S. Geological Survey Bulletin, 1157, 123 p.

Lamarck, J. B. P. A. De M. De. 1816. Histoire naturelle des animaux sans vertèbres, présentant les caracteres généraux et particuliear de ces animaux, leur distribution, leurs classes, leurs familles, leurs genres, et la citatien des principales especes qui syrattachment (7 volumes), Lamarck Echinoderms, volume 2. Paris, 568 p.

Leckie, R. M., J. I. Kirkland, and W. P. Elder. 1997. Stratigraphic framework and correlation of a principal section of the Mancos Shale (Upper Cretaceous), Mesa Verde, Colorado. New Mexico Geological Guidebook, 48th Field Conference, Mesozoic Geology and Paleontology of the Four Corners Region:163–216.

Liddell, W. D. 1975. Recent crinoid biostratinomy. Geological Society of America Abstracts with Programs, 7:1169.

Logan, W. N. 1899. Some additions to the Cretaceous invertebrates of Kansas. Kansas University Quarterly, series A, 8:87–98.

Messing, C. G., D. L. Meyer, U. E. Siebeck, L. S. Jermiin, D. I. Vaney, and G. W. A. Rouse. 2006. A modern, soft-bottom, shallow-water crinoid fauna (Echinodermata) from the Great Barrier Reef. Coral Reefs, 25:164–168.

Meyer, D. L. 1971. Post-mortem disarticulation of Recent crinoids and ophiuroids under natural conditions. Geological Society of America Abstracts with Programs, 3:645.

Meyer, D. L. 1982. Food and feeding mechanisms: Crinozoa, p. 25–42. In M. Jangoux and J. M. Lawrence (eds.), Echinoderm Nutrition. A. A. Bakema, Rotterdam.

Meyer, D. L., and C. G. Maples. 1997. Radial orientation pattern of Uintacrinus aggregations: new evidence for life at the air-water interface using surface tension. Geological Society of America Abstracts with Programs, 29:106–107.

Meyer, D. L., and C. V. Milsom. 2001. Microbial sealing in the biostratinomy of Uintacrinus Lagerstätten in the Upper Cretaceous of Kansas and Colorado, USA. Palaios, 16:535–546.

Miller, H. W. 1968. Invertebrate fauna and environment of deposition of the Niobrara Formation (Cretaceous) of Kansas. Fort Hays Studies, new series, Science Series, 8, 90 p.

Miller, H. W., G. F. Sternberg, and M. V. Walker. 1957. Uintacrinus localities in the Niobrara Formation of Kansas. Transactions of the Kansas Academy of Science, 60:163–166.

Milsom, C. V., M. J. Simms, and A. S. Gale. 1994. Phylogeny and paleobiology of Marsupites and Uintacrinus. Paleontology, 37:595–607.

Mitchell, S. F. 1995. Uintacrinus anglicus Rasmussen from the Upper Cretaceous Flamborough Chalk Formation of Yorkshire: implications for the position of the Santonian–Campanian boundary. Cretaceous Research, 16:745–756.

Mortensen, T. 1920. Studies in the development of crinoids. Papers from the Department of Marine Biology of the Carnegie Institution of Washington, 16, 94 p.

Pike, W. S., Jr. 1947. Intertonguing marine and nonmarine Upper Cretaceous deposits of New Mexico, Arizona, and southwestern Colorado. Geological Society of America Memoir, 24, 103 p.

Rasmussen, H. W. 1961. A monograph on the Cretaceous Crinoidea. Kongelige Danske Videnskabernes Selskabernes Biologiske Shrifter, 12:1–428.

Rogers, K. 1991. A Dinosaur Dynasty. The Sternberg Fossil Hunters. Mountain Press, Missoula, Montana, 302 p.

Savrda, C. E., and D. J. Bottjer. 1989. Trace fossil model for reconstruction of oxygenation histories of ancient marine bottom-waters: application to Upper Cretaceous Niobrara Formation, Colorado. Palaeogeography, Palaeoclimatology, Palaeoecology 74:49–74.

Schlotheim, E. F. Von. 1820. Die Petrefactenkunde auf ihrem jetzigen Standpunkte durch die Beschreibung seiner Sammlung versteinerter und fossiler Überreste des Thier-und Pflanzenreichs der Vorwelt erläutert. Gotha, Beckersche Buchhandlung, 437 p.

Schuchert, C. 1904. A noteworthy crinoid. Smithsonian Miscellaneous Collections, 45:450.

Seilacher, A. 1970. Begriff und Bedeutung der Fossil-Lagerstatten. Neues Jahrbuch für Geologie und Palaontologie, Monatshefte 1970:34–39.

Seilacher, A., and R. B. Hauff. 2004. Constructional morphology of pelagic crinoids. Palaios, 19:3–16.

Sokal, R. R., and F. J. Rohlf. 1981. Biometry: The Principles and Practice of Statistics in Biological Research. W. H. Freeman, New York, 859 p.

Sowerby, J. 1812–1846. The mineral conchology of Great Britain; or, Coloured figures and descriptions of those remains of testaceous animals or shells, which have been preserved at various times and depths in the earth (7 volumes). B. Meredith, London.

Springer, F. 1901. *Uintacrinus*: its structure and relations. Memoirs of the Museum of Comparative Zoology, Harvard University, 25:1–89.

Struve, W. 1957. Ein Massengrab kreidezeitlicher Seelilien: die *Uintacrinus* Platte des Senckenberg-Museums Natur und Volk, 87:361–373.

Thom, W. T., Jr., G. M. Hall, C. H. Wegemann, and G. F Moulton. 1935. Geology of Big Horn County and the Crow Indian Reservation, Montana. U.S. Geological Survey Bulletin, 856, 200 p.

Warner, G. F. 1979. Aggregation in echinoderms, p. 375–396. *In* G. Larwood and B. R. Rosen (eds.), Biology and Systematics of Colonial Organisms. Systematics Association Special Volume 11, Academic Press, London.

Warren, P. S., and M. B. B. Crockford. 1948. The occurrence of the crinoid *Uintacrinus socialis* in the Cretaceous of Alberta. Canadian Field Naturalist, 62(5):159.

Whiteaves, J. F. 1904. *Uintacrinus* and *Hemiaster* in the Vancouver Cretaceous. American Journal of Science, 18:287–289.

Williston, S. W. 1897. The Kansas Niobrara Cretaceous. University Geological Survey of Kansas, 2:237–246.

Figure 6.1. *Location of study area.*

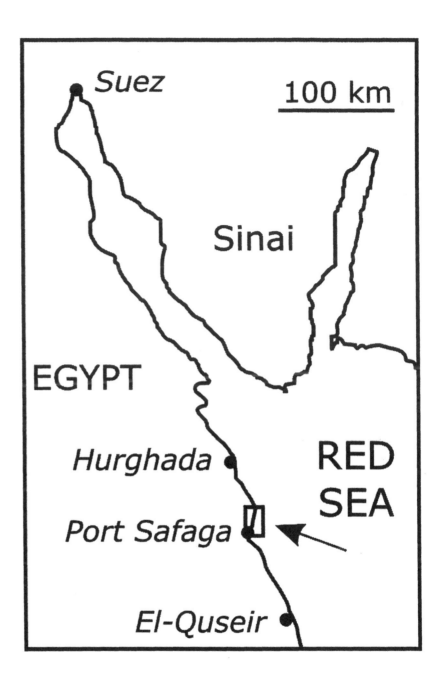

TAPHONOMY OF THE IRREGULAR ECHINOID *CLYPEASTER HUMILIS* FROM THE RED SEA: IMPLICATIONS FOR TAXONOMIC RESOLUTION ALONG TAPHONOMIC GRADES

6

James H. Nebelsick

Taphonomy describes the transition of skeletal remains from once living organisms to the preserved remains that paleontologists ultimately recover at the outcrop and store in their collections. Understanding taphonomic gradients is of essential importance for paleontologists who use fossil remains to make statements concerning such things as aspects of ecology, diversity, and evolution. Actualistic (observations can make fundamental contributions to understanding taphonomic gradients for the simple reason that the complex ecological interactions that affect taphonomic processes and features can be directly observed and measured.

Echinoderms not only have an illustrious past but also flourish today in diverse marine environments. Thus, we can conduct actualistic observations of echinoderms and directly investigate those ecological factors that affect not only their distribution but also the preservation of their remains after death. Although echinoids commonly disarticulate rapidly depending on a number of factors (see Kier, 1977; Allison, 1990; Kidwell and Baumiller, 1990; Donovan, 1991; Greenstein, 1993), clypeasteroids are comparatively robust and stay complete long enough for taphonomic factors to be observed on the surface of the test. Echinoderm taphonomy is a broad subject and has been the subject of a number of reviews by Lewis (1980), Donovan (1991), Brett et al. (1997), Ausich (2001), and most recently Nebelsick (2004).

Clypeasteroids belong to the youngest group of echinoderms to have evolved, originating in the Paleogene from cassiduloid ancestors (see Smith, 2001). They have a number of specialized features, including their characteristic flattened shape, spine miniaturization and differentiation, concentration of respiratory podia in the petals on the aboral surface, secondary unipores not restricted to ambulacra, food grooves on the oral surface, lunules in the test of many sand dollars, and a specialized Aristotle's lantern (Durham, 1955, 1966; Mooi, 1989). Stability to the test is given by internal supports in most taxa, which connect the oral and aboral sides of the test. Interlocking plates further strengthen the test. These features make clypeasteroids among the most robust of all echinoderms, allowing them to survive even in high-energy shoreface environments. As often

Introduction

Figure 6.2. *Bottom topography of the study area with main collection sites of complete* **Clypeaster humilis** *as well as the location of bulk sediment samples containing* **Clypeaster** *remains.*

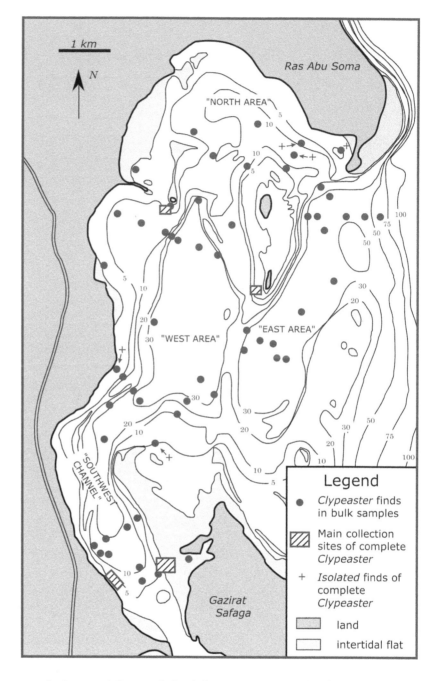

noted, these stabilizing skeletal features correspondingly increase their preservation potential (Seilacher, 1979; Nebelsick and Kampfer, 1994; Brett et al., 1997; Moffat and Bottjer, 1999; Nebelsick, 1999).

The genus *Clypeaster* Lamarck, 1801, lives today in many tropical marine environments (Ghiold and Hoffmann, 1984, 1986). They have a wide morphological variety and include flattened infaunal and domed epifaunal forms. The latter is represented by the type species *Clypeaster rosaceous* (Linnaeus, 1758), well known from shallow Caribbean waters. *Clypeaster* is an important organism in these environments. For example, they can influence sediment particle size when feeding by crushing sand-sized particles and producing finer sediments (Kampfer and Tertschnig, 1992).

Clypeaster is also one of the most common echinoids in Cenozoic shallow-water sediments. Hundreds of nominal fossil species have been described (Durham, 1966), and numerous specimens are known from fossil echinoid assemblages. Commonly, different species of *Clypeaster* occupy different habitats showing corresponding variations in size and morphology (i.e., Boggild and Rose, 1984; Poddubiuk and Rose, 1984; Nèraudeau et al., 2001; Kroh and Nebelsick, 2003).

CLYPEASTER HUMILIS IN THE STUDY AREA. *Clypeaster humilis* (Leske, 1778) is a medium-sized species that is distributed throughout the Indo–West Pacific realm (Clark and Rowe, 1971). The skeletons of *C. humilis* examined here originate from the Northern Bay of Safaga, Red Sea, Egypt (Figs. 6.1, 6.2). This is a complex area about 10 by 8 km long with coastal indentations, islands, and basins separated from one another and from the open Red Sea by submarine ridges (Piller and Pervesler, 1989; Piller and Mansour, 1990). The Northern Bay of Safaga is a shallow-water, carbonate environment and is characterized by a wide range of bottom facies. Shallower areas of the bay are dominated by coarse sands, sea grass meadows, sands with sea grass, and sands with coral patches. Coral reefs are also present and are accompanied by reef flats. However, most coral cover is represented by a coral carpet that occupies extensive areas of the seafloor. Deeper basins are occupied by carbonaceous muddy and sandy muddy sediments (Fig. 6.2). Both regular and irregular echinoids are common macrofaunal elements of this shallow-water carbonate system and have been the subject of a number of actualistic studies concerning their ecology and taphonomy (e.g., Nebelsick, 1992a, 1992b, 1999; Nebelsick and Kampfer, 1994; Nebelsick and Kowalewski, 1999).

Clypeaster is among the most common echinoids in the Northern Bay of Safaga and is found as living specimens, dead tests, and fragments. Of the four sympatric species in the study area, *Clypeaster humilis* is by far the most common and totally dominates the collected specimens. In all, 47 living specimens and 86 dead specimens were collected. *Clypeaster fervens* Koehler, 1922, *Clypeaster reticulatus* (Linnaeus, 1758), and *Clypeaster rarispinus* de Meijere, 1902, are much less common, and only a few specimens of each species were collected (Nebelsick, 1992b, 1999). Species differentiation is based on shape and size relationships of test size, petalodium length, ambitus shape, and the form of the frontal ambulacrum (Clark and Rowe, 1971).

Complete *Clypeaster humilis* specimens were found in shallow-water areas of the bay with sandy substrates (Fig. 6.1). Living *Clypeaster* were found buried just underneath the sediment surface, rarely being just visible as an outline below the surface, at maximum densities of two individuals per square meter (Nebelsick, 1992a). Dead echinoid tests were discovered on the sediment surface, partially buried, or underneath the surface. Some specimens were extracted by raking the sediment with a stout rake penetrating to a depth of approximately 5 cm into the sediment.

Figure 6.3. *Clypeaster humilis* from the Northern Bay of Safaga. Egypt. 1–3, 5–6, Aboral view. 4, Oral view. 1, Test with spines; 2, very well-preserved, denuded test; 3, test showing abrasion of surface characters, apical system depressed; 4, abraded test with encrustation (serpulids and bryozoans); note how some serpulids follow food grooves; 5, abraded test with abrasion, fragmentation, and bioerosion; apical system is missing; 6, totally abraded and corroded test. Scale bars = 1 cm.

MORPHOLOGY AND DISTRIBUTION OF CLYPEASTER HUMILIS. *Clypeaster humilis* specimens ranged in size from ~20 to 120 mm in test length but are most common at approximately 75 mm in length with a corresponding height of 15 mm. The aboral surface of the *Clypeaster* skeleton is characterized by a distinct slightly raised petalodium (Fig. 6.3.1) containing the respiratory podia, typically with up to 60 pore pairs along each ambulacrum of the petal. The apical system consists of five distinct gonopores surrounding the madreporite. The ambitus is rounded. The oral surface contains the central peristome, the ventrally positioned periproct near the posterior end of the skeleton, and straight food grooves leading to the peristome (Fig. 6.3.4). The peristome is flush with the oral surface of the test. There is a uniform distribution of small, sunken, perforated tubercles on both the oral and aboral surfaces. These tubercles are generally less than 0.5 mm in diameter and support the minute spines that cover the test during life. An important morphological feature for the preservation of the test is the presence of thick internal supports that interconnect the oral and aboral surfaces. The specialized clypeasteroid jaws complete the (larger) skeletal features of the test (see Mooi, 1989).

Transport is not regarded as a major factor in redistributing fragments in the study area (Nebelsick, 1992a, 1992b, 1999). The Red Sea is generally not affected by high-energy storms, and the study area itself is a relatively protected environment separated from the open sea by islands and under-water swells. The high differentiation of echinoid fragments within bulk samples and their close correlation to the habitat of corresponding living echinoids also suggest that transport is not an important taphonomic factor at the baywide scale of investigation (Fig. 6.2). Small-scale transport is discussed in Nebelsick (1992b) and includes transport from coral patches and coral carpet to the surrounding sediment as well as from the reefs across the reef flats and to the reef slopes. These pertain mostly to regular echinoids living in more exposed environments.

PREVIOUS STUDIES CONCERNING THE TAPHONOMY OF CLYPEASTER HUMILIS. Two previous investigations involving *Clypeaster humilis* from the study area are of importance in the interpretation of taphonomic gradients presented here. Nebelsick and Kampfer (1994) studied short-term taphonomic processes affecting test preservation of *Clypeaster humilis* and *Echinodiscus auritus* Leske, 1778, by using an experimental array of underwater caged dead specimens over a time period of 1.5 weeks. Spine disarticulation in both species commenced within 18 hours and was finished by 91 hours after the start of the experiment. After spine disarticulation, most tests remained stable, but once plate disaggregation was initiated (in three of 20 cases), the tests rapidly disarticulated into larger fragments and individual plates.

Nebelsick (1999) studied the taphonomy of *Clypeaster* fragments throughout the study area. Fragments could only be identified to genus level because of the lack of characteristic test features needed for species identification. Multivariate statistical analysis of surface preservation features (including abrasion, encrustation, and surfaces marks) of *Clypeaster* fragments led to the designation of four different taphofacies in the study area (Nebelsick, 1999). These taphofacies were related to differential exposure and sedimentation rates, ecological factors that are not readily discerned by the analysis of diversity and morphological features alone.

Taphonomic Gradient of *Clypeaster humilis* Tests

Five different preservation states along a taphonomic gradient were differentiated by using the qualitative analysis of the surface characters (Figs. 6.3–6.5). The basis of this gradient is the preservation of the following surface characters: spines, the apical system, gonopores, madreporite, petalodium, pore pairs, plate boundaries, tubercles, and the ambitus (Fig. 6.5).

Stage 1: Specimens with Spines

This stage corresponds to those animals that have just been killed, and a few dead specimens were recovered with spines still attached by epithelium (Fig. 6.3.1). The color of the animals, ranging from light to dark brown, is still recognizable. Even pedicellaria are nested between the spines. The integument covers the periproct. The teeth are in place with the jaws still articulated. It is clear that dead specimens of this stage can keep their spines for only a few days as a result of the decay of soft tissue and disarticulation of the spines (Nebelsick and Kampfer, 1994).

Figure 6.4. *Fragment preservation of **Clypeaster** sp. 1, 3, 5, Successive enlargements of a single, well-preserved plate. 2, 4, 6, Successive enlargements of two joined plates showing high rates of abrasion and encrustation of a serpulid worm tube and encrusting foraminifera. Note that the serpulid worm tube has been broken along the plate boundary, suggesting encrustation of a complete test and subsequent fragmentation. Scale bars: 1, 2 = 1 mm; 3, 4 = 400 µm; 5, 6 = 100 µm.*

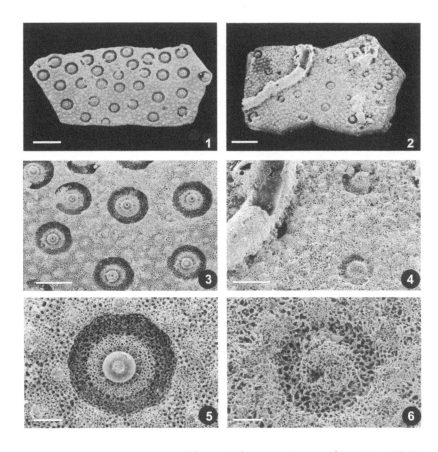

Stage 2: Very Well-preserved, Denuded Tests

CHARACTER PRESERVATION. The apical system is complete (Fig. 6.3.2). Gonopores are open and the madreporite is recognizable. The petalodium is distinct, and pore pairs are free of sediment. Ambulacral and interambulacral plate boundaries can be readily discerned. The ambitus is perfectly preserved. Pristine surface characters are present at the microscopic scale with conspicuous, sunken, perforated tubercles, and well-developed areoles (Fig. 6.4.1). Secondary tubercles and surface knobs can also be distinguished (Fig. 6.4.3). Differential stereom is readily evident (Fig. 6.4.5).

TAPHONOMIC PROCESSES. Decay of connective tissue has led to the disarticulation of spines from the surface of the test. The decay of ambulacral tube feet result in open pores of the petalodium and secondary unipores. Jaws are now loose within the test. The bare test is now white. Most specimens are still free of internal sediment.

TAXONOMIC RESOLUTION. Specimens can readily be identified at the species level. Color variations are, however, not discernible.

Stage 3a: Abraded Tests

CHARACTER PRESERVATION. The apical system can be damaged (Fig. 6.3.3, 6.3.4). Gonopores, if present, are filled with sediment. The petalodium is less distinct, and many pore pairs are plugged. Ambulacral and interambulacral plate boundaries are hardly discerned. Tubercles are abraded with damaged crenulations and mammalons. Stereom differentiation is becoming difficult to recognize at the surface. The ambitus is still well preserved.

TAPHONOMIC PROCESSES.Abrasion has led to loss of surface character resolution. First signs of encrustation and bioerosion are apparent. The test has taken a dull color. The test can be filled with sediment.

TAXONOMIC RESOLUTION.Specimens can still be identified at the species level because shape and size relationships of the test and petalodium length, ambitus shape, and frontal ambulacrum are still recognizable.

CHARACTER PRESERVATION.The apical system including the gonopores and madreporite can be completely eroded (Fig. 6.3.5). The petalodium is less distinct, and many pore pairs are plugged. Ambulacral and interambulacral plate boundaries are hardly discerned. The ambitus is still well preserved. At the microscopic scale, encrustation cover surface characters (Fig. 6.4.2), tubercles are abraded (Fig. 6.4.4), and stereom differentiation becomes difficult (Fig. 6.4.6).

TAPHONOMIC PROCESSES. Encrustation is by unilaminar bryozoans, serpulids, and encrusting foraminifera. Conspicuous holes on the test surface may be the result of boring activity. The test can be filled with sediment.

TAXONOMIC RESOLUTION. The gross morphology is still recognizable so species identification could be possible. The degree of encrustation dictates if other species-diagnostic features, such as the shape of the frontal ambulacrum and petal pores, are still recognizable.

Stage 3b: Encrusted— Bioeroded Tests

CHARACTER PRESERVATION. Corroded test (Fig. 6.3.6). All surface characters are completely missing. The ambitus is corroded. The tests are noticeably heavier and are filled with sediment.

TAPHONOMIC PROCESSES. These tests are heavily corroded. Early diagenesis has lead to the filling of stereom and lithification of infilling sediment.

TAXONOMIC RESOLUTION. Species and genus identification, as such, is no longer possible without detailed knowledge of the clypeasteroid present in the study area. General form and height are still discernible, but the corroded ambitus changes the length and width parameters of the specimen. All details concerning the petalodium and pore pairs are destroyed.

Stage 4: Corroded Tests

Spines disappear soon after death after the decay of connective tissues and muscles that connect the spines to the test. The minute spines (< 2 mm long) are then committed to the sediment. Because the analysis of bulk sediment samples was restricted to grain sizes larger than 2 mm (Nebelsick, 1992a, 1992b), the frequency of distribution of spines in the sediment is not recorded.

The apical system with the gonopores and madreporite shows a steady degradation from stage 1 to 4. Although perfectly preserved in stage 2 (Fig.

Preservation of Surface Features along the Taphonomic Gradient

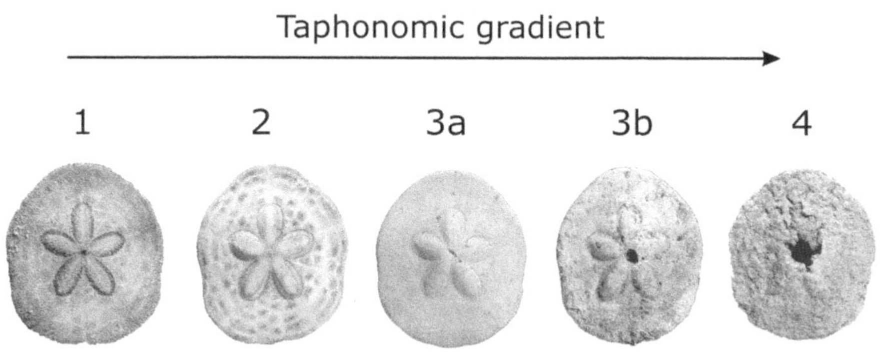

Taphonomic gradient

	1	2	3a	3b	4

Character preservation					
Apical system	0	0	1	2	2
Gonopores	0	0	1	2	2
Madreporite	0	0	1	2	2
Petalodium	0	0	1	0-1	2
Pore pairs	0	0	1	0-1	2
Plate boundaries	0	0	2	1	2
Tubercles	0	0	1	1	2
Ambitus	0	0	0	0	1
Taphonomic processes					
Denudation	0	2	2	2	2
Abrasion	0	0	1	1	2
Encrustation	0	0	1	2	0
Bioerosion	0	0	1	2	0
Corrosion	0	0	0	0	2
Diagenesis	0	0	1	1	2
Taxonomic resolution	species	species	species	species/ genus	genus/ family

Characters:
0 = complete/well preserved/free
1 = fragmented/partially preserved/partially plugged
2 = destroyed/missing/completely plugged

Taphonomic features:
0 = not present
1 = present
2 = dominant

Figure 6.5. *Taphonomic gradient, character preservation, taphonomic features, and taxonomic resolution for* **Clypeaster humilis** *test from the Northern Bay of Safaga, Red Sea, Egypt.*

6.3.2), it can completely disappear by stage 3 and 4. The gonopores are open in stage 2 but become plugged by sediment or cement by stage 3 (Fig. 6.3.3). The apical system is obviously not as stable as the rest of the test as observed in the continuous degradation from Figures 6.3.2 to 6.3.3 to 6.3.5 to 6.3.6.

The petalodium is distinct in stage 1 and especially in stage 2 specimens, as the pore pairs are free of sediment and thus result in dark holes in a stark contrast to the bright plate surfaces on the aboral surface of the *Clypeaster* skeleton. By stage 3, however, the petalodium becomes less prominent as the ambulacral pore pairs are plugged by sediment or cement (Fig. 6.2.3). The minute ambulacral small pores at the base of the petals next to the apical system also become progressively indistinct (Fig. 6.3.3). Because they represent depression on the upper side of the test, the pore pair rows are commonly preferred sites of encrustation, which then mask the pore pairs (Fig. 6.3.5). The slight outline of the petals is barely recognizable in the stage 4 specimens (Fig. 6.3.6).

Although plate boundaries are readily visible in living and stage 1 specimens, they become even more distinct in the denuded stage 2 examples. Plate boundaries are noticeably lighter in color, whereas the interior of the plates is darker, leading to a distinct pattern on the surface of the skeleton (Fig. 6.3.2). Plate boundaries become indistinct by stage 3, although they can just be recognized. For example, in Figure 6.3.5, the plate boundaries are slightly depressed compared to the plate centers. The progressive degradation of the test surface by abrasion and corrosion preclude any recognition of plate boundaries in stage 4 (Fig. 6.3.6).

The fate of the tubercles can barely be discerned at a macroscopic scale. They are still occupied by spines in stage 1. In stage 2, the tubercles are well defined and distinct; by stage 3, they start becoming indistinct, and they disappear totally by stage 4. These changes can best be seen at a microscopic level in Figures 6.4.1, 6.4.3, and 6.4.5. Well-preserved test surfaces have distinct sunken tubercles with deepened areoles, a distinct crenulated boss, and prominent, perforated mammelons. These three structures are distinct because they are constructed of different stereom types, including coarse labyrinthic stereom of the sunken areole and galleried stereom corresponding to the muscle attachment areas. As illustrated in Figure 6.4.5, the mammelon is constructed of a sparsely perforated dense stereom. The area between the tubercles is characterized by secondary tubercles and protuberances. The stereom is completely free of sediment and cementation.

An abraded and encrusted specimen is depicted in Figures 6.4.2, 6.4.4, and 6.4.6. The tubercles are not only abraded but also covered by an encrusting serpulid worm tube (itself abraded) and encrusting foraminifera. Although primary tubercles are abraded, protuberances between the tubercles can still be recognized (Fig. 6.3.2). As seen in Figure 6.4.4, the primary tubercles are less distinct, and the areoles are partially filled with sediment particles. Abrasion of the tubercle is typically accompanied by the shearing off of the mammelon. The pore space between the stereom becomes reduced due probably to diagenetic accretion of cement on the struts of the stereom (Fig. 6.3.6).

Other features include the ambitus of the clypeasteroid skeletal, which remains stable throughout the taphonomic gradient until stage 5, when the perimeter of the skeleton starts to become corroded. The oral surface has the same general tendencies as the aboral surface. Interestingly, the serpulid worm tubes follow the food grooves in stage 3a (Fig. 6.3.4). By stage 4, the food grooves become indistinguishable, and the boundaries of the periproct and peristome also become indistinct.

Discussion

There are obviously two end members of the taphonomic gradient described here, with spine-covered skeletons on the one hand and totally corroded specimens on the other. These also represent a temporal sequence with a clear beginning and an end. Spine loss is ubiquitous in the well-oxygenated, shallow, agitated waters of the study area. The loss of surface characters through abrasion is obviously linear. However, encrusted and bioeroded specimens can have well-preserved surface characters, but conversely, abraded specimens show no or little encrustation. Thus, there is a mosaic development of taphonomic features present on the test surface for stage 3 (Fig. 6.5). Toward the end of the taphonomic pathway, all features pertaining to the test surface are completely destroyed, eradicating all evidence of encrustation and bioerosion. The general shape of the whole skeleton as such, however, remains remarkably intact.

The taphonomic gradient construed here represents a mixture of destructive and constructive taphonomic processes that are active at different stages of the gradient (Fig. 6.5). Decay occurs shortly after death. Abrasion is continuous and becomes more evident along the gradient. Encrustation and bioerosion also increase along the gradient to stage 3b but is eroded away by stage 4. Diagenesis can set in by stage 3, especially in this tropical marine environment where primary marine cementation is prevalent and continues through stage 4. Corrosion, the last process to act on the skeletons, determines the appearance of the last stage of the gradient.

Decay is destructive, causing the disarticulation of spines, disaggregation of the jaws, and a loosening of plate boundaries. Abrasion is also obviously destructive, as is bioerosion. However, encrustation and early diagenesis can serve to strengthen the test. For example, encrusting serpulids and bryozoans cross plate boundaries. Early diagenesis also fills in the stereom. Thus, stage 4 skeletons waste away and do not readily break up into fragments.

Although there is a temporal succession of events, no statements can be made about the absolute timing. Studies on the ages of differentially preserved bivalves in shallow-water settings have shown surprisingly larger age discrepancies among differently preserved specimens (e.g., Kowalewski et al., 1998). This should also be expected for echinoderm remains, although to my knowledge, no such studies have been made on echinoderms. Variations in the intensity of taphonomic processes such as different rates of abrasion, bioerosion, and encrustation can be expected within different facies. The length of residence time on the sediment surface will also affect the rate of taphonomic processes affecting the echinoid skeleton. These differences have been shown to occur for *Clypeaster* fragments recovered from bulk samples (Nebelsick, 1999) such that 1) low-energy environments and low sedimentation rates (which equal long surface residence times) lead to good surface preservation and high encrustation rates; and 2) high-energy, shallow-water environments with higher sedimentation rates lead to highly abraded specimens with low encrustation rates. However, many more specimens of *Clypeaster humilis* from different facies would be needed in order to discern such patterns among complete tests.

This study is restricted to complete specimens, although they are missing their spines. Once the specimens are fragmented, different patterns emerge (see Nebelsick, 1999). For example, the bare echinoid test offers a relatively large surface for encrusting organisms in otherwise unstable environments (Nebelsick et al., 1997). Therefore, it is not surprising that encrustation can be relatively high in shallow-water environments. However, fragments in this environment are mostly highly abraded and free of encrustation. This most likely has to do with the different hydrodynamic properties of large tests on the one hand and fragments on the other. Fragments are more likely to be more highly abraded in higher-energy environments because of their smaller size and their entrainment in wave movement. Primary encrustation on fragments (as shown by encrusters occurring directly on the disarticulated plate boundaries) also occurs but is largely restricted to deeper, quiet-water settings (in the West Area; see Fig. 1) with low sedimentation rates (see Nebelsick, 1999). This allows for long surface residence times with little disturbance that would be conducive to the settlement of an encrusting epifauna.

Another factor that is obviously important in the taphonomy of the skeletons and production of fragments is destructive predation events (see Nebelsick, 1999; Kowalewski and Nebelsick, 2003). Fish predation produces a gapping wound with jagged borders on the oral surface. The role of predation in the taphonomic scenario described above is the subject of continued investigation that should add further insights into the complexity of echinoderm preservation.

The differential stages of preservation along the taphonomic gradient will affect the taxonomic resolution of identification of fossil specimens. The level of taxonomic identifications depends on the loss of diagnostic taxonomic characters along the taphonomic gradient. In stages 1 and 2, species identification is readily identifiable. Identification at stage 3a and 3b depends on which characters happen to be preserved (i.e., not destroyed by abrasion or bioerosion) or exposed (i.e., not covered by encrustation). It also depends on how well the species from the study area are known with respect to their morphologic features and phenotypic variations. The preservation of stage 4 echinoids (Fig. 6.3.6) obviously precludes direct identification at the species and even genus level, but the order Clypeastereroida can still be identified. No other echinoderms have such a flattened form with internal supports, so that they can be recognized in even the most corroded specimens. However, it will be difficult to make any closer determination at the family and suborder levels. Even if the test is totally corroded, their designation as echinoderms is possible because of the unique structure of the echinoderm stereom at a microscopic level (Smith, 1980, 1984, 1990).

This study demonstrates that fossil skeletons have to be removed from the taphonomic cycle in order to be preserved. Theoretically, all the studied specimens would have ended up as stage 4 skeletons if they remained on the surface long enough, and even then they could corrode completely away. This study also reiterates the complexity and interplay of different factors affecting the preservation of echinoderm skeletons. Although there are definite end members, the taphonomic process in between these end members can differentially affect the preservation of the skeletons. It has been argued that using fragments can increase our knowledge of the distribution of echinoderms (Gordon and Donovan, 1992; Nebelsick, 1992a, 1992b; Donovan, 2001, 2003; Kroh, 2005): the fragments can more closely preserve the long-term settlement patterns because living echinoid distributions are notoriously patchy. This study shows that both complete skeletons and fragments should also be included in taphonomic analysis in order to obtain a more complete picture of the ecological factors affecting preservation at different scales of observation.

Acknowledgments

I thank the staff of the Institute of Geosciences University of Tübingen for their support, especially W. Gerber for the photographs. I sincerely thank W. I. Ausich and C. Schneider for their careful reviews.

References

Allison, P. A. 1990. Variation in rates and decay and disarticulation of Echinodermata: implications for the application of actualistic data. Palaios, 5:432–330.

Ausich, W. I. 2001. Echinoderm taphonomy, p. 171–227. In M. Jangoux and J. M. Lawrence (eds.), Echinoderm Studies 6. Balkema, Lisse.

Boggild, G. R., and E. P. F. Rose. 1984. Mid-Tertiary echinoid biofacies as palaeoenvironmental indices. Annales Gèologique des Pays Hellènique, 22:57–67.

Brett, C. E., H. A. Moffat, and W. L. Taylor. 1997. Echinoderm taphonomy, taphofacies, and Lagerstätten, p. 147–190. In J. A. Waters and C. G. Maples (eds.), Geobiology of Echinoderms. Paleontological Society Papers, 3.

Clark, A. M., and F. W. E. Rowe. 1971. Monograph of the Shallow-Water Indo-West Pacific Echinoderms. British Museum, London, 238 p.

De Meijere, J. C. H. 1902. Vorläufige Beschreibung der neuen, durch die Siboga-Expedition gesammelten Echiniden. Tijdschrift van de Nederlansche Dierkundige Vereeniging Leiden, 2(8):1–16.

Donovan, S. K. 1991. The taphonomy of echinoderms: calcareous multi-element skeletons in the marine environment, p. 241–269. In S. K. Donovan (ed.), The Processes of Fossilisation. Belhaven Press, London.

Donovan, S. K. 2001. Evolution of Caribbean echinoderms during the Cenozoic: moving towards a complete picture using all of the fossils. Palaeogeography, Palaeoclimatology, Palaeoecology, 166:177–192.

Donovan, S. K. 2003. Completeness of a fossil record: the Pleistocene echinoids of the Antilles. Lethaia, 36:1–7.

Durham, J. W. 1955. Classification of clypeasteroid echinoids. California University Publications in Geological Science, 31:73–198.

Durham, J. W. 1966. Clypeasteroids, p. U450–U491. In R. C. Moore (ed.), Treatise on Invertebrate Paleontology. Pt. U. Echinodermata 3. Geological Society of America and University of Kansas Press, Lawrence.

Ghiold, J., and A. Hoffman. 1984. Clypeasteroid echinoids and historical biogeography. Neues Jahrbuch für Geologie und Paläontologie Monatshefte, 1984:529–538.

Ghiold, J., and A. Hoffman. 1986. Biogeography and biogeographic history of clypeasteroid echinoids. Journal of Biogeography, 13:183–206.

Gordon, C. M., and S. K. Donovan. 1992. Disarticulated echinoid ossicles in paleoecology and taphonomy: the last interglacial Falmouth Formation of Jamaica. Palaios, 7:157–166.

Greenstein, B. J. 1993. Is the fossil record of regular echinoids really so poor? A comparison of living and subfossil assemblages. Palaios, 8:597–601.

Kampfer, S., and W. Tertschnig. 1992. Feeding biology of Clypeaster rosaceus (Echinoidea, Clypeasteroidea) and its impact on shallow lagoon sediments,

p. 197–200. *In* L. Scalera-Liaci and C. Canicatti (eds.), Echinoderm Research 1991: Proceedings of the 3rd European Echinoderm Conference, Lecce, Italy, 1991. A. A. Balkema, Rotterdam.

Kidwell, S. M., and T. Baumiller. 1990. Experimental disintegration of regular echinoids: roles of temperature, oxygen and decay thresholds. Paleobiology, 16:247–271.

Kier, P. M. 1977. The poor fossil record of the regular echinoids. Paleobiology, 3:168–174.

Koehler, R. 1922. Echinides du musèe indien à Calcutta. 11. Clypeastridès et Cassidulidès. Echinoderma of the Indian Museum, part IX, Echinoidea (II). Calcutta, 161 p.

Kowalewski, M., G. A. Goodfriend, and K. W. Flessa. 1998. The high-resolution estimates of temporal mixing in shell beds: the evils and virtues of time-averaging. Paleobiology, 24:287–304.

Kowalewski, M., and J. H. Nebelsick. 2003. Predation on Recent and fossil echinoids, p. 279–302. *In* P. H. Kelley, M. Kowalewski, and T. A. Hansen (eds.), Predator-Prey Interactions in the Fossil Record. Topics in Geobiology Series, 20. Kluwer Academic/Plenum Publishers, New York.

Kroh, A. 2005. Catalogus Fossilium Austriae. Band 2. *Echinoidea neogenica*. Österreichische Akademie der Wissenschaften, Vienna, 210 p.

Kroh, A., and J. H. Nebelsick. 2003. Echinoid assemblages as a tool for palaeoenvironmental reconstruction—an example from the Early Miocene of Egypt. Palaeogeography, Palaeoclimatology, Palaeoecology, 201:157–177.

Lamarck, J. B. P. A. DE M. 1801. Système des animaux sans vertèbres, ou table général des classes, des ordres, et des genres et ces animaux. Paris, chez Deterville, 432 p.

Leske, N. G. 1778. Additamenta ad Jacobi Theodori Klein Naturalem Dispositionem Echinodermatum et Lucubratiunculam de Aculeis Echinorum Marinorum. Officina Gleditschiana, Lipsiae, 278 p.

Lewis, R. 1980. Taphonomy, p. 27–39. *In* T. W. Broadhead and J. A. Waters (eds.), Echinoderms: Notes for a Short Course. University of Tennessee Studies in Geology, 3. University of Tennessee Press, Knoxville.

Linnaeus, C. 1758. Systema naturae (10th edition). Stockholm, Laurentii Salvii. 824 p.

Moffat, H. A., and D. J. Bottjer. 1999. Echinoid concentration beds: two examples from the stratigraphic spectrum. Palaeogeography, Palaeoclimatology, Palaeoecology, 149:329–348.

Mooi, R. 1989. Living and fossil genera of the Clypeasteroida (Echinoidea: Echinodermata): an illustrated key and annotated checklist. Smithsonian Contributions to Zoology, 488:1–51.

Nebelsick, J. H. 1992a. The Northern Bay of Safaga (Red Sea, Egypt): an actuo-palaeontological approach. III Distribution of echinoids. Beiträge zur Paläontologie von Österreich, 17:5–79.

Nebelsick, J. H. 1992b. The use of fragments in deducing echinoid distribution by fragment identification in Northern Bay of Safaga; Red Sea, Egypt. Palaios, 7:316–328.

Nebelsick J. H. 1999. Taphonomic signatures and taphofacies distribution as recorded by *Clypeaster* fragments from the Red Sea. Lethaia, 32:241–252.

Nebelsick, J. H. 2004. Taphonomy of echinoderms: introduction and outlook, p. 471–477. *In* T. Heinzeller, and J. H. Nebelsick (eds.), Echinoderms München: Proceedings of the 11th International Echinoderm Meeting. Taylor and Francis, London.

Nebelsick, J. H., and S. Kampfer. 1994. Taphonomy of *Clypeaster humilis* and

Echinodiscus auritus from the Red Sea, p. 803–808. *In* B. David, A. Guille, J.-P. Fèral, and M. Roux (eds.), Echinoderms through Time. A. A. Balkema, Rotterdam.

Nebelsick, J. H., and M. Kowalewski. 1999. Drilling predation on recent Clypeasteroid echinoids from the Red Sea. Palaios, 14:127–144.

Nebelsick, J. H., B. Schmid, and M. Stachowitsch. 1997. The encrustation of fossil and recent sea-urchin tests: ecological and taphonomic significance. Lethaia, 30:271–284.

Nèraudeau, D., E. Goubert, D. Lacour, and J. M. Rouchy. 2001. Changing biodiversity of Mediterranean irregular echinoids from the Messinian to present-day. Palaeogeography, Palaeoclimatology, Palaeoecology, 175:43–60.

Piller, W., and A. M. Mansour. 1990. The Northern Bay of Safaga (Red Sea, Egypt): an actuopalaeontological approach, II. Sediment analysis and sedimentary facies. Beiträge zur Paläontologie von Österreich, 16:1–102.

Piller, W., and P. Pervesler. 1989. The Northern Bay of Safaga (Red Sea, Egypt): an actuopalaeontological approach, II. Topography and bottom facies. Beiträge zur Paläontologie von Österreich, 15:103–147.

Poddubiuk, R. H., and E. P. F. Rose. 1984. Relationships between mid-Tertiary echinoid faunas from the Central Mediterranean and eastern Caribbean and their palaeobiogeographic significance. Annales Gèologique des Pays Hellènique, 31:115–127.

Seilacher, A. 1979. Constructional morphology of sand dollars. Paleobiology, 5:191–221.

Smith, A. B. 1980. Stereom microstructure of the echinoid test. Special Papers in Palaeontology, 25:1–81.

Smith, A. B. 1984. Echinoid Palaeobiology. George Allen and Unwin, London, 191 p.

Smith, A. B. 1990. Biomineralization in echinoderms, p. 413–443. *In* J. G. Carter (ed.), Skeletal Biominerilization. Patterns, Processes and Evolutionary Trends, volume I. Van Nostrand Reinhold, New York.

Smith, A. B. 2001. Probing the cassiduloid origins of clypeasteroid echinoids using stratigraphically restricted parsimony analysis. Paleobiology, 27:392–404.

PART 2. EVOLUTIONARY PALEOECOLOGY

INTRODUCTION TO PART 2

William I. Ausich and Gary D. Webster

The Ecological Theater and the Evolutionary Play (Hutchinson, 1965) is realized in the emerging field of evolutionary paleoecology, and the metaphor reaffirms the important role of ecological-time processes for evolutionary history. Bottjer and Allmon (2001, p. 1–2) defined evolutionary paleoecology as "a loosely connected skein of research programs that focus on the environmental and ecological context for long-term (e.g., macroevolutionary) changes seen in the fossil record." Evolutionary paleoecology combines the phylogeny and paleoenvironmental history of a clade to understand the theater in which the evolutionary play occurred. If possible, some processes responsible for the evolutionary history can also be identified and may lead to a broader, predictive model to understand the history of life on Earth.

Stalked, epifaunal echinoderms dominated many Paleozoic suspension-feeding communities beginning during the Ordovician. Epifaunal suspension feeding arose among echinoderms during the Cambrian. Dornbos (Chapter 7) establishes the ecological structure of these early echinoderms and traces their history to younger times. Ausich and Kammer (Chapter 8) examine the evolutionary history of an important middle Paleozoic family, the Periechocrinidae. After establishing the phylogeny of this family, they consider the paleobiogeographic and paleoenvironmental context that led to the evolutionary history of the Periechocrinidae, including the evolution of two new genera immediately before the clade becomes extinct.

References

Bottjer, D. J., and W. D. Allmon. 2001. Evolutionary Paleoecology. Columbia University Press, New York, 357 p.

Hutchinson, G. E. 1965. The Ecological Theater and the Evolutionary Play. Yale University Press, New Haven, Connecticut, 139 p.

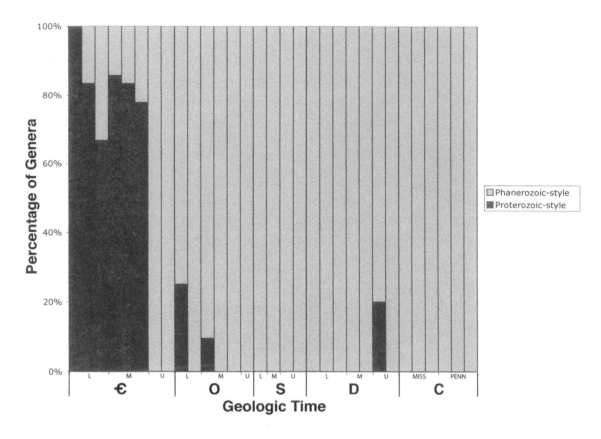

Figure 7.1. *Stacked-bar histogram showing percentage of genera adapted to typical Proterozoic-style soft substrates (black) and typical Phanerozoic-style soft substrates (gray) within each stage of the Cambrian through the Carboniferous. Sample size for each bar of histogram, from left to right: 2, 6, 3, 7, 6, 9, 2, 6, 4, 8, 11, 8, 25, 16, 4, 6, 5, 2, 2, 2, 3, 2, 4, 5, 3, 7, 7, 6, 2, 4, 2. The two initial stages (Nemakit–Daldynian and Tommotian) and final stage (Trempealeauan) of the Cambrian contain no data and thus are not included in this histogram. Data from Dornbos (2006, tables 1–5) (see Appendix 7.1 for additional references used for data presented in this and other figures).*

TIERING HISTORY OF EARLY EPIFAUNAL SUSPENSION-FEEDING ECHINODERMS

7

Stephen Q. Dornbos

Introduction

Echinoderms provide some of the earliest Phanerozoic evidence for fully skeletonized metazoans. Their skeletal remains are preserved as early as the Atdabanian Stage of the Lower Cambrian, where the first trilobite fossils are also found. Since that time, echinoderm fossils have been excellent recorders of their paleoecology, including their division of suspended trophic resources into epifaunal tiers.

Crinoids are known for achieving the highest levels of epifaunal tiering, reaching as high as 1 m above the seafloor (Bottjer and Ausich, 1986). Of interest in this particular study are epifaunal sessile suspension-feeding echinoderms that originated during the Cambrian Period, the edrioasteroids, eocrinoids, and helicoplacoids. Despite being an important part of the Cambrian radiation, and in the cases of the edrioasteroids and eocrinoids, the Ordovician radiation, these echinoderm groups never attained the tiering prowess of the crinoids (Bottjer and Ausich, 1986). Perhaps as a result of the lack of high-level tiering taxa, a good understanding of their tiering history has never been attained.

The goal of this study is the reconstruction of the tiering patterns of these early sessile epifaunal echinoderm groups throughout their entire geologic history by examining the tiering levels inhabited by each known genus at stage-level stratigraphic resolution. The resulting data will reveal evolutionary paleoecological trends in these echinoderm groups from their shared Cambrian origins to the Carboniferous demise of the edrioasteroids. These trends should reveal the impact of important Paleozoic evolutionary events such as the Cambrian radiation, Ordovician radiation, end-Ordovician mass extinction, and Late Devonian mass extinction on the tiering history of these echinoderms.

Previous Research

ECHINODERM EPIFAUNAL TIERING. Ausich and Bottjer (1982, 1985) and Bottjer and Ausich (1986) reconstructed the tiering history of epifaunal suspension-feeding echinoderms in soft substrate settings on the subtidal shelf. They defined the following four epifaunal tiers that are also utilized in this study: 0 to +5 cm, +5 to +10/15/20 cm, +20 to +50 cm, and +50 to +100 cm (Bottjer and Ausich, 1986). Epifaunal echinoderms only reached the +5 to +10 cm tier during the Cambrian (Bottjer and Ausich, 1986). With the evolution of crinoids during the Ordovician radiation, echino-

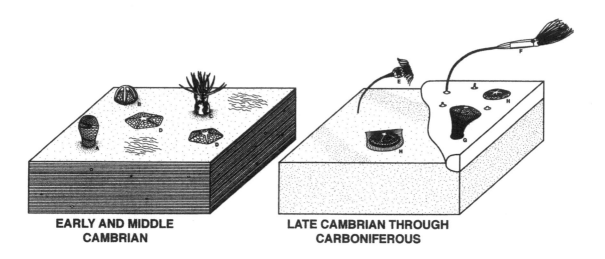

EARLY AND MIDDLE CAMBRIAN   **LATE CAMBRIAN THROUGH CARBONIFEROUS**

Figure 7.2. *Box diagrams depicting dominant substrate adaptations of sessile benthic suspension-feeding echinoderms and dominant soft substrate characteristics during Early and Middle Cambrian (left) and from Late Cambrian through Carboniferous (right). Left box:* **A**, *shallow sediment sticker* **Helicoplacus** *(Dornbos and Bottjer, 2000);* **B**, *suctorial sediment attacher* **Totiglobus** *(Bell and Sprinkle, 1978; Domke and Dornbos, 2005);* **C**, *shallow sediment sticker* **Lichenoides** *(Parsley and Prokop, 2004); and* **D**, *suctorial sediment attacher* **Stromatocystites** *(Smith and Jell, 1990). Some Early and Middle Cambrian eocrinoids were hard substrate attachers but are not shown in this reconstruction because they were not dominant diversity-wise during this time interval (see Fig. 7.1). Note low levels of horizontal bioturbation and wrinkle structures on microbial-mat bound substrate. Right box:* **E**, *holdfast strategist* **Macrocystella** *(Paul, 1968, based on reconstruction in Dzik and Orlowski, 1993);* **F**, *hard substrate attacher* **Haimacystis** *(Sumrall et al., 2001);* **G**, *hard substrate attacher* **Hypsiclavus** *(Sumrall, 1996); and* **H**, *generalized hard substrate attaching pyrgate edrioasteroids. Note well-bioturbated soft sediment and attachment to skeletal material and carbonate hardgrounds (right). The box diagrams are not intended to show echinoderms that lived at the same time or all possible substrate adaptations, just the dominant substrate adaptations during specified time intervals. Data from Dornbos (2006).*

derms reached the +50 to +100 cm tier and sustained those heights above the seafloor until the end-Permian mass extinction (Ausich and Bottjer, 2001). During the Early Triassic aftermath of this mass extinction, crinoids were reduced to one species that perhaps only reached the +20 to +50 cm tier (Schubert et al., 1992; Ausich and Bottjer, 2001).

Crinoids reoccupied the +50 to +100 cm tier when the recovery from the end-Permian mass extinction was complete (Ausich and Bottjer, 2001). At the end of the Jurassic, stalked crinoids moved off of the shelf into oceanic water depths (Bottjer and Ausich, 1986). Since that time, the +50 to +100 cm tier has been largely unoccupied in shallow-water communities (Bottjer and Ausich, 1986).

EARLY ECHINODERM SUBSTRATE ADAPTATIONS. Increasing bioturbation depth and intensity during the Cambrian radiation created a transition in dominant subtidal soft substrates from firm Proterozoic-style soft substrates without a well-developed mixed layer to soupier Phanerozoic-style soft substrates with a well-developed mixed layer (e.g., Bottjer et al., 2000). The term *soft substrates* here refers to unlithified sediments, which can have variable consistencies, from extremely firm to soupy, depending on factors such as bioturbation intensity and seafloor microbial mat development. During the Cambrian increase in bioturbation depth and intensity, microbial mats were relegated to settings with inhibited metazoan activity but made resurgences during later ecological crises, such as the aftermaths of the end-Ordovician and end-Permian mass extinctions (e.g., Schubert and Bottjer, 1992; Pruss et al., 2004; Sheehan and Harris, 2004). Previous work shows that this substrate change strongly influenced the evolution of Cambrian benthic suspension-feeding echinoderms (Bottjer et al., 2000; Dornbos and Bottjer, 2000, 2001; Lefebvre and Fatka, 2003; Clausen, 2004; Parsley and Prokop, 2004; Dornbos, 2006).

The substrate adaptations of genera in the three classes of sessile suspension-feeding epifaunal echinoderms with definitive Cambrian origins (edrioasteroids, eocrinoids, and helicoplacoids) have been examined throughout their fossil records (Dornbos, 2006). This examination revealed that genera adapted to typical Proterozoic-style soft substrates as sediment attachers, sediment resters, and shallow sediment stickers were taxonomically dominant during the Early and Middle Cambrian (Figs. 7.1, 7.2) (Dornbos, 2006). From the Late Cambrian through the Pennsylvanian, genera adapted to typical Phanerozoic-style soft substrates, most as hard substrate attachers, were taxonomically dominant (Figs. 7.1, 7.2) (Dornbos, 2006). This trend in substrate adaptations is consistent with an evolutionary response to this Cambrian substrate transition because these echinoderms had to adapt to the soupier substrates created by increasing bioturbation intensity.

Methods

The epifaunal tiers inhabited by the 94 known edrioasteroid, eocrinoid, and helicoplacoid genera from the Lower Cambrian through the Carboniferous were determined on the basis of descriptions and figures in the literature. This database represents current knowledge and will change with future fossil descriptions. The epifaunal tiers of Ausich and Bottjer (1982) and Bottjer and Ausich (1986) were used: +0 to +5 cm, +5 to +10/15/20 cm, +20 to +50 cm, and +50 to +100 cm. Sepkoski's (2002) compendium of marine genera was supplemented and amended to establish the stage range of each genus. The interpreted substrate adaptation of each genus is from Dornbos (2006).

All of these data were compiled to produce plots of the epifaunal tiers inhabited by these 94 genera through geologic time. One plot displays the class-level taxonomic classification (Edrioasteroidea, Eocrinoidea, or Helicoplacoidea) and another the general substrate adaptation type (Proterozoic style or Phanerozoic style; see Dornbos, 2006) of each genus. These substrate adaptation categorizations are based on interpretations of the adaptive morphological features of these genera used to interact directly with the substrate. All genera were placed into one of seven paleoecological categories developed by Thayer (1975, 1983) and Dornbos et al. (2005): (1) snowshoe strategists, (2) iceberg strategists, (3) holdfast strategists, (4) hard substrate attachers, (5) sediment resters, (6) shallow sediment stickers,

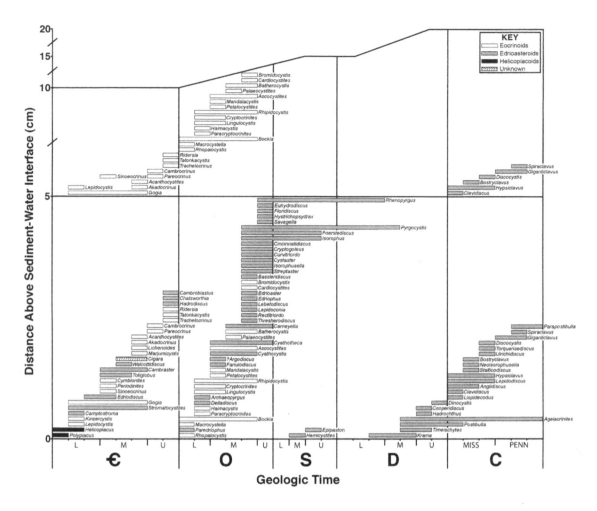

Figure 7.3. *Tiering diagram showing genera occupying each epifaunal tier, their stratigraphic range, and their taxonomic classification. Vertical arrangement of genera within tiers is not meaningful.*

Results

and (7) sediment attachers. Genera in the first four paleoecological categories are interpreted to be adapted to typical Phanerozoic-style soft substrates because they had adaptations to soupy sediments with a well-developed mixed layer (Dornbos et al., 2005). Genera in the last three categories are interpreted to be adapted to typical Proterozoic-style soft substrates because they had adaptations to firm sediments that lack a well-developed mixed layer (Dornbos et al., 2005).

GENERAL TIERING PATTERN. The Early and Middle Cambrian were dominated by inhabitants of the 0 to +5 cm tier (Fig. 7.3), most of which were adapted to firm Proterozoic-style substrates (Fig. 7.4). Occupants of the +5 to +10 cm tier, all of which were eocrinoids, comprised only 0%–33% of genera during this interval (Figs. 7.3, 7.5). Eocrinoids expanded in this tier during the Late Cambrian, causing inhabitants of this tier to comprise 50%–100% of genera (Figs. 7.3, 7.5). These Late Cambrian +5 to +10 tiering taxa were all adapted to soupy Phanerozoic-style soft substrates (Fig. 7.4).

This dominance of +5 to +10/15 cm tier occupants continued into the heart of the Ordovician radiation, the Early and Middle Ordovician (Figs.

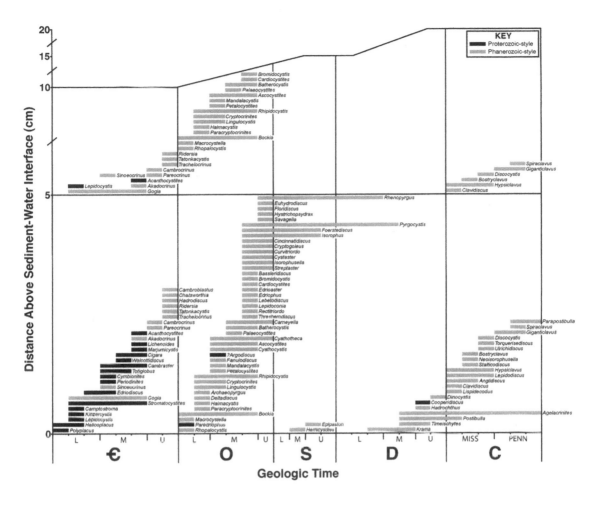

Figure 7.4. *Tiering diagram showing genera occupying each epifaunal tier, their stratigraphic range, and their substrate adaptations. Proterozoic-style indicates that genus had adaptations to typical Proterozoic-style soft substrates. Phanerozoic-style indicates that genus had adaptations to typical Phanerozoic-style soft substrates. See Dornbos (2006) for interpreted substrate adaptations for each genus. Vertical arrangement of genera within tiers is not meaningful.*

7.3, 7.5). They comprised more than 60% of genera until the final stage of the Middle Ordovician, when they declined to 24% because of the radiation of edrioasteroids in the 0 to +5 cm tier (Fig. 7.5). As in the Cambrian, all of these +5 to +10/15 cm tier occupants were eocrinoids, with edrioasteroids restricted to the 0 to +5 cm tier (Fig. 7.3). During the latest Middle Ordovician and Late Ordovician, +5 to +10/15 cm tier occupants decreased until they were no longer represented in these echinoderm groups (Fig. 7.5). This decline is the result of the extinction of eocrinoids during the latter part of the Ordovician (Fig. 7.3), perhaps as a result of the end-Ordovician mass extinction. Although their fossil record does not extend to the end of the Ordovician, it is at least clear that eocrinoids did not survive beyond this extinction event.

Of the sessile suspension-feeding echinoderm classes with Cambrian origins, only the edrioasteroids survived beyond the end-Ordovician mass extinction (Fig. 7.3). They were restricted to the 0 to +5 cm tier throughout the Silurian and Devonian (Figs. 7.3, 7.5). During most of this time, they were primarily adapted to soupy Phanerozoic-style soft substrates as hard substrate attachers (Fig. 7.4).

Edrioasteroids reached the +5 to +20 cm tier in the Carboniferous with the evolution of several clavate forms that could extend their theca as

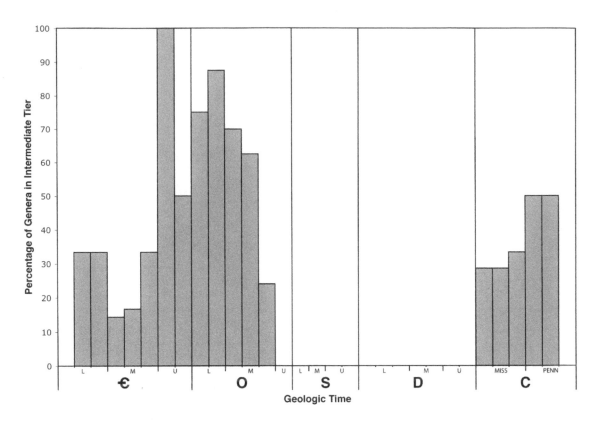

Figure 7.5. *Histogram showing the percentage of genera that reached the +5 to +10/15/20 cm tier during the history of the helicoplacoids, edrioasteroids, and eocrinoids. Sample size for each bar of histogram, from left to right: 2, 6, 3, 8, 6, 9, 2, 6, 4, 8, 11, 8, 25, 16, 4, 6, 5, 2, 2, 2, 3, 2, 4, 5, 3, 7, 7, 6, 2, 4, 2.*

high as 5 cm into the water column (Figs. 7.3, 7.5). All of these clavate forms were hard substrate attachers (Fig. 7.4). Occupants of the +5 to +20 cm tier comprised 50% of genera during the first two stages of the Pennsylvanian, but the last known edrioasteroid genus *Parapostibulla* Sumrall et al., 2001, was restricted to the 0 to +5 cm tier (Fig. 7.5).

HELICOPLACOIDS. The Early Cambrian helicoplacoids lived as shallow sediment stickers with their pointed lower end inserted into the sediment (Durham, 1993; Dornbos and Bottjer, 2000, 2001; Sprinkle and Wilbur, 2005). A recent report of helicoplacoids attached to skeletal material (Wilbur, 2004) is unconvincing because it appears to simply be the fortuitous preservation of a helicoplacoid on top of a trilobite molt with no real evidence for attachment. Even if true, the small size of the helicoplacoid specimen in question combined with the solid evidence for adult helicoplacoids living as shallow sediment stickers makes it likely that only juvenile helicoplacoids could have lived attached to hard substrates. Abundant direct evidence demonstrates that adult helicoplacoids were shallow sediment stickers (Durham, 1993; Dornbos and Bottjer, 2000, 2001; Sprinkle and Wilbur, 2005).

Although they could expand and contract their theca, helicoplacoids were restricted to the 0 to +5 cm tier (Fig. 7.3). Their maximum body height is known to be as much as 7 cm (Durham, 1993), but with the lower third of their body inserted into the sediment, they would not have attained heights greater than 5 cm above the seafloor. As such, they typify the 0 to +5 cm tier occupants of the Early and Middle Cambrian that were adapted to living on firm, often microbial mat bound, sediments (Figs. 7.4, 7.5) (Dornbos, 2006).

EOCRINOIDS. *Gogia* is the most diverse and abundant Cambrian eocrinoid. Along with *Lepidocystis*, *Gogia* is the first eocrinoid to reach the +5 to +10 cm tier, doing so in the Botomian Stage of the Lower Cambrian (Fig. 7.3). With the extinction of *Lepidocystis*, *Gogia* was the sole occupant of the +5 to +10 cm tier until the late Middle Cambrian and Upper Cambrian, when other eocrinoids reach this tier (Fig. 7.3).

Most *Gogia* species attached to skeletal material (e.g., Sprinkle, 1973), but other substrate interactions may have been possible for some species (Dornbos, 2006). Despite the abundance of Early and Middle Cambrian eocrinoid genera adapted to firm Proterozoic-style soft substrates, only two such genera, *Lepidocystis* and *Acanthocystites*, reached the +5 to +10 cm tier (Figs. 7.3, 7.4). The dominance of 0 to +5 cm tier occupants adapted to firm, often microbial mat bound, substrates contrasts sharply with large Ediacaran forms adapted to similar substrates that could reach an epifaunal tier as high as 120 cm (Clapham and Narbonne, 2002). Perhaps the substrate transition underway during the Cambrian did not provide the stability necessary for epifaunal echinoderms adapted to typical Proterozoic-style soft substrates to reach the +5 to +10 cm tier in great abundance.

As discussed above, eocrinoids became even more common in the +5 to +10/15 cm tier during the Ordovician radiation (Fig. 7.3). Just as during the Cambrian, they accounted for all +5 to +10/15 cm tier occupants in these echinoderm classes during the Ordovician (Fig. 7.3). Despite being an important part of the Ordovician radiation, eocrinoids are unknown from the Upper Ordovician fossil record (Fig. 7.3). Their extinction left edrioasteroids as the only remaining class of epifaunal echinoderms with Cambrian origins and began a long absence from the +5 to +10/15/20 cm tier for these echinoderm classes (Figs. 7.3, 7.5).

EDRIOASTEROIDS. Edrioasteroids were inhabitants of the 0 to +5 cm tier through most of their history. Specializing in this tier probably allowed them to avoid competition with eocrinoids, blastozoans, and crinoids for suspended trophic resources in the higher tiers. Forms adapted to firm Proterozoic-style soft substrates, such as *Stromatocystites*, were the dominant Early and Middle Cambrian edrioasteroids (Figs. 7.2 to 7.4). Forms adapted to hard substrate attachment dominated thereafter, with a handful of exceptions (Dornbos, 2006).

It was not until the evolution of the advanced clavate edrioasteroids during the Carboniferous that edrioasteroids reached the +5 to +20 cm tier (Fig. 7.3). Such forms could extend their aboral theca to lengths over 5 cm, allowing edrioasteroids to exploit epifaunal ecospace that was previously off limits to them (Sumrall, 1996).

Discussion

The tiering history of these early echinoderm groups was shaped by the major evolutionary events of the Paleozoic: the Cambrian radiation, Ordovician radiation, end-Ordovician mass extinction, and Late Devonian mass extinction. During the Cambrian radiation, helicoplacoids and edrioasteroids were among the earliest fully skeletonized metazoans. Eocrinoids reached the +5 to +10 cm tier during the Cambrian (Fig. 7.3), ranking them among the highest epifaunal tierers of the time (Bottjer and Ausich, 1986).

The Ordovician radiation saw a remarkable diversification in eocrinoids and edrioasteroids, as well as increased inhabitation of the +5 to +10/15 cm tier by eocrinoids (Fig. 7.3). In contrast with the Cambrian, eocrinoids were no longer among the highest epifaunal tierers, being overshadowed in this respect by the crinoids and cystoids (Bottjer and Ausich, 1986). Although edrioasteroids made an important contribution to the biodiversity of the Ordovician radiation, they still did not break free of the 0 to +5 cm tier (Fig. 7.3).

The end-Ordovician mass extinction left no echinoderm classes with Cambrian origins anywhere above the 0 to +5 cm tier (Fig. 7.3). Edrioasteroids were the sole remaining Cambrian echinoderm group, and they would not exploit higher epifaunal tiers until the Carboniferous (Fig. 7.3). Although echinoderms with Cambrian origins were not present above the 0 to +5 cm tier, the +5 to +15/20 cm tier was dominated by the calceocrinid crinoids during the Silurian and most of the Devonian (Ausich, 1986). The radiation of fenestrate bryozoans during the Devonian may have displaced calceocrinids from dominance of this epifaunal tier (Ausich, 1986).

From the end-Ordovician mass extinction until the Carboniferous, edrioasteroids remained strictly in the 0 to +5 cm tier (Fig. 7.3). The Carboniferous expansion of clavate edrioasteroids into the +5 to +20 cm tier probably had its origin in the Late Devonian mass extinction. The earliest of the clavate edrioasteroids, *Cooperidiscus* (Sumrall, 1996), evolved as a sediment-attaching form during or shortly after the Late Devonian mass extinction (Dornbos, 2006). *Cooperdiscus* gave rise to the advanced clavate edrioasteroids that exploited the +5 to +20 cm tier during the Carboniferous, reaching higher above the seafloor than any edrioasteroid group ever had or would (Sumrall, 1996). If indeed calceocrinid crinoids were significantly supplanted in the +5 to +20 cm tier by fenestrate bryozoans during the Devonian (Ausich, 1986), then clavate edrioasteroids must have been able to overcome competition with fenestrate bryozoans to inhabit this tier. Perhaps clavate edrioasteroids were able to adapt to dense aggregations of fenestrate bryozoans by having a more upright life position than the calceocrinids, which lived with their stems prostrate along the seafloor (Ausich, 1986). In addition, clavate edrioasteroids may have been able to survive the current baffling of fenestrate bryozoans because, unlike calceocrinids (Ausich, 1986), they did not rely on unidirectional currents to move through the filtration apparatus. Their domal, disk-shaped feeding surfaces may have actually been well adapted for survival in slower, more multidirectional currents.

Having evolved during the Late Devonian mass extinction or in its immediate aftermath and giving rise to a lineage that radiated after the event, *Cooperidiscus* can be identified as a crisis progenitor (Kauffman and Harries, 1996). In addition, the shift in substrate adaptations from sediment attachment in *Cooperidiscus* to hard substrate attachment in all subsequent clavate edrioasteroids recapitulates on a smaller scale the epifaunal echinoderm response to the Cambrian substrate revolution (Bottjer et al., 2000; Dornbos, 2006).

Epifaunal suspension-feeding echinoderm classes with Cambrian origins reached the +5 to +10/15/20 cm tier during the Cambrian radiation, the Ordovician radiation, and after the Late Devonian mass extinction. During the Cambrian and Ordovician, among these echinoderm classes, only eocrinoids reached the +5 to +10/15 cm tier. Helicoplacoids were restricted to the 0 to +5 cm tier throughout their Early Cambrian existence. Edrioasteroids were restricted to the 0 to +5 cm tier during the Cambrian and Ordovician.

After the end-Ordovician mass extinction, only the edrioasteroids existed, and they still were limited to the 0 to +5 cm tier. They remained solely in that tier until after the Late Devonian mass extinction, when the clavate edrioasteroids reached the +5 to +10/15 cm tier. The clavate edrioasteroids originated from the Late Devonian crisis progenitor *Cooperidiscus*, thus owing their expansion into the +5 to +10/15 cm tier to the Late Devonian mass extinction itself.

It is possible to reconstruct the tier inhabitation of these echinoderm groups throughout their history because of their intricate, well-preserved skeletons. The resulting tier inhabitation patterns show how the major evolutionary events of the Paleozoic controlled the evolutionary paleoecology of these echinoderms. The Cambrian and Ordovician radiations predictably resulted in expansions in epifaunal tier occupation. Mass extinctions had varying effects: the end-Ordovician mass extinction curtailed epifaunal tiering, whereas the Late Devonian mass extinction ultimately spurred expansion into a higher tier.

Conclusions

Thanks to W. I. Ausich for both helpful discussions and the invitation to contribute to this book. I am grateful to J.-P. Lin and R. L. Parsley for providing helpful and comprehensive reviews of drafts. A grant from the UWM Graduate School Research Committee supported this research.

Acknowledgments

References

Ausich, W. I. 1986. Palaeoecology and history of the Calceocrinidae (Palaeozoic Crinoidea). Palaeontology, 29:85–99.

Ausich, W. I., and D. J. Bottjer. 1982. Tiering in suspension-feeding communities on soft substrata throughout the Phanerozoic. Science, 216:173–174.

Ausich, W. I., and D. J. Bottjer. 1985. Phanerozoic tiering in suspension-feeding communities on soft substrata: implications for diversity, p. 255–274. *In* J. W. Valentine (ed.), Phanerozoic Diversity Patterns. Princeton University Press, Princeton, New Jersey.

Ausich, W. I., and D. J. Bottjer. 2001. Sessile invertebrates, p. 388–391. *In* D. E. G. Briggs, and P. R. Crowther (eds.), Palaeobiology II. Blackwell Scientific Publications, Oxford.

Bell, B. M., and J. Sprinkle. 1978. *Totiglobus*, an unusual edrioasteroid from the Middle Cambrian of Nevada. Journal of Paleontology, 52:243–266.

Bottjer, D. J., and W. I. Ausich. 1986. Phanerozoic development of tiering in soft substrata suspension-feeding communities. Paleobiology, 12:400–420.

Bottjer, D. J., J. W. Hagadorn, and S. Q. Dornbos. 2000. The Cabrian substrate revolution. GSA Today, 10:1–7.

Clapham, M. E., and G. M. Narbonne. 2002. Ediacaran epifaunal tiering. Geology, 30:627–630.

Clausen, S. 2004. New Early Cambrian eocrinoids from the Iberian Chains (NE Spain) and their role in nonreefal benthic communities. Eclogae Geologicae Helvetiae, 97:371–379.

Domke, K. L., and S. Q. Dornbos. 2005. Paleoecology of the Middle Cambrian edrioasteroid echinoderm *Totiglobus:* implications for unusual Cambrian body plans. North American Paleontological Convention, Program and Abstracts, Paleobios, 25(2):37.

Dornbos, S. Q. 2006. Evolutionary palaeoecology of early epifaunal echinoderms: response to increasing bioturbation levels during the Cambrian radiation. Palaeogeography, Palaeoclimatology, Palaeoecology, 237(2–4):225–239.

Dornbos, S. Q., and D. J. Bottjer. 2000. Evolutionary paleoecology of the earliest echinoderms: helicoplacoids and the Cambrian substrate revolution. Geology, 28:839–842.

Dornbos, S. Q., and D. J. Bottjer. 2001. Taphonomy and environmental distribution of helicoplacoid echinoderms. Palaios, 16:197–204.

Dornbos, S. Q., D. J. Bottjer, and J. Y. Chen. 2005. Paleoecology of benthic metazoans in the Early Cambrian Maotianshan Shale biota and Middle Cambrian Burgess Shale biota: evidence for the Cambrian substrate revolution. Palaeogeography, Palaeoclimatology, Palaeoecology, 220:47–67.

Durham, J. W. 1993. Observations on the Early Cambrian helicoplacoid echinoderms. Journal of Paleontology, 67:590–604.

Dzik, J., and S. Orlowski. 1993. The Late Cambrian eocrinoid *Cambrocrinus:* Palaeontologica Polonica, 38:21–34.

Kauffman, E. G., and P. J. Harries. 1996. The importance of crisis progenitors in recovery from mass extinction, p. 15–39. *In* M. B. Hart (ed.), Biotic Recovery from Mass Extinction Events. Geological Society Special Publication, 102.

Lefebvre, B., and O. Fatka. 2003. Palaeogeographical and palaeoecological aspects of the Cambro–Ordovician radiation of echinoderms in Gondwanan Africa and peri-Gondwanan Europe. Palaeogeography, Palaeoclimatology, Palaeoecology, 195:73–97.

Parsley, R. L., and R. J. Prokop. 2004. Functional morphology and paleoecology of some sessile Middle Cambrian echinoderms from the Barrandian region of Bohemia. Bulletin of Geosciences, 79:147–156.

Paul, C. R. C. 1968. *Macrocystella* Callaway, the earliest glyptocystitid cystoid. Palaeontology, 11:580–600.

Pruss, S. B., M. L. Fraiser, and D. J. Bottjer. 2004. Proliferation of Early Triassic wrinkle structures: implications for environmental stress following the end-Permian mass extinction. Geology, 32:461–464.

Schubert, J. K., and D. J. Bottjer. 1992. Early Triassic stromatolites as post-mass-extinction disaster forms. Geology, 20:883–886.

Schubert, J. K., D. J. Bottjer, and M. J. Simms. 1992. Paleobiology of the oldest known articulate crinoid. Lethaia, 25:97–110.

Sepkoski, J. J. 2002. A compendium of fossil marine animal genera. Bulletins of American Paleontology, 363:1–590.

Sheehan, P. M., and M. T. Harris. 2004. Microbialite resurgence after the Late Ordovician extinction. Nature, 430:75–78.

Smith, A. B., and P. A. Jell. 1990. Cambrian edrioasteroids from Australia and the origin of starfishes. Memoirs of the Queensland Museum, 28:715–778.

Sprinkle, J. 1973. Morphology and evolution of blastozoan echinoderms. Harvard University Museum of Comparative Zoology Special Publication, Cambridge, 283 p.

Sprinkle, J., and B. C. Wilbur. 2005. Deconstructing helicoplacoids: reinterpreting the most enigmatic Cambrian echinoderms. Geological Journal, 40:281–293.

Sumrall, C. D. 1996. Late Paleozoic edrioasteroids (Echinodermata) from the North American mid-continent. Journal of Paleontology, 70:969–985.

Sumrall, C. D., J. Sprinkle, and T. E. Guensburg. 2001. Comparison of flattened blastozoan echinoderms: insights from the new Early Ordovician eocrinoid *Haimacystis rozhnovi*. Journal of Paleontology, 75:985–992.

Thayer, C. W. 1975. Morphologic adaptations of benthic invertebrates to soft substrata. Journal of Marine Research, 33:177–189.

Thayer, C. W. 1983. Sediment-mediated biological disturbance and the evolution of marine benthos, p. 479–625. *In* M. J. S. Tevesz, and P. L. McCall (eds.), Biotic Interactions in Recent and Fossil Communities. Plenum Press, New York.

Wilbur, B. C. 2004. The ties that bind: attachment structure homologies in Early Cambrian echinoderms. Geological Society of America Annual Meeting, Abstracts with Programs, 36(5):521.

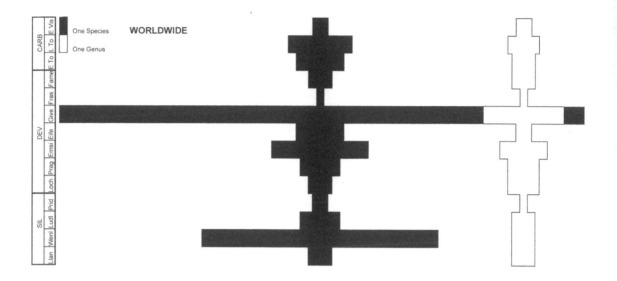

Figure 8.1. *Spindle diagram of Periechocrinidae species and genus richness; no range-through taxa are included.*

EVOLUTION AND EXTINCTION OF A PALEOZOIC CRINOID CLADE: PHYLOGENETICS, PALEOGEOGRAPHY, AND ENVIRONMENTAL DISTRIBUTION OF THE PERIECHOCRINIDS

8

William I. Ausich and Thomas W. Kammer

The periechocrinidae was a long-ranging (Early Silurian to Mississippian), cosmopolitan family that originated on Laurentia as part of the Early Silurian diversification of the Middle Paleozoic crinoid evolutionary fauna (Baumiller, 1994; Ausich et al., 1994). This family belongs to the monobathrid camerates within the suborder Compsocrinina (see Ubaghs, 1978a). Although present on Avalonia, Baltica, Gondwana, Kazakhstan, and Laurentia, on the basis of the known fossil record, the major diversifications of periechocrinids occurred on Laurentia and Gondwana during the Silurian and Devonian (Figs. 8.1, 8.2). Reported here are two new genera, which are surprising because they originated during the final phase (Mississippian) of the family's history and on the former Avalonian terrane of easternmost Euramerica (Laurentia plus Avalonia). Thus, the final originations of Periechocrinidae evolution was on a paleocontinental block and in facies where it had experienced only modest success. This curious evolutionary pattern prompted a thorough examination of the phylogenetic, paleoenvironmental, and paleogeographic evolution of this important middle Paleozoic family.

Introduction

Character analyses used here follow the stepwise approach to the development of a phylogeny (Ausich, 1998a, 1998b). In the present case, temporal and paleogeographic subsets of the Periechocrinidae are evaluated to develop a more realistic phylogeny than possible when evaluating the entire data set. This method partially eliminates the impact of asynchronous convergent, iterative, and paedomorphic evolution within the Periechocrinidae. This is necessary because a cladistic analysis of all genera in this Early Silurian to Mississippian family has several problematic aspects, as described below.

Parsimony-based character analyses were performed by PAUP 3.1.1. For analyses presented in this study, all characters were unordered and equally weighted. The search methods were heuristic, with random stepwise addition. Search results are presented as 50% majority-rule trees or as single trees. Rohlf's consistency index is given for majority-rule trees. Simple consistency indices (CI), retention indices (RI), and rescaled consis-

Methods

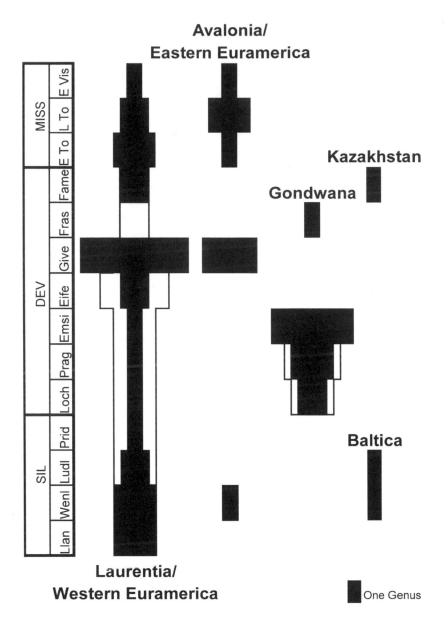

Figure 8.2. *Generic richness of periechocrinids on various paleogeographic terranes. Black indicates actual known occurrences of genera; white includes range-through genera.*

tency indices (RC) are given, as indicated in output from PAUP analyses. Details for each cladogram are presented below.

Data on the Periechocrinidae are taken directly from the literature (summary in Webster, 2003) with minimal reevaluation, except for the Carboniferous of England and *Periechocrinus prumiensis geometricus* (Schultze, 1866), which is too poorly preserved for generic assignment. On the basis of our previous experience, some of the species, especially in genera with a high species richness, should become junior synonyms when these genera are fully reevaluated systematically. Further, generic assignments may also require eventual revision, which are beyond the scope of the present study. However, regardless of these potential issues, the current

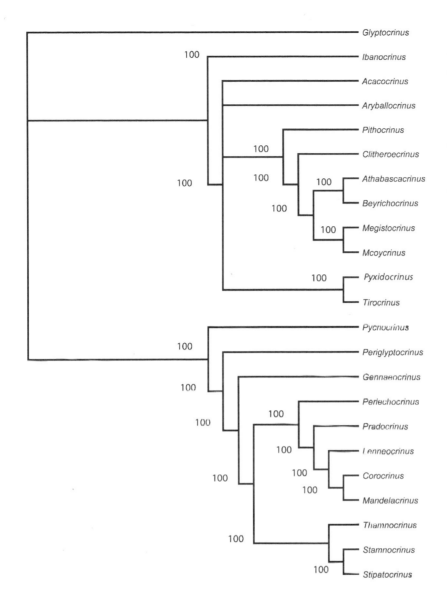

Figure 8.3. *Fifty percent majority-rule tree of all periechocrinids; outgroup* **Glyptocrinus**, **Periglyptocrinus**, *and* **Pycnocrinus**. *Twenty-two characters were parsimony informative, derived from two equally parsimonious trees of length 84. Rohlf's CI = 0.844, CI = 0.393, RI = 0.628, and RC = 0.247.*

data do give a relative indication of genus and species richness and of paleogeographic occurrence through time for the Periechocrinidae.

Characters for each genus are based on the type species of the genus. A list of characters, character states, and characters for each genus are given in Appendices 8.1 and 8.2.

Only genus-level diagnostic characters for the Periechocrinidae are used in these analyses. As argued in Ausich (1998a, 1998b), species-level characters are not used because of the likelihood that these will add noise to an analysis that is seeking to uncover the underlying structure of the phylogeny of genera. Examples of species-level characters for the Periechocrinidae are calyx plate sculpturing, the degree of definition of the median ray ridge, the exact fixed brachial (rather than the brachitaxis) where arms become free, and the number of posterior interradial plates in contact with the tegmen. Character selection for the analyses presented here include

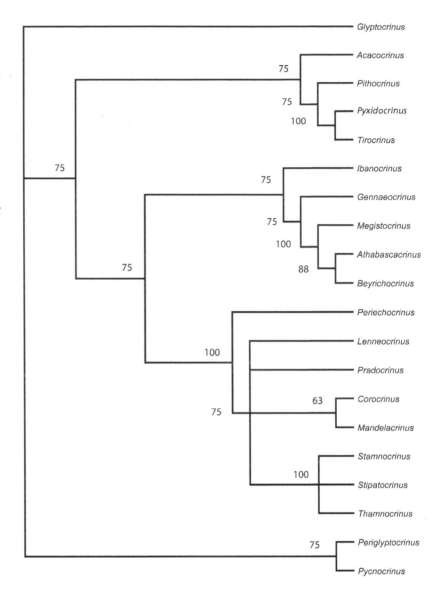

Figure 8.4. *Fifty percent majority-rule tree of all pre-Mississippian periechocrinids; outgroup* **Glyptocrinus, Periglyptocrinus, Pycnocrinus.** *Twenty-two characters were parsimony informative; derived from eight equally parsimonious trees of length 80. Rohlf's CI = 0.859, CI = 0.412, RI = 0.598, and RC = 0.247.*

those considered to represent basic architectural features of the arms, calyx, and column. Thus, 22 characters with 59 character states were used to analyze the 20 genera of the Periechocrinidae. Additional genus-level diagnostic characters (Appendix 8.1) were eliminated during preliminary analyses because these characters are known for too few taxa.

The choice of an outgroup was problematic. The Late Ordovician tanaocrinids have been taken to be the ancestral clade to the Periechocrinidae (Ubaghs, 1978b; Witzke and Strimple, 1981; Ausich, 1987). However, analyses that use *Compsocrinus* Miller, 1883, and *Canistrocrinus* Wachsmuth and Springer, 1885, resulted in extremely poorly resolved trees that yielded unrealistic results. The three basal periechocrinids (*Acacocrinus* Wachsmuth and Springer, 1897; *Ibanocrinus* Ausich, 1987; and *Tirocrinus* Ausich, 1987) are distinctly older than other members of the family, and these were also used as the outgroup for a series of analyses. The results

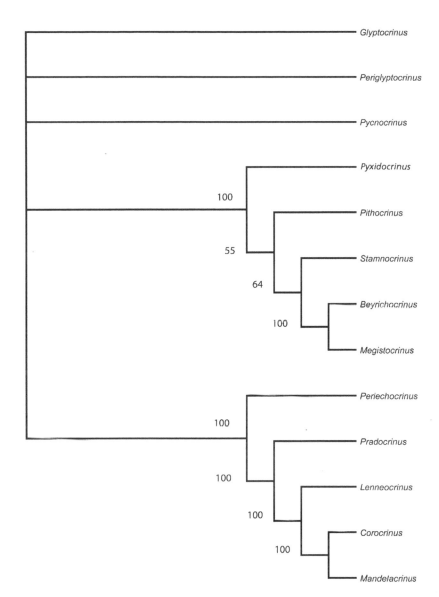

Figure 8.5 *Fifty percent majority-rule tree of all Gondwana periechocrinids; outgroup **Glyptocrinus, Periglyptocrinus** and **Pycnocrinus**. Nineteen characters were parsimony informative; derived from 22 equally parsimonious trees of length 53; Rohlf's CI = 0.513, CI = 0.585, RI = 0.651, and RC = 0.381.*

that used these basal periechocrinids were significantly better resolved, and although they were not without problems, they yielded much more realistic cladograms. A true outgroup is preferred, so another common Ordovician family, the Glyptocrinidae, was utilized. Analyses with the three glyptocrinid genera gave results that largely parallel those resulting from the outgroup composed of *Acacocrinus, Ibanocrinus,* and *Tirocrinus,* and the results are significantly more robust than those that use the Tanaocrinidae. Thus, for this study, we present PAUP results from analyses derived from the Glyptocrinidae as the outgroup. These results question the standard assumption for the origination of the Periechocrinidae, but that must be addressed in future studies.

Results from various character analyses on periechocrinid genera are presented in Figures 8.3 to 8.7. Figure 8.3 is a character analysis of all 20 periechocrinids, and the outgroup is the Ordovician Glyptocrinidae (*Glyp-*

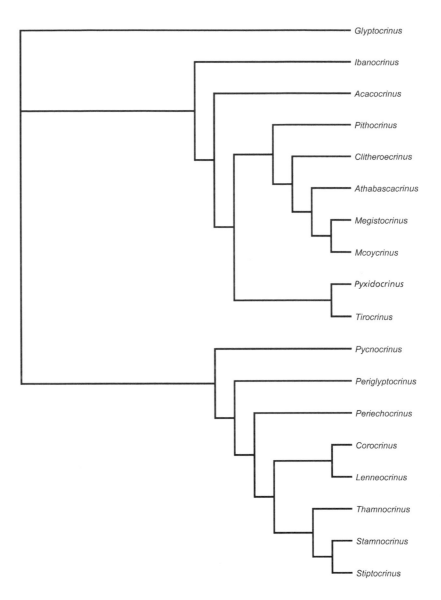

Figure 8.6 *Fifty percent majority-rule tree of all Laurentian-Avalonian-Euramerican pre-Carboniferous periechocrinids (except **Aryballocrinus** and **Gennaeocrinus**); outgroup **Glyptocrinus**, **Periglyptocrinus** and **Pycnocrinus**. Nineteen characters were parsimony informative; resulted in a single tree of length 68. Rohlf's CI = 0.458, RI = 0.615, and RC = 0.280.*

tocrinus Hall, 1847; *Periglyptocrinus* Wachsmuth and Springer, 1897; and *Pycnocrinus* Miller, 1883). Twenty-two characters were parsimony informative. The 50% majority-rule tree is derived from two equally parsimonious trees of length 84, and the tree has the following values: Rohlf's CI = 0.844, CI = 0.393, RI = 0.628, and RC = 0.247.

Character analysis of all pre-Carboniferous periechocrinids is given in Figure 8.4. Twenty genera were analyzed, and the outgroup is the Glyptocrinidae. Twenty-two characters were parsimony informative. The 50% majority-rule tree is derived from eight equally parsimonious trees of length 80, and the tree has the following values: Rohlf's CI = 0.859, CI = 0.412, RI = 0.598, and RC = 0.247.

Figure 8.5 is the cladogram for character analysis of all 10 Gondwanan genera with three glyptocrinids as the outgroup (total of 13 genera). Nineteen characters were parsimony informative. This analysis yielded 22 most parsi-

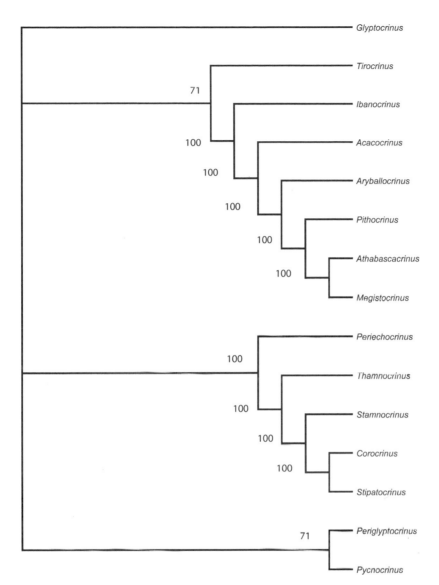

Figure 8.7. *Fifty percent majority-rule tree of all Laurentian and Western Euramerican periechocrinids with the exception of **Gennaeocrinus**; outgroup **Glyptocrinus**, **Periglyptocrinus** and **Pycnocrinus**. Nineteen characters were parsimony informative; derived from seven equally parsimonious trees of length 61. Rohlf's CI = 0.744, CI = 0.508, RI = 0.595, and RC = 0.302.*

monious trees with a length of 53, and the 50% majority-rule tree had the following values: Rohlf's CI – 0.513, CI = 0.585, RI = 0.651, and RC = 0.381.

All Euramerica, pre-Mississippian genera (15 total), with the exception of *Aryballocrinus* Breimer, 1962, and *Gennaeocrinus* Wachsmuth and Springer, 1881, are analyzed and presented in Figure 8.6 (as explained below, elimination of these two genera commonly greatly improves the resolution of the analyses). The Glyptocrinidae is the outgroup. Nineteen characters were parsimony informative. The analysis resulted in a single tree of length 68, and the tree has the following values: Rohlf's CI = 0.458, RI = 0.615, and RC = 0.280.

Figure 8.7 is from character analysis of all Laurentian–western Euramerican periechocrinids with the exception of *Gennaeocrinus*. The outgroup is the Glyptocrinidae, so 15 genera are included in the analysis. Nineteen characters were parsimony informative. The 50% majority-rule tree is derived

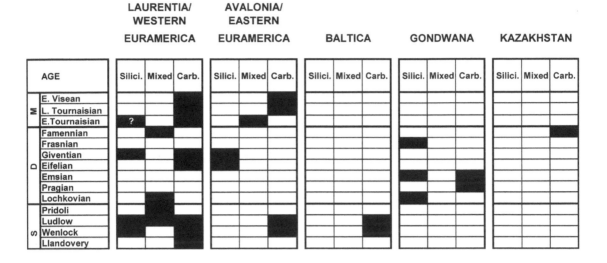

Figure 8.8. *Temporal pattern of major environmental preferences of Periechocrinidae on each paleogeographic terrane.*

from seven equally parsimonious trees of length 61, and the tree has the following values: Rohlf's CI = 0.744, CI = 0.508, RI = 0.595, and RC = 0.302.

Paleozoic paleogeographic models follow standard interpretations (<http://www.scotese.com/earth.htm>) with recognition of Avalonia, Baltica, Gondwana, Kazakhstan, and Laurentia continental areas; Euramerica comprises the combined Laurentia and Avalonia. Paleogeographic definition issues for this study are the division of Germany during the Devonian and the separation of Euramerican faunas from the former Laurentian and Avalonian terranes. Herein, we follow Franke (2000), Robardet (2002, 2003), and Kříž et al. (2003) by recognizing Germany as divided by the Devonian Rheic Ocean, separating western and northern Germany (including, for example, faunas from the Coblenzian), which were part of Avalonia, from other regions in western and southern Germany that were part of Gondwana.

Stratigraphic nomenclature follows Heckel and Clayton (2005) that outlines recent changes adopted by the Subcommission on Carboniferous Stratigraphy. Accordingly, the Carboniferous Period is divided into two subsystems, the Mississippian and Pennsylvanian.

Evolutionary History of the Periechocrinids

TEMPORAL, GEOGRAPHIC, AND PALEOENVIRONMENTAL OCCURRENCE. The oldest members of this family are from the Llandovery (Aeronian) reef-associated carbonate setting of the Brassfield Formation (Ohio) on Laurentia (Ausich, 1987; Schneider and Ausich, 2002), including *Acacocrinus, Ibanocrinus,* and *Tirocrinus* (Fig. 8.3). During the Wenlock, periechocrinids became dispersed to Avalonia and Baltica. On Laurentia, generic turnover resulted from extinction of two Llandovery genera and origination of two new Wenlock genera. The dominant genus was *Periechocrinus* Morris, 1843, which underwent significant species diversification during the Wenlock and Ludlow on Laurentia and principally during the Wenlock on Baltica and Avalonia. Periechocrinids became extinct on both Avalonia and Baltica before the end of the Ludlow.

Although the Llandovery periechocrinids were associated with reefs, those of the Wenlock, Ludlow, and Pridoli were adapted to a wide array of habitats (Fig. 8.8). Laurentian species of *Periechocrinus* occurred in reef-associated and other carbonates (e.g., Racine Dolomite of Illinois and Laurel Limestone of Indiana, respectively), siliciclastic facies (e.g., Waldron Shale of Indiana and Rochester Shale of New York), and in mixed siliciclastic-carbonate facies (e.g., Brownsport Formation of Tennessee). Avalonian species of *Periechocrinus* are only in the Much Wenlock Limestone Formation (England), and on Baltica, it occurred in several reef horizons on the Isle of Gotland, Sweden. *Stiptocrinus* Kirk, 1946, also occurred in a variety of carbonate facies during the Wenlock (Indiana) through the Pridoli (Tennessee).

Devonian periechocrinids are only known from Euramerica, Gondwana, and Kazakhstan. Euramerica will be discussed as western Euramerica (former Laurentia terrane) and eastern Euramerica (former Avalonia terrane). Periechocrinids migrated to Gondwana, with the first occurrences being *Corocrinus* Goldring, 1923, and *Mandelacrinus* Jell and Theron, 1999, in the siliciclastic Gydo Formation of South Africa (Jell and Theron, 1999). *Megistocrinus* Owen and Shumard, 1852, was also reported from the Early Devonian of New Zealand (Prokop, 1970). On Gondwana, periechocrinids again adapted to both carbonate and siliciclastic facies. The Gondwana genus richness peaked during the Emsian with six genera (Fig. 8.2).

On western Euramerica, Devonian genus richness was low and static until the Eifelian, peaking during the Givetian (Fig. 8.2). In contrast, species richness slowly increased through the Emsian, with periechocrinids poorly known during the Eifelian. During the Givetian, they attained a maximum in both genus and species richness. In addition, the Givetian represents the maximum species richness for the family as a whole. As during the Silurian, Devonian periechocrinids from western Euramerica adapted to a variety of facies. *Gennaeocrinus* and *Megistocrinus* have the highest species richness, and species of both genera occur in both carbonate and siliciclastic facies (e.g., carbonate: Columbus Limestone of Ohio, Alpena Limestone of Michigan, Cedar Valley Limestone of Iowa; siliciclastic: Moscow Shale of New York, Arkona Shale of Canada, Silica Shale of Ohio). Although species of *Megistocrinus* occurred in both carbonate and siliciclastic settings, this genus was predominantly in carbonate paleoenvironments.

Four genera—*Corocrinus*; *Gennaeocrinus*; *Lenneocrinus* Jaekel, 1918; and *Pyxidocrinus* Müller, 1855—first appeared during the Devonian of eastern Euramerica. These are all from localities in western Germany, were confined to the Givetian, and occur only in siliciclastic facies. Periechocrinids are unknown from the Frasnian of Laurentia, but during the Famennian, two genera, *Gennaeocrinus* and *Megistocrinus*, occurred in mixed siliciclastic and carbonate facies of the Sappington Formation of Montana (Gutschick and Rodriguez, 1979).

The only Devonian occurrence outside of Euramerica and Gondwana was *Athabascacrinus* Laudon et al., 1952, from Kazakhstan (Waters et al., 2003), which was endemic to Kazakhstan during the Famennian (Fig. 8.2). As presently understood, for the Devonian, *Corocrinus*; *Lenneocrinus*;

Megistocrinus; Pithocrinus Kirk, 1945; *Pyxidocrinus;* and *Stamnocrinus* Breimer, 1962, are recognized from both Euramerica and Gondwana. *Beyrichocrinus* Waagen and Jahn, 1899; *Mandelacrinus;* and *Pradocrinus* de Verneuil, 1850, were endemic to Gondwana. *Acacocrinus; Thamnocrinus* Goldring, 1923; and *Stiptocrinus* were endemic to Euramerica. During the Devonian, western Euramerica had nine genera and eastern Euramerica had four genera. Only two of these, *Corocrinus* and *Gennaeocrinus,* were common between these two regions.

The Frasnian–Famennian extinction crisis greatly affected crinoid richness (Baumiller, 1994), but it did not change the overall clade-dominance patterns among major crinoid groups (Kammer and Ausich, 2006). Periechocrinids are poorly known through this extinction crisis, with one Frasnian and three Famennian genera known. The major crisis for periechocrinid genus and species richness occurred between the Givetian and the Frasnian, similar to other crinoids, blastoids, and other groups. Peak diversity for the entire clade occurred during the Givetian, after which it dropped precipitously. The Frasnian had the lowest levels of genus and species richness for the periechocrinids, with richness levels rising from the Famennian through the late Tournaisian.

During the Mississippian, three genera occurred in western Euramerica (Kentucky, Iowa, and Missouri): *Athabascacrinus* (early Tournaisian); *Megistocrinus* (early to late Tournaisian); and *Aryballocrinus,* Breimer, 1962 (early Tournaisian to early Viséan). Three genera also occurred in eastern Euramerica on the former western Avalonian terrane (Ireland and England), including *Aryballocrinus* (early to late Tournaisian); *Mcoycrinus* new genus (late Tournaisian); and *Clitheroecrinus* new genus (late Tournaisian to early Viséan) (Fig. 8.2). On Euramerica, periechocrinids principally adapted to carbonate ramp settings, such as the Burlington Limestone, and the two new genera, *Mcoycrinus* and *Clitheroecrinus,* occur associated with Waulsortian mudmound facies, which were deeper-water carbonate facies (Lees and Miller, 1985). The only exception is *Aryballocrinus,* which occurred on the mixed carbonate-siliciclastic ramp of the Hook Head Formation in Ireland (Ausich and Sevastopulo, 2001) and also in the Nada Member of the Borden Formation, which is a siliciclastic facies (Lee et al., 2005). The mudstones of the Nada Member anomalously contain many elements of the Burlington Limestone fauna. However, as suggested by Lee et al. (2005), the Nada fauna does not represent typical delta platform siliciclastic deposition because this occurrence represents conditions associated with a high abundance of glauconite and phosphate.

PHYLOGENY. Character analysis of all genera yielded two most parsimonious trees with two polytomies and agreement in all other branches (Fig. 8.3). This analysis has three Laurentian Ordovician glyptocrinids (*Glyptocrinus, Periglyptocrinus,* and *Pycnocrinus*) as the outgroup, and are all basal on the cladogram. The three Llandovery taxa (*Acacocrinus, Ibanocrinus,* and *Tirocrinus*) are also from Laurentia and are all low in the topology of half of the cladogram. Several aspects of this tree topology are noteworthy. First, *Gennaeocrinus* (Devonian) and *Aryballocrinus* (Mississippian) are seated deeply in the topology, which is phylogenetically problematic.

This is typical of results generated in this study, and elimination of these two taxa commonly significantly reduces the number of most parsimonious trees resulting from analysis. For example, here, if *Aryballocrinus* is eliminated from the analysis, a single tree (not shown) is generated, which otherwise is identical to Figure 8.3. As noted on some analyses below, *Aryballocrinus* and/or *Gennaeocrinus* are eliminated. Compelling aspects of this tree are that all Mississippian periechocrinids are closely linked. All except *Aryballocrinus* are within a single clade, although these are also grouped with the Pragian genus *Beyrichocrinus* Waagen and Hahn, which is endemic to Gondwana.

Aspects of Figure 8.3 that are problematic include 1) the position of *Beyrichocrinus* (listed above); 2) the close and deeply rooted linkage of *Tirocrinus* (Llandovery) to *Pyxidocrinus* (Emsian to Eifelian); 3) *Corocrinus* and *Mandelacrinus*, which first appeared during the Lockhovian of Gondwana, linked high within the topology; 4) *Stiptocrinus* (Wenlock through Pragian) positioned high in the topology; 5) neither *Stiptocrinus* nor *Periechocrinus* (with first occurrences during the Wenlock of Laurentia) linked with any of the Llandovery forms also from Laurentia; and 6) neither major group with the association of genera endemic to a particular paleogeographic terrane. Because of these issues, a single comprehensive cladistic analysis is unsatisfactory, and additional analyses are performed to examine subsets of the complete data.

An analysis was run (not shown) with only periechocrinid genera and with the three Llandovery forms designated as the outgroup. Similar to the topology of Figure 8.3, this cladogram linked Mississippian genera to *Beyrichocrinus*. However, other genera were not divided into two major clades, and relationships among post-Llandovery and pre-Mississippian genera are different.

Character analysis of all pre-Mississippian genera yielded eight most parsimonious trees (Fig. 8.4). The three glyptocrinid outgroup genera are on the basal polytomy, from which also branches the clade of all pre-Mississippian periechocrinids. Three of the four oldest periechocrinids (*Acacocrinus*, *Ibanocrinus*, and *Periechocrinus*) are deeply rooted in the cladogram, whereas the Llandovery *Tirocrinus* is not. In both Figures 8.3 and 8.4, *Tirocrinus* and the Devonian *Pyxidocrinus* are closely linked, which undoubtedly represents a case of convergence by *Pyxidocrinus* during the Devonian. In this analysis, the pre-Mississippian periechocrinids are subdivided into three groupings. The first is a group of Silurian and Devonian forms with *Periechocrinus*, the most deeply rooted genus. Two of the three genera (*Pradocrinus* and *Mandelacrinus*) are endemic to Gondwana with other Silurian and Devonian forms from Avalonia, Baltica, Gondwana, and Laurentia. Second, a group of genera, discussed below, led to Mississippian periechocrinids. This group has *Ibanocrinus* as the most deeply rooted genus and also includes the Devonian taxa *Gennaeocrinus*, *Megistocrinus*, *Beyrichocrinus*, and *Athabascacrinus*. In aggregate, these genera occur on all Devonian terranes. Third is the small group with anomalous linking of *Tirocrinus* and *Pyxidocrinus*.

Figure 8.5 is the analysis of all Gondwana taxa with, again, the three glyptocrinid genera as the outgroup. *Periechocrinus* does not occur on Gondwana, but it is included in the analysis because it was paleogeographically the most widespread genus during the Silurian. It is rooted deeply in the topology of this cladogram. This analysis yielded 22 most parsimonious trees, but two major groups are clearly delineated. The grouping of *Corocrinus, Mandelacrinus, Lenneocrinus,* and *Pradocrinus* is consistent with previous analyses (Figs. 8.3, 8.4). Similarly, the association of *Beyrichocrinus* and *Megistocrinus* is consistent with other analyses, but this second cluster also includes *Stamnocrinus, Pithocrinus,* and *Pyxidocrinus,* which are not closely associated with *Beyrichocrinus* and *Megistocrinus* in previous analyses.

Analysis of all Euramerican pre-Mississippian genera with the three glyptocrinids as the outgroup yielded 23 most parsimonious trees (not shown); however, both *Aryballocrinus* and *Gennaeocrinus* are deeply rooted in these cladograms. If these two genera are eliminated from the analysis, then a single tree outcome results (Fig. 8.6). The outgroup and two of the Llandovery genera are most deeply rooted along with *Periechocrinus,* the cosmopolitan Wenlock genus. Two major groupings exist. Mississippian crinoids are clustered together along with *Pithocrinus, Pyxidocrinus,* and *Tirocrinus.* The second cluster includes only Silurian and Devonian forms with *Periechocrinus,* the most deeply rooted genus, except for outgroup genera.

If only Laurentia and western Euramerica genera are considered, the resulting cladogram (not shown) had 112 most parsimonious trees, with the basal polytomy having the outgroup genera (*Glyptocrinus, Periglyptocrinus,* and *Pycnocrinus*), two from the Llandovery (*Ibanocrinus* and *Tirocrinus*) and *Gennaeocrinus* from the Pragian to Famennian. *Gennaeocrinus* is clearly rooted too deeply, and if this genus is removed from the analysis, seven most parsimonious trees result (Fig. 8.7). With minor differences, Figures 8.6 and 8.7 delineate the same topology.

The various iterations of Periechocrinidae character analysis yield a diverse array of results, and two observations are noted. First, *Aryballocrinus* (Mississippian) and *Gennaeocrinus* (Pragian to Famennian) typically are independently rooted much more deeply in the cladogram topology than expected on the basis of other data. These two genera evolved morphologies with more primitive character states. For example, *Aryballocrinus* has a fairly consistent array of primitive character states, especially the high, bowl-shaped calyx; absent basal concavity; thin calyx plates; radial plates higher than wide; first primibrachial higher than wide; primaxil pentagonal; and CD interray proximal plating 1–2.

Second, in contrast, certain pairs of genera are always closely linked and positioned high within the cladogram topology, even though these pairs have a low probability of being closely related phylogenetically. These pairs are *Beyrichocrinus* (Pragian) with *Athabascacrinus* (Famennnian to early Tournaisian), and *Tirocrinus* (Llandovery) with *Pyxidocrinus* (Emsian to Givetian). These must be cases of convergent evolution.

EVOLUTION OF THE PERIECHOCRINIDAE. As currently understood, the Periechocrinidae evolved on Laurentia as part of the Early Silurian radiation of monobathrid camerates associated with the diversification of the Middle Paleozoic crinoid evolutionary fauna. The oldest crinoids are *Acacocrinus*, *Ibanocrinus*, and *Tirocrinus*. These crinoids were adapted to shallow, reef-associated carbonate facies. In the various character analyses used here, *Ibanocrinus* is consistently the most primitive taxon for the Periechocrinidae (except in Fig. 8.7). In contrast, solely on the basis of Llandovery genera, Ausich (1987) concluded that *Tirocrinus* was the most primitive, and *Acacocrinus* and *Ibanocrinus* were more derived. These conflicting views may result because *Tirocrinus* is closely, and presumably inappropriately, linked with *Pyxidocrinus* (Fig. 8.3), as discussed above, which may be distorting the relationships among Llandovery forms. When *Pyrixocrinus* is not included, *Tirocrinus* plots in the most primitive position (Fig. 8.7).

The Wenlock was a time of periechocrinid dispersal. *Periechocrinus* and *Stiptocrinus* evolved on Laurentia during the Wenlock adapting to a wide diversity of facies. *Periechocrinus* also migrated to Avalonia and Baltica, where species were specialized for specific facies on each terrane. Perhaps because this genus was more stenotopic on both Avalonia and Baltica, *Periechocrinus*, and consequently the Periechocrinidae, became extinct on these terranes by the close of the Ludlow. Periechocrinids did repopulate Avalonia during the Devonian, but these crinoids never reappeared on Baltica. *Stiptocrinus* was probably derived from *Periechocrinus* or a similar form.

The history of Devonian periechocrinids is complex. This was the peak of their evolutionary success, as gauged by genus and species richness and adaptation to many facies. Devonian periechocrinids had wide geographic distribution occurring on Laurentia-Euramerica, Gondwana and Kazakhstan, and demonstrated considerable faunal interchange between Laurentia-Euramerica and Gondwana.

Taking systematic and occurrence data at face value, the oldest Devonian periechocrinids are Lockhovian. On western Euramerica, *Stiptocrinus* ranged into the Devonian from Ludlow and Pridoli forms. *Megistocrinus*, *Corocrinus*, and *Mandelacrinus* first appeared on Gondwana during the Lockhovian, presumably by origination through migration. *Corocrinus* and *Mandelacrinus* are closely linked in every analysis completed (Figs. 8.3–8.5); however, *Megistocrinus* is consistently in a clade separate from *Corocrinus* and *Mandelacrinus*. *Gennaeocrinus*, which first occurred during the Pragian of western Euramerica, is problematic as noted above, and because it is variously associated with the two groupings defined by *Megistocrinus* and *Corocrinus-Mandelacrinus*. However, because it is linked to *Megistocrinus* in Figure 8.4, it may have been derived from *Megistocrinus* during a migration event to Euramerica. *Megistocrinus* appears to be derived from Llandovery forms, whereas *Corocrinus-Mandelacrinus* are probably derived from *Periechocrinus* or *Stiptocrinus*.

From these beginnings, two primary radiation events occurred during the Devonian: an Emsian event on Gondwana, and a Givetian expansion on both eastern and western Euramerica. On Gondwana, it is probable that the Emsian radiation was entirely endemic. *Beyrichocrinus*, *Pithocrinus*, and *Pyxidocrinus* evolved within the *Megistocrinus* lineage, whereas the *Corocrinus-Mandelacrinus* lineage gave rise to *Pradocrinus* and *Stamnocrinus*. This Emsian radiation was short lived, and these genera do not occur later on Gondwana. Emsian periechocrinids occurred in both siliciclastic and carbonate settings, but this paleoenvironmental diversity was insufficient for them to radiate further on Gondwana.

Stamnocrinus and *Megistocrinus* presumably migrated to Euramerica from Gondwana during the Eifelian, but peak richness of the Periechocrinidae on Euramerica was during the Givetian. This radiation occurred on both eastern and western Euramerica (Figs. 8.1, 8.2). Givetian periechocrinids on western Euramerica are a combination of reoccurrences after Lazarus gaps, migrations from Gondwana, and endemic radiation. The former include *Acacocrinus*, *Megistocrinus*, and *Periechocrinus*. *Corocrinus* and *Pithocrinus* presumably migrated from Gondwana to western Euramerica, and *Thamnocrinus* evolved through migration from within the *Corocrinus-Mandelacrinus* lineage, presumably from *Stamnocrinus* or *Periechocrinus*.

Periechocrinids reappeared on eastern Euramerica (former Avalonia) during the Givetian, presumably with *Corocrinus* and *Pyxidocrinus* migrating from Gondwana, *Gennaeocrinus* migrating from western Euramerica, and *Lenneocrinus* origination through migration from the *Corocrinus-Mandelacrinus* lineage. The ancestor of *Lenneocrinus* is unclear, but it is always closely associated with the *Corocrinus* and *Mandelacrinus* (Figs. 8.3–8.6).

Late Devonian occurrences on Gondwana and Kazakhstan are anomalies. Presumably, *Lenneocrinus* migrated to Gondwana from eastern Euramerica, and *Athabascacrinus* originated from the *Megistocrinus* lineage through migration. *Athabascacrinus* is commonly closely associated with *Megistocrinus* on cladograms (Figs. 8.3, 8.4, 8.6) and may very well be derived from this cosmopolitan form.

This complex and not completely resolved Devonian history is the result of apparently common migration among terranes. As periechocrinids were expanding into a wide diversity of facies, genera were migrating and were also originating, both endemically and through migration events.

In contrast, the final diversification of the periechocrinids is more straightforward. Mississippian periechocrinids evolved into carbonate facies, and the most probable ancestor of these crinoids was *Megistocrinus*. *Athabascacrinus* migrated from Kazakhstan to appear during the Mississippian of Euramerica. The exception is *Aryballocrinus*, which, as discussed above, is commonly rooted anomalously on cladograms (Fig. 8.3), but which also evolved into mixed siliciclastic-carbonate paleoenvironments or into unusual siliciclastic facies. However, phylogenetically, it must be rooted within the *Megistocrinus* lineage.

On Euramerica, periechocrinids evolved presumably onto carbonate ramps at middle-shelf facies. This includes *Megistocrinus* and *Aryballocrinus* in the Burlington Limestone of the Mississippi River Valley and *Athabascacrinus* in the Banff Formation of the Canadian Rocky Mountains. *Mcoycrinus* and *Clitheroecrinus* adapted to Waulsortian mudmounds on the toe of the slope in Lancashire, England (Lees and Miller, 1985). During the Mississippian, migration routes existed between eastern and western Euramerica, but a considerable degree of endemism also existed (Lane and Sevastopulo, 1987). The periechocrinids displayed a similar trend. Eastern and western Euramerica each have three periechocrinids, with one genus shared and two endemic. This latest periechocrinid radiation was unsuccessful and was also the terminus of the family. Similar to the Silurian on Baltica and Avalonia, the Mississippian occurrences were confined to carbonates, so they had relatively little paleoenvironmental diversity.

DISCUSSION. The Periechocrinidae had a long and successful history but did not survive the Mississippian. The Devonian–Mississippian was a dynamic time for crinoids, including the affects of the rise of durophagous predators (Signor and Brett, 1984) and the adaptive radiation that led to the Age of Crinoids (Kammer and Ausich, 2006). Several factors may have been significant for periechocrinid evolution.

First, the adaptive radiation of durophagous predators beginning by at least the Early Devonian is believed to have affected many benthic invertebrates (Signor and Brett, 1984). Crinoids were not an exception. Meyer and Ausich (1983) and Signor and Brett (1984) hypothesized that increased levels of predation led to crinoids that were more spinose, had thicker calyx plates, were reduced in calyx size, and had more compact, boxlike calyx design. This increased predation was further documented by Baumiller and Gahn (2004). Waters and Maples (1991) hypothesized that the changes in dominance of pinnulate crinoids, from camerates to advanced cladids, occurred due to this increase in predation. In general, the primitive periechocrinid condition and the *Corocrinus-Mandelacrinus* lineage had morphologies that would presumably have been more susceptible to predation. Alternatively, the *Megistocrinus* lineage in general had shorter, more compact, and thick-plated calyxes.

In contrast, the Mississippian *Aryballocrinus* is exceptional in having a higher, more thinly plated calyx. Predation pressure may therefore have been an important factor in the morphological evolution of periechocrinids. However, no periechocrinids developed spines. The genera that survived to the Mississippian, with the exception of *Aryballocrinus*, possessed a morphology that should have afforded them some predation resistance.

Second, Mississippian crinoids adapted to a wide array of paleoenvironments (Lane, 1971; Ausich et al., 1979; Kammer and Ausich, 1987), with individual clades displaying preferences for different environmental conditions. The majority of Devonian and Mississippian camerate crinoids displayed a distinct preference for carbonate paleoenvironments, although in many parts of Euramerica Mississippian, these became increasingly more siliciclastic. This was hypothesized to have caused a shift in faunal composition (Lane, 1971; Ausich et al., 1994). With these paleoenvironmental preferences and changing paleoenvironmental conditions, Kammer et al. (1997, 1998) demonstrated that more eurytopic crinoids had longer durations.

Figure 8.8 is a general depiction of the paleoenvironmental occurrences of the periechocrinids through time and space. Periechocrinid success appears to be correlated with paleoenvironmental disparity—that is, the temporal duration of these genera was shorter on paleogeographic terranes where periechocrinids were adapted to only a restricted range of paleoenvironments. For example, periechocrinids dispersed from Laurentia to Avalonia and Baltica during the Silurian. On Laurentia, periechocrinids were part of communities in carbonate, siliciclastic, and mixed carbonate-siliciclastic paleoenvironments, and the clade persisted. In contrast, periechocrinids occurred only in carbonate facies on Avalonia and Baltica, and the clade did not persist after the Ludlow. Similarly, Mississippian periechocrinids existed only in carbonate paleoenvironments and did not persist beyond the early Viséan. The exception to this trend was the Devonian of Gondwana, where, despite periechocrinids living in both siliciclastic and carbonate paleoenvironments, they did not persist past the Emsian, although they reappeared as a result of the migration of *Lenneocrinus* to Gondwana during the Frasnian.

Periechocrinids reached a pinnacle of success during the Silurian and Devonian, although they persisted as a relatively minor clade into the Mississippian. Regardless, post-Devonian trends in Periechocrinidae evolutionary history, although minor in genus and species richness, paralleled those of the more dominant Mississippian clades. For example, morphological trends of calyx design for predation resistance is similar to other crinoid clades from the Silurian to the Devonian (Meyer and Ausich, 1983; Signor and Brett, 1984), and both new Mississippian genera from England display these morphologies

The major radiation of camerate crinoids during the Mississippian radiations of the Age of Crinoids occurred among the Actinocrinitidae, Batocrinidae, Coelocrinidae, Dichocrinidae, and Platycrinitidae in western Laurentia and the Actinocrinitidae, Amphoracrinidae, Dichocrinidae, and Platycrinitidae in eastern Laurentia. This radiation was largely the expansion of camerates into the newly formed, expansive carbonate ramp settings that formed after the Frasnian–Famennian extinction of reefs (Kammer and Ausich, 2006). The final Mississippian radiation of periechocrinids mirrored this major expansion of camerates, both in its timing and origination within largely carbonate facies. Finally, the early Viséan extinction of the periechocrinids correlates to the major extinctions among camerates and other important Mississippian clades.

Systematic Paleontology

Terminology follows Ubaghs (1978c), and supergeneric taxonomy follows Moore and Teichert (1978) with modifications by Ausich (1998b). Specimens are housed in the National History Museum, London (BMNH), the National Museum of Scotland (NMS), and the British Geological Survey (BGS). Measurements are in millimeters; * indicates incomplete measurement.

Class CRINOIDEA Miller, 1821
Subclass CAMERATA Wachsmuth and Springer, 1885
Order MONOBATHRIDA Moore and Laudon, 1943

Suborder COMPSOCRININA Ubaghs, 1978a
Superfamily PERIECHOCRINACEA Bronn, 1849
Family PERIECHOCRINIDAE Bronn, 1849

EMENDED DIAGNOSIS. Typically elongate, many-plated calyxes (may be reduced in both features); first primibrachial tetragonal to hexagonal in early forms and hexagonal in later forms; two to five fixed secundibrachials; primanal hexagonal or heptagonal with two or three plates above; tegmen many plated, ambulacrals and orals rarely distinct; 10–40 arms that are uniserial or biserial and simple or branching (modified from Ausich, 1987, p. 554).

Genus MCOYCRINUS new genus

TYPE SPECIES. *Actinocrinus globosus* Phillips, 1836, by monotypy.

DIAGNOSIS. Calyx shape low bowl, basal concavity present, no calyx lobation, thick calyx plates, median ray ridges absent, basal plate circlet low, radial plate wider than high, orientation of radial plate nearly horizontal, radial plates equal in size, radial circlet only open in CD interray, fixed first primibrachial nearly as high as wide, primaxil pentagonal, second primibrachial axillary, fixed brachials branch isotomously, fixed rays symmetrical, highest fixed brachial in the tertibrachitaxis, numerous interradial plates, regular interray plating not in a biseries, regular interrays not depressed, CD interray proximal plating P-3, anitaxial ridge absent, CD interray wider than regular interrays, tegmen high, tegmen robust, tegmen not depressed interradially, ambulacrals and orals are not differentiated on the tegmen, anal tube present, anus positioned centrally, character of the free arms unknown, and gonoporoids absent.

DESCRIPTION. See species description below.

ETYMOLOGY. This name recognizes Frederick M'Coy, a pioneer in early crinoid studies in Ireland and England.

OCCURRENCE. *Mcoycrinus* is regarded as being only from the lower Chadian (upper Tournaisian, Tn3c) of England.

DISCUSSION. Both *Mcoycrinus* n. gen. and *Clitheroecrinus* n. gen. have the following characters that ally them with the Periechocrinidae: hexagonal first primibrachial, two to five fixed secundibrachials, interradials connected to tegmen, posterior interray wide, primanal followed by two or three plates, many-plated tegmen, and 10 to 40 biserial or uniserial arms. Previously, only *Aryballocrinus*, *Athabascacrinus*, and *Megistocrinus* were recognized from the Lower Mississippian. Diagnostic characters for these Lower Mississippian (Lower Carboniferous) genera are distinguished as follows. *Aryballocrinus* has a medium bowl-shaped calyx, basal concavity absent, calyx lobation absent, thin calyx plates, median ray ridge either present or absent, vertical radial plate orientation, first primibrachial higher than wide, primaxil pentagonal in shape, second primibrachial the primaxil, highest fixed brachial within the secundibrachitaxis, few regular interray plates, proximal CD interray plating P-3, anitaxial ridge present or absent, posterior interray much wider than normal interrays, tegmen height low, incompetent tegmen strength, unknown whether tegmen is depressed interradially, unknown whether tegmen has ambulacrals and orals differentiated, anal tube absent, and eccentric positioned anal opening.

Athabascacrinus has a low cone-shaped calyx, basal concavity present, slight calyx lobation, thick calyx plates, median ray ridge absent, horizontal radial plate orientation, first primibrachial wider that high, primaxil pentagonal in shape, first or second primibrachial the primaxil, highest fixed brachial within secundibrachitaxis or tertibrachitaxis, numerous regular interray plates, proximal CD interray plating P-3, anitaxial ridge absent, posterior interray wider than normal interrays, tegmen height low, robust tegmen strength, tegmen depressed interradially, tegmen with ambulacrals and orals differentiated, anal tube absent, and eccentric positioned anal opening. *Megistocrinus* has a low cone- to medium bowl-shaped calyx, basal concavity present, calyx lobation absent, thick calyx plates, median ray ridge absent, horizontal radial plate orientation, first primibrachial approximately as high as wide, primaxil pentagonal to heptagonal in shape, second primibrachial the primaxil, highest fixed brachial within secundibrachitaxis or tertibrachitaxis, numerous regular interray plates, proximal CD interray plating P-3, anitaxial ridge absent, posterior interray much wider than normal interrays, tegmen height low, robust tegmen strength, tegmen depressed interradially, tegmen with ambulacrals and orals differentiated, anal tube absent, and eccentric positioned anal opening.

Mcoycrinus has a low bowl-shaped calyx, basal concavity present, calyx lobation absent, thick calyx plates, median ray ridge absent, horizontal radial plate orientation, first primibrachial approximately as high as wide, primaxil pentagonal in shape, second primibrachial the primaxil, highest fixed brachial within the tertibrachitaxis, numerous regular interray plates, proximal CD interray plating P-3, anitaxial ridge absent, posterior interray wider than normal interrays, tegmen height high, robust tegmen strength, tegmen not depressed interradially, tegmen without ambulacrals and orals differentiated, anal tube present, and centrally positioned anal opening.

Clitheroecrinus has a low bowl-shaped calyx, basal concavity present, calyx lobation absent, thick calyx plates, median ray ridge absent, horizontal radial plate orientation, first primibrachial wider than high, primaxil pentagonal in shape, second primibrachial the primaxil, highest fixed brachial within the secundibrachitaxis or tertibrachitaxis, few regular interray plates, proximal CD interray plating P-2 anitaxial ridge absent, posterior interray wider than normal interrays, tegmen height unknown, tegmen strength unknown, unknown whether tegmen is depressed interradially, unknown whether tegmen has ambulacrals and orals differentiated, occurrence of anal tube unknown, and position of anal opening unknown. All periechocrinids are compared in Appendices 8.1 and 8.2.

MCOYCRINUS GLOBOSUS (Phillips, 1836)
Figures 8.9.1, 8.9.2, 8.9.5–8.9.9, 8.10.1
Actinocrinus globosus PHILLIPS, 1836, p. 206, pl. 4, figs. 26, 29; M'COY, 1844, p. 128; BRONN, 1848, p. 14; D'ORBIGNY, 1850, p. 155.
Periechocrinites globosus AUSTIN AND AUSTIN, 1842 (*non* Phillips, 1836), p. 110 (nomen nudum).
Periechocrinites globosus AUSTIN AND AUSTIN, 1843 (*non* Phillips, 1836), p. 204 (nomen nudum).

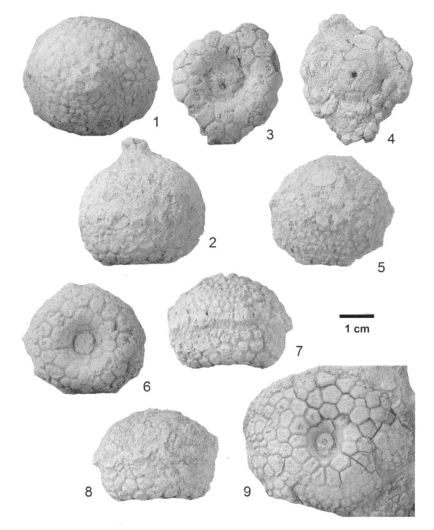

Figure 8.9. *New English periechocrinids; all to same scale, with scale bar = 1 cm. **1, 2, 5–9, Mcoycrinus globosus** (Phillips); **1, 2**, BMNH E6985, **1**, tegmen view; **2**, A-ray lateral view (note distinct anal tube from tegmen); **5–8**, BMNH E6986, lectotype; **5**, tegmen view of calyx; **6**, basal view of calyx with basal concavity; **7**, CD interray lateral view of calyx; **8**, A-ray lateral view of calyx; **9**, GSM 73326, basal view of calyx, posterior at bottom of photo. **3, 4, Clitheroecrinus wrighti** n. gen. et sp., NMS G.1958.1.1678, holotype; **3**, basal view of partial calyx (note basal concavity, posterior, at bottom of photo); **4**, interior view of the basal plating of the calyx.*

1 cm

Actinocrinus globulus Phillips, 1836 [*sic*]. DUJARDIN AND HUPÉ, 1862, p. 139.

Megistocrinus globosus (Phillips, 1836). WACHSMUTH AND SPRINGER, 1885, p. 112 (386); BASSLER AND MOODEY, 1943, p. 551; WRIGHT, 1955a (in part), p. 191–192, pl. 48, figs. 4, 8 (*non* pl. 48, figs. 1–3); WEBSTER, 1973, p. 169 (in part); DONOVAN, 1992, p. 45, fig. 7, no. L; WEBSTER, 2003 (in part); DONOVAN ET AL., 2003, p. 1, pl. 1, fig. 3.

New Genus A AUSICH AND KAMMER, 2006, p. 101.

DIAGNOSIS. See generic diagnosis above.

DESCRIPTION. Calyx and tegmen together a globose shape (Fig. 8.9.2). Calyx large, low bowl shape with basal concavity (Fig. 8.9.6, 8.9.9), calyx plates smooth with impressed sutures. Basal plates three, small, confined to basal concavity, and completely covered by proximal columnal. Radial circlet interrupted by primanal (Fig. 8.10.1). Radial plates five, as high as wide or wider than high, confined within basal concavity. Normal interrays widest at mid height (Fig. 8.9.8), plating 1-2-3-3-, narrowly in contact with

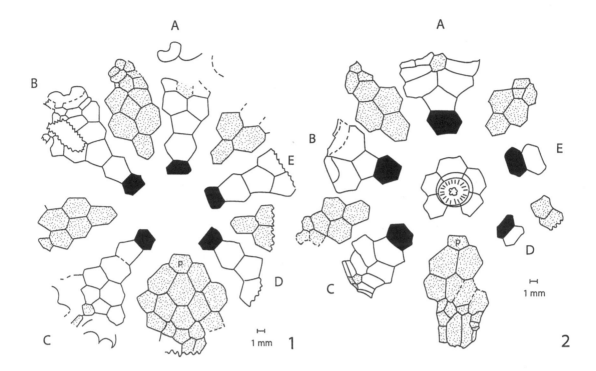

Figure 8.10. *Camera lucida plate diagrams for new periechocrinid genera. 1, Mcoycrinus globosus (Phillips). 2, Clitheroecrinus wrighti n. gen. et sp. Radial plates with black pattern; interradial plates with stippled pattern; P indicates primanal; A, B, C, D, E, indicate ray designations.*

tegmen. CD interray wider than others (Fig. 8.9.7), primanal between C and D radials, plating P-3-5-5-4- (Fig. 8.10.1). Fixed ray plates lacking a ray ridge; first primibrachial fixed, hexagonal, as high as wide; second primibrachial axillary, pentagonal; first secundibrachial fixed, axillary, hexagonal or heptagonal. Last fixed brachial approximately third tertibrachial. A few fixed intrabrachials medially between tertibrachials and arm facets of each ray. Tegmen inverted bowl shape, higher than calyx; numerous, irregular, intermediate-size plates (Fig. 8.9.5). Stout, central anal tube (Fig. 8.9.1, 8.9.2). Twenty arm openings; facets for arms large, subcircular with adoral groove. Free arms unknown. Proximalmost columnal circular, holomeric, with narrow crenularium, and pentalobate lumen.

TYPES. Lectotype, BMNH E6986 (Fig. 8.9.5–8.9.8). The label of this specimen is labeled as Bolland, Yorkshire. Bolland is not a locality per se but is now regarded as a region in Lancashire, north and west of Clitheroe.

ADDITIONAL MATERIAL. BMNH E6985 (Fig. 8.9.1, 8.9.2) (Bolland area, Lancashire), BMNH E6987 (label indicates Yorkshire?, which today could be either Yorkshire or Lancashire), and BGS 73326 (Derbyshire).

OCCURRENCE. As indicated above, only vague locality information accompanies specimens of *Mcoycrinus globosus*. We regard it to be from the lower Chadian (Tournaisian, Tn3c) Waulsortian mudmound facies of the Bowland area, Lancashire.

DISCUSSION. The label of specimen GSM 73326 is labeled as the lectotype. Phillips (1836) only illustrated a single specimen and gave no other indication that other specimens were available. The specimen designated as lectotype (Wright, 1955a) is clearly the specimen illustrated by Phillips (1836, pl. 4, figs. 26, 29).

The morphology of *Mcoycrinus globosus* is well documented. Two specimens (BMNH E6985 and BMNH E6986) are complete calyxes with tegmen. GSM 73326 is nearly a complete calyx, and BMNH E6987 is a large, partial calyx that has been prepared from the inside to reveal the proximal calyx plates.

We follow Wright (1955a) by placing *Periechocrinites globosus* Austin and Austin, 1842, in *Mcoycrinus globosus* (Phillips); however, it should be noted that we regard both usages of this name by Austin and Austin to be unavailable names that should be designated as nomina nuda.

MEASUREMENTS. BMNH E 6986 (lectotype): calyx height, 11.2; tegmen height, 14.7; width at base of calyx, 20.2; calyx width at arm openings, 31.6; exposed radial plate height, 3.3; radial plate width, 4.3; exposed primanal height, 3.7; primanal width, 4.3. BMNH E6985: calyx height, 12.0; tegmen height, 19.8; anal tube height, 4.2*; width at base of calyx, 21.5; calyx width at arm openings, 32.2; exposed radial plate height, 5.0; radial plate width, 5.3; exposed primanal height, 3.7; primanal width, 4.3. BGS 73326: calyx width, 45.0*.

Genus CLITHEROECRINUS new genus

TYPE SPECIES. *Clitheroecrinus wrighti* new species, by monotypy.

DIAGNOSIS. Calyx shape low bowl, basal concavity present, no calyx lobation, thick calyx plates, median ray ridges absent, basal plate circlet low, radial plate wider than high, orientation of radial plate nearly horizontal, radial plates equal in size, radial circlet only open in CD interray, fixed first primibrachial wider than high, primaxil pentagonal, second primibrachial axillary, fixed brachials branch isotomously, fixed rays symmetrical, highest fixed brachial in the secundibrachitaxis or tertibrachitaxis, few interradial plates, regular interray plating not a biseries, regular interrays not depressed, CD interray proximal plating P-2, anitaxial ridge absent, CD interray wider than regular interrays, characters of the tegmen unknown, characters of the free arms unknown, and gonoporoids absent.

DESCRIPTION. See species description below.

ETYMOLOGY. Clitheroe, Lancashire, is a center for Waulsortian terrane that is famous for its crinoid faunas. The holotype of this new genus and species is from the Bellman Quarry, Clitheroe.

OCCURRENCE. *Clitheroecrinus* is from either the lower Chadian (Tournaisian, Tn3c) or upper Chadian (Viséan, V1a) of England.

DISCUSSION. In his discussion of *Megistocrinus globosus*, Wright (1955a) stated that his new specimens from Bellman Quarry (here *Clitheroecrinus wrighti*) may not be conspecific with the Derbyshire specimens (here *Mcoycrinus globosus*) (including the lectotype) and that he did not study specimens of the latter. For comparisons, see discussion of *Mcoycrinus* above.

CLITHEROECRINUS WRIGHTI new species
Figures 8.9.3, 8.9.4, 8.10.2
Megistocrinus globosus? (Phillips, 1836). WRIGHT, 1955a (in part), p.

191–192, pl. 48, figs. 1–3 (*non* pl. 48, figs. 4, 8).

New Genus B AUSICH AND KAMMER, 2006, p. 101

DIAGNOSIS. See generic diagnosis above.

DESCRIPTION. Calyx large, probably low bowl shape with basal concavity (Fig. 8.9.3, 8.9.4), calyx plates smooth with impressed sutures. Basal plates small, confined to basal concavity, and completely covered by proximal columnal. Radial circlet interrupted by primanal. Radial plates five, wider than high, proximal radials within the basal concavity (Fig. 8.9.3). Normal interrays widest at mid-height, plating 1-2-1-2-1-, in contact with tegmen. CD interray wider than others, primanal between C and D radials, plating P-2-4-4-4-, plates above CD may be elongate (Fig. 8.10.2). Fixed ray plates lacking a ray ridge; first primibrachial fixed, hexagonal, as high as wide; second primibrachial axillary, pentagonal. Last fixed brachial in secundibrachitaxis or tertibrachitaxis. One or more fixed intrabrachials medially between tertibrachials of each ray. Tegmen unknown. Eleven or more free arms; features of facets for the arms and free arms unknown. Most proximal columnal circular, holomeric, with narrow crenularium, and pentalobate lumen.

ETYMOLOGY. This name recognizes James Wright, who contributed much to our understanding of Lower Carboniferous crinoids from the United Kingdom and Ireland.

HOLOTYPE. NMS G.1958.1.1678 is from the Bellman Quarry, Clitheroe, Lancashire

OCCURRENCE. This new crinoid is known from a single specimen from the Bellmanpark Quarry, Clitheroe, Lancashire, England. Strata in this quarry include the lower Chadian (Tournaisian, Tn3c) Bellmanpark Limestone Member of the Clitheroe Limestone that is separated by an erosional surface from the lower portions of the Limekiln Wood Limestone Member of the Hodder Formation, upper Chadian (Viséan, V1a) (Miller and Grayson, 1972; Cossey et al., 2004).

DISCUSSION. A key feature of this fossil is the number of free arms. Unfortunately, this aspect of the crinoid is not well preserved. The C ray has 10 arms with thin secundibrachials leading to the free arms; however, one half-ray in the A ray has two secundibrachials, with the second one axillary. The thin secundibrachials do not exist in this half-ray. Other rays are too poorly preserved to be certain.

MEASUREMENTS. NMS G.1958.1.1678 (holotype): calyx height, 10.1; width at base of calyx, 14.9; calyx width approximately at arm openings, 22.5*; exposed radial plate height, 1.83; radial plate width, 3.9; exposed primanal height, 1.7; primanal width, 3.4.

Acknowledgments

For making specimens available for study, we thank A. B. Smith and D. N. Lewis (British Museum Natural History, London), L. I. Anderson (National Museum of Scotland, Edinburgh), and M. Howe (British Geological Survey, Keyworth). A. Munnecke and A. Stigall helped with paleogeographic issues, and B. Heath helped with typing. S. K. Donovan, J. A. Waters, and G. D. Webster suggested important improvements. This research was supported by the National Science Foundation (EAR-02059068 and EAR-0206307).

References

Ausich, W. I. 1987. Brassfield Compsocrinina (Lower Silurian crinoids) from Ohio. Journal of Paleontology, 61:552–562.

Ausich, W. I. 1998a. Early phylogeny and subclass division of the Crinoidea (Phylum Echinodermata). Journal of Paleontology, 72:499–510.

Ausich, W. I. 1998b. Phylogeny of Arenig to Caradoc crinoids (Phylum Echinodermata) and suprageneric classification of the Crinoidea. University of Kansas Paleontological Contributions, new series, 9, 36 p.

Ausich, W. I., and T. W. Kammer. 2006. Stratigraphical and geographical distribution of Mississippian (Lower Carboniferous) Crinoidea from England and Wales. Proceedings of the Yorkshire Geological Society, 56:91–109.

Ausich, W. I., and G. D. Sevastopulo. 2001. The Lower Carboniferous (Tournaisian) crinoids from Hook Head, County Wexford, Ireland. Palaeontographical Society Monograph, 155(617), 137 p.

Ausich, W. I., T. W. Kammer, and T. K. Baumiller. 1994. Demise of the Middle Paleozoic crinoid fauna: a single extinction event or a rapid faunal turnover? Paleobiology, 20:345–361.

Ausich, W. I., T. W. Kammer, and N. G. Lane. 1979. Fossil communities of the Borden (Mississippian) delta in Indiana and northern Kentucky. Journal of Paleontology, 53:1182–1196.

Austin, T., Sr., and T. Austin Jr. 1842. Proposed arrangement of the Echinodermata, particularly as regards the Crinoidea, and a subdivision of the class Adelostella (Echinidae) Annals and Magazine of Natural History, series 1, 10:106–113.

Austin, T., Sr., and T. Austin, Jr. 1843. Description of several new genera and species of Crinoidea. Annals and Magazine of Natural History, series 1, 11:195–207.

Bassler, R. S., and M. W. Moodey. 1943. Bibliographic and faunal index of Paleozoic pelmatozoan echinoderms. Geological Society of America Special Paper, 45, 734 p.

Baumiller, T. K. 1994. Patterns of dominance and extinction in the record of Paleozoic crinoids, p. 193–198. In B. David, A. Guille, J. P. Féral, and M. Roux (eds.), Echinoderms through Time. A. A. Balkema, Rotterdam.

Baumiller, T. K., and F. J. Gahn. 2004. Testing predation-driven evolution with Paleozoic crinoid arm regeneration. Science, 305:1453–1455.

Breimer, A. 1962. A monograph on Spanish Palaeozoic Crinoidea. Overdruk uit Leidse Geologische Mededelingen, Deel, 27, 190 p.

Bronn, H. G. 1848. Neues Jahrbuch für Mineralogie, Geologie und Paläontologie, p. 355.

Bronn, H. G. 1849. Index palaeontologicus, unter Mitwirkung der Herren Prof. H. R. Göppert und H. von Meyer. Handbuch einer Geschichte der Natur, Nomenclator paleontologicus, 5(2):776–1381.

Cossey, P. J., N. J. Riley, A. E. Adams, and J. Miller. 2004. Craven Basin, p. 257–302. In P. J. Cossey, A. E. Adams, M. A. Purnell, M. J., Whiteley, M. A. Whyte, and V. P. Wright (eds.). British Lower Carboniferous Stratigraphy. Joint Nature Conservation Committee, Peterborough, England.

Donovan, S. K. 1992. A field guide to the fossil echinoderms of Coplow, Bellman and Salt Hill Quarries, Clitheroe, Lancashire. North West Geologist, 2:33–54.

Donovan, S. K., D. N. Lewis, and P. Crabb. 2003. Lower Carboniferous echinoderms of north-west England. Palaeontological Association Fold-out Fossils, 1, 12 p.

D'Orbigny, A. D., 1850. Prodrome du paléontologie stratigraphique universelle des animaux mollusques et rayonnés faisant suite au cours élémentaire de paleontologie et de géologie stratigraphique, volume 1. Victor Masson, Paris, 392 p.

Dujardin, F., and L.-H. Hupé. 1862. Histoire Naturelle des Zoophytes, échino-dermes. Library Encyclopédidque. Paris, de Roret, Roret, 8 volumes, 628 p.

Franke, W. 2000. The mid-European segment of the Variscides: tectonostrati-graphic units, terrane boundaries and plate evolution, p. 35–61. *In* W. Franke, V. Haak, O. Oncken, and D. Tanner (eds.), Oceanic Processes: Quantification and Modelling in the Varsican Belt. Geological Society, London, Special Publications, 179.

Goldring, W. 1923. The Devonian crinoids of the state of New York. New York State Museum Memoir, 16, 670 p.

Gutschick, R. S., and J. Rodriguez. 1979. Biostratigraphy of the Pilot Shale (De-vonian–Mississippian) and contemporaneous strata in Utah, Nevada, and Montana. Brigham Young University Geology Studies, 26(1):37–63.

Hall, J. 1847. Palaeontology of New York, volume 1, Containing descriptions of the organic remains of the lower division of the New-York system (equiva-lent of the Lower Silurian rocks of Europe). Natural History of New York, Albany, New York, 6, 338 p.

Heckel, P., and G. Clayton. 2005. Official names of the Carboniferous System. Geology Today, 21:213–214.

Jaekel, O. 1918. Phylogenie und System der Pelmatozoen. Paläeontologische Zeitschrift, 3(1): 128 p.

Jell, P. A., and J. N. Theron. 1999. Early Devonian echinoderms from South Africa. Memoirs of the Queensland Museum, 43(1):115–199.

Kammer, T. W., and W. I. Ausich. 1987. Aerosol suspension feeding and current velocities: distributional controls for late Osagean crinoids. Paleobiology, 13:379–395.

Kammer, T. W., and W. I. Ausich. 2006. The "Age of Crinoids": a Mississippian biodiversity spike coincident with widespread carbonate ramps. Palaios, 21:238–248.

Kammer, T. W., W. I. Ausich, and T. K. Baumiller. 1997. Species longevity as a function of niche breadth: evidence from fossil crinoids. Geology, 25:219–222.

Kammer, T. W., W. I. Ausich, and T. K. Baumiller. 1998. Evolutionary signifi-cance of differential species longevity in Osagean–Meramecian (Mississip-pian) crinoid clades. Paleobiology, 24:155–176.

Kirk, E. 1945. Four new genera of camerate crinoids from the Devonian. Ameri-can Journal of Science, 243:341–345.

Kirk, E. 1946. *Stiptocrinus*, a new camerate crinoid genus from the Silurian. Journal of the Washington Academy of Science, 36:33–36.

Kříž, J., J. M. Degardin, A. Ferretti, W. Hansch, J. C. Gutiérrez Marco, F. Paris, J. M. Piçarra D-Almeida, M. Robardet, H. P. Schönlaub, and E. Serpagli. 2003. Silurian stratigraphy and paleogeography of Gondwana and Peruni-can Europe, p. 105–178. *In* E. Landing and M. E. Johnson (eds.), Silurian Land and Seas Paleogeography outside Laurentia. New York State Museum Bulletin, 493.

Lane, N. G. 1971. Crinoids and reefs. Proceedings of the North American Pale-ontological Convention, September 1969, part J, 1430–1443.

Lane, N. G., and G. D. Sevastopulo. 1987. Stratigraphic distribution of Mississippian camerate crinoid genera from North America and western Europe. Courier Forschugs-institute Senckenberg, 98:199–206.

Laudon, L. R., J. M. Parks, and A. C. Spreng. 1952. Mississippian crinoid fauna from the Banff Formation Sunwapta Pass, Alberta. Journal of Paleontology, 26:544–575.

Lee, K. G., W. I. Ausich, and T. W. Kammer. 2005. Crinoids from the Nada Member of the Borden Formation (Lower Mississippian) in eastern Kentucky. Journal of Paleontology, 79:337–355.

Lees, A., and J. Miller. 1985. Facies variation on Waulsortian buildups, part 2. Mid-Dinantian buildups from Europe and North America. Geological Journal, 20:159–180.

M'Coy, F. 1844. Echinodermata, p. 171–183. In R. Griffith (ed.), A Synopsis of the Characters of the Carboniferous Limestone Fossils of Ireland. Dublin University Press, Dublin.

Meyer, D. L., and W. I. Ausich. 1983. Biotic interaction among Recent and among fossil crinoids, p. 377–427. In M. J. S. Tevesz and P. L. McCall (eds.), Biotic Interactions in Recent and Fossil Benthic Communities. Plenum, New York.

Miller, J. E., and R. F. Grayson 1972. Origin and structure of the lower Viséan "reef" limestones near Clitheroe, Lancashire. Proceedings of the Yorkshire Geological Society, 38:607–638.

Miller, J. S. 1821. A Natural History of the Crinoidea or Lily-Shaped Animals, with Observations on the Genera *Asteria*, *Euryale*, *Comatula*, and *Marsupites*. Bryan & Co, Bristol, 150 p.

Miller, S. A. 1883. *Glyptocrinus* redefined and restricted, *Gaurocrinus*, *Pycnocrinus* and *Compsocrinus* established, and two new species described. Journal of the Cincinnati Society of Natural History, 6:217–234.

Moore, R. C., and L. R. Laudon. 1943. Evolution and classification of Paleozoic crinoids. Geological Society of America Special Paper, 46, 153 p.

Moore, R. C., and C. Teichert (eds.). 1978. Treatise on Invertebrate Paleontology. Pt. T. Echinodermata 2. Geological Society of America and University of Kansas Press, Lawrence, 1027 p.

Morris, J. 1843. A catalogue of British fossils. Comprising all the genera and species hitherto described; with reference to their geological distribution and to the localities in which they have been found (first edition). John Van Voorst, London, 222 p.

Müller, J. 1855. Über die Echinodermen in der Umgegend von Coblenz und in dem Eifeler Kalke, p. 79–85. In F. Zeiler and P. Wirtgen, Bemerkungen über die Petrefacten der altern devonischen Gebirge am Rheine, insbesondere uber die in die Umgegend von Coblenz vorkommenden Arten Verhandlungen des Naturhistorischen Vereins der Preussischen Rheinlande und Westfalen, 12.

Owen, D. D., and B. F. Shumard. 1852. Descriptions of one new genus and twenty-two new species of Crinoidea from the Subcarboniferous limestone of Iowa, p. 587–598. In D. D. Owen, Report of a Geological Survey of Wisconsin, Iowa, and Minnesota.

Phillips, J. 1836. Illustration of the geology of Yorkshire, or a description of the strata and organic remains. Part 2, The Mountain Limestone districts. John Murray, London, 253 p.

Prokop, R. J. 1970. Crinoidea from the Reefton Group (Lower Devonian), New Zealand. Transactions of the Royal Society of New Zealand, Earth Sciences, 8:41–43.

Robardet, M. 2002. Alternative approach to the Variscan Belt in southwestern

Europe: preorogenic paleobiogeographical constraints, p. 1–15. *In* J. R. Martínez Catalán, R. D. Hatcher Jr., R. Arenas, and F. Díaz García (eds.), Varscian-Appalachian Dynamics: The Building of the Late Paleozoic Basement. Geological Society of America Special Paper, 364.

Robardet, M. 2003. The Amorica "microplate": fact or fiction? Critical review of the concept and contradictory paleobiogeographic data. Palaeogeography, Palaeoclimatology, Palaeoecology, 195:125–148.

Schneider, K. A., and W. I. Ausich. 2002. Paleoecology of framebuilders in Early Silurian reefs (Brassfield Formation, southwestern Ohio). Palaios, 17:237–248.

Schultze, L. 1866 (advance publication). Monographie der Echinodermen des Eifler Kalkes. Denkschriften der Kaiserlich Akademie der Wissenschaften Mathematisch-Naturwissen schaftlichen Classe, Vienna, 26:113–230.

Signor, P. W., and C. E. Brett. 1984. The mid-Paleozoic precursor to the Mesozoic marine revolution. Paleobiology, 10:229–245.

Ubaghs, G. 1978a. Suborder Compsocrinina Ubaghs, new suborder, p. T440–T452. *In* R. C. Moore and C. Teichert (eds.), Treatise on Invertebrate Paleontology. Pt. T. Echinodermata 2. Geological Society of America and University of Kansas Press, Lawrence.

Ubaghs, G. 1978b. Evolution of camerate crinoids, p. T281–T294. *In* R. C. Moore and C. Teichert (eds.), Treatise on Invertebrate Paleontology, pt. T. Echinodermata 2. Geological Society of America and University of Kansas Press, Lawrence.

Ubaghs, G. 1978c. Skeletal morphology of fossil crinoids, p. T58–T216. *In* R. C. Moore and C. Teichert (eds.), Treatise on Invertebrate Paleontology, pt. T. Echinodermata 2. Geological Society of America and University of Kansas Press, Lawrence.

Verneuil, E. De. 1850. Note sur les fossiles dévoniens du district de Sabero (Léon). Bulletin de la Société Géologique de France, series 2, 7:155–186.

Waagen, W., and J. Jahn. 1899. Famille des Crinoides, p. 1–216. *In* J. Barrande, Systeme Silurien du centre de La Bohême, Classe des Echinodermes, 7(2). Rivnác, Prague; Gerhard, Leipzig.

Wachsmuth, C., and F. Springer. 1880–1886. Revision of the Palaeocrinoidea. Proceedings of the Academy of Natural Sciences of Philadelphia. Pt. I, The families Ichthyocrinidae and Cyathocrinidae (1880), p. 226–378, pl. 15–17 (separate repaged p. 1–153, pl. 1–3). Pt. II, Family Sphaeroidocrinidae, with the sub-families Platycrinidae, Rhodocrinidae, and Actinocrinidae (1881), p. 177–411, pl. 17–19 (separate repaged, p. 1–237, pl. 17–19). Pt. III, Sec. 1, Discussion of the classification and relations of the brachiate crinoids, and conclusion of the generic descriptions (1885), p. 225–364, pl. 4–9 (separate repaged, 1–138, pl. 4–9). Pt. III, Sec. 2, Discussion of the classification and relations of the brachiate crinoids, and conclusion of the generic descriptions (1886), p. 64–226 (separate repaged to continue with section 1, 139–302).

Wachsmuth, C., and F. Springer. 1897. The North American Crinoidea Camerata. Harvard College Museum of Comparative Zoology Memoir, 20–21, 897 p.

Waters, J. A., and C. G. Maples. 1991. Mississippian pelmatozoan community reorganization: a predator-mediated faunal change. Paleobiology, 17:400–410.

Waters, J. A., C. G. Maples, N. G. Lane, S. Marcus, Liao Zhou-Ting, Liu Lujun, Hou Hong-Fei, and Wang Jin-Xing. 2003. A quadrupling of Famennian pelmatozoan diversity: new Late Devonian blastoids and crinoids from Northwest China. Journal of Paleontology, 77:922–948.

Webster, G. D. 1973. Bibliography and index of Paleozoic crinoids, 1942–1968. Geological Society of America Memoir, 137, 341 p.

Webster, G. D. 2003. Bibliography and index of Paleozoic crinoids, coronates, and hemistreptocrinids 1758–1999. Geological Society of America Special Paper, 363 <http://crinoid.gsajournals.org/crinoidmod>.

Witzke, B. J., and H. L. Strimple. 1981. Early Silurian crinoids of eastern Iowa. Proceedings of the Iowa Academy of Science, 88:101–137.

Wright, J. 1950–1960. A monograph of the British Carboniferous Crinoidea. Palaeontographical Society, London. [1950, v. 1, pt. 1, i–xxx + 1–24, pl. 1–7, 4 text-figs.; 1951a, v. 1, pt. 2, 25–46, pl. 8–12, 10 text-figs.; 1951b, v. 1, pt. 3, 47–102, pl. 13–31, 27 text-figs.; 1952, v. 1, pt. 4, 103–148, pl. 32–40, 40 text-figs.; 1954, v. 1, pt. 5, 149–190, pl. 41–47, 27 text-figs.; 1955a, v. 2, pt. 1, 191–254, pl. 48–63, 16 text-figs.; 1955b, v. 2, pt. 2, 255–272, pl. 64–67, 2 text-figs.; 1956, v. 2, pt. 3, 273–306, pl. 68–75, 2 text-figs.; 1958, v. 2, pt. 4, 307–328, pl. 76–81, 4 text-figs.; 1960, v. 2, pt. 5, 329–347, pl. A, B.].

PART 3. MORPHOLOGY FOR REFINED PHYLOGENETIC STUDIES

INTRODUCTION TO PART 3

William I. Ausich and Gary D. Webster

The negative side of the bewildering number of polygonal plates is maintaining a consistent and useful terminology that has meaning. Terms are not passive labels for bumps and knobs on an organism; rather, terminology is a vehicle to understanding the biology of an organism, which in turn is needed to track the details of morphological change through time and to shed light on the evolutionary history of a group. Terminology leads to an appreciation for separating grades of evolution from homologous clades, and, of course, evolutionary innovation of characters is the fundamental basis for modern phylogenetic reconstructions.

For characters to be meaningful for monophyletic reconstructions, terms within and between groups must consistently reflect homologous relationships. The topology of plates around the oral region of pelmatozoans are thought to be evolutionarily conservative, and Sumrall (Chapter 11) reexamines homology among glyptocystitoid rhombiferans. Arms are of primary significance among crinoids because they reflect the filtration fan presented for feeding via aerosol suspension feeding. Webster and Maples (Chapter 10) reexamine the terminology associated with crinoid arms and the temporal distribution of arm characteristics. Kammer (Chapter 9) examines morphological change in the posterior of Mississippian cladid crinoids. This region of the aboral cup is considered to be evolutionary conservative and thus has primary significance for recognizing homologies. Kammer proposes that heterochronic processes are responsible for the convergent trends that exist among clades of cladid crinoids.

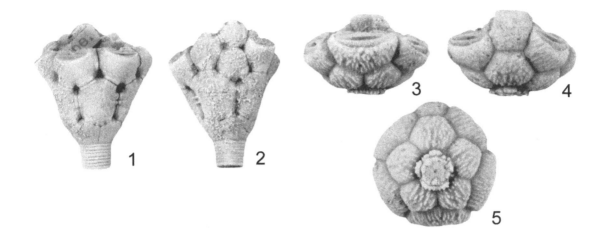

Figure 9.1. *1–4, Examples of aboral cup plate arrangements. 1, 2, Anterior and posterior views of **Hypselocrinus? subimpressus** (Meek and Worthen, 1861) from the Burlington Limestone at Burlington, Iowa; MCZ 104482, holotype of **Poteriocrinus subimpressus**; original magnification, ×1.0. 1, Three circlets of plates in this high cup: infrabasals, basals, and radials. 2, Three anal plates in the posterior. 3–5, Anterior, posterior, and aboral views of **Acylocrinus striatus** (Meek and Worthen, 1869) from the Burlington Limestone at Burlington, Iowa; MCZ 103750, holotype of **Poteriocrinus (Scaphiocrinus) striatus**; original magnification, ×2.0. 3, In this low cup, the basals and radials are exposed, but the infrabasals are nearly hidden by the stem attachment. 4, One anal plate in the posterior.*

PAEDOMORPHOSIS AS AN ADAPTIVE RESPONSE IN PINNULATE CLADID CRINOIDS FROM THE BURLINGTON LIMESTONE (MISSISSIPPIAN, OSAGEAN) OF THE MISSISSIPPI VALLEY

9

Thomas W. Kammer

Introduction

Well-preserved crinoid fossils offer an excellent opportunity to infer how developmental, or ontogenetic, changes may have played a role in their adaptation to local environments. Crinoid development involves both the addition of plates and shape changes in plates during ontogeny (Mortensen, 1920; Springer, 1920; Breimer, 1978; Brower et al., 1978; Mladenov and Chia, 1983; Lahaye and Jangoux, 1987; Nakano et al., 2003). By using known developmental sequences in modern crinoids as an analog, it may be possible to infer the effects of developmental evolution in Paleozoic crinoids.

MISSISSIPPIAN PINNULATE CLADID CRINOIDS. Crinoid generic richness was at its maximum during the Tournaisian and Viséan stages of the Mississippian (Kammer and Ausich, 2006). Much of the Tournaisian generic richness is associated with the late Tournaisian (early Osagean) Burlington Limestone of the Mississippi Valley (Iowa, Illinois, and Missouri), which contains the most diverse and abundant concentration of crinoids in the geologic record. Gahn (2002) listed 86 crinoid genera that include approximately 300 species. Among these are 27 genera of the order Cladida with pinnules (small, nonbranching appendages attached to each arm plate) on their arms, also known as the advanced cladids (Kammer and Ausich, 1992, 1993, 1994) or the suborder Poteriocrinina of Moore et al. (1978). The advanced cladids were a diverse group during the Mississippian (Kammer and Ausich, 2006) and were usually associated with increased current velocities to take advantage of their pinnules as food-gathering devices (Kammer and Ausich, 1987; Kammer et al., 1997, 1998).

Advanced cladids display a wide range of crown morphologies ranging from high to low aboral cups and from few to many arms, arranged on five rays (Kammer and Ausich, 1992, 1993, 1994). Their cup always contains three circlets of plates, infrabasals, basals, and radials, the latter of which supported the arms (Fig. 9.1). Also present in the cup are extra plates in the posterior that support the anal sac, termed the anal plates. Typically there are three anal plates within the cup, less commonly only one (Fig. 9.1), or rarely two. Among 30 genera from the late Osagean–early Meramecian, 77% had three anal plates, 7% had two anal plates, and 17% had one anal plate (Kammer and Ausich, 1992, 1993, 1994). Among 27 genera of advanced cladids from the early Osagean Burlington Limestone, 26% (n = 7) had only one plate, which is an unusually high amount. After investigating these genera with a single anal plate, it was noted they shared many characters indicative of an overall reduction in plate number and morphological complexity. This reduction in plate number and complexity produced morphologies that are inferred to have been paedomorphic by analogy with developmental patterns in modern crinoids.

PAEDOMORPHOSIS AND HETEROCHRONY. Paedomorphosis refers to retention of ancestral juvenile morphology in adult morphology. The study of paedomorphosis is part of the larger field of heterochrony involving timing changes in the appearance of characters during development (McNamara, 1986, 1995, 2002; McKinney, 1988, 1999; McNamara and McKinney, 2005; Webster and Zelditch, 2005). The opposite of paedomorphosis is peramorphosis, where the ancestral adult morphology is reached during a juvenile stage in development as the organism develops farther than its ancestor. Peramorphosis will not be considered further.

The study of heterochrony has been plagued with a proliferation of terms (Gould, 1977, 2000), but the basic terms outlined by McNamara (1986) are still widely used. Three different processes are thought to lead to paedomorphosis: progenesis, neoteny, and postdisplacement. Progenesis results from early sexual maturation that produces smaller adults with ancestral juvenile characters. Neoteny involves both reduction and retardation of development, leading to a large adult with juvenile morphology. Postdisplacement refers to timing changes only in particular organs or structures rather than the whole organism, so that only some features retain ancestral juvenile characteristics.

Considering recent conceptual breakthroughs in the understanding of evolutionary developmental genetics, or evo-devo (Carroll, 2005), the three processes of progenesis, neoteny, and postdisplacement may not be actual processes but rather common end-member states produced by the action of genetic switches during development. The amazing complexity and number of genetic switches associated with the developmental *Hox* genes allows evolution to tinker with development in an unlimited number of ways and not simply produce the end-member states recognized in the literature of heterochrony. Whereas the term *paedomorphosis* is a useful concept for recognizing juvenile features in adult morphology, the developmental processes leading to paedomorphosis are probably more complex than the three processes defined above. As such, actual morphologies should not necessarily be categorized as resulting from simply one of these three alternatives. Rather, we should attempt to understand the possible origins and adaptiveness of the mosaic of characters possessed by any particular species. Recognition of ancestral juvenile characters is then an indication of developmental timing changes producing adaptive morphologies.

CRINOID DEVELOPMENT. Modern crinoids, subclass Articulata, are descended from the advanced cladids (Simms and Sevastopulo, 1993). Thus, their developmental stages represent some combination of features that are both homologous and analogous with those of the advanced cladids, and as such, they provide a useful model for understanding early development not fossilized in advanced cladids. Significant papers on development in modern crinoids include Mortensen (1920), Springer (1920), Breimer (1978), Mladenov and Chia (1983), Lahaye and Jangoux (1987), and Nakano et al. (2003).

There are common themes in the development of modern crinoids demonstrated by a variety of species. The embryo develops in the egg until the membrane ruptures, releasing the doliolaria, or free-living larva. After a period lasting from hours to days, the larva settles on a substrate and metamorphoses to the stalked cystidean stage, which may last several weeks. At this stage, the only plates present are the basals, orals, terminal stem plate, and a few columnals. Infrabasals are generally absent, having been lost in most groups of the Articulata (Rasmussen, 1978). Additional plates appear in the following order as the pentacrinoid stage is reached: radials and radianal, primibrachials (first arm plates), the first axillary (branching plate), then additional brachials and axillaries, and finally pinnules. The pentacrinoid stage ends in the stalkless comatulids when the crown breaks free of the stalk. In contrast, the isocrinids retain the stalk and reach adulthood in the pentacrinoid stage. The theca with basals and orals always forms before the radials, which appear before the arms, which grow from the base up by addition of brachials. This specific growth sequence indicates activation of a series of genetic switches probably controlled by *Hox* genes, as observed in several phyla where there is a regular and predictable order of development in a proximal to distal direction (Carroll, 2005). Because the cup develops first before the arms, it is easy to visualize a wide variety of arm branching styles developed on the same cup morphology. Thus, cup morphology in advanced cladids is presumably a primary homology that should perhaps have greater importance in phylogenetic reconstructions than arm branching patterns. Similarities in arm branching between genera with substantially different cups may simply be homoplasy because there is a limited range of patterns in which arms may branch.

Brower (2005) provided a detailed ontogenetic study of the Ordovician disparid crinoid *Cincinnaticrinus varibrachialus* Warn and Strimple, 1977. Preserved specimens range from juveniles with only 10 arms (two branches per ray) to adults with 80 arms (16 branches per ray), clearly demonstrating the addition of plates and increasing morphologic complexity during ontogeny. On the basis of this ontogenetic series, it is easy to imagine how heterochronic changes could readily alter the number and arrangement of arm branches in adult crinoids by limiting the number of new brachial plates and/or the number of branching plates (axillaries).

POSTERIOR PLATES IN ADVANCED CLADID CRINOIDS. The aboral cup of advanced cladid crinoids usually does not exhibit true pentameral, or even bilateral, symmetry. Instead, the circlets of plates are interrupted by the insertion of anal plates in the posterior of the cup (Fig. 9.1). However, if only one anal plate is present, the cup exhibits true bilateral symmetry (Figs. 9.1.3–9.1.5). In general, the anal plates of the cup are continuous

Abrotocrinus*	Coeliocrinus*	Histocrinus*	Parascytalocrinus
Acylocrinus*†	Corythocrinus*†	Holcocrinus*†	Pelecocrinus*
Adinocrinus	Cosmetocrinus	Hylodecrinus*	Poteriocrinites
Aphelecrinus*	Cromyocrinus	Hypselocrinus*	Sarocrinus*
Armenocrinus†	Culmicrinus	Lanecrinus*	Scytalocrinus*
Ascetocrinus*	Cydrocrinus*	Lebetocrinus†	Springericrinus*
Aulocrinus	Decadocrinus*	Lekocrinus	Stinocrinus
Blothrocrinus*	Dinotocrinus	Linocrinus*	Tropiocrinus*†
Bollandocrinus	Eratocrinus*	Nactocrinus*†	Ulrichicrinus
Bursacrinus*†	Gilmocrinus*	Ophiurocrinus	Worthenocrinus
Cercidocrinus*	Graphiocrinus*†	Pachylocrinus*	

Note.—See Webster (2003) for author and literature citations.

† Genera with one anal plate (n = 9).

* Burlington Limestone genera (n = 27).

Table 9.1. *Pinnulate cladid genera (n = 43). Osagen and early Meramecian stages, eastern North America.*

with and support the plates in the overlying anal sac. The number and arrangement of the anal plates is widely variable between genera (Webster and Maples, 2006). In advanced cladids, there are typically three anal plates termed (from aboral to oral direction) radianal, anal X, and right-tube plate. Webster et al. (2004) preferred the nonhomologous descriptive terms primanal, secundanal, and tertanal, respectively, which are named by counting the plates from the bottom up. They argue that the radianal and anal X plates are often incorrectly identified when only one or two anal plates are present in the cup, so that the use of these terms has been inconsistent. In particular, there has been a great deal of uncertainty as to whether the radianal or anal X is the correct term when only one anal plate is present. In the present study, the number of anal plates that are present is of concern, rather than the terms for the individual plates or plate homologies.

Data and Methods

GENERA AND SPECIES WITH ONLY ONE ANAL PLATE IN CUP. A total of 43 genera of advanced cladid crinoids occur in Osagean and early Meramecian rocks of the eastern United States, including the early Osagean Burlington Limestone of the Mississippi Valley; the late Osagean Keokuk Limestone of the Mississippi Valley, the Borden Group of Indiana, and equivalent rocks in Kentucky; and the early Meramecian Harrodsburg Limestone of Indiana and the Warsaw Formation of the Mississippi Valley (Kammer and Ausich, 1992, 1993, 1994) (Table 9.1).

Data for the Burlington Limestone, which includes 27 genera, were compiled for the present study. Among these 43 genera, nine genera possess only a single anal plate. Species in these nine genera include: *Acylocrinus mcadamsi* (Meek and Worthen, 1873) (Figs. 9.2.5, 9.2.6), *A. striatus* (Meek and Worthen, 1869), *A. tumidus* Kirk, 1947, *Armenocrinus neglectus* (Miller and Gurley, 1896), *A. tenuidactylus* (Worthen, 1882), *Bursacrinus wachs-*

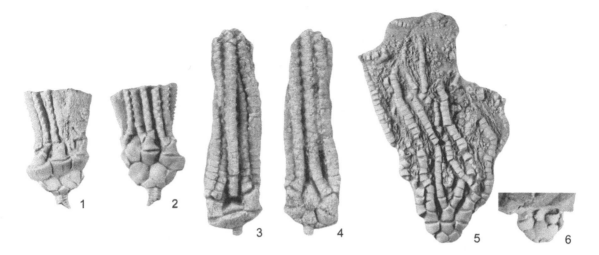

muthi (Meek and Worthen, 1861), *Corythocrinus fragilis* Kirk, 1946, *C. tenuis* Kirk, 1946, *Graphiocrinus? whitei* (Meek and Worthen, 1869), *Holcocrinus spinobrachiatus* (Hall, 1861) (Figs. 9.2.1, 9.2.2), *H. wachsmuthi* (Meek and Worthen, 1861), *Lebetocrinus grandis* Kirk, 1940, *Nactocrinus antiquus* (Meek and Worthen, 1869), *N. nitidus* Kirk, 1947, *Tropiocrinus rudis* (Meek and Worthen, 1869), and *T. simplex* (Hall, 1858) (Figs. 9.2.3, 9.2.4) (see Webster, 2003, for synonymy lists of these species). *Tropiocrinus* Kirk, 1947, was synonymized with *Acylocrinus* Kirk, 1947, by Moore et al. (1978), but *Tropiocrinus* is a distinct genus with only nine arms, rather than 10, and a low or saucer-shaped cup, rather than a bowl-shaped cup as in *Acylocrinus*.

Eleven species occur only in the early Osagean Burlington Limestone and include *A. striatus*, *A. tumidus*, *B. wachsmuthi*, *C. tenuis*, *G.? whitei*, *H. spinobrachiatus*, *H. wachsmuthi*, *N. antiquus*, *N. nitidus*, *T. rudis*, and *T. simplex*. The remaining five species—*A. mcadamsi*, *A. neglectus*, *A. tenuidactylus*, *C. fragilis*, and *L. grandis*—occur in the late Osagean and early Meramecian.

CHARACTERS ASSOCIATED WITH A SINGLE ANAL PLATE. A matrix of 31 characters was compiled to determine those characters shared by the nine genera and the remaining 34 genera. Nine of the 31 characters had the same character state for most of the genera with a single anal plate. The characters, character state definitions, and character matrix for these nine characters in all 43 genera are included in Appendix 9.1.

Those character states present in all or most single-anal-plate genera include the following: 1, high cup; 2, smooth cup plates; 3, single anal plate; 4, plenary radial facets (arm facet extends across the full width of the radial); 5, primibrachial one axillary (arms branch on the first arm plate); 6, A-ray arm atomous (nonbranching) or with extra primibrachials before the first axillary; 7, nine or 10 arms maximum; 8, cuneate (wedge-shaped) or subcuneate brachials; and 9, round column. Although these characters are in all or most genera with single anal plates, this does not necessarily indicate that these characters were linked to possession of a single anal plate because these characters may be common in the remaining advanced cladid genera. However, if any characters could be shown to be linked to

Figure 9.2. *1–6, Examples of crinoids with only one anal plate. 1, 2,* Anterior and posterior views of **Holcocrinus spinobrachiatus** *(Hall, 1861) from the Burlington Limestone at Burlington, Iowa; USNM S2861; original magnification, ×1.0. 3, 4,* Anterior and posterior views of **Tropiocrinus simplex** *(Hall, 1858) from the Burlington Limestone at Burlington, Iowa; note the atomous A ray; UIX-806, holotype of **Poteriocrinus (Scaphiocrinus) simplex**; original magnification, ×1.5. 5, 6,* Anterior and posterior views of **Acylocrinus mcadamsi** *(Meek and Worthen, 1873) from the Edwardsville Formation at Crawfordsville, Indiana; USNM S2857, holotype of **Scaphiocrinus mcadamsi**; original magnification, ×1.0.*

Characteristic	Character							
	1	2	4	5	6	7	8	9
Genera with 1 anal plate (n = 9)	6	8	9	7	8	6	6	7
Genera with > 1 anal plate (n = 34)	14	30	26	19	18	11	24	26
Expected frequency of each character	47%	88%	81%	60%	60%	40%	70%	77%
Observed frequency in genera with 1 anal plate	67%	89%	100%	78%	89%	67%	67%	78%
Cumulative binomial distribution	94%	67%	100%	93%	99%	98%	54%	66%
Critical value at 90% cumulative distribution	6	9	9	7	7	5	8	8
≥ 90% cumulative value	Yes	No	Yes	Yes	Yes	Yes	No	No

Note.—Microsoft Excel used to calculate probabilities (BINOMDIST) and critical values (CRITBI-NOMEM). The 90%, rather than 95%, cumulative distribution is used because of the low number of trials (n = 9). Characters as follows: 1, high cup; 2, cup plates smooth; 4, plenary radial facets; 5, primibrachial one axillary; 6, A-ray arm atomous or with extra primibrachials; 7, nine or ten arms total; 8, cuneate or subcuneate brachials; 9, column round. See Appendix for character data matrix.

Table 9.2. *Binomial distribution probabilities for majority character occurrences in pinnulate cladid genera with only one anal plate, Osagean and early Meramecian stages, eastern North America.*

possession of a single anal plate, this plexus of characters could then be inferred to have been repeatedly selected by natural selection in several genera. If so, the reasons for such selection might then be inferred. This suite of nine characters would seem to indicate relatively simple crowns, with fewer cup plates and arms, suggestive of paedomorphism.

CHARACTER ANALYSIS USING THE BINOMIAL DISTRIBUTION. In order to test for those character states that might be linked to cups with a single anal plate, the cumulative binomial distribution probability was calculated for each character by using the function BINOMDIST in Microsoft Excel. The binomial distribution is used to evaluate the probability of only two possible outcomes (Dowdy and Wearden, 1983)—in this case, the presence or absence of a particular character state. In order to calculate the cumulative binomial distribution probability for each character state, it is necessary to know the frequency of that character state in the sample of interest as well as in the remainder of the group the sample was drawn from. Combining these two frequencies gives the expected frequency for each character state that is then compared with the observed frequency in the sample group (Tables 9.2 and 9.3). If the observed frequency in the sample group exceeds some chosen critical value of the cumulative binomial distribution—say 90% (rather than 95%) because the number of trials is less than 10—then the high frequency of that particular character state may be considered statistically significant. The chance that the observed frequency could be even higher would be 1.0 minus the cumulative binomial distribution. This is a one-tailed test because only one end—high frequencies—of the character distributions are considered.

The function CRITBINOMEM in Excel was used to determine those character states whose frequency met or exceeded the critical value. For example, six out of nine genera with single anal plates have high cups (Table 9.2); the expected frequency is 47% (20 of 43 genera), whereas the ob-

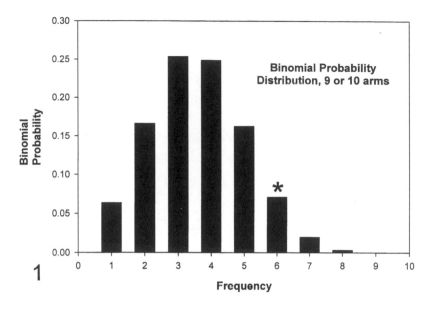

1

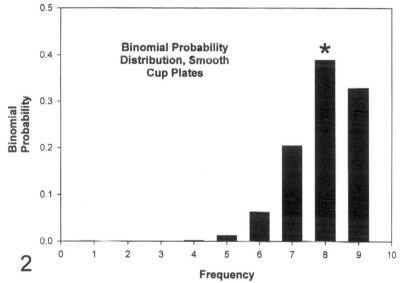

2

Figure 9.3. *Examples of binomial probability distribution plots for the data in Tables 9.2 and 9.3. 1, Plot for the occurrence of nine or 10 arms in Osagean–early Meramecian crinoid genera with only one anal plate. The asterisk indicates the cumulative observed frequency of six genera, which is 98%, and significantly higher than expected. The null hypothesis that these two characters are random with respect to one another may be rejected at the 2% significance level. 2, Plot for the occurrence of smooth cup plates in Osagean–early Meramecian crinoid genera with only one anal plate. The asterisk indicates the cumulative observed frequency of eight genera, which is 67%. The null hypothesis that these two characters are random with respect to one another cannot be rejected because the cumulative binomial distribution does not exceed 90%, which results from the high expected frequency and would only be possible if all nine genera had smooth plates.*

served frequency is 67%, which is associated with a cumulative binomial distribution of 94%; thus, there is only a 6% chance that the frequency of high cups could be greater, which suggests the observed frequency is not random. By analogy, this is the same as saying there is only a 6% chance of randomly tossing more than seven heads in 10 coin tosses where the expected frequency is 50%, and the cumulative binomial distribution probability is 94%. The null hypothesis that a given character is randomly distributed with respect to the possession of a single anal plate may be rejected if its observed frequency meets or exceeds the critical value at the 90% cumulative binomial distribution (Tables 9.2 and 9.3).

Graphic examples of the binomial probability distribution and the cumulative binomial distribution for two characters from Table 9.2 are shown in Figure 9.3. For one character, the null hypothesis of randomness can be

Characteristic	Character							
	1	2	4	5	6	7	8	9
Genera with 1 anal plate (n = 7)	4	7	7	6	6	6	4	6
Genera with > 1 anal plate (n = 20)	11	17	13	12	10	5	17	16
Expected frequency of each character	56%	89%	74%	67%	59%	41%	78%	81%
Observed frequency in genera with 1 anal plate	57%	100%	100%	86%	86%	86%	57%	86%
Cumulative binomial distribution	67%	100%	100%	94%	97%	100%	19%	76%
Critical value at 90% cumulative distribution	6	7	7	6	6	5	7	7
≥ 90% cumulative value	No	Yes	Yes	Yes	Yes	Yes	No	No

Note.—Microsoft Excel used to calculate probabilities (BINOMDIST) and critical values (CRITBI-NOMEM). The 90%, rather than 95%, cumulative distribution is used because of the low number of trials (n = 7). Characters the same as Table 9.2. See Appendix for character data matrix.

Table 9.3. *Binomial distribution probabilities for majority character occurrences in the subset (Table 10.2) of pinnulate cladid genera with only one anal plate from just the Burlington Limestone.*

rejected at the 2% significance level (Fig. 9.3.1), whereas for the other character the null hypothesis of randomness cannot be rejected (Fig. 9.3.2).

Two groups of data were analyzed by binomial distribution. The first group consisted of all 43 genera of Osagean and early Meramecian advanced cladids (Tables 9.1 and 9.2). The second group consisted of only the subset of 27 genera from the Burlington Limestone (Tables 9.1 and 9.3). In the first group, five character states were found to have a nonrandom association with possession of a single anal plate in the cup. These included: a high cup, plenary radial facets, primibrachial one axillary, A-ray arm atomous or with extra primibrachials, and nine or 10 arms total (Table 9.2). In the second group, a slightly different set of five characters have a nonrandom association with possession of a single anal plate in the cup. These included smooth cup plates, plenary radial facets, primibrachial one axillary, A-ray atomous or with extra primibrachials, and nine or 10 arms total (Table 9.3). The results of the two analyzes together indicate four character states that are consistently associated with only one anal plate in the cup: plenary radial facets, primibrachial one axillary, A-ray atomous or with extra primibrachials, and nine or 10 arms total (characters 4–7, Tables 9.2 and 9.3).

As a check to ensure that these character states were uniquely correlated with possession of a single anal plate, the cumulative binomial distributions were calculated for the two groups (all 43 genera and the 27 genera of the Burlington Limestone) of advanced cladids with more than one anal plate in the cup (Tables 9.4 and 9.5). The cumulative binomial distributions were generally low, usually less than 50%, indicating that the observed frequencies were generally less than the expected frequencies for these character states. The notable exception was for cuneate or subcuneate brachials, which was 85% for the Burlington fauna.

Discussion

The plexus of four character states consistently shared by advanced cladid genera with single anal plates describes relatively simple crowns, suggest-

Characteristic	Character							
	1	2	4	5	6	7	8	9
Genera with 1 anal plate (n = 9)	6	8	9	7	8	6	6	7
Genera with > 1 anal plate (n = 34)	14	30	26	19	18	11	24	26
Expected frequency of each character	47%	88%	81%	60%	60%	40%	70%	77%
Observed frequency in genera with > 1 anal plate	41%	88%	76%	56%	53%	32%	71%	76%
Cumulative binomial distribution	33%	57%	29%	35%	23%	25%	61%	55%
Critical value at 90% cumulative distribution	20	32	30	24	24	17	27	29
≥ 90% cumulative value	No	No	No	No	No	No	No	No

Note.—Microsoft Excel used to calculate probabilities (BINOMDIST) and critical values (CRITBI-NOMEM). Characters are the same as Table 9.2. See Appendix for character data matrix.

Table 9.4. *Binomial distribution probabilities for character occurrences in pinnulate cladid genera with greater than one anal plate, Osagean and early Meramecian stages, eastern North America.*

ing underdevelopment and paedomorphosis. These simple crowns with only nine or 10 arms are close to the minimum number of arms necessary for survival. The minimum of five arms is known in only three advanced cladid genera from the Osagean–early Meramecian: *Cromyocrinus* Trautschold, 1867, *Gilmocrinus* Laudon, 1933, and *Ophiurocrinus* Jaekel, 1918, all of which have three anals in the cup and may have followed a different developmental path that also produced paedomorphosis, or at least crown simplification.

The reasons for the linkage of plenary radial facets and branching on the first primibrachial with nine or 10 arms are not clear. Many advanced cladids with plenary facets have more than 10 arms, although the possession of nine or 10 arms is common in this group (Moore et al., 1978). However the cumulative binomial distribution for those genera with more than one anal plate and only nine or 10 arms is only 25% (Table 9.4). Plenary radial facets are also commonly associated with branching on primibrachial one, although many genera branch on higher primibrachials (Moore et al., 1978). Advanced cladids with angustary radial facets do not appear to ever branch on the first primibrachial (Moore et al., 1978).

The A-ray arm is commonly different from other arms in advanced cladids. In 16 of the 43 genera from the Osagean–early Meramecian, the A ray either has extra primibrachials below the first axillary plate, or the arm is atomous (Kammer and Ausich, 1992, 1993, 1994) (Fig. 9.2.3). This character has a 99% cumulative binomial distribution in the group with a single anal plate (Table 9.2), but only a 23% cumulative distribution in the group with more than one anal plate (Table 9.4). The functional reason for a different A-ray arm is unknown, but its origin could be related to the perfect bilateral symmetry along the A ray and the C-D interray axis in the single-anal-plate group.

Why might these advanced cladids have reduced their anal plates from three to one, skipping two anals in nearly all cases? (A few genera may have two or three anals, e.g., *Culmicrinus* Jaekel, 1918, and *Eratocrinus* Kirk, 1938.) This appears to be the result of plate reduction during development. Did the plates not form at all in the cup, or were they pushed up into the anal sac? All genera that possess only one anal plate in this study did not have a variable number of anals in the cup. In other words, all individuals of each of these genera have only a single anal plate. However, Wright (1951) reported wide variation in the number of anal plates in the advanced cladid *Phanocrinus calyx* (M'Coy, 1849), with individuals exhibiting one, two, or three anals, with the latter most common. The anal plates did seem to be reduced in number by being pushed upward out of the cup (Wright, 1951, pl. 18). This suggests that the number of anal plates is a developmentally labile feature that could be modified during development. Beyond simplifying development, it is difficult to imagine what the functional advantage might be in reducing the number of anal plates.

Among the Osagean–early Meramecian genera with a single anal plate, seven of the nine genera, including 11 of 16 species, occur in the Burlington Limestone. Of these 16 species, only 11 have just nine or 10 arms; the rest have 18–40 arms. Of these 11 species, 10 occur in the Burlington Limestone, and one occurs in the late Osagean Edwardsville Formation of Indiana, *Acylocrinus mcadamsi*.

Another feature shared by advanced cladids in the Burlington Limestone is their small size relative to the other crinoids, the camerates, flexibles, and nonpinnulate cladids. Specific data on size has not been compiled for the present study, but casual inspection of museum collections of Burlington Limestone crinoids (e.g., Springer Collection, USNM; Field Museum, Chicago; Museum of Comparative Zoology, Harvard) clearly shows this pattern. In general, but with some obvious exceptions, the maximum volume (as a metric for size) of advanced cladid specimens is commonly one order, or as much as two orders, of magnitude smaller than for the maximum volume of other crinoids. This can also be verified from illustrations of Burlington crinoids: camerates (Wachsmuth and Springer, 1897), flexibles (Springer, 1920), nonpinnulate or primitive cladids (Gahn and Kammer, 2002; Kammer and Gahn, 2003), and advanced cladids (Hall, 1858; Meek and Worthen, 1866, 1868, 1873; Kirk, 1938, 1940, 1941, 1943, 1945, 1946, 1947). The smaller size of advanced cladids is judged to be real, but is it a taphonomic artifact? If advanced cladids had grown to larger sizes than the crowns present in museum collections, these larger specimens would have been easier to find and should be in collections, or at least their larger cups, which are more resistant to disarticulation than the arms, should be in museum collections. Larger crinoids should not necessarily have been more subject to disarticulation, because probably all advanced cladids with preserved arms represent live burial from storm sedimentation (Gahn and Baumiller, 2004). If anything, taphonomic bias should eliminate smaller specimens from the fossil record because of their fragility (Kidwell and Holland, 2002), which is generally true for the smaller juveniles. However, Gahn and Baumiller (2004, p. 28) noted, "Generally, crinoids with larger body sizes and relatively thin, large plates show a greater propensity for disarticulation than crinoids possessing smaller body sizes and thick, small plates." As a first approximation, there is no clear pattern that advanced cladids have thinner plates than either primitive cladids or platycrinitid camerates, both of which can be relatively large in Burlington

Limestone faunas. Thus, although it does not appear that the smaller size in general for advanced cladids is a taphonomic artifact, further study is certainly warranted, although it is beyond the scope of the present study.

The overall simplification of crown morphology and the general reduction in size of the genera with a single anal plate may have been related to onset of early maturity. The simpler morphology and smaller size are consistent with paedomorphosis that results from progenesis (Gould, 1977; McNamara, 1986). Progenesis refers to an adaptive strategy that is based on sexual precociousness leading to rapid reproduction, which is the primary object of selection, with paedomorphosis an incidental by-product (Gould, 1977).

Microcrinoids, which are defined as having a theca less than 2 mm in height, some of which can be argued were small adults, could be an extreme example of progenesis (Lane and Sevastopulo, 1982; Lane et al., 1985; Sevastopulo, 2005). In contrast, *Parazophocrinus callosus* Strimple, 1963, a codiacrinid, as are many microcrinoids, has typical microcrinoid morphology with an armless theca and prominent oral plates, but the holotype is approximately 3 cm in diameter. Ausich (2003) interpreted this crinoid as an example of neoteny on the basis of the large juvenile morphology. Eckert (1987) interpreted the morphology of the Silurian disparid *Homocrinus parvus* Hall, 1852, from the Rochester Shale to be progenetic on the basis of its unusually small size, greatly elongated brachials, and absence of an anal tube. He interpreted this to be an adaptation for early maturity and reproduction in order to survive in an environment with episodic and frequent, high rates of sedimentation. Sprinkle and Bell (1978) reported paedomorphosis, most likely by progenesis, in cyathocystid edrioasteroids from the Ordovician and Devonian, which they interpreted as adaptation to current-swept, nearshore, shallow-water environments.

Why did the Burlington Limestone have so many species with a single anal plate with only nine or 10 arms? These paedomorphic morphologies may have evolved in response to substrate conditions during Burlington Limestone deposition. Typical Burlington Limestone lithology consists of packstones and grainstones composed of coarse bioclastic debris overwhelmingly dominated by crinoidal debris (Fig. 9.4) (Ausich, 1997, 1999), although poorly fossiliferous mudstones and wackestones are also present (Witzke et al., 1990). The low amount of fine-grained matrix in the packstones and grainstones indicates current winnowing on the shallow Burlington Shelf or ramp (Lane, 1978; Witzke et al., 1990; Witzke and Bunker, 2002). The Burlington Shelf was approximately 20° south of the Osagean paleoequator (Witzke and Bunker, 1996). Thus, the coarse bioclastic substrate during Burlington Limestone deposition may have been mobile at some regular frequency, perhaps from storms (Marsaglia and Klein, 1983; Carbonate Petrology Seminar, 1987; Barron, 1989). Gahn (2002, p. 60–61) reported that "well-articulated crinoid crowns are often found buried by coarse crinoidal grainstone. Furthermore, the graded and low-angle cross-stratified crinoidal limestones are indicative of storm-generated sedimentary processes" and that many grainstones are amalgamated storm beds. An example of such burial is shown in Figure 9.5.

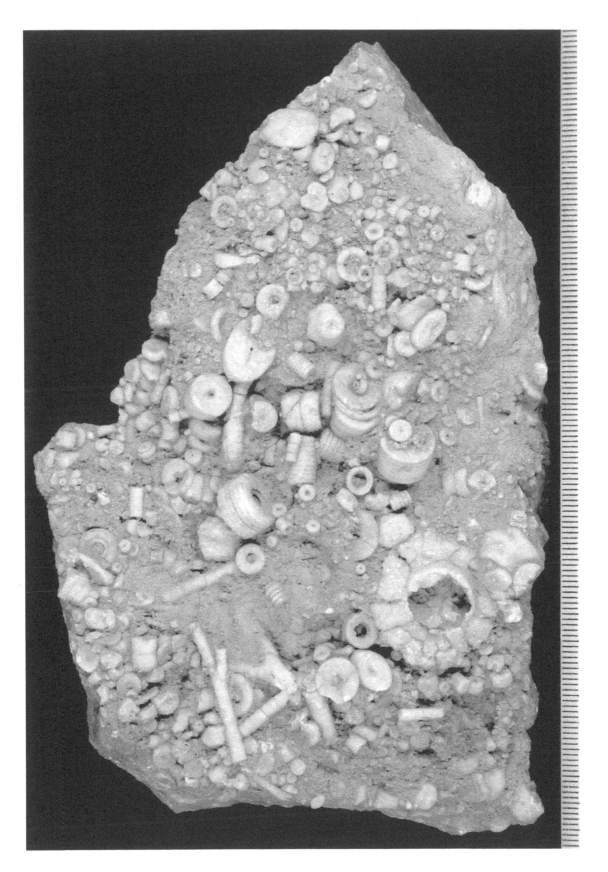

Figure 9.5. *Example of a well-articulated advanced cladid crinoid crown buried in coarse crinoidal grainstone of the Burlington Limestone.* **Sarocrinus varsoviensis** *(Worthen, 1882), non-type, USNM S8872; ×3.0.*

Figure 9.4. *Example of Burlington Limestone encrinital lithology exhibiting coarse bioclastic grains; USNM 530469 from Burlington, Iowa; millimeter scale on right, ×1.0. Photo courtesy of Forest Gahn, National Museum of Natural History, Smithsonian Institution.*

Characteristic	Character							
	1	2	4	5	6	7	8	9
Genera with 1 anal plate (n = 7)	4	7	7	6	6	6	4	6
Genera with > 1 anal plate (n = 20)	11	17	13	12	10	5	17	16
Expected frequency of each character	56%	89%	74%	67%	59%	41%	78%	81%
Observed frequency in genera with > 1 anal plate	55%	85%	65%	60%	50%	25%	85%	80%
Cumulative binomial distribution	57%	39%	24%	34%	27%	11%	85%	52%
Critical value at 90% cumulative distribution	14	19	17	16	15	11	18	18
≥ 90% cumulative value	No	No	No	No	No	No	No	No

Note.—Microsoft Excel used to calculate probabilities (BINOMDIST) and critical values (CRITBI-NOMEM). Characters the same as Table 9.2.. See Appendix for character data matrix.

Table 9.5. *Binomial distribution probabilities for character occurences in the subject (Table 9.4) of pinnulate cladid genera with more than one anal plate from just the Burlington Limestone.*

The small size of Burlington Limestone advanced cladids in general and the paedomorphic morphology of the group with single anal plates specifically may have been a strategy for rapid reproduction to cope with the mobile, unstable substrate that may easily have destroyed juvenile advanced cladids. Gould (1977) argued that frequent disruption in unstable environments selects for rapid reproduction. The large camerates, flexibles, and nonpinnulate cladids are all much larger and may have used a strategy of rapid growth in order to reach sizes that made them less susceptible to damage by substrate mobility. These different adaptive strategies may be the reason for the clear size dichotomy between advanced cladids and all other crinoids in the Burlington Limestone. Advanced cladids in younger late Osagean and early Meramecian rocks are commonly much larger than those of the Burlington Limestone (Kammer and Ausich, 1992, 1993, 1994).

Finally, could all of these genera with a single anal plate simply be an adaptive radiation within a single lineage characterized by the presence of one anal plate? The nine genera with a single anal plate (Table 9.1) are distributed within five families in four superfamilies as defined by Moore et al. (1978) in the *Treatise on Invertebrate Paleontology* (Table 9.6). Admittedly the classification of cladids in the *Treatise* is in need of revision (McIntosh, 2001), and a modern phylogenetic analysis, which is beyond the scope of this chapter, would probably rearrange these genera into different families. Nonetheless, the distribution of these genera among five families suggests that the evolution of cups with a single anal plate may have occurred independently in several lineages responding to similar selection pressures. Thus, the seven genera with single anal plates found together in the Burlington Limestone are probably a greater reflection of homoplasy than homology.

The results of this study have implications for the study of evolution, phylogenetics, and paleoecology in crinoids. Homoplasy is probably rampant in cladid crinoids, as well as other crinoid subclasses. This homoplasy can result from developmental evolution through heterochrony in multiple lineages producing either paedomorphosis or peramorphosis, relative to ancestors, as a response to environmental pressures. In order to sort out

| Scytalocrinacea, Corythocrinidae—*Corythocrinus* |
| Agassizocrinacea, Bursacrinidae—*Bursacrinus, Nactocrinus, Lebetocrinus* |
| Ampelocrinidae—*Armenocrinus* |
| Decadocrinacea, Decadocrinidae—*Acylocrinus, Tropiocrinus* |
| Erisocrinacea, Graphiocrinidae—*Graphiocrinus, Holcocrinus* |

Table 9.6. *Classification of the nine genera of pinnulate cladids with a single anal plate based on the* **Treatise on Invertebrate Paleontology** *(Moore et al., 1978).*

Note—The distribution across five families in four superfamilies suggests that possession of a single anal plate is more related to homoplasy than homology, although the cladid crinoid genera are in need of an updated, comprehensive phylogenetic analysis (McIntosh, 2001).

true phylogenetic relationships among crinoids when choosing from most parsimonious trees, it may be useful to consider the possible effects of heterochrony, just as it is useful to consider stratigraphic range (Ausich and Harvey, 1997).

Conclusions

The crinoid fauna of the Burlington Limestone includes seven genera of advanced cladids, *Acylocrinus, Bursacrinus* Meek and Worthen, 1861, *Corythocrinus* Kirk, 1946, *Graphiocrinus?* de Koninck and Le Hon, 1854, *Holcocrinus* Kirk, 1945, *Nactocrinus* Kirk, 1947, and *Tropiocrinus*, united in the possession of a single anal plate in the aboral cup. The cumulative binomial distribution test indicates four other character states are highly associated with this feature: plenary radial facets, primibrachial one axillary, A-ray atomous or with extra primibrachials, and nine or 10 arms total. Together, this plexus of character states, when considered in light of crinoid development, indicates overall plate reduction and relatively simple morphology associated with arrested development, or paedomorphosis. Along with the generally small size of these crinoids, this suggests early maturation related to rapid reproduction, or progenesis. Rapid reproduction may have been an adaptive strategy for coping with the mobile, unstable, coarse bioclastic substrate of the Burlington Limestone.

Acknowledgments

I thank the following individuals for access to collections under their supervision: J. Thompson, U.S. National Museum of Natural History (USNM); F. Collier, Museum of Comparative Zoology (MCZ), Harvard University; and S. Lidgard, Field Museum of Natural History. This research was partially supported by the National Science Foundation (EAR-0206307). F. Gahn, U.S. National Museum of Natural History, provided the photograph for Figure 9.4. G. D. Sevastopulo and D. L. Meyer provided helpful reviews that improved the chapter.

References

Ausich, W. I. 1997. Regional encrinites: a vanished lithofacies, p. 509–519. *In* C. E. Brett and G. C. Baird (eds.), Paleontological Events: Stratigraphic, Ecologic, and Evolutionary Implications. Columbia University Press, New York.

Ausich, W. I. 1999. Lower Mississippian Burlington Limestone along the Missis-
sippi River Valley in Iowa, Illinois, and Missouri, USA, p. 139–144. *In* H.
Hess, W. I. Ausich, C. E. Brett, and M. J. Simms (eds.), Fossil Crinoids.
Cambridge University Press, Cambridge.

Ausich, W. I. 2003. Developmental breakdown during the early evolution of the
Codiacrinidae: *Parazophocrinus callosus* Strimple, 1963 (Class Crinoidea).
Journal of Paleontology, 77:471–475.

Ausich, W. I., and E. W. Harvey. 1997. Phylogeny of calceocrinid crinoids (Paleozoic: Echinodermata): biogeography and mosaic evolution. Journal of
Paleontology, 71:299–305.

Barron, E. J. 1989. Severe storms during Earth history. Geological Society of
America Bulletin, 101:601–612.

Breimer, A. 1978. General morphology, Recent crinoids, p. T9–T58. *In* R. C.
Moore and C. Teichert (eds.), Treatise on Invertebrate Paleontology, pt. T.
Echinodermata 2. Geological Society of America and University of Kansas
Press, Lawrence.

Brower, J. C. 2005. The paleobiology and ontogeny of *Cincinnaticrinus varibrachialus* Warn and Strimple, 1977 from the Middle Ordovician (Shermanian)
Walcott-Rust Quarry of New York. Journal of Paleontology, 79:152–174.

Brower, J. C., N. G. Lane, and H. W. Rasmussen. 1978. Postlarval ontogeny of
fossil crinoids, p. T244–T274. *In* R. C. Moore and C. Teichert (eds.), Treatise on Invertebrate Paleontology, pt. T. Echinodermata 2. Geological Society of America and University of Kansas Press, Lawrence.

Carbonate Petrology Seminar. 1987. Ramp Creek and Harrodsburg Limestones;
a shoaling-upward sequence with storm-produced features in southern Indiana, USA. Sedimentary Geology, 52:207–226.

Carroll, S. B. 2005. Endless Forms Most Beautiful: The New Science of Evo
Devo and the Making of the Animal Kingdom. W. W. Norton & Co., New
York, 350 p.

De Koninck, L. G., and H. Le Hon. 1854. Recherches sur les crinoides du terrain
carbonifère de la Belgique. Academie Royal de Belgique, Memoir 28, 215 p.

Dowdy, S., and S. Weardon. 1983. Statistics for Research. John Wiley & Sons,
New York, 537 p.

Eckert, J. D. 1987. Heterochrony (progenesis) in the Silurian crinoid *Homocrinus*
Hall. Canadian Journal of Earth Sciences, 24:2568–2571.

Gahn, F. J. 2002. Crinoid and blastoid biozonation and biodiversity in the Early
Mississippian (Osagean) Burlington Limestone, p. 53–74. *In* Pleistocene,
Mississippian, and Devonian Stratigraphy of the Burlington, Iowa, Area.
Iowa Geological Survey, Guidebook Series 23.

Gahn, F. J., and T. K. Baumiller. 2004. A bootstrap analysis for comparative taphonomy applied to Early Mississippian (Kinderhookian) crinoids from the
Wassonville cycle of Iowa. Palaios, 19:17–38.

Gahn, F. J., and T. W. Kammer. 2002. The cladid crinoid *Barycrinus* from the
Burlington Limestone (Early Osagean) and the phylogenetics of Mississippian botryocrinids. Journal of Paleontology, 76:123–133.

Gould, S. J. 1977. Ontogeny and Phylogeny. Belknap Press of Harvard University
Press, Cambridge, Massachusetts, 501 p.

Gould, S. J. 2000. Of coiled oysters and big brains: how to rescue the terminology of heterochrony, now gone astray. Evolution and Development,
2:241–248.

Hall, J. 1852. Palaeontology of New York. Volume 2, containing descriptions of
the organ remains of the lower middle division of the New York System.
Albany, New York, 362 p.

Hall, J. 1858. Report on the Geological Survey of Iowa embracing the results of investigations made during portions of the years 1855, 1856, 1857. Geological Survey of Iowa, volume 1, parts 1 and 2, 724 p.

Hall, J. 1861. Descriptions of new species of Crinoidea from the Carboniferous rocks of the Mississippi Valley. Boston Society of Natural History Journal, 7:261–328.

Jaekel, O. 1918. Phylogenie und System der Pelmatozoen. Paläontologisches Zeitschrift, 3(1):1–128.

Kammer, T. W., and W. I. Ausich. 1987. Aerosol suspension feeding and current velocities: distributional controls for late Osagean crinoids. Paleobiology, 13:379–395.

Kammer, T. W., and W. I. Ausich. 1992. Advanced cladid crinoids from the middle Mississippian of the east-central United States: primitive-grade calyces. Journal of Paleontology, 66:461–480.

Kammer, T. W., and W. I. Ausich. 1993. Advanced cladid crinoids from the middle Mississippian of the east-central United States: intermediate-grade calyces. Journal of Paleontology, 67:614–639.

Kammer, T. W., and W. I. Ausich. 1994. Advanced cladid crinoids from the middle Mississippian of the east-central United States: advanced-grade calyces. Journal of Paleontology, 68:339–351.

Kammer, T. W., and W. I. Ausich. 2006. The "Age of Crinoids": a Mississippian biodiversity spike coincident with widespread carbonate ramps. Palaios, 21:238–248.

Kammer, T. W., and F. J. Gahn. 2003. Primitive cladid crinoids from the early Osagean Burlington Limestone and the phylogenetics of Mississippian species of Cyathocrinites. Journal of Paleontology, 77:121–138.

Kammer, T. W., T. K. Baumiller, and W. I. Ausich. 1997. Species longevity as a function of niche breadth: evidence from fossil crinoids. Geology, 25:219–222.

Kammer, T. W., T. K. Baumiller, and W. I. Ausich. 1998. Evolutionary significance of differential species longevity in Osagean–Meramecian (Mississippian) crinoid clades. Paleobiology, 24:155–176.

Kidwell, S. M., and S. M. Holland. 2002. The quality of the fossil record: implications for evolutionary analyses. Annual Review of Ecology and Systematics, 33:561–588.

Kirk, E. 1938. Five new genera of Carboniferous Crinoidea Inadunata. Journal of the Washington Academy of Science, 28:158–172.

Kirk, E. 1940. Seven new genera of Carboniferous Crinoidea Inadunata. Journal of the Washington Academy of Science, 30:321–334.

Kirk, E. 1941. Four new genera of Carboniferous Crinoidea Inadunata. Journal of Paleontology, 16:382–386.

Kirk, E. 1943. Zygotocrinus, a new fossil inadunate crinoid genus. American Journal of Science, 241:640–646.

Kirk, E. 1945. Holcocrinus, a new inadunate crinoid genus from the Lower Mississippian. American Journal of Science, 243:517–521.

Kirk, E. 1946. Corythocrinus, a new inadunate crinoid genus from the Lower Mississippian. Journal of Paleontology, 20:269–274.

Kirk, E. 1947. Three new genera of inadunate crinoids from the Lower Mississippian. American Journal of Science, 245:287–303.

Lahaye, M. C., and M. Jangoux. 1987. The skeleton of the stalked stages of the comatulid crinoid Antedon bifida (Echinodermata). Zoomorphology, 107:58–65.

Lane, H. R. 1978. The Burlington shelf (Mississippian, north-central United States). Geologica et Paleontologica, 12:165–176.

Lane, N. G., and G. D. Sevastopulo. 1982. Microcrinoids from the Middle Pennsylvanian of Indiana. Journal of Paleontology, 56:103–115.

Lane, N. G., G. D. Sevastopulo, and H. L. Strimple. 1985. *Amphipsalidocrinus:* a monocyclic camerate microcrinoid. Journal of Paleontology, 59:79–84.

Laudon, L. R. 1933. The stratigraphy and paleontology of the Gilmore City Formation of Iowa. University of Iowa Studies, 15(2), 74 p.

Marsaglia, K. M., and G. De V. Klein. 1983. The paleogeography of Paleozoic and Mesozoic storm depositional systems. Journal of Geology, 91:117–142.

McIntosh, G. C. 2001. Devonian cladid crinoids: families Glossocrinidae Goldring, 1923, and Rutkowskicrinidae new family. Journal of Paleontology, 75:783–807.

McKinney, M. L. (ed.). 1988. Heterochrony in Evolution: A Multidisciplinary Approach. Plenum Press, New York, 348 p.

McKinney, M. L. 1999. Heterochrony: beyond words. Paleobiology, 25:149–153.

McNamara, K. J. 1986. A guide to the nomenclature of heterochrony. Journal of Paleontology, 60:4–13.

McNamara, K. J. (ed.). 1995. Evolutionary Change and Heterochrony. John Wiley & Sons, Chichester, 286 p.

McNamara, K. J. 2002. Changing times, changing places: heterochrony and heterotopy. Paleobiology, 28:551–558.

McNamara, K. J., and M. L. McKinney. 2005. Heterochrony, disparity, and macroevolution. Paleobiology, 32(Supplement 2), 17–26.

M'Coy, F. 1849. On some new Paleozoic Echinodermata. Annals and Magazine of Natural History (series 2), 3:244–254.

Meek, F. B., and A. H. Worthen. 1861. Descriptions of new Paleozoic fossils from Illinois and Iowa. Academy of Natural Sciences, Philadelphia, Proceedings for 1861:128–148.

Meek, F. B., and A. H. Worthen. 1866. Descriptions of invertebrates from the Carboniferous system. Illinois Geological Survey, 2:143–411.

Meek, F. B., and A. H. Worthen. 1868. Palaeontology of Illinois. Illinois Geological Survey, 3:289–565.

Meek, F. B., and A. H. Worthen. 1869. Descriptions of new Carboniferous fossils from the western states. Academy of Natural Sciences, Philadelphia, Proceedings, 22:137–172.

Meek, F. B., and A. H. Worthen. 1873. Descriptions of invertebrates from Carboniferous system. Illinois Geological Survey, 5:321–619.

Miller, S. A., and W. F. E. Gurley. 1896. New species of Echinodermata and a new crustacean from the Palaeozoic rocks. Illinois State Museum Bulletin, 10:1–91.

Mladenov, P. V., and F. S. Chia. 1983. Development, settling behavior, metamorphosis, and pentacrinoid feeding and growth of the feather star *Florometra serratissima*. Marine Biology, 73:309–323.

Moore, R. C., N. G. Lane, and H. L. Strimple. 1978. Order Cladida, p. T578–T755. *In* R. C. Moore and C. Teichert (eds.), Treatise on Invertebrate Paleontology, pt. T. Echinodermata 2. Geological Society of America and University of Kansas Press, Lawrence.

Mortensen, T. 1920. Studies in the Development of Crinoids. Papers from the Department of Marine Biology, Carnegie Institution of Washington, 16, 94 p.

Nakano, H., T. Hibino, T. Oji, Y. Hara, and S. Amemiya. 2003. Larval stages of a living sea lily (stalked crinoid echinoderm). Nature, 421:158–160.

Rasmussen, H. W. 1978. Articulata, p. T813–T928. *In* R. C. Moore and C. Teichert (eds.), Treatise on Invertebrate Paleontology, pt. T. Echinodermata 2. Geological Society of America and University of Kansas Press, Lawrence.

Sevastopulo, G. D. 2005. Paleobiology of Carboniferous microcrinoids. Geological Society of America Abstracts with Programs, 37(7):62.

Simms, M. J., and G. D. Sevastopulo. 1993. The origin of articulate crinoids. Palaeontology, 36:91–109.

Springer, F. 1920. The Crinoidea Flexibilia. Smithsonian Institution Publication, 2501, 486 p.

Sprinkle, J., and B. M. Bell. 1978. Paedomorphosis in edrioasteroid echinoderms. Paleobiology, 4:82–88.

Strimple, H. L. 1963. Crinoids of the Hunton Group. Oklahoma Geological Survey Bulletin, 100, 169 p.

Trautschold, H. 1867. Einige crinoideen und andere Thierreste des Jungeren Bergkalks im Gouvernment Moskau. Bulletin de la Societie Imperial Naturalistes de Moscou, 15(3–4), 49 p.

Wachsmuth, C., and F. Springer. 1897. The North American Crinoidea Camerata. Harvard College Museum of Comparative Zoology, Memoir, 20, 21, 897 p.

Warn, J. M., and H. L. Strimple. 1977. The disparid inadunate superfamilies Homocrinacea and Cincinnaticrinacea (Echinodermata: Crinoidea), Ordovician–Silurian, North America. Bulletins of American Paleontology, 72:1–138.

Webster, G. D. 2003. Bibliography and Index of Paleozoic Crinoids, Coronates, and Hemistreptocrinids 1758–1999. Geological Society of America Special Paper, 363 <http://crinoid.gsajournals.org/crinoidmod>.

Webster, G. D., and C. G. Maples. 2006. Cladid crinoid (Echinodermata) anal conditions: a terminology problem and proposed solution. Palaeontology, 49:1–26.

Webster, G. D., C. G. Maples, G. D. Sevastopulo, T. Frest, and J. A. Waters. 2004. Carboniferous (Viséan–Moscovian) echinoderms from the Bechair Basin area of western Algeria. Bulletins of American Paleontology, 368, 98 p.

Webster, M., and M. L. Zelditch. 2005. Evolutionary modifications of ontogeny: heterochrony and beyond. Paleobiology, 31:354–372.

Witzke, B. J., and B. J. Bunker. 1996. Relative sea-level changes during Middle Ordovician through Mississippian deposition in the Iowa area, North American craton, p. 307–330. In B. J. Witzke, G. A. Ludvigson, and J. Day (eds.), Paleozoic Sequence Stratigraphy: Views from the North American Craton. Geological Society of America Special Paper, 306.

Witzke, B. J., and B. J. Bunker. 2002. Bedrock geology in the Burlington area, southest Iowa, p. 23–51. In Pleistocene, Mississippian, and Devonian Stratigraphy of the Burlington, Iowa, Area. Iowa Geological Survey Guidebook Series, 23.

Witzke, B. J., R. M. McCay, B. J. Bunker, and F. J. Woodson. 1990. Stratigraphy and paleoenvironments of Mississippian strata in Keokuk and Washington Counties, southeast Iowa. Department of Natural Resources, Geological Survey Bureau. Guidebook Series, 10, 105 p.

Worthen, A. H. 1882. Descriptions of fifty-four new species of crinoids from the Lower Carboniferous limestones and Coal Measures of Illinois and Indiana. Illinois State Museum of Natural History Bulletin, 1:3–38.

Wright, J. 1951. A monograph of the British Carboniferous Crinoidea. Monograph of the Palaeontographical Society, 1(3):47.

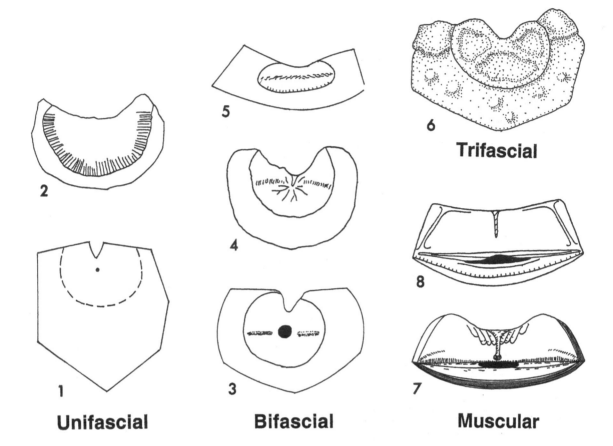

Unifascial　　**Bifascial**　　**Muscular**

Trifascial

Figure 10.1. *Illustrations of the four types of radial facets recognized in the cladid crinoids. 1–2, Unifascial. 1, **Eoparisocrinus siluricus** Springer, 1926. 2, **Rhabdocrinus scotocarbonarius** (Wright, 1937). 3–5, Bifascial. 3, **Vasocrinus valens** Lyon, 1857. 4, **Barycrinus spurius** (Hall, 1858). 5, Unnamed radial from the Famennian of the Saalfeld Germany area. 6, Trifascial, **Cyathocrinites multibrachiatus** (Lyon and Casseday, 1859), after Van Sant (1964). 7–8, Muscular. 7, Diagrammatic, after Moore and Plummer (1940). 8, **Moapacrinus rotundatus** Lane and Webster, 1966.*

CLADID CRINOID RADIAL FACETS, BRACHIALS, AND ARM APPENDAGES: A TERMINOLOGY SOLUTION FOR STUDIES OF LINEAGE, CLASSIFICATION, AND PALEOENVIRONMENT

10

Gary D. Webster and Christopher G. Maples

During our past investigations of Devonian to Permian crinoids, we became increasingly convinced that the radial facet morphologies, brachial types, and arm appendages are of great significance for understanding evolution and lineages within the cladids. We also became convinced that all of these features have been undervalued in the characters used in the classification of the cladids and that their functional morphology is inadequately understood. As a result, we began an investigation of these cladid features and their stratigraphic occurrences.

Study of the cladid crinoids resulted in a classification (Moore and Teichert, 1978) that is mostly morphology based. The morphology used did not include some elements that we recognize as important; therefore, we consider the classification to inadequately represent evolutionary lineages. Moreover, most morphology-based classifications are biased strongly toward characters of the dorsal cup, without proper consideration of the radial facets, arm appendages, brachials, and stem morphology. Part of this bias results from the lack of exposure of the radial facets on crowns where the articulated proximal brachials cover the radial facets, and enclosed arms preclude observation of the arm appendages. Another part of the bias is the lack of descriptions by early workers of the details of the radial facet if they were well exposed. Also, some of the bias is a result of the lack of the preservation of the arms and stems on some genera while the significance of the facet morphology was overlooked in favor of other, more easily observable morphological features of the cup.

Problems with the higher-level classification of the crinoids as given in the *Treatise* (Moore and Teichert, 1978) have resulted in revised classifications as proposed by Simms and Sevastopulo (1993) and Ausich (1998). Modifications of the classification at various levels appear regularly as new discoveries and reevaluation of earlier collections result in recognition of the need to revise earlier classifications. One example is the discovery of an isocrinid in the Permian and recognition of the order Ampelocrinida by Webster and Jell (1999) for the Paleozoic "stem-group articulates" of Simms and Sevastopulo (1993). A second example is the transfer of *Neoprotencrinus* Knapp, 1969, from the Protencrinidae Knapp, 1969, to the

Introduction

Erisocrinidae Wachsmuth and Springer, 1886, by Webster and Kues (2006).

In this study, the dendrocrinids are subdivided into three groups: primitive, transitional, and advanced. This subdivision is based on the morphology of the radial facets as explained below and is not used in the sense of clades except for the transitional dendrocrinids, nor is it implying specific evolution because there is time overlap among the three groups. Primitive dendrocrinids include the superfamilies Dendrocrinacea Wachsmuth and Springer, 1886; Mastigocrinacea Jaekel, 1918; and Merocrinacea S. A. Miller, 1890, as used by Moore et al. (1978), and the family Thalamocrinidae Miller and Gurley, 1895, following McIntosh and Brett (1988). This includes part of the "stem-group cladids" of Simms and Sevastopulo (1993) and the Dendrocrinida Bather, 1899, as used by Ausich (1998). With two exceptions, the primitive dendrocrinids have unifascial and bifascial radial facets. The Glossocrinacea Webster et al., 2003, are referred to as the transitional dendrocrinids because they include taxa with bifascial or muscular articulation. We believe that they gave rise to most of the advanced dendrocrinids. Advanced dendrocrinids bear pinnules, have muscular articulation, and include superfamilies, except the Poteriocrinacea Austin and Austin, 1842, previously included in the Poteriocrinina Jaekel, 1918.

As indicated above, there are several superfamilies recognized in the primitive dendrocrinids; likewise, there are several superfamilies recognized in the advanced dendrocrinids. Polyphyletic evolution of the radial facet, as well as other morphologic characters, is recognized within both the primitive and advanced dendrocrinids. Until evolutionary lineages are established in each of these groups including the character of the radial facet, we consider the current cladid classifications to be incomplete.

The study is based on the type species of 399 genera of the cladid crinoids (Appendices 10.1–10.4). Data for the radial facets, brachials, and arm appendages were compiled from the original descriptions and illustrations of the type species for each genus. For a few genera, a more recent publication was also referred to where the types were reillustrated or new information provided that supplemented the original description of the type species. These references are also cited in Appendices 10.1–10.4, and if dated after 1999 are fully cited in the references. All pre-1999 references in the appendices are given in Webster (2003).

Genera included are 46 cyathocrinids (excluding the codiacrinids, sphaerocrinids, and petalocrinids; Appendix 10.1), 67 primitive dendrocrinids (Appendix 10.2), 22 transitional dendrocrinids (Appendix 10.3), and 264 advanced dendrocrinids (Appendix 10.4). The geologic age for the type species of each genus is that given in the original publication, adjusted for modern age assignments. For example, most genera assigned a Middle Ordovician age in older timescales for North America are now considered Late Ordovician and follow the timescale of Webby et al. (2004).

In addition to problems with the classification, there have been considerable differences in the use of some of the terms describing the cladids. For example, Webster and Maples (2006) recently pointed out that the "primitive condition" of three anal plates in the cup is actually a transitional plate arrangement (although the most common in Paleozoic cladids) between genera with four or five anals and genera with two, and one or no anals. They also recognized eight subarrangements of the three anal plates that may be of evolutionary significance. Thus, the previous use of the descriptor "primitive condition" to imply three anal plates without qualification is inappropriate. Likewise, the terms *ramule*, *armlet*, and *pinnule* have been applied with different meanings by various authors over the years. What one author has called a ramule, another might consider a pinnule. For example, *Imitatocrinus* Schmidt, 1934 (p. 104), was described as having pinnules, whereas they are called ramules by Moore et al. (1978, p. 615). Springer (1911, p. 40) referred to the armlets as ramules with branches when he described *Ottawacrinus billingsi* (=*Grenprisa billingsi*). Definitions of these terms in the *Treatise* glossary (Moore and Teichert, 1978) are fairly clear, but they have not been followed consistently in generic diagnoses within the *Treatise* and require modification.

As we compiled our data, we recognized that some genera are incorrectly classified at the family level and that some families may contain two or more clades that might be recognized at the family level. We believe that this study and likely some future studies will lead to an improved use of the crinoids for zonation and correlation purposes. In addition, it may lead to a better understanding of the functional use of some of the features and may be of significance for a better understanding of cladid lineages.

The purpose of this paper is fivefold: 1) to summarize the stratigraphic distribution of the four major types of cladid radial facets, the six types of cladid brachials, and three types of arm appendages in the cyathocrinids and dendrocrinids; 2) to compare the stratigraphic distribution of the types of radial facets with the types of brachials; 3) to compare the stratigraphic distribution of the types of radial facets with the type of arm appendage; 4) to propose terminology to correct inconsistencies and provide stability in future usage of the radial facets, brachials, and arm appendages; and 5) to suggest future studies to further evaluate the full utilization of these characters.

Radial Facets

Investigations and summaries of several workers studying crinoids provide the guidelines for descriptions of the cladid radial facets. Bather (1893) made minimal descriptions and illustrated some radial facets of Silurian cladids of Gotland. Moore (1939, fig. 9) and Moore and Plummer (1940, fig. 4) defined much of the major morphology of the more complex muscular radial facets of the cladid crinoids. Van Sant (1964) provided some of the best descriptions of the morphology of the radial facets of a few of the Paleozoic genera in each of the camerates, cladids, and flexible crinoids. In addition, he stressed the trifascial nature of the facets with a transverse ridge separating the outer ligament area from the two muscle fields. Macurda and Meyer (1975) provided the first scanning electron micrographs that showed the microstructure of the radial facets and other parts of the crinoid endoskeleton. Ubaghs (1978) presented an excellent summary of the known morphology of the radial facets, brachials, and arm appendages.

Following Moore and Plummer (1940), more detailed descriptions of the radial facet morphology have been given by Strimple (1962), Lane and Webster (1966), Strimple and Watkins (1969), among others. The illustrations of the radial facets by McIntosh (1979, 1984, among others) showed excellent detail of the radial facets. Brower (1992, 2002, among others) has given measurements and ratios of the facet width–radial width. He also described the ontogeny of specimens and plate growth in Ordovician crinoids from the midcontinent (Brower, 1992), noting that they may change with growth.

As applied in generic diagnoses in the *Treatise* (Moore and Teichert, 1978), cladid radial facets are grouped into three general types: angustary, peneplenary, and plenary. Plenary facets occupy the full width of the distal end of the radial. Angustary facets do not extend the full width of the distal end of the radial. They may be on a pedestal or extended distal end of the radial, at or not at its widest point, and occupy less width than the peneplenary facet. In the past, there has been no precisely defined usage of what constitutes the angustary versus peneplenary condition. Webster et al. (2004) recommended using a ratio of radial facet width to radial width as less than 0.7 for the angustary condition. The peneplenary facet would then be between 0.7 and full width of the widest part of the radial, with the plenary condition remaining as the full width facet. This recommendation, although arbitrary, quantifies the *Treatise* definition (Moore and Teichert, 1978, p. 239) of "occupying most but not all of distal extremity of plate." A measurable ratio should provide stability in terminology for usage consistency and comparative studies.

The width ratio of the radial facet may change with growth, becoming angustary from peneplenary or peneplenary from angustary, because radial size and radial facet width are not consistent throughout the growth in some species. Brower (1992, p. 115) listed radial facets as one of the aboral cup variables in the ontogeny of the cupulocrinids. Brower (2002, p. 999) reported relatively wider radial facets in adult specimens than in immature specimens of *Quintuplexacrinus oswegoensis* (Meek and Worthen, 1868) resulting in gradual constriction of the interradial areas with growth. A slab of the Lodgepole Formation from the Little Snowy Mountains of central Montana in the GDW research collections contains more than 50 storm-killed crowns, including growth stages of the camerate crinoid *Platycrinites bozemanensis* (Miller and Gurley, 1897). The two smallest immature crowns have peneplenary radial facet ratios of 0.83 and 0.71, whereas three larger adult crowns have angustary facet ratios of 0.69, 0.63, and 0.45.

In a few crinoids, the radial facet may be angustary in some rays and peneplenary in other rays on a single specimen (*Mollocrinus poculum* Wanner, 1916, pl. 101, fig. 6b). In some Famennian cladids from Germany currently under study by us, the peneplenary facets of the C- and D-ray radials are narrower than the angustary facets of the E-, A-, and B-ray radials. In the cyathocrinids and primitive dendrocrinids, the evolutionary tendency is from angustary to peneplenary and probably occurred polyphyletically. In the advanced dendrocrinids, however, the angustary and peneplenary conditions were a result of modification of the cup. For example, in the Pirasocrinidae Moore and Laudon (1943), the cup became discoidal and the radial facets are subvertical around the sides of the cup, resulting in interradial notches between the peneplenary radial facets (Van Sant, 1964). The angustary radial facets on genera of the Mollocrinidae Wanner (1916) are developed on the inward-curved radials of a globose cup.

Radial facets have an immense amount of morphological disparity in their surface structure (Fig. 10.1). Van Sant (1964) reviewed the historical development of the recognition and terminology applied to radial facet morphologies and recognized the general evolutionary trend from simple to most complex. He recognized unifascial, bifascial, and trifascial ligamentary articulations in the cyathocrinids and dendrocrinids and referred to muscular articulation in more advanced taxa. Ubaghs (1978, p. 161–174, figs. 133–146) described and illustrated muscular and ligamentary articulations with emphasis on their occurrences in the crinoid brachials and pinnulars. He described three types of ligamentary articulations, which he designated synostosial (=unifascial of Van Sant, 1964), bifascial, and trifascial; and he recognized the five fossae and ligament pit on the muscular articulations. Although the evolutionary trend is considered to be from the simplest (unifascial) to the most complex (muscular) radial facet, evolutionary lineages of the radial facets have not been established in the cyathocrinids or dendrocrinids.

We recognize four general radial facet morphologic conditions (Fig. 10.1) in the cladids, each of which may occur on angustary, peneplenary, or plenary type facets.

1. Unifascial radial facets (Van Sant, 1964; = synostosial Ubaghs, 1978) are smooth, flat (*Levicyathocrinites acinotubus* [Angelin, 1878]), or more commonly a gently to moderately concave fossa lacking other morphological features (Fig. 10.1.1) except the outline, which may vary considerably as recognized by applied descriptive terms, such as *circular, subcircular, oval, elongate,* and *horseshoe shaped*. Most of these simple radial facets are circular, subcircular, or horseshoe shaped. More complex forms (Fig. 10.1.2) may contain crenulae along the outer margins (*Rhabdocrinus scotocarbonarius* [Wright, 1937]). They have ligamentary synostosial articulation (Ubaghs, 1978). They are generally angustary or peneplenary, rarely plenary.

2. Bifascial radial facets (Van Sant, 1964; Ubaghs, 1978) have a transverse ridge that divides the facet into outer and inner fossae, which without the transverse ridge would be one continuous fossa. The inner fossa is only partly divided by the ambulacral groove and thus considered a single fossa. Bifascial facets include two types: a) a transverse ridge (normally curved, e.g., *Barycrinus* Wachsmuth and Worthen in Wachsmuth, 1868, which may be a low rounded ridge that extends partly or entirely across the width of the facet (e.g., *Levicyathocrinites monilifer* [Angelin, 1878]); the fossae lack any other ridges or grooves (Fig. 10.1.3); and b) a transverse ridge and ridges and grooves along the

Period	Epoch	Unifascial	Bifascial	Trifascial	Unknown
Permian	Mid	—	—	—	1
	Early	—	3	—	—
	Late	—	—	—	—
Pennsylvanian	Mid	—	2	—	—
	Early	—	—	—	—
	Late	—	—	—	—
Mississippian	Mid	1	1	—	4
	Early	1	1	—	—
	Late	—	—	—	—
Devonian	Mid	7	3	3	3
	Early	1	1	—	—
	Late	1	—	—	2
Silurian	Mid	1	1	—	2
	Early	1	—	—	1
Ordovician	Late	4	—	—	1

Table 10.1. *Stratigraphic distribution of type of radial facet in cyathocrinid genera.*

outer margin (e.g., *Poteriocrinites crassus* J. S. Miller, 1821), radiating from the center of the transverse ridge (e.g., *Barycrinus spurius* [Hall, 1858]), or along the lateral sides of the facet (Fig. 10.1.4, 10.1.5). Some bifascial facets have a central entoneural canal (e.g., *Vasocrinus* Lyon, 1857). More complex bifascial facets have more elevated transverse ridges with or without splitting into multiple ridges on the lateral ends, and some developed a series of short ridges and grooves (described as denticulate) normal to or slightly oblique to the linear trend of the transverse ridge. Bifascial radial facets are generally angustary or peneplenary, rarely plenary. They may have evolved polyphyletically in both the cyathocrinids and dendrocrinids from unifascial radial facets (Tables 10.1, 10.2).

3. Trifascial radial facets (Van Sant, 1964) have a transverse ridge separating the outer fossa from the two inner fossae that are partially separated by the intermuscular notch or intermuscular furrow, with or without a central pit. The three fossae are each distinct depressions. With weathering, they may be difficult to distinguish from bifascial facets. They are most commonly developed on the axillary brachials, are less commonly recognized on the radials, and are thought to have evolved from the bifascial facet.

4. Muscular radial facets (Van Sant, 1964; Ubaghs, 1978) have five fossae and are easily recognized by the presence of a ligament pit on the outer side of the transverse ridge (Fig. 10.1.6, 10.1.7). They have additional morphologic features on both the outer and inner marginal areas (Moore, 1939, fig. 9; Moore and Plummer, 1940, fig. 4). Ubaghs (1978) noted that the two fossae of the lateral furrows on the inner marginal area are poorly developed in the cladids. Thus, they commonly appear more like a trifascial facet. The muscular facet devel-

Period	Epoch	Unifascial	Bifascial	Muscular	Unknown
Mississippian	Late	—	—	1	—
	Mid	1	—	1	2
	Early	1	1	—	1
Devonian	Late	1	—	—	4
	Mid	5	2	—	1
	Early	3	1	—	7
Silurian	Late	—	—	—	3
	Mid	6	1	—	4
	Early	—	—	—	1
Ordovician	Late	8	1	—	7
	Mid	—	—	—	2
	Early	—	—	—	2

oped during the Devonian and Mississippian in the Glossocrinacea and is also recognized on two Mississippian primitive dendrocrinids (*Springericrinus* Jaekel, 1918; *Hebohenocrinus* Webster et al., 2004). These radial facets are dominantly plenary (e.g., *Decadocrinus* Wachsmuth and Springer, 1880), may be secondarily peneplenary (e.g., *Retusocrinus* Knapp, 1969), and rarely are secondarily angustary (e.g., *Strongylocrinus* Wanner, 1916).

Table 10.2. *Stratigraphic distribution of type of radial facet in primitive dendrocrinid genera.*

In the cyathocrinids (Table 10.1), the unifascial facet is present on 17 (37%) of the 46 genera and first recognized in the Late Ordovician extending into the Middle Mississippian. The bifascial facet on 12 (26%) of the 46 genera is first recognized in the Middle Silurian and ranges into the Early Permian. The trifascial facet is only recognized in the Middle Devonian in three genera (7%). The facet type is unknown on 14 genera (30%) of the cyathocrinids. Muscular facets are not recognized in the cyathocrinids. The evolutionary pattern of unifascial→bifascial→trifascial is supported by the stratigraphic first occurrences. Lineage studies have not been established, and polyphyletic evolution of the bifascial facet may have occurred within the cyathocrinids.

In the primitive dendrocrinids (Table 10.2), both the unifascial and bifascial facets are first recognized in the Late Ordovician and range into the Mississippian. The facet types are unknown on the Early and Middle Ordovician genera, probably extending the stratigraphic range downward for one or both of these type facets. Unifacial facets are present on 25 (37%) of the 67 genera, bifascial facets on six genera (9%), muscular facets on two genera (3%), and unknown facet types on 34 genera (51%) of the primitive dendrocrinids. Although trifascial facets are unrecognized among the primitive dendrocrinids, it is possible that they are present on some of the unknown genera. The bifascial facets may have been derived polyphyletically from the unifascial facets, and lineages have not been established. It is uncertain whether the muscular facets developed from the bifascial or trifascial facets in the dendrocrinids, but they are not known until the Middle Mississippian. Primitive dendrocrinids are unknown thereafter.

Only bifascial (four of 22 genera, 18%) and muscular (11 of 22 genera, 50%) radial facets have been identified on genera of the transitional dendrocrinids (Glossocrinacea) with the facet type of seven genera (32%) unknown (Table 10.3). The bifascial facet is known only from the Devonian, and the muscular facet ranges from the Late Devonian–Late Mississippian. We consider the Glosssocrinacea to be the transitional dendrocrinids, wherein the muscular radial facet became established in the cladids. We also consider the transitional dendrocrinids to be the progenitors of most of the advanced cladids, all of which have muscular facets and bear pinnules (Appendix 10.4). *Poteriocrinites* J. S. Miller, 1821, was previously thought to be an advanced cladid, but the type species, *P. crassus* J. S. Miller, 1821, has a bifascial radial facet with a transverse ridge and denticulation along the outer margin. It lacks a ligament pit and other details of the muscular radial facet. Thus, it is here considered to be a primitive dendrocrinid.

The lack of recognized trifascial facets in the transitional dendrocrinids (Table 10.3) suggests that the muscular facet evolved from the bifascial facet in that clade unless some of the unknowns are trifascial facet bearers. We recognize significant variations within the morphologic details of the muscular radial facet that may be of family significance, including differences within the outer marginal area, intermuscular furrow and central pit, and muscle fields. For example, all subfeatures of the outer marginal area, as described by Moore and Plummer (1940), are well developed in the wide outer marginal area of most genera of the Cromyocrinidae Bather (1890), whereas some of the subfeatures are not recognized in the narrow outer marginal area of genera of the Aphelecrinidae Strimple, 1967. The functional significance of these subfeatures for control of the arms while feeding or movement of the crinoid remains to be resolved. Details of the subfeature differences remain to be worked out; they should be of value in lineage studies and may be of value in classification and paleoecology of the cladids.

Brachials

Bather (1900, p. 115, 116, fig. 22) described the evolutionary trend from uniserial to biserial brachial arrangement in the crinoids and recognized three cuneate and two biserial types. Grabau (1903) noted that the biserial arms in some cladids have uniserial brachials both proximally and distally with biserial brachials in the medial part of the arm showing primary uniseriality of the brachials as supported by Ubaghs (1978). Bather, Grabau, and Ubaghs did not distinguish the wedge biserial brachial from the two chisel biserial brachials described below. We recognize seven basic conditions within the brachials of the arms of the cladids. Four conditions are recognized in uniserial arms, and three conditions are recognized in biserial arms (Fig. 10.2).

UNISERIAL ARMS. Arms in which the brachials are arranged in a single column and brachials extend entirely across the arm. Two major types are recognized.

Period	Epoch	Bifascial	Muscular	Unknown
	Late	—	1	—
Mississippian	Mid	—	1	—
	Early	—	4	—
	Late	1	5	4
Devonian	Mid	2	—	1
	Early	1	—	2

Table 10.3. *Stratigraphic distribution of type of radial facet in transitional dendrocrinid genera.*

1. Rectilinear brachial: form that in lateral view has a quadrate shape with right-angle corners and may be equidimensional, longer than wide, or wider than long (Fig. 10.2.1). Where present the articular facet of the arm appendage may modify the shape or angle of that distal corner of the brachial.

2. Cuneate brachial: form that in lateral view has a quadrate-wedge shape with two obtuse-angle and two acute-angle corners and may be longer than wide or wider than long. Three subtypes are recognized: a) slightly cuneate: brachial with the proximal and distal sides less than 5° from parallel (Fig. 10.2.2); b) moderately cuneate: brachial with the proximal and distal sides between 5° and 20° from parallel (Fig. 10.2.3); and c) strongly cuneate: brachial with the proximal and distal sides between 20° and 40° from parallel. The quadrate form approaches, but never attains, a triangular shape in lateral view (Fig. 10.2.4).

BISERIAL ARMS. In biserial arms, the brachials are arranged in two columns interlocking medially. Individual brachials, except axillaries and brachials in the proximal or distal extremities of the arm, do not extend entirely across the arm. Two major types are recognized.

1. Cuneate wedge: brachial with an apparent isosceles triangle shape in lateral view with the proximal and distal facets forming the two equal sides of the triangle. Individual brachials are much wider than long and typically extend nearly the full width of the arm (Fig. 10.2.5).

2. Chisel brachial: brachial with a pentagonal outline resembling a chisel in lateral view with the proximal and distal facets at or nearly at right angles to the lateral facet and one each of the two inner facets interlocking with one of the inner facets of the two laterally adjacent brachials. Two major types are recognized, one with two subtypes: a) Rounded chisel: brachial with the aboral surface moderately to strongly rounded and is approximately as long as wide (Fig. 10.2.6). In cross section, the arm is nearly circular aborally and cut by the V-shaped ambulacral groove adorally. b) Flat chisel: brachial with the aboral surface flat or slightly convex and is generally much wider than long (Fig. 10.2.7). In cross section, the arm is flat or gently convex aborally and has a pentagonal outline cut by the V-shaped ambulacral groove adorally.

Figure 10.2. *Illustrations of the seven types of brachials recognized in cladid crinoids. 1–4, Uniserial: 1, rectilinear; 2, weakly cuneate; 3, moderately cuneate; 4, strongly cuneate. 5–7, Biserial: 5, wedge; 6, round chisel; 7, flat chisel.*

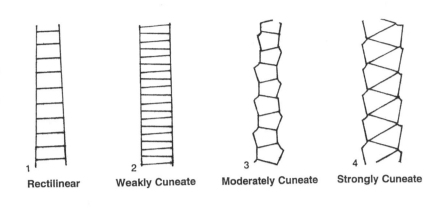

UNISERIAL BRACHIALS

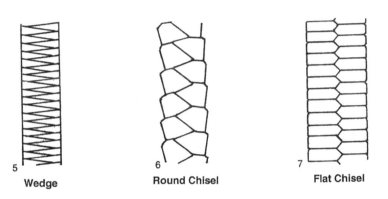

BISERIAL BRACHIALS

Where known in the cyathocrinids (33 of 46 genera, 72%), the brachials are mostly rectilinear (Table 10.4). Only two Permian cyathocrinid genera (*Ceratocrinus* Wanner, 1937, and *Necopinocrinus* Webster and Jell, 1999) are recognized with slightly cuneate brachials. Where known in the 67 genera of primitive dendrocrinids, 57 genera (85%) have rectilinear brachials, seven (10%) have slightly cuneate brachials, one (*Eopinnacrinus* Brower and Venius, 1982) has moderately cuneate brachials, and two (3%) are unknown (Table 10.5). Genera of the transitional dendrocrinids are equally divided between forms with rectilinear and cuneate brachials (Table 10.6). The cuneate brachials are dominated by the slightly cuneate forms (six genera) with three genera with moderately cuneate brachials and two genera with strongly cuneate brachials.

Both uniserial and biserial brachials are present in the 264 genera of the advanced dendrocrinids (Table 10.7). Uniserial brachials (159 genera, 60%) are more common than biserial brachials (41 genera, 15%) in the advanced dendrocrinids, and the brachial type is unknown for 65 genera

(25%). In the uniserial forms, cuneate brachials (97 genera) are more common than rectilinear brachials (62 genera), but the rectilinear brachial is more common than each of the three types of cuneate brachial. Slightly cuneate (50 genera) and moderately cuneate (35 genera) brachials are much more common than strongly cuneate brachials (12 genera). Wedge biserial brachials (10 genera) are less common than either rounded chisel brachials (17 genera) or flat chisel brachials (13 genera).

Nearly 25% (65 of 264 genera) of the advanced dendrocrinids are known from cups without the arms. If the family assignments of these 65 genera are correct, only five to seven should have biserial brachials, and most of the others probably would have gently to moderately cuneate brachials. However, it is doubtful that they are all correctly assigned at the family level.

It should be noted that taxa with rectilinear brachials increase gradually in numerical abundance throughout the Devonian to Permian in the advanced dendrocrinids (Table 10.7). Both the slightly and moderately cuneate brachial-bearing taxa reached their greatest numbers in the Mississippian and declined thereafter. Wedge biserial brachials reached their greatest numbers in the Mississippian, whereas chisel biserial brachials reached their greatest numbers in the Pennsylvanian. Webster and Lane (1967) noted that cladid genera with biserial brachials greatly decreased in the Permian. It should be noted that if species other than the type species had been included in this study, there would be a greater number of genera with biserial arms recognized in the Permian because the type species of some genera are restricted to the Pennsylvanian. However, biserial arm-bearing genera would still have a decline from the number recognized in the Pennsylvanian.

Brachial type has been given inadequate consideration in the classification and lineage studies of the cladids. If genera with cuneate brachials are to be included within a family in which all other genera have rectilinear brachials (compare genera within the Scytalocrinidae Moore and Laudon, 1943, or Blothrocrinidae Moore and Laudon, 1943), it should be demonstrated that the form with cuneate brachials is derived from one of those with rectilinear brachials.

This discussion concerns the definition and inconsistent usage of the terms *ramules, armlets,* and *pinnules.* Consistent usage of these terms is a must if they are to be of value in the classification, phylogenetic analysis, evolution, paleoecology, and lineage studies within the cladids and may have similar applications in the camerates, disparids, and articulates.

The evolutionary sequence from arms lacking minor appendages to those with minor appendages has occurred repeatedly in the crinoids as recognized by Van Sant (1964). An excellent discussion of the various types, developments, and structure of pinnules (broadly applied including ramules and armlets) is given by Ubaghs (1978). He ended that discussion (p. 159) by noting, "One may therefore question whether all crinoid appendages designated as pinnules have the same origin and morphological significance." It is clear from his discussion and illustrations that arm appendages evolved repeatedly and independently within each of the various

Cladid Arm Appendages

Period	Epoch	Rectilinear	Slightly Cuneate	Unknown
Permian	Mid	—	1	—
	Early	2	1	—
	Late	—	—	—
Pennsylvanian	Mid	2	—	—
	Early	—	—	—
	Late	—	—	—
Mississippian	Mid	6	—	—
	Early	1	—	1
	Late	—	—	—
Devonian	Mid	10	—	6
	Early	1	—	1
	Late	3	—	—
Silurian	Mid	2	—	2
	Early	1	—	1
Ordovician	Late	3	—	2

Table 10.4. *Stratigraphic distribution of type of brachial in Cyathocrinina genera.*

Table 10.5. *Stratigraphic distribution of type of brachial in primitive Dendrocrinina genera.*

major crinoid groups (except flexibles) and that most of these homologous appendages served the same general purpose as part of the food-gathering process. The exceptions usually involve modification of or specialization of only some of the pinnules, including examples in which proximal pinnulars serve other purposes (e.g., brood pouches in some modern crinoids) and the tip of the terminal pinnular may develop a protective structure (spines on *Mrakibocrinus* Webster et al., 2005).

The evolutionary significance of the minor arm branches is implied in the classification of the cladids as the poteriocrinids were partly recognized on the presence of pinnules, which may have a polyphyletic origin. Many cyathocrinids and dendrocrinids lack minor branches of the arms,

Period	Epoch	Rectilinear	Slightly Cuneate	Moderately Cuneate	Unknown
Mississippian	Late	1	—	—	—
	Mid	3	1	—	—
	Early	3	—	—	—
Devonian	Late	4	1	—	—
	Mid	6	1	—	1
	Early	11	—	—	—
Silurian	Late	3	—	—	—
	Mid	8	2	—	1
	Early	—	1	—	—
Ordovician	Late	14	1	1	—
	Mid	2	—	—	—
	Early	2	—	—	—

Period	Epoch	Rectilinear	Slightly Cuneate	Moderately Cuneate	Strongly Cuneate
Mississippian	Late	—	—	1	—
	Mid	—	—	1	—
	Early	—	2	1	1
Devonian	Late	6	3	—	1
	Mid	3	—	—	—
	Early	2	1	—	—

whereas others have ramules or armlets, as given in generic diagnostic characters in the *Treatise* (Moore and Teichert, 1978). However, inconsistencies are recognized with this approach because, among others, *Imitatocrinus* is classified as a botryocrinid but is described as having ramules on secundibrachials (Moore and Teichert, 1978, fig. 399-4, show pinnules on alternate sides of each brachial) and *Rhenocrinus* Jaekel, 1906, is described as having armlets but classified in the poteriocrinids (Moore and Teichert, 1978), whereas the type species actually has pinnules. With reevaluation of classifications within the cladids, the use of pinnules as a morphological character with suborder-level significance was rejected by Webster and Hafley (in Webster et al., 1999); they reassigned the Cupressocrinitidae to the cyathocrinid superfamily Gasterocomacea. They considered the cupressocrinitids to be a lineage within the gasterocomids, not the Poteriocrinina. Webster et al. (2003) recognized some early Glossocrinacea as having ramules, whereas later taxa have pinnules. They suggested that the Scytalocrinacea (all of which have pinnules) were evolved from the Glossocrinacea. The two Late Ordovician dendrocrinid genera assigned to the Metabolocrinidae Jaekel, 1918, have pinnules (bifascial radial facets known only on *Eopinnacrinus*), but no younger descendants have been reported.

Cover plates are exceedingly small plates that line the ambulacral groove along the arm, ramules, and armlets (Ubaghs, 1978) and are recognized in all major groups of the cladids. They are seldom preserved on crowns with flared or partly extended arms; thus, they are commonly overlooked and not reported. Cover plate attachment facets on the brachials, although small, are commonly preserved and show that they were present. Enclosed crowns may conceal cover plates or appendages and are listed as unknown for the cyathocrinids, primitive dendrocrinids, and transitional dendrocrinids herein (Tables 10.8–10.10). Cover plates and arm appendages are not listed for the advanced dendrocrinids because they are all judged to have pinnules even where genera are defined on cups or enclosed crowns.

A ramule is defined by Moore and Teichert (1978, p. 241) as "bifurcating or nonbifurcating minor branch of arms, differing from pinnule in less regular occurrence and in some crinoids by presence of pinnules on it." Armlet is referred to ramule (Moore and Teichert, 1978, p. 231). Pinnule is defined (Moore and Teichert, 1978, p. 240) as "generally slender, unbifurcated, uniserial branchlet of arm, typically borne on alternate sides of successive brachials except hypozygals and axillaries."

Table 10.6. *Stratigraphic distribution of type of brachial in transitional dendrocrind genera.*

Period	Epoch	Rectilinear	Slightly Cuneate	Moderately Cuneate	Strongly Cuneate	Flat Chisel	Rounded Chisel	Wedge Biserial	Unknown
Permian	Mid	10	—	—	1	—	—	—	22
	Early	7	3	3	1	1	—	1	8
	Late	12	4	1	2	8	4	1	13
Pennsylvanian	Mid	10	7	5	2	6	7	2	7
	Early	3	2	—	1	—	1	1	7
	Late	6	12	4	2	2	—	3	2
Mississippian	Mid	4	12	18	2	—	1	2	6
	Early	6	9	2	1	—	—	—	—
	Late	2	1	2	—	—	—	—	—
Devonian	Mid	1	—	—	—	—	—	—	—
	Early	1	—	—	—	—	—	—	—

In practice, the definitions of unbranched ramules and pinnules have not always been followed. In particular, coarse versus slender is a subjective interpretation, and the regular occurrence on every other brachial versus irregular occurrence has been used to differentiate pinnule and ramule. For example, "coarse" minor branches, borne on alternate sides of successive proximal brachials and both sides of successive distal brachials of *Cyliocrinus* Jaekel, 1918, are considered ramules (Moore and Teichert, 1978, p. 628, fig. 410.2a). "Long, slender" appendages borne on alternate sides of every brachial of *Imitatocrinus* Schmidt, 1934, are called ramules (Moore and Teichert, 1978, p. 615, fig. 399.4).

The subjective distinction between coarse and slender creates a problem when recognizing ramules and pinnules. If the definitions are to include coarse and slender, then a repeatable, clearly understood measure of coarse versus slender must be defined in order to determine when to stop calling the structure a ramule and recognize it as a pinnule. Because unbranched appendages occur in varying sizes that include short and stout, long and stout, very slender, long and slender, and wide base versus narrow base, we recommend defining the terms on their occurrence.

Ramules (Fig. 10.3) are defined herein as unbranched minor appendages of the arm, regardless of length and girth, either occurring irregularly along the arm, or having a regular occurrence of every fourth or greater number of brachials distally (i.e., *Goniocrinus sculptilis* Miller and Gurley 1891, pl. 6, fig. 4). Pinnules (Fig. 10.3) are here defined as unbranched minor appendages of the arm, regardless of length and girth, occurring in a regular pattern on the arms, typically borne on alternate sides of successive brachials except hypozygals and axillaries. We consider the regular pattern to include the same or opposite sides of adjacent brachials or every second or rarely even every third or fourth brachial when consistent throughout the length of the arm. The application of "when consistent throughout the length of the arm" is the major distinction that we recognize between pinnules and ramules. Hyperpinnulation is the occurrence of multiple pinnules on one brachial.

Armlet commonly has been used to describe the minor, regular (usually every third or greater number of brachials) or irregular branching on the arms and is basically considered a ramule with branches or pinnules as implied in the *Treatise*'s (Moore and Teichert, 1978, p. 231, 241) definition of a ramule. Armlet, as applied, typically is much smaller than the arm from which it branches and is easily differentiated with its small secondary branches. The trunk of an armlet has been called a ramule that branches, although the arms of *Eifelocrinus* Wanner, 1916, are described as "heterotomous armlets alternate sides of every second brachial; armlets composed of long, stout main branch and side ramules given off distally from opposite sides of every second plate" by Moore and Teichert (1978, p. 623, fig. 406.5), and the arms of *Pellecrinus* Kirk, 1929, as "armlets similarly bearing ramules alternately on opposite sides of each second brachial" by Moore and Teichert (1978, p. 583, fig. 374.3b). These armlets are so small compared with the size of the arms that they would not be considered normal arms, as is typical of other armlets either regular or irregular in occur-

Table 10.7. *Stratigraphic distribution of type of brachial of advanced Dendrocrinina genera.*

Period	Epoch	Cover Plates	Ramules	Ramules/ Armlets	Armlets	Pinnules	Unknown
Permian	Mid	1	—	—	—	—	—
	Early	3	—	—	—	—	—
	Late	—	—	—	—	—	—
Pennsylvanian	Mid	1	1	—	—	—	—
	Early	—	—	—	—	—	—
	Late	—	—	—	—	—	—
Mississippian	Mid	3	1	1	1	—	—
	Early	1	—	—	—	—	1
	Late	—	—	—	—	—	—
Devonian	Mid	8	—	—	—	2	6
	Early	—	1	—	—	—	1
	Late	2	—	—	—	—	1
Silurian	Mid	2	—	—	—	—	2
	Early	1	—	—	—	—	1
Ordovician	Late	2	—	—	—	—	3

Table 10.8. *Stratigraphic distribution of cladid arm appendages in Cyathocrinina genera.*

Table 10.9. *Stratigraphic distribution of cladid arm appendages in primitive Dendrocrinina genera.*

rence. In the cladids, armlets generally have been uniformly recognized as such, although the terms applied to their parts have not. Therefore, we define armlet (Fig. 10.3) as a minor secondary appendage of the arm, composed of a trunk and aculets, occurring at irregular or regular spacing of every third or greater number of brachials (Fig. 10.3). Aculets (Latin, "needle" or "pin") are the secondary branches on the trunk of an armlet. This distinguishes armlet from ramule and aculet from ramule and pinnule; the latter three we recognize as minor unbranched appendages of the arms. Two cyathocrinids (*Barycrinus spurius* [Hall, 1858] and *Parabarycrinus camplanatus* Chen and Yao, 1993) and one primitive dendrocrinid (*Bot-*

Period	Epoch	Cover Plates	Ramules	Armlets	Pinnules	Unknown
Mississippian	Late	—	—	—	—	1
	Mid	—	2	—	2	—
	Early	—	—	—	1	2
Devonian	Late	1	1	1	2	—
	Mid	1	—	2	2	3
	Early	3	2	2	1	3
Silurian	Late	1	—	—	—	2
	Mid	2	—	2	3	4
	Early	1	—	—	—	—
Ordovician	Late	9	1	1	3	2
	Mid	1	—	—	1	—
	Early	2	—	—	—	—

Period	Epoch	Ramules	Armlets	Pinnules	Unknown
Mississippian	Late	—	—	1	—
	Mid	—	—	1	—
	Early	—	—	3	1
Devonian	Late	2	—	8	—
	Mid	—	—	3	—
	Early	1	1	1	—

ryocrinus ramosissimus Angelin, 1878) have armlets and ramules on the same arm. Ramules, aculets, and pinnules serve the same general purpose in the food-gathering system, but their precise functions may be quite different and need further investigation.

The stratigraphic distribution of the types of brachials of the advanced dendrocrinids, all of which bear pinnules, shows that they reached their apex in the Pennsylvanian (Table 10.7). The stratigraphic distributions of the arm appendages in the cyathocrinids, primitive dendrocrinids, and glossocrinaceans are given in Tables 10.8–10.10.

In the 46 genera of cyathocrinids (Table 10.8, Appendix 10.1), 24 have cover plates, seven have arm appendages, and 15 are unknown. Three genera with ramules are recognized, one each in the Early Devonian, Middle Mississippian, and Middle Pennsylvanian. The Devonian and Pennsylvanian genera are euspirocrinids, but their lineage is not established. The Mississippian genus is a barycrinid. The Middle Mississippian *Pellecrinus hexadactylus* (Lyon and Casseday, 1860) has both ramules and armlets on the same arm and is assigned to the Barycrinidae Jaekel, 1918, as is the armlet-bearing Middle Mississippian *Meniscocrinus magnitubus* Kammer and Ausich, 1996. Two Middle Devonian genera with pinnules are cupressocrinids. Although the ramules occur before the first occurrence of pinnules and armlets (Table 10.8), they are in a different family and probably evolved independently, not in an evolutionary lineage. The three genera with ramules may represent

Table 10.10. *Stratigraphic distribution of cladid arm appendages in transitional dendrocrinid genera.*

Ramules

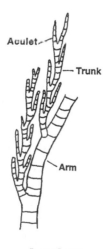

Armlets

Aculet
Trunk
Arm

Pinnules

Figure 10.3. *Illustration of the three types of arm appendages (ramules, armlets, pinnules) in the cladid crinoids.*

polyphyletic evolution even though the Early Devonian and Middle Pennsylvanian taxa are both euspirocrinids.

In the 67 genera of primitive dendrocrinids (Table 10.9, Appendix 10.2), 21 have cover plates, six ramules, eight armlets, 15 pinnules, and 17 are unknown. Pinnules are first recognized in the Middle Ordovician, whereas ramules and armlets are first recognized in the Late Ordovician. Genera with cover plates are reported from the Early Ordovician–Late Devonian. Pinnulate genera have the second best record with gaps in the Early Silurian, Late Devonian, and Middle Mississippian. Ramules have a spotty record with none known in the Silurian, Middle Devonian, and Early Mississippian. Armlets lack representatives in the Early and Late Silurian. Lineages are not established in any of these groups, and the spotty record suggests that independent and polyphyletic evolution also occurred in the primitive dendrocrinids.

Arm appendages in the 22 genera of transitional dendrocrinids (Table 10.10, Appendix 10.3) are dominated by pinnules (17 genera), only three with ramules, one with armlets, and one unknown. Ramules, armlets, and pinnules are recognized, one each on the three different Early Devonian genera, and may have independently evolved from Silurian species of the glossocrinid *Nassovacrinus* Jaekel, 1918, or another primitive dendrocrinid from which the glossocrinids were derived. However, lineages have not been established. Pinnules are recognized throughout the range of the glossocrinids, whereas armlets are unknown after the Early Devonian occurrence and ramules are only known on two Late Devonian genera.

Where known, all advanced dendrocrinids have pinnules (Appendix 10.4).

Stratigraphic Distribution of Combined Morphologic Features

As the results of the investigations of the radial facets, brachials, and arm appendages were compiled, it became apparent that the stratigraphic distribution patterns of each of these morphologic features within the cyathocrinids (Tables 10.1, 10.4, 10.8), primitive dendrocrinids (Tables 10.2, 10.5, 10.9), and glossocrinids (Tables 10.3, 10.6, 10.10) have some similarities. Tables were then constructed to compare the stratigraphic distribution of the combination of the type of radial facet and brachial type and radial facet and type of arm appendage for the cyathocrinids (Tables 10.11, 10.12) and primitive dendrocrinids (Tables 10.13–10.15) and the facet and brachial types for the transitional dendrocrinids (Table 10.16).

Tables 10.11 and 10.12 reemphasize the dominance of the unifascial radial facet and rectilinear brachials in the cyathocrinids until the Permian, when the weakly cuneate brachials are present. In addition, they show that the greatest diversification in the types of radial facets, brachial types, and arm appendages is coincident with the Middle Devonian cyathocrinid appearance of 14 genera. After a gap in the Late Devonian, the cyathocrinids recovered slightly in the Early Mississippian and show another evolutionary peak in the Middle Mississippian (appearance of six new genera). After the Middle Mississippian, the cyathocrinids are poorly known, and genera with the bifascial radial facet lacking arm appendages are the dominant forms. There are no cyathocrinids known with arm appendages, and unifascial radial facets although the arms are unknown on nine genera with unifascial facets (Table 10.12). Of the six genera of cyathocrinids with known arm appendages, four are present on bifascial facets, one on trifascial facets, and one on unknown facets (Table 10.12). This suggests that the bifascial and trifascial facets were necessary for the function of the arm with appendages. The functional relationships remain to be resolved.

Table 10.13 shows that the unifascial facet and rectilinear brachials dominate the primitive dendrocrinids, although the facet type is unknown for 30 genera with rectilinear brachials. However, it also shows that the bifascial facet (six genera) is only known with rectilinear brachials and that the muscular facets are known with rectilinear (one genus) and slightly cuneate brachials (one genus). It should also be noted that the combination of the bifascial facet and rectilinear brachial is present in each of the four periods of greatest introduction of new genera of primitive dendrocrinids. These are the Late Ordovician (16 genera), Middle Silurian (11 genera), Early Devonian (11 genera), and Middle Devonian (eight genera).

Disregarding the four genera for which the type of radial facet is unknown, only six genera of primitive dendrocrinids with unifascial facets have arm appendages (Table 10.14). Among these genera, pinnules (four genera) occur (Late Ordovician) before armlets (one genus) and ramules (one genus) in the Early Devonian. This is the reverse of what might be expected if the evolutionary sequence of arm appendages is ramules→ ramules/armlets→armlets→pinnules. This suggests that each of these types of arm appendages evolved independently. There may be two exceptions to this. First is the evolutionary sequence with the development of ramules on the lower parts and armlets on the distal parts of the arms of *Botryocrinus* Angelin, 1878. Second, ramules may have evolved into pinnules by occurring on every second to fourth brachial distally. If progeny of these crinoids have a regular occurrence of the appendage, they would be considered pinnules by the revised definition; however, no evolutionary lineage has been proposed for such development.

A similar occurrence of the arm appendages is present in the primitive dendrocrinid genera (five) with bifascial radial facets (Table 10.14). Pinnules (three genera) are first recognized in the Middle Silurian, and armlets (two genera) are not known until the Middle Devonian. However, among genera for which the type of radial facet is unknown (Table 15), pinnules (nine genera) are first recognized in the Middle Ordovician (ranging into the Early Mississippian), whereas ramules and armlets are first recognized in the Late Ordovician. Other ramule-bearing genera are known in the Early and Late Devonian and Middle Mississippian for which the radial facet type is unknown. Additional armlet-bearing genera are known in the Middle Silurian and Early and Late Devonian for which the facet type is unknown. It should be noted that only one primitive dendrocrinid with bifascial facets lacks arm appendages, disregarding the two taxa for which the arm appendages are unknown (Table 10.14). It should also be noted that 13 genera have cover plates for which the radial facet is unknown (Table 10.15).

Period	Epoch	U/R	U/Un	B/R	B/SC	B/Un	T/R	T/Un	Un/R	Un/SC	Un/Un
Permian	Mid	—	—	—	—	—	—	—	—	1	—
	Early	—	—	2	1	—	—	—	—	—	—
	Late	—	—	—	—	—	—	—	—	—	—
Pennsylvanian	Mid	—	—	2	—	—	—	—	—	—	—
	Early	—	—	—	—	—	—	—	—	—	—
	Late	—	—	—	—	—	—	—	—	—	—
Mississippian	Mid	1	—	1	—	—	—	—	4	—	—
	Early	—	2	—	—	—	—	—	—	—	—
	Late	—	—	—	—	—	—	—	—	—	—
Devonian	Mid	4	5	3	—	—	1	2	1	—	—
	Early	—	1	1	—	—	—	—	—	—	—
	Late	1	—	—	—	—	—	—	2	—	—
Silurian	Mid	2	1	—	—	1	—	—	—	—	—
	Early	1	—	—	—	—	—	—	—	—	1
Ordovician	Late	2	2	—	—	—	—	—	1	—	—

Abbreviations.—B, bifascial; R, rectilinear; SC, slightly cuneate; T, trifascial; U, unifascial; Un, unknown.

Table 10.11. *Stratigraphic distribution of combined radial facet type and type of brachial in the Cyathocrinina.*

Period	Epoch	U/CP	U/Un	B/CP	B/R	B/P	B/Un	T/P	T/Un	Un/CP	Un/RA	Un/A	Un/Un
Permian	Mid	—	—	—	—	—	—	—	—	1	—	—	—
	Early	—	—	3	—	—	—	—	—	—	—	—	—
	Late	—	—	—	—	—	—	—	—	—	—	—	—
Penn	Mid	—	—	1	1	—	—	—	—	—	—	—	—
	Early	—	—	—	—	—	—	—	—	—	—	—	—
	Late	—	—	—	—	—	—	—	—	—	—	—	—
Miss	Mid	1	—	—	1	—	—	—	—	2	1	1	—
	Early	—	1	1	—	—	—	—	—	—	—	—	—
	Late	—	—	—	—	—	—	—	—	—	—	—	—
Devonian	Mid	4	3	2	—	1	—	1	2	2	—	—	1
	Early	—	1	—	1	—	—	—	—	—	—	—	—
	Late	—	—	1	—	—	—	—	—	2	—	—	—
Silurian	Mid	—	1	—	—	—	1	—	—	2	—	—	—
	Early	1	—	—	—	—	—	—	—	—	—	—	1
Ord	Late	1	3	—	—	—	—	—	—	1	—	—	—

Abbreviations.—B, bifascial; CP, cover plates; Miss, Mississippian; Ord, Ordovician; P, pinnules; Penn, Pennsylvanian; R, ramules; RA, ramules/armlets; T, trifascial; U, unifascial; Un, unknown.

Table 10.12. *Stratigraphic distribution of combined radial facet type and type of arm appendage in the Cyathocrinina.*

Period	Epoch	U/R	U/SC	U/MC	U/Un	B/R	M/R	M/SC	Un/R	Un/SC
Mississippian	Late	—	—	—	—	—	1	—	—	—
	Mid	1	—	—	—	—	—	1	2	—
	Early	1	—	—	—	1	—	—	1	—
Devonian	Late	—	1	—	—	—	—	—	4	—
	Mid	3	—	—	1	2	—	—	1	1
	Early	3	—	—	—	1	—	—	7	—
Silurian	Late	—	—	—	—	—	—	—	3	—
	Mid	5	—	—	1	1	—	—	2	2
	Early	—	—	—	—	—	—	—	—	1
Ordovician	Late	7	—	1	—	1	—	—	6	1
	Mid	—	—	—	—	—	—	—	2	—
	Early	—	—	—	—	—	—	—	2	—

Abbreviations.—B, bifascial; M, muscular; MC, moderately cuneate; R, rectilinear; SC, slightly cuneate; U, unifascial; Un, unknown.

The glossocrinids (Table 10.16) are considered the transitional dendrocrinids because they have a large number of genera with the muscular radial facet that are present in all advanced dendrocrinid families. The muscular facet is also associated with the development of the brachials from rectilinear into strongly cuneate types. It should be noted that the radial facet type is unknown in the Early Devonian glossocrinid with slightly cuneate brachials, and that the strongly cuneate brachials are recognized in the Late Devonian before the moderately cuneate brachials in the Early Mississippian. This latter sequence is out of sync with the generally accepted increasing complexity of evolutionary lineage and suggests that the fossil record is incomplete or that these two facet types developed independently. Again, no evolutionary lineages have been proposed for the glossocrinids.

Table 10.13. *Stratigraphic distribution of combined radial facet type and type of brachial in the primitive Dendrocrinina.*

Observations and Reflections

As data were compiled for this study, some taxa were considered to be incorrectly assigned at the family and superfamily level on the basis of observations of the morphologic characters investigated or in combination with other morphologic features of the taxon. In some instances, we believe that similar studies of other morphologic characters (e.g., columnals) of the cladids may result in their being of greater use in the classification. Some of these are briefly discussed below. We certainly would not limit the issues to those included herein, but we consider them to be representative of problems within the classification of the cladids.

In the current classification of the cladids, many families contain one or more genera that do not belong in that family. A few examples that we recognize in the cyathocrinids are given to illustrate some of the problems.

Cyathocrinites J. S. Miller, 1821, has a bifascial radial facet, whereas most other genera assigned to the family Cyathocrinitidae Bassler, 1938,

Period	Epoch	U/CP	U/R	U/A	U/P	U/Un	B/A	B/P	B/Un
Mississippian	Mid	—	—	—	1	—	—	—	1
	Early	—	—	—	—	1	—	—	—
	Late	—	—	—	1	—	—	—	—
Devonian	Mid	2	—	—	—	2	2	—	—
	Early	1	1	—	—	1	—	—	1
	Late	—	—	—	—	—	—	—	—
Silurian	Mid	1	—	1	—	4	—	1	—
	Early	—	—	—	—	—	—	—	—
Ordovician	Late	5	—	—	2	1	—	—	1

Abbreviations.—A, armlets; B, bifascial; CP, cover plates; R, ramules; U, unifascial; Un, unknown.

Table 10.14. *Stratigraphic distribution of type of arm appendages on primitive Dendrocrinina genera with unifascial and bifascial type radial facets.*

have unifascial radial facets. It is uncertain whether *Cyathocrinites* evolved from one of the earlier genera or is from another lineage.

We agree with the proposal by Webster and Lane (2007) to remove *Mixocrinus* Haude and Thomas, 1992, from the cyathocrinids and include it in the family Bridgerocrinidae Webster and Lane (2007), because it has muscular radial facets among other morphologic characters of the cup and brachials relating it to the bridgerocrinids.

The arms of *Parabarycrinus* Chen and Yao, 1993, are unknown, and there are significant differences in the cup and radial facet from those of *Barycrinus*, suggesting that it should not be included in the Barycrinidae.

Ancyrocrinus Hall, 1862, may belong to the Lecythocrinidae Kirk, 1934, on the basis of the cup, brachial, and stem morphology, and the Lecythocrinidae should probably be included in the Gasterocomacea Roemer, 1854.

Some families with many genera (although one or more may not belong) have overall similar morphologic characters, but one or two morphologic features that could allow the family to be subdivided into two or three subfamilies or families.

For example, cups of the 14 genera of the Euspirocrinidae Bather, 1890, have either a conical (nine) or bowl (five) shape that does not specifically relate to the unifascial and bifascial radial facets occurring within the family. Both cup shapes have rectilinear arms without appendages, except one genus of each cup type that has ramules. Additional study is required to determine the possibility that the cup shapes reflect paleoecologic or other adaptations and whether the bowl-shaped cups are a single clade or represent polyphyletic evolution. Should the family be split into two families or subfamilies on this basis?

The 20 genera assigned to the Scytalocrinidae have significant differences in the cup shapes and brachials. Lane et al. (2001) removed three genera previously included in the Scytalocrinidae when establishing the Sostronocrinidae on the basis of the presence of 20 rather than 10 arms. Webster (1997) recognized three clades within the Scytalocrinidae on the basis of brachial shapes; we suggest that they could be recognized as separate families, pending lineage studies.

Period	Epoch	Cover Plates	Ramules	Armlets	Pinnules	Unknown
Mississippian	Mid	—	2	—	—	—
	Early	—	—	—	1	—
	Late	1	1	1	1	—
Devonian	Mid	—	—	—	2	—
	Early	2	1	2	1	1
	Late	1	—	—	—	2
Silurian	Mid	1	—	1	2	—
	Early	1	—	—	—	—
	Late	4	1	1	1	—
Ordovician	Mid	1	—	—	1	—
	Early	2	—	—	—	—

The arms are known for 13 of the 25 genera of the Pirasocrinidae Moore and Laudon, 1943. Ten of the 13 genera have uniserial arms, and three have biserial arms. Should the family be subdivided on this basis?

One of the morphologic characters utilized for the classification of the Erisocrinacea Wachsmuth and Springer (1886) was the presence or absence of one anal plate in the cup. Of the five families in the Erisocrinacea recognized by Webster and Kues (2006), two families contain genera with uniserial arms (one family with a single anal in the cup and one family with the anal interior on a notch on the C and D radials), two families contain genera with biserial arms (one family with a single anal in the cup and one family with the anal interior on a notch on the C and D radials), and one family for which the arms are unknown. It has not been demonstrated that the families with uniserial arms gave rise to the families with biserial arms, nor has it been demonstrated that the families without the external anal arose from the families with the exposed anal. The uniserial and biserial families with an external anal may have evolved from different lineages. It is also highly probable that genera with the internal anal arose polyphyletically from genera with the external anal in both lineages. The brachial type in the erisocrinids should be given greater consideration than the number of anals when determining the progenitors of the different lineages within the erisocrinids. Thus, we recommend combining the two

Table 10.15. *Stratigraphic distribution of type of arm appendages on primitive Dendrocrinina genera with unknown type radial facets.*

Table 10.16. *Stratigraphic distribution of combined radial facet type and type of brachial in the Glossocrinacea.*

Period	Epoch	B/R	M/R	M/SC	M/MC	M/StC	Un/R	Un/SC
Mississippian	Late	--	--	--	1	--	--	--
	Mid	--	--	--	1	--	--	--
	Early	--	--	2	1	1	--	--
Devonian	Late	1	1	3	--	1	4	--
	Mid	2	--	--	--	--	1	--
	Early	1	--	--	--	--	1	1

Abbreviations.—B, bifascial; M, muscular; R, rectilinear; SC, slightly cuneate; MC, moderately cuneate; StC, strongly cuneate; Un, unknown.

families with uniserial arms and the two families with biserial arms, but placing them in separate superfamilies.

Although not included in this study, the radial facets of genera of the Codiacrinacea Bather, 1890, are all of unifascial or bifascial type, and Paleozoic articulates all have muscular articulation where known.

Future Studies

Some additional considerations of the radial facets, type of brachials, and arm appendages that require further investigation within the cladids include the following.

In order to have a more complete record and enhance this preliminary analysis, the many unknown types of radial facets, brachials, and arm appendages in the cladids need to be determined wherever possible. Some of these could probably be found by searching museum collections for topotype, paratype, or additional specimens from the original study illustrating these features. Others might be determined by study of nontype species assigned to the genus. However, care must be taken that they are indeed the same genus because we recognize that a number of species are incorrectly identified at the genus level.

Is there a relationship of arm-branching patterns and radial facet types, brachial types, or arm appendages? In the closed position, the plenary radial facets lack the interradial notches that exist between the arm bases of the angustary and penplenary facets. This would deprive access to the tegmen by some predators and symbionts. In the open position, what are the functional advantages or disadvantages of the plenary radial facet compared with the angustary and peneplenary facets?

The occurrence of pinnules in the primitive dendrocrinids before ramules and armlets and at approximately the same time in the cyathocrinids implies that the different types of arm appendages evolved independently. Exceptions supporting evolution of ramules into armlets and pinnules were mentioned above and support polyphyletic evolution in the development of arm appendages.

Pinnules are most commonly associated with plenary muscular radial facets as occurring in the advanced dendrocrinids. Why are plenary muscular facets dominantly less convex aborally than the angustary and peneplenary facets? Why do they most commonly have a semitriangular to quadrate wedge shape narrowing internally rather than a horseshoe or circular shape? Is this related to strengthening of the connective tissues on the radial facet with both the development of morphologic characters of the musculature condition and widening into the plenary condition in order to provide greater control of the pinnulate arms in the advanced dendrocrinids?

As recognized by Van Sant (1964), the development of the peneplenary radial facet in several genera of the plaxocrinids and stellarocrinids is not considered an evolutionary condition derived from earlier angustary or peneplenary cladids. It is judged to be the result of the development of a basal concavity with concurrent flattening of the cup into a low bowl or a disk. Is this an adaptation to a living position on the substrate in genera with enlarged basals as recognized in the calceolispongids (Teichert, 1949;

Webster and Jell, 1999), the outflaring of the arms with a nearly vertical radial facet as present on some moderately large-stemmed stellarocrinids such as *Celonocrinus* Lane and Webster, 1966, the development of a pronounced basal impression as in *Pirasocrinus* Moore and Plummer, 1940, or all of these?

The advanced dendrocrinids evolved from the primitive dendrocrinids in the Devonian and Early Mississippian. Evolutionary lineages of advanced dendrocrinid families that began in the Devonian and Early Mississippian are currently thought by us to mostly be from the glossocrinacids. Other progenitors of some advanced dendrocrinids might be the rhenocrinids or other species currently assigned to *Poteriocrinites* that have muscular articulation. A systematic revision of the Poteriocrinitidae Austin and Austin, 1842, is needed to resolve part of the problem with classification of the cladids.

Feeding strategies, including tiering as recognized by Ausich (1980) and density of the filtration fans as described by Ausich (1980), Baumiller (1992) and Brower (2006) were not a part of this investigation, but the radial facets, brachials, and arm appendages are critical elements for those functions. Is there a relationship of the type of arm appendage and position in tiering levels? The recognition of the width of the base of the pinnule as a current indicator (Baumiller and Meyer, 2000) suggests the possibility that with additional study this concept could have broader application within the size and distribution of ramules, armlets, and pinnules. Baumiller (1992) demonstrated that fine-mesh filter feeders required a higher current velocity to push sufficient food particles through the pinnulate mesh than coarse-mesh filter feeders. What is the effect of the current velocity difference between coarse and fine pinnules, both with fine-mesh filters or coarse-mesh filters? The differences between ramule and armlet-bearing taxa have not been investigated. What is the relationship of coarse ramules, armlets, and pinnules (regardless of width of basal ossicles) versus thin ramules, armlets, and pinnules to current strength in unidirectional and multidirectional currents? Is there a relationship between cladid attachment structure or substrate and radial facet, brachial, or arm appendage types?

Conclusions

Cyathocrinids are cladids with unifascial or bifascial radial facets and rectilinear brachials lacking arm appendages, with several exceptions. A few cyathocrinids have trifascial facets with pinnules; and a few have ramules, ramules and armlets, or armlets, which probably evolved polyphyletically. Muscular articulating radial facets are not recognized in the cyathocrinids.

Primitive dendrocrinids are dominated by forms with unifascial radial facets and rectilinear brachials, and they may or may not have arm appendages. Pinnules are recognized in the primitive dendrocrinids in the Middle Ordovician before the appearance of ramules and armlets in the Late Ordovician. Muscular articulating radial facets are first recognized in the Middle Mississippian in the primitive dendrocrinids. Slightly cuneate and moderately cuneate brachials are first recognized in the primitive dendrocrinids in the Late Ordovician.

Transitional dendrocrinids are considered a cladid clade (Glossocrinacea), and most have bifascial or muscular radial facets, have rectilinear or slightly cuneate brachials, and bear pinnules. A few transitional dendrocrinids have moderate or strongly cuneate brachials, and a few have ramules or armlets. The transitional dendrocrinids are considered to be the progenitors of most of the advanced dendrocrinids.

The advanced dendrocrinids all have muscular articulating radial facets that are dominated by rectilinear brachials, and all have pinnules. All seven brachial types are recognized in the advanced dendrocrinids, wherein the uniserial brachials evolved into two types of biserial brachials. Peneplenary and angustary radial facets in the advanced dendrocrinids are considered to be secondarily derived as a result of cup modifications.

Polyphyletic evolution of the unifascial to bifascial radial facet is believed to have occurred in both the cyathocrinids and primitive dendrocrinids. It is also believed that polyphyletic evolution occurred in development of arm appendages in the cyathocrinids and primitive dendrocrinds. Evolution of the biserial brachials from the uniserial brachials may be of polyphyletic origin in the advanced dendrocrinids.

Radial facet, brachial, and arm appendage types are undervalued in the classification of the cladid crinoids and should be incorporated into any functional, paleoecologic, systematic, or lineage study of the cladids.

Acknowledgments

Discussions with N. G. Lane clarified some of our ideas and definitions for terms as proposed herein. The reviews of C. Brett, F. Gahn, and W. Ausich helped us make significant improvements to the chapter and are gratefully acknowledged.

References

Angelin, N. P. 1878. Iconographia Crinoideorum: in stratis Sueciae Siluricis fossilium. Samson and Wallin, Holmiae, 62 p., 29 pl.

Ausich, W. I. 1980. A model for niche differentiation in Lower Mississippian crinoid communities. Journal of Paleontology, 54:273–288.

Ausich, W. I. 1998. Phylogeny of Arenig to Caradoc crinoids (Phylum Echinodermata) and suprageneric classification of the Crinoidea. University of Kansas Paleontological Contributions, n.s., 9, 36 p.

Ausich, W. I., and G. D. Sevastopulo. 2001. The Lower Carboniferous (Tournaisian) crinoids from Hook Head, County Wexford, Ireland. The Palaeographical Society Monograph, 216, 136 p., 13 pls.; supplement 1:1–34; supplement 2:1–11.

Ausich, W. I., M. D. Gil Cid, and P. Domínguez Alonso. 2002. Ordovician [Dobrotivian (Llandeillian Stage) to Ashgill] crinoids (phylum Echinodermata) from the Montes de Toledo and Sierra Morena, Spain with implications for paleogeography of Peri-Gondwana. Journal of Paleontology, 76:975–992.

Austin, T., and T. Austin. 1842. XVIII.—Proposed arrangement of the Echinodermata, particularly as regards the Crinoidea, and a subdivision of the Class Adelostella (Echinidae). Annals and Magazine of Natural History, series 1, 10(63):106–113.

Bassler, R. S. 1938. Pelmatozoa Palaeozoica. *In* W. Quenstedt (ed.), Fossilium catalogus, I: Animalia. Part 83. W. Junk, s'Gravenhage, 194 p.

Bather, F. A. 1890. British fossil crinoids. II. The classification of the Inadunata. Annals and Magazine of Natural History, series 6, 5:310–334, 373–388, 485–486, pl. 14, 15.

Bather, F. A. 1893. The Crinoidea of Gotland. Part I. The Crinoidea Inadunata. Kongliga Svenska Vetenskaps-Akademiens Handlingar, 25(2), 200 p., 10 pls.

Bather, F. A. 1899. A phylogenetic classification of the Pelmatozoa. British Association for the Advancement of Science (1898):916–923.

Bather, F. A. 1900. Part III. The Echinoderma. The Pelmatozoa. Crinoidea, p. 94–204. Assisted by J. W. Gregory and E. S. Goodrich. *In* E. R. Lankester (ed.), A Treatise on Zoology. Adam and Charles Black, London.

Baumiller, T. K. 1992. Survivorship analysis of Paleozoic Crinoidea: effect of filter morphology on evolutionary rates. Paleobiology, 19:304–321.

Baumiller, T. K., and D. L. Meyer. 2000. Crinoid postures in oscillating current: insights into some biomechanical aspects of passive filter feeding. Geological Society of America, Abstracts with Programs, 32(7):372.

Bohaty, J. 2005. Doppellagige Kronenplaten: ein neues anatomisches Merkmal paläozoicher Crinoiden und Revision der Familie Cupressocrinitidae (Devon). Paläontologische Zeitschrift, 79:201–225.

Brower, J. C. 1992. Cupulocrinid crinoids from the Middle Ordovician (Galena Group, Dunleith Formation) of northern Iowa and southern Minnesota. Journal of Paleontology, 66:99–128.

Brower, J. C. 2002. *Quintuplexacrinus*, a new cladid crinoid genus from the Upper Ordovician Maquoketa Formation of the northern midcontinent of the United States. Journal of Paleontology, 76:993–1006.

Brower, J. C. 2006. Ontogeny of the food-gathering system in Ordovician crinoids. Journal of Paleontology, 80:430–446.

Brower, J. C., and J. Veinus. 1982. Long-armed cladid inadunates, p. 129–144. *In* J. Sprinkle. (ed.), Echinoderm faunas from the Bromide Formation (Middle Ordovician) of Oklahoma. University of Kansas Paleontological Contributions, Monograph, 1.

Chen, Z-T., and J-H. Yao. 1993. Palaeozoic echinoderm fossils of western Yunnan, China. Geological Publishing House, Beijing, 102 p., 16 pl.

Grabau, A. W. 1903. Notes on the development of the biserial arm in certain crinoids. American Journal of Science, series 4, 16:289–300.

Hall, J. 1858. Chapter 8. Palaeontology of Iowa, p. 473–724. *In* J. Hall and J. D. Whitney (eds.), Report of the Geological Survey of the state of Iowa: embracing the results of investigations made during portions of the years 1855, 56 & 57. Palaeontology, 1(II).

Hall, J. 1862. Preliminary notice of some of the species of Crinoidea known in the Upper Helderberg and Hamilton groups of New York. New York State Cabinet of Natural History, 15th Annual Report, p. 87–125, 1 pl.

Haude, R., and E. Thomas. 1992. Die unter-karbonischen Crinoiden von "Kohleiche" bei Wuppertal, p. 307–361. *In* E. Thomas (ed.), Oberdevon und Unterkarbon von Aprath im Bergischen Land (Nördliches Rheinisches Schiefergebirge). Verlag Sven von Loga, Köln.

Jaekel, O. 1906. *In* W. E. Schmidt, Der oberste Lenneschiefer zwischen Letmathe und Iserlohn. Zeitschrift der Deutschen Geologischen Gesellschaft, 57:544.

Jaekel, O. 1918. Phylogenie und System der Pelmatozoen. Paläontologische Zeitschrift, 3, 128 p.

Kammer, T. W., and W. I. Ausich. 1996. Primitive cladid crinoids from upper

Osagean–lower Meramecian (Mississippian) rocks of east-central United States. Journal of Paleontology, 70:835–866.

Kirk, E. 1929. The fossil crinoid genus *Vasocrinus* Lyon. Proceedings of the U.S. National Museum, 74(15):1–16.

Kirk, E. 1934. *Corynecrinus*, a new Devonian crinoid genus. Proceedings of the U.S. National Museum, 83(2972):1–7, 1 pl.

Knapp, W. D. 1969. Declinida, a new order of late Paleozoic inadunate crinoids. Journal of Paleontology, 43:340–391.

Lane, N. G., C. G. Maples, and J. A. Waters. 2001. Revision of Late Devonian (Famennian) and some Early Carboniferous (Tournaisian) crinoids and blastoids from the type Devonian area of North Devon. Palaeontology, 44:1043–1080.

Lane, N. G., and G. D. Webster. 1966. New Permian crinoid fauna from southern Nevada. University of California Publications in Geological Sciences, 63, 60 p., 13 pl.

Lyon, S. S. 1857. P. 467–497, pl. 1–5. *In* S. S. Lyon, E. T. Cox, and L. Lesquereux (eds.), Palaeontological Report. Geological Report of Kentucky, 3.

Lyon, S. S., and S. A. Casseday. 1859. Description of nine new species of crinoidea from the Subcarboniferous rocks of Indiana and Kentucky. American Journal of Science and Arts, series 2, 28:233–246.

Lyon, S. S., and S. A. Casseday. 1860. Description of nine new species of crinoidea from the Subcarboniferous rocks of Indiana and Kentucky. American Journal of Science and Arts, series 2, 29:68–79.

Macurda, D. B., and D. L. Meyer. 1975. The microstructure of the crinoid endoskeleton. University of Kansas Paleontological Contributions, 74, 22 p., 30 pl.

McIntosh, G. C. 1979. Abnormal specimens of the Middle Devonian crinoid *Bactrocrinites* and their effect on the taxonomy of the genus. Journal of Paleontology, 53:18–28.

McIntosh, G. C. 1984. Devonian cladid inadunate crinoids: family Botryocrinidae Bather, 1899. Journal of Paleontology, 58:1260–1281.

McIntosh, G. C. 2001. Devonian cladid crinoids: families Glossocrinidae Goldring, 1923, and Rutkowskicrinidae new family. Journal of Paleontology, 75:783–807.

McIntosh, G. C., and C. E. Brett. 1988. Occurrence of the cladid inadunate crinoid *Thalamocrinus* in the Silurian (Wenlockian) of New York and Ontario. Royal Ontario Museum, Life Sciences Contributions, 149:1–17, 2 pl.

Meek, F. B. and A. H. Worthen. 1868. Palaeontology of Illinois. Illinois Geological Survey, 3:289–565, pl. 1–20.

Miller, J. S. 1821. A natural history of the Crinoidea, or lily-shaped animals; with observations on the genera, *Asteria, Euryale, Comatula* and *Marsupites*. Bryan & Co., Bristol, England, 150 p.

Miller, S. A. 1890. The structure, classification and arrangement of American Palaeozoic crinoids into families. American Geologist, 6:275–286, 340–357.

Miller, S. A., and W. F. E. Gurley. 1891. Description of some new genera and species of Echinodermata from the Coal Measures and Subcarboniferous rocks of Indiana, Missouri, and Iowa. Indiana Department of Geology and Natural Resources, 16th Annual Report:327–373, pl. 1–10.

Miller, S. A., and W. F. E. Gurley. 1895. New and interesting species of Palaeozoic fossils. Illinois State Museum Bulletin, 7:1–89, 5 pl.

Miller, S. A., and W. F. E. Gurley. 1897. New species of crinoids, cephalopods, and other Palaeozoic fossils. Illinois State Museum Bulletin, 12:1–69, 5 pl.

Moore, R. C. 1939. The use of fragmentary crinoidal remains in stratigraphic paleontology. Denison University Bulletin Journal of the Scientific Laboratories, (1938), 33:165–250, pl. 1–4.

Moore, R. C.,and L. R. Laudon. 1943. Evolution and classification of Paleozoic crinoids. Geological Society of America Special Paper, 46:1–151, 14 pl.

Moore, R. C., and F. B. Plummer. 1940. Crinoids from the Upper Carboniferous and Permian strata in Texas. University of Texas Publication, 3945, 468 p., 21 pl.

Moore, R. C., and C. Teichert (eds.). 1978. Treatise on Invertebrate Paleontology, pt. T. Echinodermata 2, 3 volumes. Geological Society of America and University of Kansas Press, Lawrence, 1027 p.

Moore, R. C., N. G. Lane, and H. L. Strimple. 1978. Suborder Dendrocrinina Bather, 1899, p. 606–630. In R. C. Moore and C. Teichert (eds.), Treatise on Invertebrate Paleontology, pt. T. Echinodermata 2. Geological Society of America and University of Kansas Press, Lawrence.

Roemer, C. F. 1852–54. Erste Periode, Kohlen-Gebirge, p. 210–291, pl. 4. In Lethaea Geognostica, volume 2, H. G. Bronn, 1851–1856, third edit. E. Schweizerbart, Stuttgart.

Schmidt, W. E. 1934. Die Crinoideen des Rheinischen Devons, I. Teil; Die Crinoideen des Hunsrückschiefers. Abhandlung der Preussischen Geologischen Landesanstalt, 163:1–149, 34 pl.

Simms, M. J., and G. D. Sevastopulo. 1993. The origin of articulate crinoids. Palaeontology, 36:91–109.

Springer, F. 1911. On a Trenton echinoderm fauna. Canada Department Mines, Memoir, 15-P, 70 p., 5 pl.

Springer, F. 1926. American Silurian crinoids. Smithsonian Institution Publication, 287, 239 p., 33 pl.

Strimple, H. L. 1962. Crinoids from the Oologah Formation. Oklahoma Geological Survey Circular, 60, 75 p.

Strimple, H. L. 1967. Aphelecrinidae, a new family of inadunate crinoids. Oklahoma Geology Notes, 27:81–85, 1 pl.

Strimple, H. L. and W. T. Watkins. 1969. Carboniferous crinoids of Texas with stratigraphic implications. Palaeontographica Americana, 6(40):141–275.

Stukalina, G. A. 2000. Siluriiskie krinoidei Sibirskoi Platformy [Silurian crinoids of the Siberian Platform]. Rossiiskaya Akademiua Nauk, Trudy Paleontologicheskogo Instituta, 278:1–108, 24 pl.

Teichert, C. 1949. Permian crinoid Calceolispongia. Geological Society of America, Memoir, 34, 132 p., 26 pl.

Ubaghs, G. 1978. Skeletal morphology of fossil crinoids, p. 58–216. In R. C. Moore and C. Teichert (eds.), Treatise on Invertebrate Paleontology, pt. T. Echinodermata 2. Geological Society of America and University of Kansas Press, Lawrence.

Van Sant, J. F. 1964. Crawfordsville crinoids, p. 34–136. In J. F. Van Sant and N. G. Lane, Crawfordsville (Indiana) crinoid studies. University of Kansas Paleontological Contributions, Echinodermata Article, 7.

Wachsmuth, C. 1868. Notes on some points in the structure and habits of Paleozoic Crinoidea. Proceedings of the Academy of Natural Sciences of Philadelphia, 19:323–334.

Wachsmuth, C., and F. Springer. 1880–1886. Revision of the Palaeocrinoidea. Pt. I. The families Ichthyocrinidae and Cyathocrinidae (1880), p. 226–378, pl. 15–17. Pt. II. Family Sphaeroidocrinidae, with the sub-families Platycrinidae, Rhodocrinidae, and Actinocrinidae (1881), p. 177–411, pl. 17–19. Pt. III, sec. 1. Discussion of the classification and relations of the brachiate crinoids, and conclusion of the generic descriptions (1885), p. 225–364, pl.

4–9. Pt. III, sec. 2. Discussion of the classification and relations of the brachiate crinoids, and conclusion of the generic descriptions (1886), p. 64–226. Proceedings of the Academy of Natural Sciences, Philadelphia.

Wanner, J. 1916. Die Permischen echinodermen von Timor, I. Teil. Palaontologie von Timor, 11:1–329, pl. 94–114.

Wanner, J. 1937. Neue beiträge zur kenntnis der Permischen echinodermen von Timor, VIII-XIII. Palaeontographica, Supplement 4, IV Abteilungen, Abschnitt 1, 212 p., 14 pl.

Webby, D. B., R. A. Cooper, and S. M. Bergstrom. 2004. Stratigraphic framework and time slices, p. 41–47. *In* D. B. Webby, F. Paris, M. L. Droser, and I. G. Percival (eds.), The Great Odovician Biodiversification Event. Columbia University Press, New York.

Webster, G. D. 1997. Lower Carboniferous echinoderms from northern Utah and western Wyoming. Utah Geological Survey Bulletin 128, Paleontology Series, 1, 65 p.

Webster, G. D. 2003. Bibliography and index of Paleozoic crinoids, coronates, and hemistreptocrinoids, 1758–1999. Geological Society of America, Special Paper, 363, 2335 p. <http://crinoid.gsajournals.org/crinoidmod/>.

Webster, G. D., and P. A. Jell. 1999. New Permian crinoids from eastern Australia. Memoirs of the Queensland Museum, 43:279–340.

Webster, G. D., and B. S. Kues. 2006. Pennsylvanian crinoids of New Mexico. New Mexico Geology, 28(1):1–36.

Webster, G. D., and N. G. Lane. 1967. Additional Permian crinoids from southern Nevada. University of Kansas Paleontological Contributions, Paper, 27:1–32, 8 pl.

Webster, G. D., and N. G. Lane. 2007. New Permian crinoids from the Battleship Wash patch reef in southern Nevada. Journal of Paleontology 81:953–967.

Webster, G. D., and C. G. Maples. 2006. Cladid crinoid (Echinodermata) anal conditions: a terminology problem and proposed solution. Palaeontology, 49:187–212.

Webster, G. D., D. J. Hafley, D. B. Blake, and A. Glass. 1999. Crinoids and stelleroids (Echinodermata) from the Broken Rib Member, Dyer Formation (Late Devonian, Famennian) of the White River Plateau, Colorado. Journal of Paleontology, 73:461–486.

Webster, G. D., C. G. Maples, R. Mawson, and M. Dastanpour. 2003. A cladid-dominated Early Mississippian crinoid and conodont fauna from Kerman Province, Iran and revision of the glossocrinids and rhenocrinids. Journal of Paleontology, Memoir, 60, 77(3 supp.), 35 p.

Webster, G. D., C. G. Maples, G. D. Sevastopulo, T. Frest, and J. A. Waters. 2004. Carboniferous (Viséan–Moscovian) echinoderms from the Béchar Basin area of western Algeria. Bulletins of American Paleontology, 368, 98 p.

Webster, G. D., R. T. Becker, and C. G. Maples. 2005. Biostratigraphy, paleoecology, and taxonomny of Devonian (Emsian and Famennian) crinoids from southeastern Morocco. Journal of Paleontology, 79:1052–1071.

Wright, J. 1937. Scottish Carboniferous crinoids. Geological Magazine, 74:385–411, pls. 13–16.

Figure 11.1. *Plate terminology used for the peristomial border. **1**, Glyptocystitoid **Lepadocystis moorei** UC 57349. Note that the first brachiole facets are shared between the first primary floor plates and the adjacent proximal oral plate. **2**, Diploporan **Protocrinites fragum** Eichwald, 1856, PIN 257/671. **3**, Crinoid **Hybocrinus crinerensis** Strimple and Watkins, 1949 (modified from Sprinkle, 1982a). Oral plates (O1–O7) with light shading. Most proximal primary floor plates of shared ambulacra (P0) and most proximal primary floor plate on A ambulacrum (P1). Primary floor plates (P1–Pn) and secondary floor plates (S1–Sn) are not shaded. A–E are the distal ambulacra, and BC and DE are the shared ambulacra.*

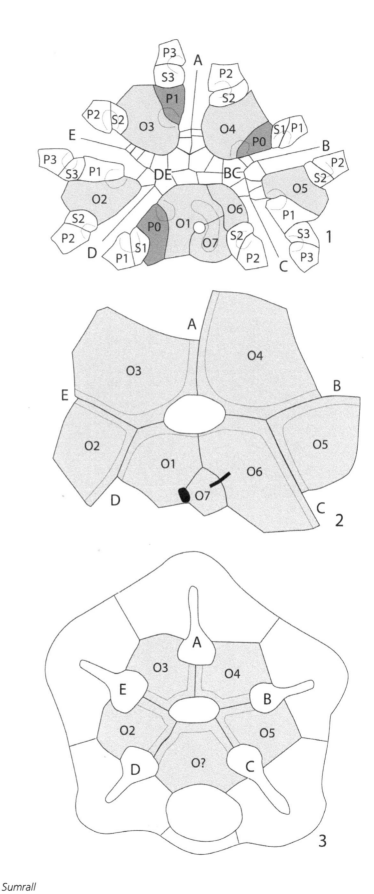

THE ORIGIN OF LOVÉN'S LAW IN GLYPTOCYSTITOID RHOMBIFERANS AND ITS BEARING ON THE PLATE HOMOLOGY AND HETEROCHRONIC EVOLUTION OF THE HEMICOSMITOID PERISTOMIAL BORDER

11

Colin D. Sumrall

Homology is a contentious issue in echinoderm paleontology, as recent articles discussing the cup plating of crinoids illustrates (Simms, 1994; Ausich, 1996, and later; Guensburg and Sprinkle, 2003). Yet no issue is more important for the understanding of the phylogenetic history of a clade, for without robust and accurate interpretations of homology, phylogenetic reconstruction becomes a meaningless exercise. Echinoderms bear a highly complex, mesodermal skeleton that is geologically stable and therefore provides high-resolution data on which hypotheses of homology can be examined. This study investigates the development of the oral plating and proximal ambulacral floor plates of glyptocystitoid rhombiferans to argue for homology of elements across the clade and other blastozoans. It shows that Lovén's Law (Lovén, 1874) (similar plating of the A, C, and E ambulacra that is different from the similarly plated B and D ambulacra) results from the ontogenetic insertion of the lateral oral plates O2 and O5. Then this homology and developmental trajectory will be used to reinterpret the oral plating of the closely related triradial peristomial border in the hemicosmitoid rhombiferan *Hemicosmites* von Buch, 1840, through paedomorphic ambulacral reduction (Sumrall and Wray, 2007).

Hypotheses of heterochrony, evolution by changes in the timing of the developmental trajectory, are commonly argued for groups of fossil echinoderms (Sprinkle and Bell, 1978; Broadhead, 1981, 1987, 1988; McKinney, 1984; Waters et al., 1985, Sumrall, 1993, 2000; Broadhead and Sumrall, 2003; Sumrall and Wray, 2007). These studies typically attempt to demonstrate the existence of one or more types of heterochrony or drastic changes to the ambulacral system. Hypotheses of heterochrony require knowledge of both the ancestral ontogenetic sequence and the phylogenetic position of the organisms in question. The ancestral ontogenetic sequence allows a detailed understanding of the plesiomorphic ontogenetic trajectory hypothesized to have changed evolutionarily and the identification of homologous elements. In contrast, a phylogenetic hypothesis provides the polarity of evolutionary change.

Introduction

Phylogenetically, glyptocystitoids and hemicosmitoids are highly derived blastozoans, closely related to trachelocrinids, blastoids, and coronoids (Paul, 1988; Sumrall, 1997). This clade is characterized by having double biserial plating of the ambulacral floor plates with brachioles placed along the sutures between pairs of primary and secondary floor plates. Furthermore, plesiomorphically, these advanced blastozoans have three oral plates in the CD interambulacrum, but this number can be reduced to one or two, as in many blastoids and coronoids (Beaver et al., 1967; Brett et al., 1983). This plating of the peristomial border is plesiomorphic for the clade: a similar arrangement of interradially positioned oral plates with multiple oral plates in the CD interambulacrum is also known for paracrinoids, parablastoids, some diploporan groups such as *Protocrinites* Eichwald, 1840, and plesiomorphic crinoids (Fig. 11.1).

Terminology

In this chapter, the ambulacra are labeled A through E following the Carpenter (1884) system. The shared BC and shared DE ambulacra are the portions of the lateral ambulacra proximal to the bifurcation point. Oral plates O1–O7 are the interradially positioned thecal plates forming or adjacent to the peristomial opening that bear the proximal portions of the ambulacra along their adjacent sutures. Identity of these plates follows the terminology developed for glyptocystitoids by Kesling (1967). Primary floor plates P1–Pn are the large ambulacral floor plates that form the distal half of each brachiole facet. Secondary floor plates S1–Sn are smaller floor plates that form the proximal half of all but the proximalmost brachiole facet in each ambulacrum. P0 are primary ambulacral floor plates borne on the shared BC and shared DE ambulacra that share a brachiole facet across the O1 and O4 sutures. Brachioles are small biserial feeding appendages that attached to brachiole facets that are shared between primary and secondary floor plates or the first primary floor plate and an oral plate in each ambulacrum (Fig. 11.1.1).

General Features of Early Stemmed Echinoderm Oral and Ambulacral Areas

Primitively, the ambulacral system of stemmed echinoderms is developed into an arrangement known as 2-1-2 symmetry (Sprinkle, 1973). The 2-1-2 ambulacral arrangement is characterized by having three ambulacra radiating from the centrally located peristomial opening, two of which branch distally, forming the typical pentaradiate symmetry. In the triradial portion of the ambulacral system, the A ambulacrum projects anteriorly (i.e., away from the anus in standard orientation), and the shared BC and shared DE ambulacra project to the right and left, respectively. Bifurcation of the shared ambulacra forms the distal B, C, D, and E ambulacra. Although some have argued phylogenetically that echinoderms arose from unirayed taxa (Jefferies, 1986; Smith, 2005), these hypotheses are largely based on misinterpretation of "homalozoan" appendage anatomy and inverted character polarity. Regardless, these issues are not relevant to this chapter because advanced blastozoans are unequivocally derived from pentaradiate taxa (Paul, 1988; Sumrall, 1997).

Many advanced blastozoans (sensu Sumrall, 1997) have homologous oral plates, including crinoids (Fig. 11.1.3), glyptocystitoids (Fig. 11.1.1), hemicosmitoids (radials in part), blastoids (deltoids), some diploporans (Fig. 11.1.2), parablastoids, and several others. In all cases, these oral plates are positioned interradially, form the border of the peristomial opening, and bear the proximal portion of the ambulacra along their sutures without underlying floor plates. In many taxa, especially those with a strongly developed 2-1-2 ambulacral arrangement, the lateral oral plates O2 and O5 do not share in the peristomial border (Fig. 11.1.2), whereas in other taxa with a more weakly developed 2-1-2, O2 and O5 share the peristomial border (Fig. 11.1.3). Furthermore, in many taxa there are extra plates in the CD interambulacrum that here are called O6 and O7 following Kesling (1967). These accessory orals are differently expressed in some clades and variously lost phylogenetically in others (Fig. 11.1).

Ontogenetically the 2-1-2 ambulacral system develops in three stages. The shared ambulacra develop, forming a marked peristomial elongation present in most early echinoderms. Next, the A ambulacrum develops, forming the triradial portion of the proximal ambulacra. Finally, the lateral ambulacra bifurcate, completing the five-part symmetry of the echinoderm ambulacral system (Bell, 1976; Sumrall, 2000; Sumrall and Wray, 2007). Consequently, such echinoderms pass through bilateral, triradial, and pentaradial phases of the ambulacral system during development.

Sumrall (2000) and Sumrall and Wray (2007) documented that paedomorphic loss of any combination of these three stages of the ambulacral developmental trajectory can result in ambulacral systems that bear one through four ambulacra and that these changes occur repeatedly throughout echinoderm phylogeny. These changes include both large-scale changes, as in some of the homalozoan clades where a reduced number of ambulacra is the norm, as well as small subclades, such as advanced callocystitid glyptocystitoids, where a reduced number of ambulacra (four) is typical but does not greatly modify the overriding body plan. In this chapter, paedomorphic ambulacral reduction is shown for one unusual clade (Hemicosmitidae Jaekel, 1918) where paedomorphic loss of the lateral bifurcation stage of development allows for a new interpretation of oral plate homology.

DEVELOPMENT. Two studies (Sumrall and Sprinkle, 1999; Sumrall and Schumacher, 2002) have documented the three-part developmental trajectory in glyptocystitoid rhombiferans. In *Lepadocystis moorei* (Meek, 1871), a specimen with a thecal height of 1.1 mm was shown to lack two sets of thecal plates, including the five-plate radial circlet and two plates from the oral circlet, O2 and O5 (Sumrall and Sprinkle, 1999). A slightly larger specimen, measuring 1.7 mm in thecal height, had a small open radial plate circlet and five rudimentary ambulacra interpreted as terminal brachiole facets, but the plating of the summit was somewhat obscure. A clear and unambiguous O2 plate was present, indicating that the placement of the lateral oral plates occurs later in ontogeny than the proximal peristomial bordering oral plates O1, O3, O4, O6, and O7 (Sumrall and Sprinkle,

Glyptocystitoids

1999). No proximal ambulacral floor plates were present in either specimen, but this may be a result of inadequate preservation and not indicative of a lack of development of these plates.

Sumrall and Schumacher (2002) described a 5-mm-high specimen of *Cheirocystis fultonensis* that had incomplete development of the ambulacral system. This specimen had extremely well-developed, laterally shared ambulacra resulting from proportionately large peristomial bordering oral plates O1, O3, O4, O6, and O7 and disproportionately small bifurcating orals O2 and O5. This was thought to reflect the late development of the bifurcating oral plates. This specimen also has the distal ambulacra limited to a single brachiole facet in each of the five ambulacra, although plating is insufficiently preserved to interpret details.

Other glyptocystitoids also have single terminal brachioles that can be interpreted to have resulted from paedomorphic evolution through developmental loss of more distal brachiole facets, including *Tyrridocystis* Broadhead and Strimple, 1978, *Sprinkleocystis*, Broadhead and Sumrall, 2003, and pleurocystitids. Although the brachioles of pleurocystitids have been interpreted as erect ambulacra (Parsley, 1970; Broadhead, 1974), this seems unlikely because of the biserial (brachiole-like) rather than double biserial (floor plate–like) plating of these structures and their lack of brachioles, which would be expected if they were of floor plate origin.

An unusual feature of nearly all glyptocystitoids is the branching pattern of the brachioles on the ambulacra. In most cases, the A, C, and E ambulacra bear the first brachiole on the left side of the food groove and alternate right to left distally if viewed proximally to distally (Fig. 11.1.1). The B and D ambulacra, however, have the first two brachioles on the left side before alternating distally (Paul, 1984) (Fig. 11.1.1). These differences follow Lovén's Law (Lovén, 1874) that states the A, C, and E ambulacra are differently plated than the B and D ambulacra. Also noteworthy is the first brachiole in each of the ambulacra is borne on the suture between the first left primary ambulacral floor plate and the oral plate on its left side (Fig. 11.1.1). The remaining brachioles are borne along sutures between pairs of primary and secondary floor plates. However, there are exceptions. In *Cheirocystis fultonensis*, all of the brachioles branch along the left side of the ambulacrum (Sumrall and Schumacher, 2002); and in Devonian forms with a split hydropore, such as *Strobilocystites* White, 1876, and *Lipsanocystis* Ehlers and Leighley, 1922, the B and D ambulacra lack this first left brachiole facet. Finally, the peristomial opening is not shared by all of the oral plates. The peristomial bordering oral plates are located in the AB, CD, and DE interambulacra, whereas O2 and O5 located in the DE and BD interambulacra do not bear the edge of the peristome.

ONTOGENETIC MODEL. The simplest explanation for the anomalous first left brachiole on the B and D ambulacrum is that these are the brachioles of the shared BC and shared DE ambulacra. Early during ontogeny, the glyptocystitoid summit is plated with only O1, O3, O4, O6, and O7 (Sumrall and Sprinkle, 1999). Three rudimentary ambulacra are borne along the O3/O4 (A ambulacrum), O4/O1–O6 (shared BC ambulacrum), and O1/O3 (shared DE ambulacrum) sutures (Fig. 11.2.1). Brachioles are

Figure 11.2. *Ontogenetic model for the development of the glyptocystitoid peristomial border.* **1**, *Early stage of development with five peristomial bordering oral plates (O1, O3, O4, O6, O7) and three ambulacra;* **2**, *insertion of brachioles and first primary floor plates on the A (P1), shared BC and shared DE ambulacra (P0) (dark shading);* **3**, *insertion of the lateral bifurcating oral (O2 and O5) plates to the right of the P0 plates of the shared ambulacra forming five ambulacra;* **4**, *development of the distal ambulacra. Note that U-shaped brachiole facets are mounted on pairs of primary and secondary floor plates and that the first brachiole is on the left of each ambulacrum.* **5**, *Camera lucida drawing of* **Lepadocystis moorei** *UC 57349 with plates labeled and shaded the same as the model.*

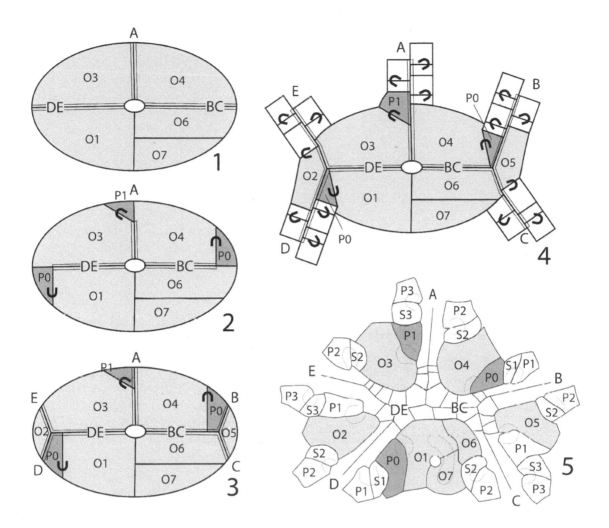

added on the left side of each of these ambulacra by the insertion of the first primary floor plate in the A ambulacrum (P1) and the first primary ambulacral floor plate of the shared BC and DE ambulacra (P0) (Fig. 11.2.2). Next, O2 and O5 are inserted along the right side of each of the P0 plates (viewed proximally to distally), resulting in the bifurcation of the lateral shared ambulacra (Fig. 11.2.3). This forms the proximal portions of the distal B, C, D, and E ambulacra along the O4–P0/O5, O5/O6, O1–P0/O2, and O2/O3 sutures, respectively (Fig. 11.2.3). The distal ambulacra develop as additional primary-secondary floor plate pairs are added along the growing tip of each ambulacrum beginning on the left side (Fig. 11.2.4). With maturity, the facet for the brachiole of the shared ambulacrum shifts slightly distally, placing them along the proximalmost B and D ambulacra. The specimen figured shows the brachiole facet of the BC shared ambulacrum still emptying into the shared ambulacrum (Fig. 11.2.5).

DISCUSSION. This developmental model can explain the morphology of nearly all glyptocystitoid rhombiferans despite being based on a relatively derived callocystitid *Lepadocystis* Carpenter, 1891. Although callocystitids are derived, *Lepadocystis* is the oldest known member of the clade and has many features common to other glyptocystitoid groups, including the retention of the A ambulacrum, a relatively large number of pectinirhombs, and a retention of all thecal plates except for R5. Certain glyptocystitoids, including some echinoencrinitids, pleurocystitids, and some callocystitids, have reduced summits that have not been examined in detail for this study. *Cheirocystis fultonensis* lacks all brachiole facets on the right side of the ambulacra. It is unclear whether this represents a loss of half of the floor plate complement or a shift of the right side plates to the left side of the ambulacra. *Hesperocystis deckeri* Sinclair, 1945, elaborates the side food grooves of the ambulacra by developing clusters up to three brachiole facets (Sprinkle, 1982b). Even so, the placement of the first two brachiole clusters of the B and D ambulacra are on the left, consistent with the developmental model.

Hemicosmitoids

GENERAL FEATURES. Hemicosmitoids are a clade of stemmed echinoderms characterized by a theca with hemicosmitoid plating four basals (three in Thomacystidae Paul, 1969), six infralaterals, eight or nine laterals, a variable complement of "radials," cryptorhombs, and erect double biserial ambulacra that bear short biserial brachioles (Sprinkle, 1975; Bockelie, 1979; Paul, 1984). This group has unusual threefold symmetry in the ambulacral system and the thecal plating. This clade ranges from Middle Ordovician through Early Devonian and consists of three groups: Hemicosmitidae; Caryocrinitidae Bernard, 1895; and Thomacystidae (Paul, 1984).

Hemicosmitidae, typically thought to be the most primitive group, is characterized by four basals, six infralaterals, nine laterals, and nine radials (the peristomial bordering circlet described here) (Kesling, 1967; Bockelie, 1979; Paul, 1984). Three of these radials (R3, R6, and R9) are interradial, whereas six (R1, R2, R4, R5, R7, and R8) are paired perradially (Fig. 11.3). Furthermore, small accessory plates called wedge plates (Bockelie, 1979) are placed consistently among the nine orals (Fig. 11.3). The triangular peristomial opening is covered with small biserial cover plates, and three erect ambulacra bearing brachioles are borne on large facets at the corners of the triangular peristome.

Bockelie (1979) suggested that the number and position of wedge plates are somewhat variable. However, in all specimens where these wedge plates were observed, three occur to the left of the large arm facets, and two are associated with the hydropore opening. Observations for this study found wedge plates present in most specimens of several species, and where they were not observed, the specimens were either abraded or otherwise poorly preserved. Wedge plates are difficult to find in juvenile specimens, probably because of their small size, but these plates become increasingly more obvious through ontogeny. These wedge plates are also associated with a brachiole facet that is shared between the wedge plate and the adjacent radially positioned radial plate to its right. Bockelie (1979) speculated that wedge plates were important for increasing the size of the peristome.

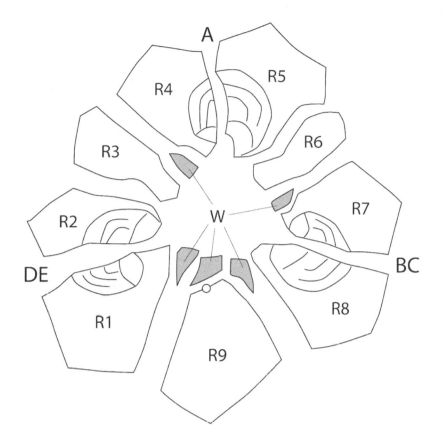

Figure 11.3. *Plating of summit of hemicosmitid* **Hemicosmites** *sp. PIN4125/76. R1–R9 are radials; W are wedge plates (plate designations after Bockelie, 1979).*

Caryocrinitidae is similar in thecal plating, except details of the peristomial area cannot be determined because the cover plates over the peristome are enlarged into a tegmen. Erect arms arise from several facets around the margin of the summit clustered in three areas, reflecting the triradial nature of the ambulacral system (Sprinkle, 1975).

Thomacystidae is different from other hemicosmitids both in terms of thecal plating and the peristomial border (Paul, 1984). This group, known from a handful of specimens, has three basals, six infralaterals, eight laterals, and five radials (orals) separating four ambulacra B, C, D, and E.

The remainder of the discussion will concentrate on the peristomial border of the hemicosmitid *Hemicosmites*. Orientation of the summit is based on three factors. 1) The triradial nature of the ambulacral system as it exits the peristome (a remnant of the 2-1-2 symmetry) indicates that the A, shared BC, and shared DE ambulacra are present (Sprinkle, 1973; Bell, 1976). 2) The hydropore primitively is located in the CD interambulacral area. 3) The periproct primitively is located in the CD interarea. This indicates that the ambulacrum opposite the hydropore and anus is A, the one to the right is shared BC, and the one to the left is shared DE.

ONTOGENETIC MODEL. The simplest explanation for the unusual triradiate peristomial border of *Hemicosmites* is that it evolved by paedomorphic loss of the bifurcation of the lateral shared ambulacra into the distal B, C, D, and E ambulacra. Early during ontogeny, the summit of *Hemicosmites* is plated with only O1, O3, O4, O6, and O7 (Fig. 11.4.1). Three rudi-

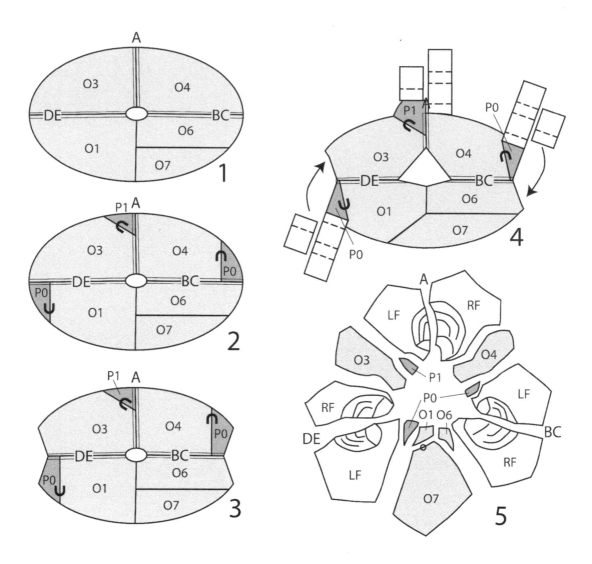

mentary ambulacra are borne along the O3/O4 (A ambulacrum), O4/O1–O6 (shared BC ambulacrum), and O1/O3 (shared DE ambulacrum) sutures (Fig. 11.4.1). Brachioles are added on the left side of each of these ambulacra by the insertion of the first primary floor plate of the A ambulacrum (P1) and the first primary floor plate of the shared ambulacra (P0), i.e., three of the wedge plates (Fig. 11.4.2). To this point, the development is the same as in glyptocystitoids (Fig 11.2). However, unlike glyptocystitoids, O2 and O5 fail to develop and are lost through paedomorphic ambulacral reduction. Loss of these plates causes a failure of the lateral ambulacral bifurcation resulting in three ambulacra at maturity: A, shared BC, and shared DE (Fig. 11.4.3). The proximal portions of the distal ambulacra develop as additional primary-secondary floor plate pairs are added along the growing tip of each ambulacrum (Fig. 11.4.4). However, these plates fuse such that the plates of the right and left sides of each ambulacrum develop into paired perradial plates forming the radial edges of each of the three ambulacra. The erect, brachiole-bearing portions of the ambulacra mount on these fused floor plates with maturity.

DISCUSSION. This developmental model is conjectural but is fully testable if a complete developmental series were discovered. It is based on a modification of the developmental sequence of the closely related glyptocystitoids that uses the paedomorphic ambulacral reduction model of Sumrall and Wray (2007). Although it is clear that the paired arm facet plates are modified floor plates fused and incorporated into the peristomial border, it is uncertain from which ambulacrum they derive. On the right, for example, they may be derived either from the B or C ambulacrum or from a combination of left plates from the B ambulacrum and right plates from the C ambulacra. All options are equivocal at present. The peristomial border of Caryocrinitidae remains unknown because the peristome is covered by large peristomial cover plates.

The peristomial border of *Thomacystis* Paul, 1969, is uninterruptible with the paedomorphic ambulacral reduction model. Although the symmetry and position of the gonopore and anus suggest the four ambulacra are B, C, D, and E (Paul, 1984), the oral plates present cannot be reconciled with this interpretation. In all glyptocystitoids that are missing the A ambulacrum, the O3, O4 suture, which plesiomorphically bears the proximal portion of the A ambulacrum, is still present. However, in *Thomacystis*, either O3 or O4 is absent, resulting in only a single plate separating the E and B ambulacrum. This radically different development of the ambulacral system suggests that Thomacystidae branched from other hemicosmitids before the evolution of the triradial ambulacral system of Hemicosmitidae and Caryocrinitidae.

The position of the initial brachiole facet of each of the three ambulacra of *Hemicosmites* is slightly shifted from what one would expect. In glyptocystitoids, the first brachiole facet on each ambulacrum crosses the first primary oral plate suture. In *Hemicosmites*, this brachiole facet crosses sutures of the first primary floor plate and the fused arm plates. Brachiole facets are known to be differently plated in different groups, including multiple facets per floor plate pairs (Sprinkle, 1982b); some clades of diploporans have brachiole facets on thecal plates without underlying floor plates, as in *Glyptosphaerites* Müller, 1854. It is suggested that this anomalous placement results from a slight positional shift to keep the brachiole more in line with the main ambulacral food groove.

The anomalous placement of the first two brachioles on the left side of the B and D ambulacra of glyptocystitoids results from the proximal facet being the brachiole of the shared BC and DE ambulacra. These brachioles are ontogenetically shifted to the B and D ambulacra when the shared ambulacra bifurcate as O2 and O5 are added to the right of these plates. Then the distal ambulacra grow, adding pairs of floor plates beginning on the left.

The model for the development of the peristomial border of *Hemicosmites* clarifies the plating and homology of its unusual triradial summit. Three of the five small, wedge-shaped plates in the peristomial border of

Homology and Conclusions

Figure 11.4. *Ontogenetic model for development of hemicosmitoid peristomial border.* **1**, *Early stage of development with five peristomial bordering oral plates (O1, O3, O4, O6, O7) and three ambulacra;* **2**, *insertion of brachioles and first primary floor plates on the A (P1), shared BC and shared DE ambulacra (P0) (dark shading);* **3**, *failure to develop O2 and O5, resulting in the retention of only the three juvenile ambulacra;* **4**, *addition and perradial fusion of the first floor plates. To reach full maturity, the peristome must enlarge so that the fused floor plates are in contact.* **5**, *mature morphology of* **Hemicosmites** *sp. PIN4125/76 shaded and labeled identically to the model.*

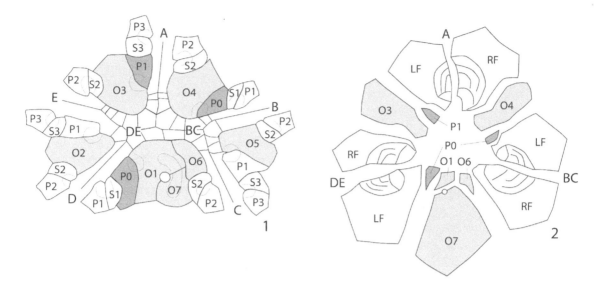

Figure 11.5. *Comparison of plating of glyptocystitoid* **Lepadocystis** *and hemicosmitoid* **Hemicosmites**. *Dark shading indicates the first primary floor plate in the A ambulacrum and the primary floor plates of the shared ambulacra. Light shading indicates the oral plates. Unshaded plates are primary and secondary floor plates of the distal ambulacra. RF and LF are right and left fused floor plates. A–E, BC, and DE are the ambulacral identities.*

Hemicosmites are interpreted as homologues of the first primary ambulacral floor plates in the A, shared BC, and shared DE ambulacra. The remaining two wedge plates in the CD interambulacrum are interpreted as homologues of O1 and O6 (Fig. 11.5). The three large radial plates that lie interradially, R9, R3, and R6 are interpreted as homologues of the proximal oral plates O7, O3, and O4, respectively. Note that in glyptocystitoids, O7 is much smaller, but the position as distal to O1 and O6, and the presence of the hydropore along its suture with O1 are consistent (Fig. 11.5). The paired perradial, arm facet–bearing radials R1, R2, R4, R5, R7, and R8 are interpreted as fused proximal ambulacral floor plates of the D and or E, A, and B and or C ambulacra, respectively.

High-resolution data from the information-rich echinoderm skeleton can be useful in understanding evolutionary patterns. Unlike the irregular thecal plating of more plesiomorphic echinoderms, such as edrioasteroids and eocrinoids, the plating of the ambulacral system has an exact elemental homology scheme that can be recovered in closely related taxa, such as those discussed here. Both glyptocystitoids and hemicosmitoids follow Lovén's Law in the placement of brachioles resulting from the developmental insertion of primary ambulacral floor plates on which the single pair of shared ambulacral brachioles mount on the left side. In glyptocystitoids, development of the distal ambulacra with alternate brachioles starting on the left results in ambulacra B and D bearing the first two brachiole facets on the left. Failure of the distal ambulacra to bifurcate in hemicosmitoids results in an apparent loss of any manifestation of Lovén's Law, except for its footprint, the primary ambulacral floor plates of the shared ambulacra.

The model for the origin of the peristomial bordering plates in hemicosmitoids is fully testable. It requires a fortuitous discovery of a growth series of small, postmetamorphosis *Hemicosmites* that show the development of the summit. Similar studies for glyptocystitoids (Sumrall and Sprinkle, 1999) and blastoids (Sevastopulo, 2005) have shown great research potential.

This chapter was greatly improved through numerous discussions with J. Sprinkle, University of Texas; K. S. Koverman, University of Chicago; J. A. Waters, Appalachian State University; and G. A. Wray, Duke University. S. V. Rozhnov, Palaeontological Institute, Moscow; and P. Wagner, Field Museum, Chicago, provided access to specimens of *Hemicosmites* and *Lepadocystis* for this study. T. E. Guensburg and S. Q. Dornbos provided comments that improved the chapter. This study was funded in part by a grant from the National Geographic Society.

Acknowledgments

References

Ausich, W. I. 1996. Crinoid plate circlet homologies. Journal of Paleontology, 70:955–964.

Beaver, H. H., R. O. Fay, B. D. Macurda Jr., R. C. Moore, and J. Wanner. 1967. Blastoids, p. S298–S255. *In* R. C. Moore (ed.), Treatise on Invertebrate Paleontology. Pt. U. Echinodermata 1. Geological Society of America and University of Kansas Press, Lawrence.

Bell, B. M. 1976. Phylogenetic implications of ontogenetic development in the class Edrioasteroidea (Echinodermata). Journal of Paleontology, 50:1001–1019.

Bernard, F. 1895. Eléments de paléontologie. Paris, 1168 p.

Bockelie, J. F. 1979. Taxonomy, functional morphology and paleoecology of the Ordovician cystoid family Hemicosmitidae. Palaeontology, 22:363–406.

Brett, C. E., T. J. Fresl, J. Sprinkle, and C. R. Clement. 1983. Coronoidea: a new class of blastozoan echinoderms based on taxonomic reevaluation of *Stephanocrinus*. Journal of Paleontology, 57:627–651.

Broadhead, T. W. 1974. Reevaluation of the morphology of *Amecystis laevis* (Raymond) Journal of Paleontology, 48:670–673.

Broadhead, T. W. 1981. Carboniferous camerate crinoid subfamily Dichocrininae. Palaeontographica A, 176:81–157.

Broadhead, T. W. 1987. Heterochrony and the achievement of the multibrachiate grade in camerate crinoids. Paleobiology, 13:177–186.

Broadhead, T. W. 1988. Heterochrony—a pervasive influence in the evolution of Paleozoic crinoids, p. 115–123. *In* R. D. Burke, P. V. Mlandenov, P. Lambert, and R. L. Parsley (eds.), Echinoderm Biology. A. A. Balkema, Rotterdam.

Broadhead, T. W., and H. L. Strimple. 1978. Systematics and distribution of the Callocystitidae (Echinodermata, Rhombifera). Journal of Paleontology, 52:164–177.

Broadhead, T. W., and C. D. Sumrall. 2003. Heterochrony and paedomorphic development of *Sprinkleocystis ektopios*, new genus and species (Rhombifera, Glyptocystida) from the Middle Ordovician (Carodoc) of Tennessee. Journal of Paleontology, 77:113–120.

Buch, C. L. von. 1840. Über Spaeroniten und einige andere Geschlechter, aus welchen Crinoideen entstehen. Bericht über die zur Bekanntmachung geeigneten Verhandlungen der Koniglich Preussische Akademie der Wissenschaften zu Berlin, 1840:56–60.

Carpenter, P. H. 1884. Report on the Crinoidea—the stalked crinoids. Report on the scientific results of the voyage of the H. M. S. *Challenger*. Zoology, 11:1–440.

Carpenter, P. H. 1891. On certain points in the morphology of the Cystidea. Linnean Society of London (Zoology) Journal, 24(149):1–52.

Ehlers, G. M., and J. B. Leighley. 1922. *Lispanocysis traversensis*, a new cystoid from the Devonian of Michigan. Michigan Academy of Sciences, Arts and Letters Papers, 2:155–160.

Eichwald, E. von. 1840. Sur le systéme silurien de l'Esthonie. Journal de Medicine et d'Histoire Naturelle L'Académie de Médecine de St. Petersbourgh Journal, 1, 222 p.

Eichwald, E. von. 1856. Beitrag zur geographischen Verbreitung der fossilen Thiere Russlands. Alte Periode. Bulletin de la Societe impériale des naturalistes de Moscou, 29:88–127.

Guensburg, T. E., and J. Sprinkle. 2003. The oldest known crinoids (Early Ordovician, Utah) and a new crinoid plate homology system. Bulletins of American Paleontology, 364:1–43.

Jaekel, O. 1918. Phylogenie und system der pelmatozoan. Paläontologisches Zeitschrift, 3, 128 p.

Jefferies, R. P. S. 1986. The Ancestry of the Vertebrates. British Museum (Natural History), Cromwell, London, 376 p.

Kesling, R. V. 1967. Cystoids, p. S85–S267. In R. C. Moore (ed.) Treatise on Invertebrate Paleontology. Pt. U. Echinodermata 1. Geological Society of America and University of Kansas Press, Lawrence.

Lovén, S. 1874. Études sur les echinoidées. Kongelige Svenska Vetenskaps-kademiens Handlinger, new series, 11:1–91.

McKinney, M. L. 1984. Allometry and heterochrony in an Eocene echinoid lineage: morphological change as a by-product of size selection. Paleobiology, 10:407–419.

Meek, F. B. 1871. On some new Silurian crinoids and shells. American Journal of Science, series 3, 2:295–299.

Müller, J. H. J. 1854. Über den Bau der Echinodermen. Abhandlugen der Preussischen Akademie Wissenschaften Berlin (1853), 44:123–219.

Parsley, R. L. 1970. Revision of the North American Pleurocystitidae (Rhombifera—Cystoidea). Bulletins of American Paleontology, 58:135–213.

Paul, C. R. C. 1969. Thomacystis, a unique new hemicosmitid cystoid from Wales. Geological Magazine, 106:190–196.

Paul, C. R. C. 1984. British Ordovician cystoids, part 2. Palaeontographical Society Monograph, 136(563):65–152.

Paul, C. R. C. 1988. The phylogeny of the cystoids, p. 199–213. In C. R. C. Paul and A. B. Smith (eds.), Echinoderm Phylogeny and Evolution. Clarendon Press, Oxford.

Sevastopulo, G. D. 2005. The early ontogeny of blastoids. Geological Journal, 40:351–362.

Simms, M. J. 1994. Reinterpretation of thecal plate homology and phylogeny in the class Crinoidea. Lethaia, 26:303–312.

Sinclair, G. W. 1945. Some Ordovician echinoderms from Oklahoma. American Midland Naturalists, 34:707–716.

Smith, A. B. 2005. The pre-radial history of echinoderms. Geological Journal, 40:255–280.

Sprinkle, J. 1973. Morphology and evolution of blastozoan echinoderms. Harvard University Museum of Comparative Zoology, Special Publication, 283 p.

Sprinkle, J. 1975. The "arms" of Caryocrinites, a rhombiferan cystoid convergent on crinoids. Journal of Paleontology, 49:1062–1073.

Sprinkle, J. 1982a. Hybocrinus, p. 119–128. In J. Sprinkle (ed.) Echinoderm Faunas from the Bromide Formation (Middle Ordovician) of Oklahoma. University of Kansas, Lawrence.

Sprinkle, J. 1982b. Cylindrical and globular rhombiferans, p. 231–273. In J.

Sprinkle (ed.) Echinoderm Faunas from the Bromide Formation (Middle Ordovician) of Oklahoma. University of Kansas, Lawrence.

Sprinkle, J., and B. M. Bell. 1978. Paedomorphosis in edrioasteroid echinoderms. Paleobiology, 4:42–48.

Strimple, H. L., and W. T. Watkins. 1949. *Hybocrinus crinerensis*, new species from the Ordovician of Oklahoma. American Journal of Science, 247:131–133.

Sumrall, C. D. 1993. Thecal designs in isorophinid edrioasteroids. Lethaia, 26:289–302.

Sumrall, C. D. 1997. The role of fossils in the phylogenetic reconstruction of Echinodermata, p. 267–288. *In* J. A. Waters and C. G. Maples (eds.) Geobiology of Echinoderms. Paleontological Society Papers, 3.

Sumrall, C. D. 2000. Developmental control of ambulacral reduction in fossil echinoderms. Geological Society of America, Abstracts with Programs, 33(6):A72.

Sumrall, C. D., and G. A. Schumacher. 2002. *Cheirocystis fultonensis*, a new glyptocystitid rhombiferan from the Upper Ordovician of the Cincinnati Arch. Journal of Paleontology, 76:843–851.

Sumrall, C. D., and J. Sprinkle. 1999. Early ontogeny of the glyptocystitid rhombiferan *Lepadocystis moorei*, p. 409–414. *In* M. D. C. Carnevali and F. Bonasoro (eds.), Echinoderm Research 1998. A. A. Balkema, Rotterdam.

Sumrall, C. D., and G. A. Wray. 2007. Ontogeny in the fossil record: Diversification of echinoderm body plans and the evolution of "aberrant" symmetry in Paleozoic echinoderms through paedomorphic ambulacral reduction (PAR). Paleobiology, 33:149–163.

Waters, J. A., A. S. Horowitz, and B. D. Macurda Jr. 1985. Ontogeny and phylogeny of the Carboniferous blastoid *Pentremites*. Journal of Paleontology, 59:701–712.

White, C. A. 1876. Description of new species of fossils from Paleozoic rocks of Iowa. Proceedings of the Academy of Natural Sciences of Philadelphia, 28:27–34.

PART 4. MISSISSIPPIAN IMPACTS AND BIOMARKERS

INTRODUCTION TO PART 4

William I. Ausich and Gary D. Webster

Part 4 could be titled, "What will they think of next?" and presents two innovative studies that take advantage of the unique biologic and taphonomic attributes of echinoderms. Echinoderms are composed of innumerable plates. Each is composed of high-magnesium calcite and is a single crystal in optical continuity. Furthermore, these plates are constructed as a porous arrangement of calcite bars and spaces. When alive, the spaces are filled with soft tissues, but soon after death, this pore space is occluded with calcite cement that assumes the crystallographic orientation of the echinoderm plate. Apparently, during this process, some organic molecules become entombed within the calcite and can be preserved in ancient echinoderms. O'Malley, Ausich, and Chin (Chapter 13) present an initial account of how to isolate and analyze organic molecules preserved in this way.

This and other taphonomic attributes of echinoderms are used by Miller and others (Chapter 12) to solve an important geological problem. Large meteorite impacts occurred commonly throughout the history of Earth. They leave a scar in the lithosphere from the destruction of the impact, including a confused jumble of rocks from the lithosphere that rebound into the disturbed zone. Because of the destructive nature of impacts, dating the time of impact with precision is a major challenge. The facts that echinoderms fall apart quickly after death, that cements fill internal porosity within plates, and that eventually cement binds unconsolidated sediment into a rock, leads to predictions for a temporal sequence to help, along with echinoderm and conodont biostratigraphy, to date the impact of the Weaubleau Structure (Miller et al., Chapter 12).

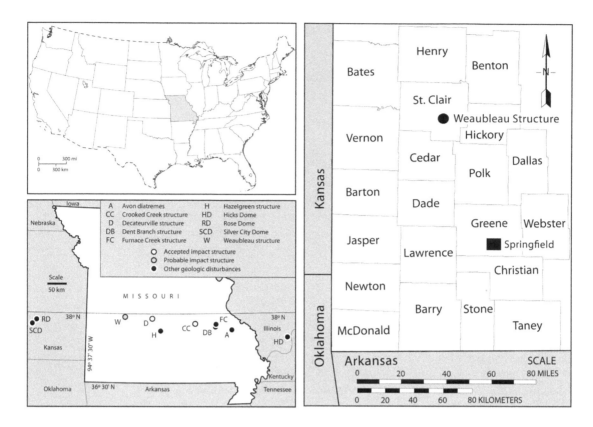

Figure 12.1. *Location of study area.* **Upper left**, *Map of United States, with Missouri shaded.* **Lower left**, *Weaubleau, Decaturville, and Crooked Creek structures and other geological disturbances along the 38th degree of north latitude in Kansas, Missouri, and Illinois. Silver City Dome, Rose Dome, and Hicks Dome are unlikely to be impact structures.* **Right**, *counties of southwest Missouri and other localities mentioned in text.*

MIXED-AGE ECHINODERMS, CONODONTS, AND OTHER FOSSILS USED TO DATE A METEORITE IMPACT, AND IMPLICATIONS FOR MISSING STRATA IN THE TYPE OSAGEAN (MISSISSIPPIAN) IN MISSOURI, USA

12

James F. Miller, Kevin R. Evans, William I. Ausich, Susan E. Bolyard, George H. Davis, Raymond L. Ethington, Charles W. Rovey II, Charles A. Sandberg, Thomas L. Thompson, and Johnny A. Waters

Introduction

Disarticulated skeletal parts of echinoderms are among the most easily recognized and abundant constituents of Paleozoic marine strata. During the Mississippian (Osagean) evolutionary peak of crinoids and blastoids, these animals' remains were major contributors to the rock record and formed the bulk of the Burlington and Keokuk limestones in Missouri, Iowa, and Illinois. Strata of the Osagean Series record the time of greatest diversity and abundance of crinoids and blastoids (Kammer and Ausich, 2006). Macurda and Meyer (1983) estimated that 99% of the Burlington Limestone is made of disarticulated crinoids and blastoids. This formation is 40 to 50 m thick, and those authors estimated that one cubic meter of the Burlington Limestone contains the remains of 15,000 crinoids. Crinoids and blastoids are useful for general age determination of Paleozoic strata, and Laudon (1937) divided the Burlington Limestone into formal zones on the basis of crinoids and blastoids. Gahn (2002) listed 84 genera and 293 species and subspecies of crinoids and 15 genera and 25 species of blastoids in the Burlington Limestone and used these echinoderms to revise previous biozonal schemes. Other fossil groups, such as conodonts, are used more commonly for detailed biozonation.

This report describes crinoid and blastoid calyxes, disarticulated echinoderm skeletal plates, conodonts, and other fossils from a breccia associated with a Mississippian-age meteorite impact in west-central Missouri. By understanding the age, distribution, and preservation of bioclasts, it is possible to estimate the depth of excavation within the impact structure. Identifiable echinoderm taxa include a mixture of species that are characteristic of both the Burlington and Keokuk limestones. Conodonts from the breccia include taxa characteristic of the Lower Ordovician Ibexian Series, the Upper Devonian, and the Lower to Middle Mississippian (Kinderhookian, Osagean, and possibly Meramecian Series). Rugose and tabu-

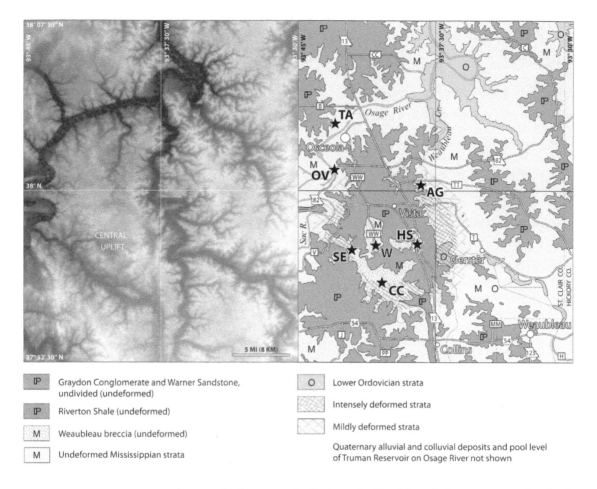

	Graydon Conglomerate and Warner Sandstone, undivided (undeformed)			O	Lower Ordovician strata
	Riverton Shale (undeformed)				Intensely deformed strata
M	Weaubleau breccia (undeformed)				Mildly deformed strata
M	Undeformed Mississippian strata				

Quaternary alluvial and colluvial deposits and pool level of Truman Reservoir on Osage River not shown

late corals also occur in the breccia. Brachiopods, bryozoans, and a single trilobite occur in residual chert remaining after dissolution of overlying Mississippian limestones. Together, data from these fossil groups allow the age of the meteorite impact to be well constrained, although the entire story cannot be deciphered by using only one of these fossil groups. An unexpected result of this research has been an explanation for the lack of upper Osagean strata in the type area of the Osagean Series, which is at the northern edge of the research area.

Weaubleau Structure

A series of anomalous geological features, collectively referred to as 38th parallel structures, extends east-west across Missouri (Fig. 12.1); all of the features are slightly south of 38° north latitude (Snyder and Gerdemann, 1965). The Decaturville structure (Offield and Pohn, 1979) and the Crooked Creek structure (Hendricks, 1954) are confirmed meteorite-impact sites and are included in the Earth Impact Data Base (2006). The Weaubleau structure, named for Weaubleau Creek near the town of Weaubleau in Saint Clair County, Missouri (Figs. 12.1, 12.2), is another of these structures and has been suspected to be an impact site (Rampino and Volk, 1996); our recent research confirms this interpretation (Evans et al., 2003a, 2003b; Davis et al., 2005). Intrusions associated with Rose Dome and Silver City Dome in eastern Kansas and with Hicks Dome in southern Illinois (Fig. 12.1) suggest that these structural features have a nonimpact origin.

The origins of the impacts (Weaubleau, Decaturville, and Crooked Creek structures) along the 38th parallel are unclear, but it has been suggested that they formed as a serial impact similar to impacts of fragments of the Shoemaker-Levy 9 comet on Jupiter in 1994 (Rampino and Volk, 1996). One problem with this interpretation is that the ages of formation of the individual structures across Missouri are not well constrained; these structures also have not been demonstrated to be of the same age or to have formed by the same processes. An alternative hypothesis is that these impact structures are of different ages, and they are oriented in a straight line by chance. The principle of Occam's razor suggests the simpler of two explanations is more likely to be correct, and the serial impact hypothesis is the simpler one.

Dulin and Elmore (2006) studied the paleomagnetism of breccias and in situ strata associated with the Weaubleau and Decaturville structures. Their data show that the Weaubleau structure formed not younger than Late Mississippian, whereas formation of the Decaturville structure was not older than Pennsylvanian. They concluded that the two structures were of different ages and could not have been part of a serial impact. Their interpretation of the age of the Weaubleau structure is consistent with our data (presented below). However, their data for the Decaturville structure are complex, and we interpret the situation differently. Outside the Decaturville structure, the in situ Lower Ordovician Jefferson City Dolomite preserves an Early Pennsylvanian paleopole (Dulin and Elmore, 2006), whereas breccias within the structure preserve a Permian paleopole. They stated that Mississippi Valley type sulfide deposits in the Decaturville impact breccia range in age from Pennsylvanian to Triassic, and these sulfide deposits formed before impact. Thus, Dulin and Elmore (2006) argued that the age of the Decaturville impact must be Pennsylvanian or Permian, but this argument hinges on the relative timing of mineralization versus impact. Sulfide mineralization at the Decaturville structure is localized to an area known as the sulfide pit in the central uplift of the structure (Offield and Pohn, 1979). These sulfides are fractured, and it is assumed that the fractures formed during impact. These are hydrothermal sulfide deposits, and brecciation or fracturing during their formation would not be unusual. We argue that it is possible that the mineral deposits and the fractures that cut them could be postimpact, thus negating the necessity of a Pennsylvanian or Permian age for the impact. However, whether the Weaubleau and Decaturville structures form part of a serial impact is not a major theme of this report.

The Weaubleau structure is located in a cratonic area that has generally undeformed strata that dip a few degrees to the west off the western flank of the Ozark Dome. NASA space shuttle imagery shows that the main area of uplifted rock in the Weaubleau structure is about 11 km in diameter (Fig. 12.2). An area about 19 km in diameter includes somewhat deformed Ordovician to Mississippian strata, and undeformed Pennsylva-

Figure 12.2. *Detailed location of Weaubleau structure and associated localities discussed in text. Straight lines delineate four 7.5-minute topographic quadrangles. Left side is a NASA space shuttle image; right side is a geologic map. Locations with stars are: AG, Ash Grove Aggregates Quarry; CC, small roadcut along Coon Creek; HS, roadcuts and measured section along Missouri Highway 13; OV, overlook site, a scenic viewpoint above confluence of the Sac and Osage Rivers; SE, small roadcut along Saint Clair County gravel road SE 450; TA, type area of Osagean Series and location of sections studied by Lane et al. (2005); W, small roadcut near south end of Saint Clair County paved road WW.*

nian strata also occur in this area. Stockdell et al. (2004) studied gravity and magnetic anomalies associated with this structure. Bouger gravity anomaly data show high-frequency gravity maxima and minima, with two gravity minima at the suspected point of impact and at a point to the northeast. A residual gravity anomaly map that removes regional anomalies highlights the gravity minimum that is thought to be related to the impact, and aeromagnetic data confirm these gravity data. Presumably, the gravity minimum is due to the presence of low-density breccia in subsurface rocks.

Beveridge (1951) mapped the geology of the Weaubleau 7.5-minute topographic quadrangle (southeast quarter of areas in Fig. 12.2), which includes only part of the Weaubleau structure. He interpreted unusual structures as thrust faults that interpose slices of Ordovician and Mississippian strata; we interpret these faults as being related to meteorite impact. Beveridge (1951) also noted in the Vista quadrangle (southwest part of areas in Fig. 12.2) an unusual conglomerate facies that is rich in lithoclasts as well as echinoderm bioclasts, and he believed this conglomerate to be part of the Burlington Limestone. Herein, we interpret this rock as part of the breccia that formed from impact processes.

Stratigraphy

PRECAMBRIAN TO LOWER ORDOVICIAN. Stratigraphic units near the Weaubleau structure include Precambrian basement and strata as young as Middle Pennsylvanian (Fig. 12.3). On the basis of a core (discussed below), the Precambrian rocks are granite of probable Paleoproterozoic age (Thompson, 1995; fig. 1). The nearest Precambrian exposures are in the middle of the Decaturville structure (Fig. 12.1), and they also crop out extensively in the core of the Ozark Dome, in the Saint Francois Mountains of southeastern Missouri. Middle to Upper Cambrian strata are mixed siliciclastics and carbonates in the eastern part of the Ozark Dome, but these strata, which also are exposed in the Decaturville structure, are not exposed in Saint Clair County. The Lower Ordovician Gasconade Dolomite also is not exposed in this county, but it crops out farther east. Megabreccia clasts of quartz sandstone, provisionally identified as deriving from the Roubidoux Formation, are exposed on the southern margin of the central uplift near Coon Creek (Fig. 12.2). The overlying Lower Ordovician Jefferson City Dolomite and Cotter Dolomite typically are mapped as a single unit that comprises the oldest contiguous exposures in this area (Figs. 12.2, 12.3); these strata are mostly cherty dolomite with minor sandstone. Thompson (1995, p. 26) stated that the Cotter Dolomite is absent in Saint Clair County, although the Jefferson City Dolomite is present. Thompson (1991) provided a comprehensive catalog of Ordovician stratigraphic units in Missouri.

MIDDLE ORDOVICIAN TO DEVONIAN. A regional unconformity occurs at the base of the Mississippian around the margins of the Ozark Dome. Middle Ordovician to Upper Devonian strata are present on the eastern and northern flanks of the dome but generally are absent on the western

Figure 12.3. *Stratigraphic columns for rocks in Southwest Missouri (left), near the Weaubleau structure (center), and in reference area for Osagean Series along the Mississippi River Valley (right). Crinoid and blastoid biozones in Burlington Limestone follow Gahn (2002). Abreviations: Cambr., Cambrian; Ord., Ordovician.*

SOUTHWEST MISSOURI		WEST-CENTRAL MISSOURI	MISSISSIPPI RIVER VALLEY			
Greene County	Polk County	Central St. Clair County	NE Missouri		SE Iowa	
Isolated exposures of Pennsylvanian channel sands		Atokan and Desmoinesian shales, conglomerates, and sandstones	WIDESPREAD PENNSYLVANIAN			
Isolated exposures of Warsaw Formation		Deposition of Parent Rock for residual chert in Graydon Congl. (Keokuk, Warsaw, Salem) ? ? ? *WEAUBLEAU BRECCIA* ? ? HIATUS ? ?	Salem and St. Louis Limestones			MERAMECIAN SERIES
			upper	WARSAW FORMATION		
			lower			
Short Creek Oolite Member			upper	KEOKUK LIMESTONE		OSAGEAN SERIES
Burlington-Keokuk Limestone (undivided)			Montrose Chert Member			
			Pelmatozoan Association III		Cedar Fork Member	BURLINGTON LIMESTONE
			Macrocrinus verneuilianus Biozone	*Pentremites elongatus* Biozone		
			Pelmatozoan Association II		Haight Creek Member	
		Burlington-Keokuk Limestone (undivided)	*Uperocrinus pyriformis* Biozone	*Globoblastus norwoodi* Biozone		
			Pelmatozoan Association I		Dolbee Creek Member	
Reeds Spring & Elsey Fms.			*Dorycrinus unicornis* Biozone	*Cryptoblastus melo* Biozone		
Pierson Limestone		Pierson Limestone				
Northview Formation	Chouteau Group	Northview Formation	McCraney Limestone			KINDERHOOKIAN SERIES
		Sedalia Formation	Chouteau Limestone		Chouteau Group	
Compton Limestone		Compton Limestone	Hannibal Shale			
Bachelor Formation		Bachelor Formation				
M. Ordovician and Devonian present as isolated erosional remnants			M. Ordovician–U. Devonian strata present			
Not exposed in western Missouri		Jefferson City-Cotter Dolomite (undivided)			IBEXIAN SERIES	ORD.
		Roubidoux Formation (sandstone and dolomite)				
		Gasconade Dolomite				
		Sandstone and dolomite			MIDDLE AND UPPER	CAMBR.
PALEOPROTEROZOIC GRANITIC BASEMENT						

flank. Presumably, strata of these ages were deposited and then eroded after the Late Devonian uplift of the Ozark Dome.

Offield and Pohn (1979) mapped the geology of the Decaturville structure (Fig. 12.1) and showed many loose blocks of the latest Ordovician Noix Oölite, which crops out near the Mississippi River. John Repetski (U.S. Geological Survey, Reston, Virginia; personal commun.) processed samples of fallback breccia at the Decaturville structure and obtained conodont elements of very Late Ordovician age. Miller processed two samples of fallback breccia collected along Missouri Highway 5 south of the village of Decaturville. Ethington, John Repetski, and Steve Leslie (University of Arkansas, Little Rock) studied the two samples and identified a diverse conodont fauna that includes taxa known from the upper part of the Plattin Group or lower part of the Decorah Group in eastern Missouri (personal commun.). These strata are assigned to the Mohawkian Series (lower Upper Ordovician). The color alteration index (CAI) of these conodont elements is 1 to 1.5 (Epstein et al., 1977), indicating that they have not been subjected to abnormally high temperatures due to burial, tectonic deformation, or hydrothermal activity. The insoluble residues of those conodont samples have abundant, well-rounded to spherical, frosted-quartz sand grains similar to grains in the Saint Peter Sandstone, the basal Mohawkian formation in Missouri. These unpublished data confirm that Upper Ordovician strata once existed in the western part of the Ozark Dome, but to date, no Upper Ordovician conodonts have been recovered from strata near the Weaubleau structure.

Two Late Devonian conodonts (discussed below) may have been derived from thin intervals of Devonian strata that are known from outcrop and in wells in the western Ozark Dome. Less than 1 m of Upper Devonian Chattanooga Shale occurs between Lower Ordovician dolomites and the Mississippian Bachelor Formation (discussed below) along U.S. Highway 65 in west-central Christian County (Fig. 12.1), Missouri (Thompson, 1986, fig. 24), about 100 km south of the Weaubleau structure. A rounded black-shale clast was recovered from the MoDOT-SMSU-Vista 1 core (discussed below); this lithology may have been derived from the Chattanooga Shale. The Missouri Geological Survey maintains a computerized database of water and oil wells in the state, and a search of this database reveals that the Upper Devonian Chattanooga Shale occurs in Greene County (two counties south of Saint Clair County; Fig. 12.1) and in Barton County (two counties southwest of Saint Clair County; Fig. 12.1). Devonian limestone is interpreted as occurring in wells in Saint Clair County and in Vernon County (next county to the west; Fig. 12.1). Many occurrences of Devonian limestone are logged in wells in Henry and Benton counties, north of Saint Clair County (Fig. 12.1). Most logs show 3 to 6 m of Devonian strata, although a few occurrences are thicker. However, many other wells in these counties have no record of Devonian strata, and no Devonian is known from outcrop in these counties. These strata are probably isolated remnants of units that were deposited over the entire area and eroded after the Late Devonian uplift of the Ozark Dome but before Mississippian deposition.

MISSISSIPPIAN. Strata assigned to the Kinderhookian and Osagean Series are exposed and complexly deformed in and near the Weaubleau structure (Fig. 12.2). Thompson (1986) provided a comprehensive catalog of Mississippian stratigraphic units in Missouri, and he included the Bachelor, Compton, Sedalia, and Northview formations in the Kinderhookian Series (Fig. 12.3). The Bachelor Formation usually is less than 1 m thick and typically consists of a basal sandstone overlain by green shale. The Compton, Sedalia, and Northview formations comprise the Chouteau Group. The Compton Formation, typically about 2 to 4 m thick, is fine crinoidal limestone that may be dolomitized locally. The Sedalia Formation is a sandy, cherty, crinoidal limestone that is 10 to 15 m thick. The Northview Formation includes dolomitic siltstone, shale, and dolomite and is only 1 to 3 m thick in Saint Clair County, but it thickens southward toward the impact area. Lane et al. (2005) measured 0.75 m of Northview Formation near the Osage River north of Osceola (locality TA on Fig. 12.2). Facies changes involving the Compton, Sedalia, and Northview formations result in lateral variations in their thicknesses. Near Stockton Reservoir, about 30 km south of the Weaubleau structure, the Northview Formation is about 25 m thick, and the Sedalia Formation is indistinguishable from the Compton Formation.

Thompson (1986) included the Pierson, Reeds Spring, Elsey, Burlington, and Keokuk formations in the Osagean Series. The Pierson Limestone is brown crinoidal limestone, and Lane et al. (2005) measured 3 m of Pierson in their section north of Osceola (Fig. 12.2). In northern Greene County and Polk County (Fig. 12.1), the Pierson Limestone contains a diverse crinoid fauna characteristic of the *Dorycrinus unicornis* Biozone of Gahn (2002) (Fig. 12.3). The Reeds Spring and Elsey formations are dense, fine-grained to crinoidal limestone with up to 50% chert, but chert decreases northward so that limestones above the Pierson Formation become indistinguishable from the Burlington Limestone (Fig. 12.3). Accordingly, the Reeds Spring and Elsey formations are not identified as far north as the Weaubleau structure. In Greene County and Polk County, the Elsey has crinoids characteristic of the *Dorycrinus unicornis* Biozone.

The Burlington Limestone and the Keokuk Limestone are cherty limestones mostly consisting of coarse-grained crinoid fragments; together, the two units are 50 to 60 m thick. Lane et al. (2005) measured 24 m of strata that they identified as Burlington Limestone near Osceola, but they identified no Keokuk Limestone. The Burlington and Keokuk formations can be distinguished easily from one another where they were named in southeastern Iowa (Fig. 12.3), but in southwestern and central Missouri, they are difficult to distinguish and usually are referred to as the Burlington-Keokuk Limestone (undivided). In Iowa and Missouri, the crinoid and blastoid faunas of the two formations are distinguishable, and these differences are useful for interpreting the age of individual exposures of the Burlington Keokuk Limestone in southwest and central Missouri. In Greene County (Fig. 12.1), the lower beds of the Burlington-Keokuk Limestone have crinoids characteristic of the *Dorycrinus unicornis* Biozone, and higher strata have crinoids characteristic of the *Uperocrinus pyriformis* Biozone. In Greene County, the upper part of the Burlington-Keokuk Limestone has crinoids and blastoids that are characteristic of the *Macrocrinus verneuilianus* Biozone of the Burlington Limestone (Gahn, 2002), and higher parts of the Burlington-Keokuk have faunas characteristic of the Keokuk Limestone in northeastern Missouri (Fig. 12.3).

The top 1 to 1.5 m of the Osagean succession is the Short Creek Oölite Member, a marker bed originally named in eastern Kansas as the top member of the Keokuk Limestone. The Short Creek Oölite also crops out in Oklahoma, Arkansas, and southwest Missouri (Thompson, 1986). This

LITHOLOGY	STRATIGRAPHIC UNIT

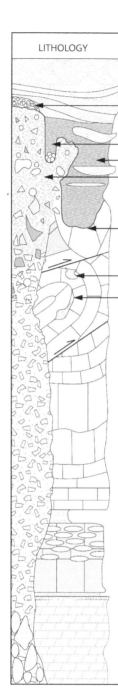

Warner Sandstone (Desmoinesian) Cross-bedded fluvial to paralic quartz sandstone

Graydon Conglomerate (Desmoinesian) Chert-clast breccia with sandstone matrix that rarely contains "round rocks" or "Weaubleau eggs" derived from weathered Weaubleau breccia

Paleokarst-fill material varies from shale to sandstone

Riverton Shale (Atokan) Valley-fill marine to non-marine shale with thin sandstone lenses

Weaubleau breccia (late Osagean to early Meramecian) Includes several types of breccia, ranging from sorted, fine-grained breccia near the top (resurge facies) to crystalline basement breccia recovered in drill core; other facies include megabreccia blocks, dilation breccia, and injection breccia

Karstified and eroded surface

Parautochthonous succession (lower Ordovician and overlying strata shown on the right side of the lithologic column) ranges from lower Ordovician to middle Mississippian

Siltstone injection body

Injection breccia

Burlington-Keokuk Limestone (undivided) (Osagean) Coarse- to fine-grained, gray, cherty crinoidal grainstone; heavily fractured, folded, and faulted around the perimeter of the Weaubleau Structure; upper part of unit in gradational contact with Weaubleau breccia; only the lower part of this unit is preserved in the vicinity of Osceola

scale
m   ft
3   10

2   5

1

0   0

Pierson Limestone (Osagean) Brown lime mudstone to crinoidal grainstone.

Northview Formation (Kinderhookian) Gray-green siltstone and dolomite

Sedalia Formation (Kinderhookian) Reddish to gray sandy limestone to dolomite with nearly 40% white to light-gray chert in the upper part

Compton Limestone (Kinderhookian) Gray to buff fine-grained crinoidal lime wackestone, locally dolomitized

Bachelor Formation (Kinderhookian) Gray sandstone and green shale with phosphatic pebbles above the Ordovician-Mississippian unconformity

Jefferson City Dolomite, (Ibexian) Buff, fine-grained dolomite, locally cherty, with a prominent unit of silicified oöid grainstone at the top, known informally as the St. Clair oölite

Figure 12.4. *Stratigraphic column for Weaubleau breccia and nearby in situ strata. Folded and thrust-faulted part of Burlington-Keokuk Limestone (undivided) based on exposures in Ash Grove Aggregates Quarry (AG on Fig. 12.2).*

Figure 12.5. *Part of roadcuts along Missouri Highway 13 in Saint Clair County (HS on Fig. 12.2) exposing Weaubleau breccia (B) and Pennsylvanian gray shale (S) (Fig. 12.4). Most vertical surfaces are weathered breccia that formed as sides of sinkholes that were filled with shale. Upper part of exposure is chert breccia (CB) assigned to Pennsylvanian Graydon Conglomerate; chert clasts are residual material derived from weathering of Osagean to Meramecian carbonates. White chert includes quasi-spherical Weaubleau eggs that weathered out of breccia and rolled into open sinkholes. Circled rock hammer shows scale. P is a pole supporting electric-power transmission wires.*

thin oöid limestone has an insoluble residue consisting primarily of tiny, doubly terminated quartz crystals that appear to be authigenic.

Thompson and Fellows (1970) followed tradition in assigning the Warsaw Formation to the Meramecian Series in their report on Kinderhookian to lowermost Meramecian strata and conodonts in southwest Missouri. Thompson (1986) also assigned the Warsaw Formation to the Meramecian Series. The Weaubleau structure has no exposure of the Warsaw Formation, and the nearest exposures are about 32 km to the southwest. Kammer et al. (1991) placed the Osagean-Meramecian series boundary within the Warsaw Formation in eastern Missouri, near Saint Louis (Fig. 12.3).

WEAUBLEAU BRECCIA. Weaubleau breccia is an informal name for a polymict carbonate breccia that formed during the impact that produced the Weaubleau structure, but formal description of this unit is not the goal of this report. Mapping the extent of this breccia is the subject of ongoing research, but the breccia crops out in many places in and near the Weaubleau structure (Fig. 12.2). A prominent exposure along Missouri Highway 13 includes roadcuts that are located 8.2 km north of Collins (locality HS on Fig. 12.2). Discontinuous exposures of breccia extend for 230 m along the side of the roadway, and individual exposures are separated by shales that fill Pennsylvanian sinkholes (Figs. 12.4, 12.5). These breccia exposures are up to 9 m thick and consist of one massive bed of carbonate breccia that consists mostly of abundant echinoderm debris and lithoclasts in a fine-grained carbonate matrix. The ratio of bioclasts to lithoclasts varies

Figure 12.6. *Surfaces of Weaubleau breccia exposed in roadcuts along Missouri Highway 13 (HS on Fig. 12.2). Left, weathered surface showing abundant bioclasts, white crinoid plates, columnals, and segments of stems. Right, unweathered surface of Weaubleau breccia exposed by highway construction, showing abundant angular lithoclasts. Abbreviations: C, white chert; D, brown dolomite, M, light-green clay mudstone. Swiss Army pocketknife is 9 cm long.*

laterally and vertically. Echinoderms include disarticulated crinoids, blastoids, and rare echinoids. Lithoclasts include abundant white, black, and rare red, angular chert clasts that clearly are not nodules (Fig. 12.6). Some chert clasts have distinctive lithologies that are exposed in a roadcut of the Jefferson City Dolomite along Missouri Highway 13 in northwestern Polk County (Fig. 12.1). Exposure there includes the informally named Saint Clair oölite, and clasts of this oöid chert occur in Weaubleau breccia at the Ash Grove Aggregates Quarry (locality AG on Fig. 12.2).

The Weaubleau breccia also contains common, angular clasts of light green to gray clay mudstone that may be derived from the Jefferson City, Bachelor, or Northview formations (Fig. 12.6). Shale beds in the former two formations usually are less than 15 cm thick, so most of the shale clasts are probably derived from the Northview Formation. Small, well-rounded pebbles and granules of quartz are similar to clasts in the basal Cambrian sandstone, where it crops out near the center of the Ozark Dome in southeast Missouri. Presumably, similar sandstone is in the subsurface near the Weaubleau structure, but wells in the area do not penetrate to the Cambrian. Large lithoclasts of dolomite and crinoidal limestone are uncommon in breccia, and most of those present are the size of pebbles or smaller. Insoluble residues from breccia sampled at these roadcuts show that the deposit is normally graded and that chert lithoclasts are more abundant and coarser near the base and become less abundant and smaller upward. Figure 12.7 shows angular chert lithoclasts and fossils that were in the insoluble residue of a conodont sample collected at the base of the exposed breccia along Highway 13.

Unusual, occasionally porous chert nodules that range in size from smaller than golf balls to larger than grapefruit (about 3–15 cm) are common in the Weaubleau breccia, but they may be as large as basketballs (about 25 cm). These nodules are somewhat spherical but typically are slightly flattened and elongated (Fig. 12.8). They weather out of the breccia and occur in regolith above the breccia and in adjacent road ditches. Locals refer to the nodules as "round rocks" or "Weaubleau eggs" and utilize them for fence construction and exterior home decoration. Many such chert nodules occur in Missouri only near the Weaubleau structure, and they are a characteristic feature of the Weaubleau breccia.

Figure 12.7. *Angular chert and silicified fossils dissolved out of a Weaubleau breccia conodont sample at base of roadcut along Highway 13. Chert clasts are mostly white and brown. A horn coral (H) and two segments of crinoid stems (CS) presumably were replaced by chert after breccia deposition, perhaps contemporaneously with formation of chert in Weaubleau eggs. Swiss Army pocketknife is 9 cm long; paper clip is 32 mm long.*

Miller broke open about 100 of these Weaubleau eggs, and most have a central clast ("yolk") of green clay mudstone or brown dolomite (Fig. 12.8). The outer parts of these nodules ("egg white") typically consist of fine- to coarse-grained clasts similar to the size of material comprising the Weaubleau breccia (including small crinoid stems), but the sediment is replaced by chert. Rarely, such nodules are elongated and contain two clasts of green mudstone, one near each end; such nodules are referred to as "double-yolked Weaubleau eggs." A few flat, elongated eggs have a single, elongated, flat mudstone core that generally is similar to the shape of the egg. A few nodules have a core of solid chert that is surrounded by somewhat porous chert. A few other nodules are geodes with hollow areas that are lined by small quartz or dolomite crystals. In a few eggs, the chert does not completely surround the central clast, and the mudstone remains in contact with carbonate breccia through a small, circular "navel." One egg encountered in a core is quite incompletely formed and has chert surrounding only one end of a mudstone clast. Microscopic examination of the boundary between breccia sediment and sediment that is replaced by chert shows many individual crinoid columnals that are partly original calcite and partly replaced by chert. Pennsylvanian sinkhole deposits along Highway 13 contain many Weaubleau eggs that apparently weathered out of the breccia and rolled into the sinkholes before shale deposition, as shown in Figure 12.5.

Our preliminary study of Weaubleau eggs led to two hypotheses regarding their origin. One hypothesis is that they formed immediately after impact by accretion of breccia sediment around sticky, wet mud clasts, and the chert formed later by diagenetic replacement of this accreted breccia

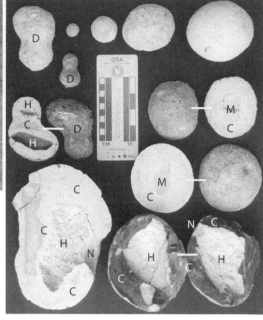

Figure 12.8. *Various sizes of Weaubleau eggs, chert nodules occurring in the Weaubleau breccia. Abbreviations: C, chert; H, holes where mudstone was present; M, clay mudstone centers of eggs.* **Left**, *Weaubleau breccia with three in situ eggs, partly broken away by highway construction along Highway 13, so some of the mudstone has weathered away. Note abundant angular chert lithoclasts in breccia.* **Right**, *eggs weathered out of breccia. Holes are where clay mudstone was removed so as to obtain conodonts. D, double-yolked eggs, which include two clay-mudstone clasts; N, navel, where clay mudstone was incompletely covered by chert and extended to outside of egg. White lines connect the halves of the same egg. Swiss Army pocketknife is 9 cm long.*

sediment. A second hypothesis is that the nodules formed long after deposition of the Weaubleau breccia, but before the Pennsylvanian. In this scenario, clay, dolomite, and chert lithoclasts occur randomly throughout the breccia sediment, and some of the sediment that surrounded such lithoclasts has been replaced by porous chert. Chert replacement was commonly localized by geochemical conditions related to lithoclast lithology. This chert developed first at the outer edge of the central lithoclast, and the egg grew outward as chert continued to replace the breccia sediment. Of the two hypotheses, the second one appears to explain more easily the irregular shapes of some eggs, the presence of double-yolked eggs, the incomplete eggs with a navel, and partial silicification of crinoid columnals at the outer margins of the eggs. However, we have not made a comprehensive study of the origin of Weaubleau eggs.

The Missouri Department of Transportation drilled the MoDOT-SMSU-Vista 1 core at the top of the roadcuts along Highway 13 (locality HS on Fig. 12.2). The upper 8.9 m of the hole passed through unconsolidated chert breccia that is assigned to the Pennsylvanian Graydon Conglomerate, and no core was recovered (see preliminary core log in Table 12.1). Drilling continued in breccia to the total depth of the hole at 75.5 m, and several breccia facies were encountered. Breccia from 8.9 to 21.9 m is similar to the material described from the roadcuts. We interpret this breccia facies to have formed when a meteorite impacted in a shallow ocean. Some lithoclasts derived from shattered bedrock units, and a large amount of loose crinoidal sediment was probably blown into the atmosphere and fell back into the ocean; other clasts may have remained in the crater as a slurry after the impact. Impact seiches rushing into the resulting crater mixed this material with other unconsolidated sediment that was on the seafloor, and all of this material was deposited as a graded resurge breccia. Comparable resurge breccias were described from the Lochne struc-

Interval in Core (m)	Lithologic Description	Interpreted Stratigraphic Assignment (Interpreted Parent Rock)
0–8.9	No core recovery	Graydon Conglomerate (chert breccia)
8.9–21.9	Crinoidal and lithoclastic breccia	Resurge breccia
21.9–27.9	Crinoidal limestone	Burlington–Keokuk or Pierson Limestone
27.9–30.3	Bedded shale dipping approx. 20 degrees	Northview Formation megaclast
30.3–31.7	Contorted shale	Northview Formation
31.7–34.3	Crinoidal limestone	Burlington–Keokuk or Pierson Limestone megaclast
34.3–47.0	Abundant large chert clasts and vuggy, porous limestone alternating with crinoidal limestone	Sedalia Formation, possibly admixed with Burlington and/or Pierson clasts
47.0–48.2	Buff, porous, crinoidal dolomite	Compton Limestone
48.2–48.6	Green mudstone	Upper part of Bachelor Formation
48.6–49.9	Buff, porous, crinoidal dolomite	Compton Limestone
49.9–50.1	Sandstone injected into buff crinoidal	Bachelor injected into Compton dolomite
50.1–50.4	Quartz sandstone	Lower part of Bachelor Formation
50.4–61.1	Buff dolomite	Jefferson City Dolomite, may include rock of older dolomite units
61.1–69.0	Granite clasts mixed with buff dolomite	Mixed Jefferson City and Paleoproterozoic Breccia basement
69.0–75.5	Cataclastic granite with carbonate cement	Paleoproterozoic basement

Note.—All recovered core is brecciated.

ture of Sweden (Lindström et al., 2005). Most of the fossils discussed below are from this facies.

Breccia from about 21.9 to 61.5 m in the Vista 1 core is fundamentally different from the resurge breccia discussed above, and this second type of breccia is identified with the generic term *fracture breccia* because we are uncertain of its specific origin. This fracture breccia was produced by meteorite impact, and it consists of a parautochthonous succession of thoroughly shattered sedimentary rock that somewhat resembles a normal succession of the local sedimentary units. The contact between resurge breccia and fracture breccia is indistinct in the core. All rock in the core is brecciated, and only a few intervals of rock that we interpret as parts of megabreccia blocks appear to be nonbrecciated.

The most unusual breccia facies is crystalline basement breccia in the lower part of the core. Small granite clasts first occur at 61.1 m below the top of the core, and the interval from 61.1 to 69 m is dolomite breccia that resembles the Jefferson City Dolomite mixed with granite clasts. The rock is entirely granite from 69.0 m to the bottom of the core at 75.5 m. Polished sections of this core reveal that this rock is shattered at the scale of individual crystals, but the shattered granite is cemented together by fine-grained carbonate material that reacts with dilute HCl. Presumably, this carbonate

Table 12.1. *Preliminary log of MoDOT-SMSU-Vista 1 core.*

cement was deposited by groundwater after the impact. Other clasts in this facies may be melt rock, but study of this granite, and the impact breccia facies in general, is in a preliminary state. Granitic clasts almost certainly were derived from Precambrian basement, which normally would be 350 to 400 m below the surface in this area. The presence of brecciated granite 69 m below the surface suggests that these roadcuts, the core, and many of the fossils discussed below are all located within the central uplift that is typical of complex craters (Melosh, 1989).

The resurge-breccia facies contain rare tiny clasts that probably were derived from granite basement. Samples of the breccia were dissolved and concentrated with heavy liquid so as to obtain conodonts. The light fraction of the insoluble residue contains abundant quartz sand grains and rare sand grains consisting of microcline and quartz that probably were derived from granite. As much as 10% of the quartz grains in some insoluble residues exhibit planar deformation fractures; such material is commonly referred to as shocked quartz. Quartz sand grains may be derived either from sandstones in the Bachelor, Cotter, or Roubidoux formations; Cambrian sandstone; Precambrian granite; or from a combination of sources. Some of the quartz sand also could have been derived from the basal Mohawkian Saint Peter Sandstone.

Relationships between the Weaubleau breccia and other stratigraphic units are not fully established, but several observations reveal basic information. The Vista 1 core includes shattered limestone that appears to be derived from an Osagean crinoidal limestone, such as the Pierson Limestone or the Burlington-Keokuk Limestone. Near where Highway 82 crosses Weaubleau Creek (Fig. 12.2), a roadcut exposes Burlington-Keokuk Limestone that apparently was faulted during meteorite impact. The Ash Grove Aggregates Quarry (locality AG on Fig. 12.2) has faulted Northview Formation and Burlington-Keokuk Limestone with breccia that is injected into fractures (Fig. 12.4). On the basis of these data, meteorite impact and deposition of the Weaubleau breccia occurred after deposition of at least part of the Burlington-Keokuk Limestone. However, the relationship of the breccia to the Warsaw Formation is unclear because the Warsaw does not crop out near the Weaubleau structure. Warsaw and younger Mississippian strata may have been deposited in this area and eroded away before deposition of the Pennsylvanian. However, Pennsylvanian strata clearly are younger than the Weaubleau breccia because sinkholes in the breccia are filled with Pennsylvanian sediments (Fig. 12.5).

Cross-cutting relationships from faulting, combined with superposition, indicate that the impact and the breccia are younger than some part of the Osagean but are older than Pennsylvanian. Even this broad age determination is better resolution than is available for the age of many meteorite impacts around the world. Fortunately, fossils from the breccia permit a much better determination of the age of the Weaubleau breccia, as discussed below.

PENNSYLVANIAN. Gentile and Thompson (2004) provided a comprehensive catalog of Pennsylvanian stratigraphic units in Missouri. In many places in Iowa, Illinois, and Missouri, the sub-Pennsylvanian unconformity is manifested as sinkholes in underlying carbonates, and these sink-holes are filled with Pennsylvanian sediments (Figs. 12.4, 12.5). Such features occur in and near the Weaubleau structure and were exposed by highway construction along Missouri Highway 13 (locality HS on Fig. 12.2), where sinkholes are developed in the Weaubleau breccia. A few kilometers farther northwest, along Missouri Highway 82, similar sinkholes occur in thrust-faulted Burlington-Keokuk Limestone. In both areas, Pennsylvanian sediments filling the sinkholes are gray shales that are assigned to the Middle Pennsylvanian Atokan Series (Figs. 12.2–12.4).

More extensive Pennsylvanian strata in the area are assigned to the Desmoinesian Graydon Conglomerate and Warner Sandstone (Fig. 12.2), which overlie deformed strata in the Weaubleau structure. In some areas, the Pennsylvanian is completely eroded away, exposing deformed Mississippian and older strata beneath (Fig. 12.2). At the exposures along Highways 13 and 82 noted above, the only Pennsylvanian strata that have not been eroded away are chert breccia that overlies the Weaubleau breccia, as well as shale that occurs as sinkhole fillings. An exposure of the chert breccia along Highway 13 was assigned to the Graydon Conglomerate by Gentile and Thompson (2004, fig. 19, p. A-28; fig. 20, p. A-29), where this breccia underlies the Warner Sandstone.

Stratigraphic relationships involving chert clasts in the Graydon Conglomerate are complex. The chert appears to be a residual deposit left after dissolutional weathering of cherty carbonate that was deposited above the Weaubleau breccia. The chert typically forms irregular, tabular clasts that are typically as large as a human hand or fist but may be as large as a 20-cm cube. Clearly, these clasts are larger than and are lithologically different from the pebble-sized chert lithoclasts that occur in the underlying Weaubleau breccia. Chert clasts in the Graydon Conglomerate may be derived from more than one formation—for example, from the Warsaw and Salem formations. Also, the chert breccia along Highway 13 was exposed during highway construction, so it is probable that chert clasts derived from different formations have been disturbed and mixed together during construction; Pennsylvanian shales are similarly disturbed. Some Weaubleau eggs are included in this residual chert near the contact with the Weaubleau breccia. This relationship indicates that an unknown amount of the Weaubleau breccia was dissolutionally weathered after the overlying carbonate was weathered away, leaving the cherty residue.

A Pennsylvanian age is assumed for this cherty residue, but it may have formed during the Late Mississippian. The age of deposition of the original carbonate above the Weaubleau breccia places an upper limit on the time of meteorite impact and breccia formation. Thus, determining the age of deposition of the limestone parent rock is more important than determining when weathering dissolved the carbonate and left behind only the chert. Chert clasts contain molds of small crinoid columnals, many bryozoa, brachiopods, and other fossils. This lithology is consistent with the Warsaw Formation and may be consistent with the upper part of the underlying Burlington-Keokuk Limestone. Jack Ray, research archeologist with the Center for Archeological Research at Missouri State University, made a comprehensive study of chert lithologies occurring in Paleozoic strata in southwest Missouri, in order to match chert Paleo-Indian artifacts with the source rock. He examined clasts from the chert breccia and concluded (personal commun.) that at least some of it is derived from the Warsaw Formation. One large block of chert has many coated and oöid grains and may be derived from the Salem Limestone (Fig. 12.3). Fossils in some of these chert clasts help determine the age of the parent rock from which the chert is derived, and these fossils are discussed below.

Fossils from the Weaubleau Breccia

Fossils from the resurge facies of the Weaubleau breccia were studied to gain insight into the timing of the impact. Most macrofossils are from roadcuts along Highway 13 (locality HS on Fig. 12.2), on the eastern side of the central uplift of the Weaubleau structure (Fig. 12.2). These fossils include abundant crinoids, less common blastoids, rare echinoid spines, five brachiopods, and 18 corals. Brachiopods include one complete rhynchonellid and the hinge areas of four spiriferids that are too badly fragmented to identify. Three crinoids and a shark tooth were recovered from a small roadcut along Saint Clair County Road SE 450, on the west side of the central uplift (locality SE on Fig. 12.2). Microfossils include conodont elements from Highway 13, from the Ash Grove Aggregates Quarry (locality AG on Fig. 12.2), from Coon Creek (locality CC on Fig. 12.2), and from County Road SE 450. Miller et al. (2005) made a preliminary report on these fossils.

CRINOIDS. Crinoids are abundant in the Weaubleau breccia and include pluricolumnals up to about 3 cm long, individual columnals, calyx plates and spines, and more or less complete calyxes that all lack anal tubes. Most of the identifiable material was collected as loose specimens that were weathered out of the fine matrix of deeply weathered Weaubleau breccia near the northern end of the Highway 13 roadcuts. Most identified specimens from these roadcuts came from an area about 3 to 5 m across (Fig. 12.9), referred to informally as Crinoid Knob. Every complete or partial calyx found over a period of 3 years was removed from exposures along Highway 13, but only a fraction of the disarticulated material (columnals, loose plates) was collected. Figure 12.10 illustrates all taxa that could be identified to species, as well as some taxa that could be identified only to genus, and also representative examples of stem segments, columnals, and tegmen spines.

Most of the intact calyxes are small. The smallest is a single *Macrocrinus gemmiformis* (Hall, 1860) that is 7 to 8 mm long (Fig. 12.10.5, 12.10.6). A complete calyx of *Aorocrinus symmetricus* (Hall, 1858) (Fig. 12.10.10) is 11 to 12 mm long; commonly this crinoid has been misidentified as *A. parvus* (Shumard, 1855). A calyx of *Synbathocrinus dentatus* Owen and Shumard, 1852 (Fig. 10.13.23), is 10 mm wide. The largest complete calyx is an *Uperocrinus pyriformis* (Shumard, 1855) (Fig. 12.10.9) that is 23 mm long, which is small for this species, and it may be a juvenile. The most common calyx is *Macrocrinus verneuilianus* (Shumard, 1855), of which eight have been found (Fig. 12.10.1–12.10.3). There also is a partial calyx of *Uperocrinus* sp. (Fig. 12.10.18). Individual tegmen spines of *Dorycrinus* Roemer, 1854, are quite

common, but typically they have broken tips (Fig. 12.10.27, 12.10.28), which is not unusual for such spines where they occur as part of in situ Burlington-Keokuk Limestone. Two sets of three articulated basal plates (not illustrated) may be from *Dorycrinus* or *Uperocrinus* Meek and Worthen, 1865. Abundant parts of *Platycrinites* J. S. Miller, 1821, occur as groups of three basal plates, as isolated radial plates, and as elliptical columnals (Fig. 12.10.11–12.10.16). Individual cup plates of actinocrinitids are identified by the stellate ornament, and some plates have characteristic large nodes (Fig. 12.10.19–12.10.21). Such plates may be from genera such as *Actinocrinites* J. S. Miller, 1821, *Strotocrinus* Meek and Worthen, 1866, or *Teleiocrinus* Wachsmuth and Springer, 1881. The fauna includes individual radial plates of *Barycrinus* Meek and Worthen, 1868 (Fig. 12.10.24). In addition, there are innumerable individual circular and elliptical columnals, calyx plates, and lengths of stem and cirri (Fig. 12.10.31, 12.10.32) that can be identified only as crinoid parts.

Five calyxes were recovered at localities other than Crinoid Knob. One calyx of *Rhodocrinites barrisi* (Hall, 1861) (Fig. 12.10.17) was chiseled out of breccia on Highway 13 about 50 m south of Crinoid Knob. A calyx of *Macrocrinus konincki* (Shumard, 1855) (Fig. 12.10.4) and a partial calyx of *Agaricocrinus* Hall, 1858, were found weathered out of breccia about 100 m farther south, near the collecting site of a series of conodont samples in a measured section (discussed below). A small roadcut along Saint Clair County Road SE 450 (locality SE on Fig. 12.2) produced three crinoids, including two calyxes of *Agaricocrinus planoconvexus* Hall, 1861 (Fig. 12.10.7, 12.10.8), and a partial calyx (three basal plates) of *Platycrinites* sp. or *Eucladocrinus* sp.; the latter is 26 mm wide and is the largest crinoid found to date in the Weaubleau breccia (Fig. 12.10.22). The three specimens of *Agaricocrinus* (from two localities) have similar size and morphology, and they may be conspecific.

Crinoids recovered from these localities are well known from the middle and upper members of the Burlington Limestone in its type area

Figure 12.9. *Deeply weathered small exposure of Weaubleau breccia (B) near north end of Highway 13 roadcuts, referred to in text as Crinoid Knob. Most of the echinoderms illustrated in Figure 12.10 were picked up above and below this small exposure after weathering free of the breccia matrix. Chert breccia clasts (CBC) are from Graydon Conglomerate. Weaubleau eggs (E), abundant echinoderm bioclasts, and red clay weather out of the Weaubleau breccia. Material left of exposure is gray Riverton Shale (S) that fills sinkholes in breccia. White handle of rock hammer is 33 cm long.*

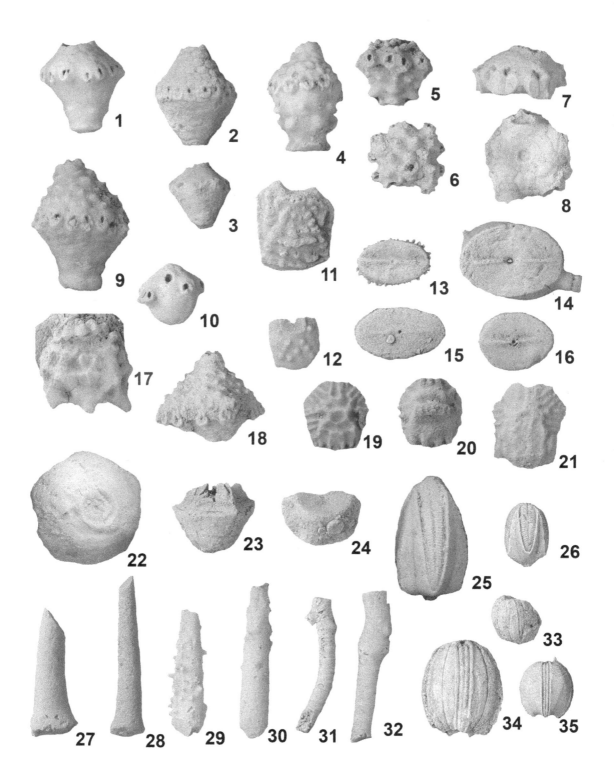

Figure 12.10. *Echinoderms from the resurge facies of the Weaubleau breccia; 1–24, 27, 28, 31, 32 are crinoids; 25, 26, 33–35 are blastoids; 29, 30 are echinoids. Specimens are reposited at the Paleontology Museum at the University of Iowa (SUI), Iowa City. Most specimens are from locality HS on Figure 12.2; 7, 8, 22 are from locality SE. 1–3,* **Macrocrinus verneuilianus** *(Shumard, 1855), lateral calyx views of three*

specimens, ×2.0; 1, SUI 102871; 2, SUI 102872; 3, SUI 102873. **4, Macrocrinus konincki** *(Shumard, 1855), lateral view of calyx, ×2.5, SUI 102874.* **5, 6, Macrocrinus gemmiformis** *(Hall, 1861), ×3.0, SUI 102875;* **5,** *lateral view of calyx;* **6,** *oral view of tegmen.* **7, 8, Agaricocrinus planoconvexus** *(Hall, 1861), ×1.5, SUI 102876;* **7,** *lateral view of calyx;* **8,** *basal view of calyx.* **9, Uperocrinus pyriformis** *(Shumard, 1855), lateral view of calyx, ×2.0, SUI 102877.* **10, Aorocrinus symmetricus** *(Hall, 1858), CD interray view of calyx, ×2.0, SUI 102878.* **11, 12,** *Isolated radial plates of* **Platycrinites** *sp., ×2.0;* **11,** *SUI 102879;* **12,** *SUI 102880.* **13–16,** *articular facets of columnals from* **Platycrinites** *sp., ×2.0;* **13,** *SUI 102881;* **14,** *SUI 102882;* **15,** *SUI 102883;* **16,** *SUI 102884.* **17, Rhodocrinites barrisi** *(Hall, 1861), lateral view of calyx, ×3.0, SUI 102885.* **18, Uperocrinus?** *sp., lateral view of calyx, ×2.0, SUI 102886.* **19–21,** *radial plates of actinocrinitids, ×2.0;* **19,** *SUI 102887;* **20,** *SUI 102888;* **21,** *SUI 102889.* **22,** *Basal view of an isolated basal circlet of* **Eucladocrinus** *sp. or* **Platycrinites** *sp., ×1.5, SUI 102890.* **23, Synbathocrinus dentatus** *Owen and Shumard, 1852, lateral view of aboral cup, original magnification, ×2.5, SUI 102891.* **24,** *Isolated radial plate of* **Barycrinus** *sp., ×2.0, SUI 102892.* **25, 26, Pentremites conoideus** *Hall, 1856, both ×2.0;* **25,** *adult, SUI 102893;* **26,** *juvenile, SUI 102894.* **27, 28,** *tegmen spines from* **Dorycrinus** *sp., both original magnification, ×2.0;* **27,** *SUI 102895;* **28,** *SUI 102896.* **29, 30,** *Echinoid spines, both ×2.0;* **29,** *SUI 102897;* **30,** *SUI 102898.* **31, 32,** *lengths of articulated cirri, both ×2.0;* **31,** *SUI 102899;* **32,** *SUI 102900.* **33, Poroblastus granulosus** *(Meek and Worthen, 1865), ×2.0, SUI 102901.* **34, 35, Globoblastus norwoodi** *(Owen and Shumard, 1850),* **34,** *SUI 102902, ×2.5;* **35,** *SUI 102903, ×2.0; note preservation of growth lines.*

in Iowa (Fig. 12.3). Gahn (2002) reported that *Agaricocrinus planoconvexus*, *Macrocrinus konincki*, and *Uperocrinus pyriformis* are common in the *Uperocrinus pyriformis* Biozone in Iowa, in the middle member of the Burlington Limestone. *Macrocrinus verneuilianus* and *Uperocrinus pyriformis* typically are in the *Macrocrinus verneuilianus* Biozone in the upper member of the Burlington Limestone in Missouri and Iowa, and *M. konincki* occurs rarely in that interval (Gahn, 2002; Fig. 12.3 herein). Wachsmuth and Springer (1897) reported *Aorocrinus parvus* (herein identified as *A. symmetricus*) from the upper part of the Burlington Limestone. The largest number of calyxes from the breccia are *Macrocrinus verneuilianus*, which is especially characteristic of the upper member of the Burlington Limestone in northeast Missouri and southwest Iowa, so this species is used to name the biozone.

Almost none of these taxa occur in either the lower part of the Burlington Limestone or in the overlying Keokuk Limestone. They do not occur in the Pierson Limestone or in the lower part of the Burlington-Keokuk Limestone in western Missouri. An exception is the rare species *Macrocrinus gemmiformis*, which occurs in the lower member (Laudon, 1937; Gahn, 2002) and ranges into the upper member of the Burlington Limestone. Lane (1958) listed *Macrocrinus carica* (Hall, 1861) as a junior synonym of *M. gemmiformis* (Hall, 1860), and Wachsmuth and Springer (1897) reported *M. carica* from the upper part of the Burlington Limestone in northwest Missouri and southeast Iowa. Material identified as *Dorycrinus*, platycrinitids, and actinocrinitids occur in both the Burlington Limestone and Keokuk Limestone (Wachsmuth and Springer, 1897).

Nearly all of these identified crinoids could have come from the Burlington-Keokuk Limestone of central Missouri, where they occur in strata that are coeval with the upper member of the Burlington Limestone (*Macrocrinus verneuilianus* Biozone) in northeastern Missouri and southeastern Iowa (Fig. 12.3). *Agaricocrinus planoconvexus* is the only species that does not occur in the upper member; Gahn (2002) reported that it occurs abundantly in the middle member of the Burlington Limestone (*Uperocrinus pyriformis* Biozone).

Fossils illustrated in Figure 12.10 are reposited at the Paleontology Museum at the University of Iowa, Iowa City. In addition, a representative collection of crinoid calyx plates, spines, columnals, and lengths of columns and cirri were placed in that collection under museum number SUI 102904. These are materials that were picked up loose after weathering out of the Weaubleau breccia along Missouri Highway 13.

BLASTOIDS. Blastoids are less common than crinoids in the Weaubleau breccia. The collection includes three *Globoblastus norwoodi* (Owen and Shumard, 1850) (Figs. 12.10.34, 12.10.35), one *Schizoblastus sayi* (Shumard, 1855) (not illustrated), one *Poroblastus granulosus* (Meek and Worthen, 1865) (Fig. 12.10.33), and five *Pentremites conoideus* Hall, 1856 (Fig. 12.10.25, 12.10.26). The first three species are characteristic taxa of the *Globoblastus norwoodi* Biozone and *Pentremites elongatus* Biozone in the Burlington Limestone in northwestern Missouri and southeastern Iowa, but all of them occur in the *P. elongatus* Biozone (Gahn, 2002; Fig. 12.3 herein). Thus, they all could have come from those parts of the Burlington-Keokuk Limestone that are coeval with the middle and upper member of the Burlington Limestone in northeastern Missouri and southeast Iowa. *Pentremites conoideus* is a younger species that occurs in situ in the upper part of the Burlington-Keokuk Limestone in Greene County (Fig. 12.1) and elsewhere in the Warsaw Formation. This species apparently occurs earlier in southwestern Missouri (upper part of the Burlington-Keokuk Limestone) than in the Mississippi River Valley, where Kammer et al. (1991) stated that its lowest occurrence, in the middle of the Warsaw Formation, characterizes the base of the Meramecian Series.

Most of the blastoids were found weathered out of the Weaubleau breccia at Crinoid Knob (Fig. 12.9). One *Globoblastus norwoodi*, however, was chiseled out of breccia matrix about 100 m to the south, and one was weathered out of breccia even farther south, near a measured section that was sampled for conodonts (discussed below). One *Pentremites conoideus* was chiseled out of breccia matrix about 10 m north of Crinoid Knob. The rock at these localities is typical resurge breccia that includes many small lithoclasts and echinoderm plates.

ECHINOIDS. Several echinoid spines that weathered out of breccia were found at Crinoid Knob (Fig. 12.9); they are illustrated in Figures 12.10.29 and 12.10.30. These fossils have minute spines protruding from the main spine. In 31 years of collecting echinoderms from Osagean limestones in southwest Missouri, Miller previously had never found echinoid spines and had found only four echinoid coronas from the Burlington-Keokuk Limestone; two are from Springfield and two are from Christian County (Fig. 12.1). Many years ago, Porter Kier (written commun.) studied two of the four specimens and identified them as *Lovenechinus missouriensis* Jackson, 1912, a species known from the Burlington Limestone and the Keokuk Limestone (Shimer and Shrock, 1944, p. 217). The echinoid spines reported herein are not identified.

IMPLICATIONS OF ECHINODERM PRESERVATION. Echinoderms from the Weaubleau breccia are primarily disarticulated but otherwise are in an excellent state of preservation (Fig. 12.10) that is comparable to in situ material weathered out of nearby exposures of Osagean crinoidal limestones (Fig. 12.11). An obvious difference between weathered surfaces of the Weaubleau breccia and of the Burlington-Keokuk Limestone is the angular lithoclasts in the breccia (compare Figs. 12.6 and 12.8 with Fig. 12.11). In most cases, specimens illustrated in Figure 12.10 were prepared simply by washing them in water, although some calyxes have been scrubbed with a toothbrush and soap. Fine ornament and small morphological details are preserved on many specimens (Fig. 12.10.35, and none of the echinoderm calcite is intensely cleaved. These specimens did not weather out of large lithoclasts in the breccia, but they were small bioclasts in the breccia. They could not have been alive at the time of impact because the fauna includes taxa that are of mixed ages that do not occur together normally.

These observations have far-reaching implications because they mean that the echinoderms, including calyxes and disarticulated fragments, must have been parts of loose, coarse-grained sediment that was not lithified at the time of impact. Most echinoderm columnals and calyx plates that are weathered out of the breccia are complete. Each echinoderm plate is an individual crystal of calcite, and if these plates had been part of indurated Burlington-Keokuk Limestone, the echinoderms would have been cleaved to bits during impact and would not have been cracked out of hard limestone as intact fossils. Immediately after impact, loose Osagean crinoidal sediment presumably was washed around by strong currents. Echinoderms of different ages (and conodonts of different ages, discussed below) were mixed together, and they also were mixed with lithoclasts of Precambrian to lower Osagean strata that were blown out of the impact crater. Finally, all of this material was deposited en masse as the resurge facies of the Weaubleau breccia. Many of the bioclasts may have been washed in from surrounding areas and partly filled the crater made during the impact.

It is noteworthy that echinoderms in the Weaubleau breccia do not include calyxes of large genera that are common in the Burlington-Keokuk Limestone, such as *Actinocrinites*, *Eutrochocrinus* Wachsmuth and Springer, 1897, *Strotocrinus*, and *Teleiocrinus*, but disarticulated plates of such large genera are present in the breccia (Fig. 12.10.19 to 12.10.21). Perhaps large fossils were sorted out by currents, or perhaps they simply did not survive the rigorous waves that transported and deposited the material now comprising the resurge breccia. In any case, only disarticulated plates of large crinoids are represented in the breccia.

In seeming contradiction to the inferred unconsolidated nature of the crinoidal sediment, some of the Burlington-Keokuk Limestone in the Weaubleau structure was faulted during impact. Such faulting, which can be seen at the Ash Grove Aggregates Quarry (locality AG on Fig. 12.2) and at roadcuts along Highway 82 (north of locality AG), implies that those strata were brittle, indurated limestone at the time of impact. The apparent contradiction can be resolved if the faulted exposures are from the lower

Figure 12.11. *Comparison of weathered surfaces of Weaubleau breccia at roadcut on Missouri Highway 13 (left) with in situ Burlington-Keokuk Limestone in Springfield (right). Clasts weathering into relief on left image include white chert (CH), red chert (R), calyx spines of the Osagean crinoid* **Dorycrinus** *(D), elliptical columnals from a platycrinitid crinoid (P), and many other white crinoid columnals and plates. On right image, the rock includes many circular crinoid columnals and stem segments, many elliptical columnals and stem segments of a platycrinitid crinoid (P), and many disarticulated crinoid plates. The fine matrix is mostly tiny crinoid skeletal plates. Similar material is mixed with lithoclasts in the Weaubleau breccia. Compare with Figure 12.6. Swiss Army pocketknife is 9 cm long.*

part of the Burlington-Keokuk Limestone. Echinoderms in the breccia are coeval with the middle and upper member of the Burlington Limestone and with the Keokuk Limestone in their type areas, which suggests that the upper parts of the Burlington-Keokuk Limestone in Saint Clair County may have been loose sediment at the time of impact, whereas the lower part of the formation already had been cemented into hard rock.

The above scenario has the startling implication that some tens of meters of crinoidal limestone, representing a considerable part of the Osagean Series, had not been cemented into rock for perhaps as many as several million years after deposition. Presumably, this interpretation might apply over a large geographic area where Osagean crinoidal sediments were deposited, although the interface between cemented versus loose sediment may have been quite irregular, at least locally and perhaps regionally. The delay between deposition of some Osagean crinoidal sediment and its induration has important implications for diagenesis. Pore waters were able to move through the uncemented sediments and to cause diagenetic changes for long periods of time. In particular, the large amounts of chert in these Osagean limestones could have formed before the pore spaces were filled with cement. Chert in Osagean limestones typically has many crinoid and blastoid parts (including rare whole calyxes), brachiopods, and horn corals that occur as external and internal molds. Opaline sponge spicules in the crinoidal sediment is a possible source of the silica for the chert.

CONODONTS. Conodont samples from the resurge facies of the Weaubleau breccia were collected at four localities, and conodonts were recovered at all four. Conodont elements from all localities are of mixed ages that represent a much greater stratigraphic range than that of the echinoderms, which were derived from strata coeval with the middle and upper members of the Burlington Limestone and from the Keokuk Limestone. Conodont preservation is good to excellent and is comparable to material recovered from nearby in situ Paleozoic rock. The CAI is 1 to 1.5, indicating that conodont elements have not been subjected to abnormally high temperatures due to burial, tectonic deformation, or hydrothermal activity (Epstein et al., 1977). Impact apparently did not adversely affect the conodonts in the resurge breccia any more than the echinoderms. Representative conodont taxa recovered from the resurge breccia are illustrated in Figure 12.12; most are the P (platform) elements that are most important for taxonomic identification. Many additional conodont elements, especially various S and M (bar and blade) elements, are not identified to genus and are not illustrated on Figure 12.12 but are recorded in Table 12.2. References to illustrations on Figure 12.12 are to all elements of a species, without distinguishing between elements from different localities. Table 12.3 gives Universal Transverse Mercator grid coordinates for conodont collection localities shown on Figure 12.2.

A diverse conodont fauna of different ages occurs in the Weaubleau breccia. The first sample of resurge breccia processed for conodonts had a mass of 7.7 kg and was selected randomly from near the top of the Ash Grove Aggregates Quarry (locality AG on Fig. 12.2; Table 12.2, Ash Grove breccia sample). Most of the sample dissolved in dilute formic acid in 2–3 days, but some dolomite clasts required nearly a week to dissolve completely. This two-step dissolution pattern occurred with some of the other samples. The initial sample yielded the coniform species *Colaptoconus quadraplicatus* (Branson and Mehl, 1933; Fig. 12.12.1 to 12.12.4 herein), which is common in the Ibexian (Lower Ordovician) Jefferson City Dolomite and also is known in the underlying Roubidoux Formation (Repetski et al., 1998, 2000; Fig. 12.13 herein). This species presumably came from dolomite clasts that dissolved slowly. Many small clasts in the breccia appear to be Jefferson City Dolomite, so this source rock is a logical possibility. Alternatively, elements of *C. quadraplicatus* might have been weathered out of Ordovician dolomite during the earliest Mississippian and redeposited in the Bachelor Formation, which then was broken into clasts during impact and included in the Weaubleau breccia.

This sample also yielded *Polygnathus communis* Branson and Mehl, 1934b (Fig. 12.12.16); *Polygnathus purus?* Voges, 1959; and a fragment of *Siphonodella* sp. (Table 12.2); these taxa indicate Kinderhookian and Osagean ages. Other recovered taxa include M elements that would be assigned to *Idioprioniodus* Gunnell, 1933 (Fig. 12.12.19 herein), and P elements of *Gnathodus texanus* Roundy, 1926 (Fig. 12.12.23 to 12.12.25); both taxa occur in uppermost Osagean to lowermost Meramecian strata (Fig. 12.13). Another species in this sample is recorded as *Hindeodus?* sp. in Table 12.2, but its identification is a matter of disagreement among authors of this report. Sandberg identified it as *Hindeodus cristulus* (Youngquist and Miller, 1949), but Thompson contends that it is not that species and may be a different genus. The name *Hindeodus?* sp. was chosen as a compromise by the senior author of this report; this taxon is illustrated by Figure 12.12.21.

Rexroad and Collinson (1965) and Thompson and Fellows (1970) found *G. texanus* in the upper part of the Keokuk Limestone and in the overlying Warsaw Formation. Miller processed 6.8 kg of crinoidal limestone from a locality in Springfield (Fig. 12.1) from near the top of the Burlington-Keokuk Limestone, about 1 m below the top of the unit and just below the base of the Short Creek Oölite. This sample produced hundreds of elements of *Gnathodus texanus* and *Hindeodus?* sp. From about 2 to 3 m higher at the

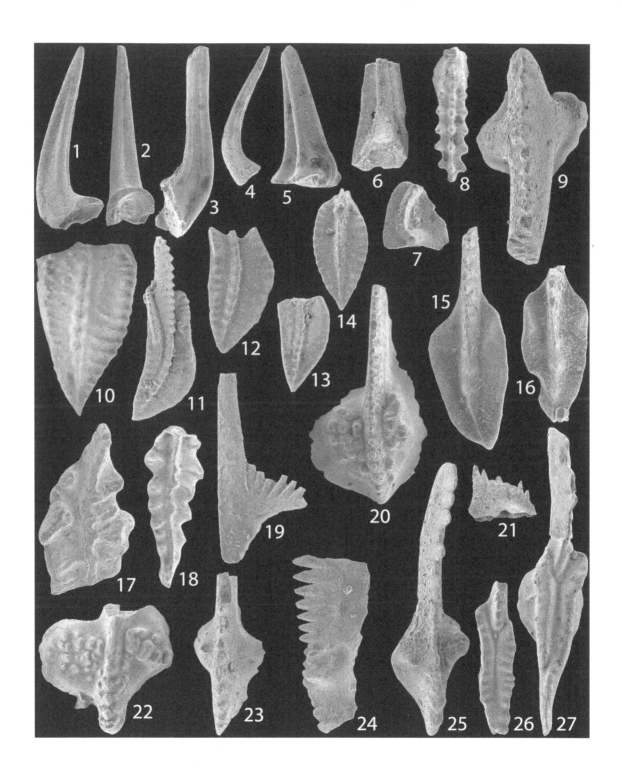

Figure 12.12. *Scanning electron microscope micrographs of conodonts from the Weaubleau breccia in Saint Clair County, Missouri. Specimens are reposited at the Paleontology Museum at the University of Iowa (SUI), Iowa City. Pa elements shown in upper view, except as indicated.* **1, 2, 4, 16, 19, 23, 24,** *Ash Grove Aggregates Quarry;* **3, 7–10, 12, 13, 15, 17, 18, 21, 22, 25–27,** *roadcuts along Missouri Highway 13, including the 8-m-thick measured section;* **5, 6, 14, 20,** *Saint Clair County Road SE 450;* **11,** *Coon Creek.* **1–6,** *Early Ordovician taxa;* **7, 8,** *Late Devonian taxa;* **9,** *taxon that may be Late Devonian or Mississippian;* **10–27,** *Mississippian taxa.* **1–4, Colaptoconus quadraplicatus** *(Branson and Mehl, 1933);* **1, 2,** *lateral-oblique and posterior views, ×87.5, SUI 102905;* **3,** *lateral view, ×72.5, SUI 102906, from 6-m level in measured section;* **4,** *lateral view, ×87.5, SUI 102907.* **5, Ulrichodina deflexa** *Furnish, 1938, anterolateral view, ×87.5, SUI 102908.* **6, Aloxoconus staufferi** *(Furnish, 1938), posterior view, ×87.5, SUI 102909.* **7, Palmatolepis** *Ulrich and Bassler, 1926, ×87.5, SUI 102910, from random sample B.* **8, Icriodus symmetricus** *Branson and Mehl, 1934a, ×87.5, SUI 102911, from random sample C (Crinoid Knob).* **9, Bispathodus stabilis** *(Branson and Mehl, 1934b), ×87.5, SUI 102912, from 6-m level in measured section.* **10, Siphonodella cooperi** *morphotype 2 of Sandberg et al. (1978), ×58, SUI 102913, 8-m level in measured section.* **11, 12, Siphonodella cooperi hassi** *Thompson and Fellows, 1970, ×87.5;* **11,** *SUI 102914;* **12,** *SUI 102915, from 2-m level in measured section.* **13, Siphonodella obsoleta** *Hass, 1959, ×87.5, SUI 102916, from core of Weaubleau egg at Highway 13.* **14, Polygnathus mehli** *Thompson, 1967, ×87.5, SUI 102917.* **15, Polygnathus purus subplanus** *Voges, 1959, ×87.5, SUI 102918, from random sample A.* **16, Polygnathus communis** *Branson and Mehl, 1934b, ×131, SUI 102919, with attached doubly terminated quartz crystal.* **17, Pseudopolygnathus multistriatus** *morphotype 1 of Lane, Sandberg, and Ziegler, 1980, upper oblique view, ×58, SUI 102920, from 2 m level in measured section.* **18, Pseudopolygnathus multistriatus** *morphotype 2 of Lane, Sandberg, and Ziegler, 1980, ×58, SUI 102921, from random sample B.* **19,** *M element of* **Idioprioniodus** *Gunnell, 1933, lateral view, SUI 102922.* **20, Protognathodus praedelicatus** *Lane, Sandberg, and Ziegler, 1980, ×87.5, SUI 102923.* **21, Hindeodus?** *sp., ×87.5, SUI 102924, from random sample C.* **22, Gnathodus punctatus** *(Cooper, 1939), ×58, SUI 102925, from 3 m level in measured section.* **23–25, Gnathodus texanus** *Roundy, 1926, all ×87.5;* **23,** *SUI 102926, 0.9 m level in measured section;* **24,** *lateral view, SUI 102927;* **25,** *SUI 102928, from 0.9-m level in measured section.* **26, 27, Taphrognathus varians** *Branson and Mehl, 1941a;* **26,** *SUI 102929, ×58, from 5 m in measured section;* **27,** *SUI 102930, ×87.5, from 6-m level in measured section.*

same locality in Springfield, a 10.2-kg sample of the Short Creek Oölite produced a dozen elements of *G. texanus*; two of *Taphrognathus varians* Branson and Mehl, 1941a (a species named from the Keokuk Limestone); and M elements of *Idioprioniodus*. *Hindeodus?* sp. is also common in a sample from 2 to 3 m above the base of the Warsaw Formation in Springfield, Missouri. Although its identity may be controversial, *Hindeodus?* sp. occurs in situ in the Keokuk and Warsaw formations, and it also occurs in several samples of the Weaubleau breccia (Table 12.2).

In summary, the first breccia sample that was processed for conodonts yielded taxa from the Lower Ordovician and the Lower and Middle Mississippian. Subsequent samples of the breccia from three other localities produced conodont faunas that include some of the same taxa found in the first sample and some additional ones, but the overall stratigraphic range of faunas recovered is quite similar at all localities.

A random sample of breccia (Table 12.2, random sample A) from the Highway 13 roadcuts (locality HS on Fig. 12.2) yielded a diverse fauna that included several of the taxa found at the quarry as well as *Polygnathus purus subplanus* Voges, 1959 (Fig. 12.12.15), and *Taphrognathus varians* (Fig. 12.12.26, 12.12.27). On the basis of these favorable preliminary results, the thickness of the exposed Weaubleau breccia was measured about 30 m

Locality	Highway 13 (m above base)*								
	0 m	0.9 m	2 m	3 m	4 m	5 m	6 m	7 m	8 m
Mississippian and Devonian Condonts									
Bispathodus stabilis (Upper Devonian to Mississippian)	1	—	—	—	—	—	2†	—	—
Elictognathus sp.	—	—	—	—	—	—	—	—	—
Gnathodus delicatus	—	—	—	—	—	—	—	—	—
Gnathodus punctatus	—	—	—	1†	—	—	—	—	—
Gnathodus texanus Pa element	2	1†	4	4	1	3	1	1	3
Gnathodus texanus Pb element	—	—	2	—	—	1	1	—	—
Gnathodus sp. aff. *G. typicus*	—	—	—	—	—	1	—	—	—
Hindeodus? sp.	—	—	—	—	—	—	—	—	—
Icriodus symmetricus (Famennian)	—	—	—	—	—	—	—	—	—
Idioprioniodus robust M element	1	—	—	1	—	—	1	1	—
Palmatolepis sp. (Famennian)	—	—	—	—	—	—	—	—	—
Polygnathus communis	1	—	—	—	—	—	—	—	—
Polygnathus purus?	—	—	1	—	—	—	—	1	—
Polygnathus purus subplanus	—	—	—	—	—	—	—	—	—
"Polygnathus" mehli	—	—	—	—	—	—	—	1	—
Polygnathus sp. fragments	—	—	—	—	—	—	—	—	—
Protognathodus praedelicatus	—	—	—	—	—	—	—	—	—
Pseudopolygnathus multistriatus M1	—	—	—	1†	—	—	—	—	—
Pseudopolygnathus multistriatus M2	—	—	—	—	—	—	—	—	—
Siphonodella cooperi M2	—	—	—	—	—	—	—	—	1†
Siphonodella cooperi hassi	—	—	1†	—	—	—	—	—	—
Siphonodella obsoleta	—	—	1	—	—	—	—	—	—
Siphonodella sp. fragments	—	—	—	—	—	1	—	—	—
Taphrognathus varians	—	—	—	—	—	1†	1†	—	—
Unidentified ramiform elements	1	1	4	2	1	—	3	1	1
Lower Ordovician and possible Upper Cambrian coniform conodonts									
Aloxoconus staufferi	—	—	—	—	—	—	—	—	—
Colaptoconus quadraplicatus	—	—	1	—	—	—	1†	—	1
Eucharodus parallelus	—	—	—	—	—	—	—	—	—
Oneotodus gracilis	—	—	—	—	—	—	—	—	—
Rossodus sp. aff. *R. manitouensis*	—	—	—	—	—	—	—	—	—
Ulrichodina deflexa	—	—	—	—	—	—	—	—	—
Utahconus n. sp.	—	—	—	—	—	—	—	—	—
Unidentified coniform elements	—	—	—	—	—	—	—	—	1
Rock dissolved (kg)	3.8	3.2	3.8	4.1	3.9	4.1	3.8	4.8	3.7

Abbreviations.—Juv, juvenile specimen; M1, morphotype 1; M2, morphotype 2.
* Includes measured section, random breccia samples, and eggs from road cuts on Highway 13. Distance is meters above base of breccia exposure in measured section.

| Random Samples | | | | Ash Grove | Ash Grove | Coon | SE 450 |
A	B	C	Eggs	Breccia	Eggs	Creek	Roadcut
—	—	—	—	—	—	—	—
—	—	1	—	—	—	—	—
—	5	—	—	—	—	—	—
—	—	—	—	—	—	—	—
6	34	45	—	16†	—	1	5
2	—	—	—	1	—	—	—
—	—	—	—	—	—	—	—
—	6	12†	—	2	—	1	2
—	—	1†	—	—	—	—	—
4	6	2	—	2†	—	—	1
—	1†	—	—	—	—	—	—
—	2	6	—	2†	—	—	—
—	—	—	—	2	—	—	—
1†	—	—	—	—	—	—	—
—	—	—	—	—	—	—	1†, 1?
—	—	9	—	—	1 juv	—	
—	—	—	—	—	—	—	1†
—	—	—	—	—	—	—	1
—	1†	1	—	—	—	—	—
—	—	—	—	—	—	—	5
—	—	3	—	—	—	1†	—
—	1	—	3†	—	—	—	—
1	—	—	—	1	—	—	—
1	—	1	—	—	—	—	1
—	77	111	—	4†	—	1	23
—	—	—	—	—	—	—	1†
1	7	7	—	6†	—	1	—
—	4	—	—	—	—	—	—
—	11	—	—	—	—	—	—
—	1	—	—	—	—	—	—
—	—	—	—	—	—	—	1†
—	2	—	—	—	—	—	—
4	4	—	—	—	—	—	1
?	?	7.7	?	7.7	?	4.1	6.9

Table 12.2. *Distribution* of conodonts from Weaubleau breccia.

† One or more specimens from this sample are illustrated on Figure 12.13. Famennian indicates highest Devonian stage. Random sample C is from Crinoid Knob, where most crinoids and blastoids were collected. Figure 14 shows ranges of many of these taxa based on in situ occurrences.

Figure 12.13. *Ranges of conodonts found in the Weaubleau breccia as documented from in situ occurrences. Species may not range through entire zone at base or top of indicated range. Zonation modified from Sandberg (2002, fig. 1). Light and dark parts of some lines indicate ranges in some parts of North America (light part of lines) may be greater than in Missouri (black part of lines). Numbers after names indicate illustrations on Figure 12. Tournaisian, Viséan, and Namurian are series used in Europe. Kinderhookian, Osagean, Meramecian, and Chesterian are series used in North America.*

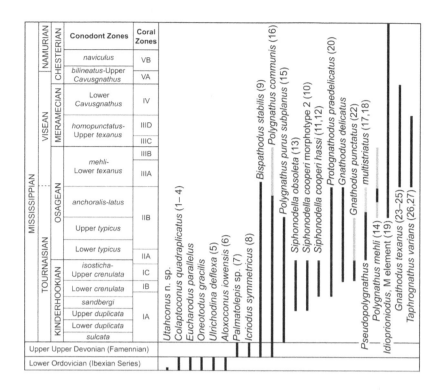

from the south end of the roadcuts, and conodont samples were collected at 1-m intervals. The results are summarized in Table 12.2, and the ranges of recovered taxa, on the basis of known in situ occurrences, are shown on Figure 12.13. Additional taxa recovered from the measured section include *Bispathodus stabilis* (Branson and Mehl, 1934b) (Fig. 12.12.9); *Colaptoconus quadraplicatus; Gnathodus punctatus* (Cooper, 1939) (Fig. 12.12.22); *Gnathodus* sp. aff. *G. typicus* Cooper, 1939; *Polygnathus purus?; Polygnathus mehli* Thompson, 1967 (Fig. 12.12.14); *Pseudopolygnathus multistriatus* morphotype 1 of Lane, Sandberg, and Ziegler, 1980 (Fig. 12.12.17); *Siphonodella cooperi* morphotype 2 of Sandberg et al. (1978) (Fig. 12.12.10); and *Siphonodella cooperi hassi* Thompson and Fellows, 1970 (Fig. 12.12.11, 12.12.12).

Two other random samples of breccia from Highway 13 produced additional taxa (Table 12.2, random samples B, C). Random sample B produced elements of *Utahconus* n. sp., a Lower Ordovician species shown on a range chart by Miller et al. (2003, fig. 7). It occurs in the middle part of the Barn Canyon Member of the House Limestone in Utah; coeval strata in Missouri are assigned to the Gasconade Dolomite. Other Lower Ordovician conodonts in the same sample probably are from the Jefferson City Dolomite and include *Eucharodus parallelus* (Branson and Mehl, 1933); *Oneotodus gracilis* (Furnish, 1938); *Rossodus* sp. aff. *R. manitouensis* Repetski and Ethington, 1983; and four unidentified coniform elements. Random sample B also contained *Gnathodus delicatus* Branson and Mehl, 1938, and a broken element of *Palmatolepis* Ulrich and Bassler, 1926 (Fig. 12.12.7). Jeff Over (personal commun.) examined the fragmentary *Palmatolepis* and identified it as a morphotype that occurs in the Famennian Stage (uppermost Devo-

Sample	UTM Coordinates of Locality	Remarks
Highway 13 measured section	443475 m E, 4201430 m N (locality HS on Fig.12.2)	Samples from 0 m to 8 m; only conodonts were found in these samples, which are from near south end of Highway 13 road cuts.
Highway 13 Random Samples A, B	Located between Measured Section and Random Sample C (locality HS on Fig. 12.2)	Only conodonts were found in these samples.
Highway 13 Random Sample C	443335 m E, 4201621 m N (locality HS on Fig. 12.2)	Also known as Crinoid Knob. Echinoids and most crinoids and blastoids on Figure 12.10 found here; also conodonts, corals, and fragmentary brachiopods.
Ash Grove Aggregates Quarry	444200 m E, 4206500 m N (locality AG on Fig. 12.2)	Only conodonts in sample. Location is approximate because rock at the exact sample locality has been removed by quarrying operations.
Coon Creek	440713 m E, 4198808 m N (locality CC on Fig. 12.2)	Only conodonts were found at this locality.
County Road SE 450	438320 m E, 4201023 m N (locality SE on Fig. 12.2)	Crinoids, conodonts, and a shark tooth were found at this locality.
County Highway W	440473 m E, 4201865 m N (locality W on Fig. 12.2)	A single tabulate coral was found at this locality.

Note.—See also Figure 12.2 and Table 12.2. All UTM coordinates are in Zone 15 and are ±5 m. All localities are in Saint Clair County, Missouri.

Table 12.3. *Locations of fossil collections.*

nian). Random sample C at Highway 13 also contained *Elictognathus* Cooper, 1939; *Icriodus symmetricus* Branson and Mehl, 1934a (Fig. 12.12.8); *Pseudopolygnathus multistriatus* morphotype 2 of Lane, Sandberg, and Ziegler, 1980 (Fig. 12.12.18); and *Siphonodella cooperi hassi*.

Two taxa found at Highway 13 have stratigraphic ranges that extend from Famennian to Osagean strata (Fig. 12.13). *Polygnathus communis* occurs from Famennian to upper Osagean strata, and several elements (Fig. 12.12.16, Table 12.2) are an Osagean morphotype. *Bispathodus stabilis* is known from the Famennian to middle Osagean (Fig. 12.12.9; Table 12.2), and the specific morphotype is similar to specimens from the Famennian.

Two taxa, *Icriodus symmetricus* and *Palmatolepis* sp., are of Late Devonian (Famennian) age. There are two possible explanations for these conodont occurrences, which are comparable to the occurrence of Mohawkian conodonts in breccia at the Decaturville structure, as discussed above. One explanation is that Upper Devonian strata were deposited in Saint Clair County, but most were eroded away after the Late Devonian uplift of the Ozark Dome. Such erosional remnants are known from Saint Clair County and bordering counties. Erosional remnants of Devonian strata may have been buried by Mississippian strata and incorporated into the breccia at the time of meteorite impact. Alternatively, it is possible that the Devonian conodont elements were reworked from Devonian strata into the basal Mississippian Bachelor Formation, and clasts of the Bachelor later were incorporated into the Weaubleau breccia.

The third locality from which conodonts were recovered is Coon Creek (locality CC on Fig. 12.2). A single sample produced *Colaptoconus quadraplicatus, Gnathodus texanus, Hindeodus?* sp., and *Siphonodella cooperi hassi* (Fig. 12.12.11) (Table 12.2).

The fourth locality from which conodonts were recovered is a roadcut along County Road SE 450 (SE on Fig. 12.2). A single sample (Table 12.2) included the Lower Ordovician species *Aloxoconus staufferi* (Furnish, 1938) (Fig. 12.12.6); *Ulrichodina deflexa* Furnish, 1938 (Fig. 12.12.5); and an unidentified coniform element. This sample also included the Mississippian taxa *Gnathodus texanus, Hindeodus?* sp.; *Idioprioniodus* sp.; *Polygnathus mehli; Protognathodus praedelicatus* Lane, Sandberg, and Ziegler, 1980 (Fig. 12.12.20); *Pseudopolygnathus multistriatus;* and *Siphonodella cooperi.*

Most of the conodont taxa recovered from the Weaubleau breccia are known elsewhere from Ibexian (Lower Ordovician) and Kinderhookian, Osagean, and lowest Meramecian (Mississippian) strata (Fig. 12.13). Presumably these taxa are derived from both consolidated and unconsolidated sediments that were incorporated into the unbedded Weaubleau breccia. There is no discernible pattern to the associations of species in samples collected at 1-m intervals at Highway 13, and young and old species appear to be mixed randomly. The Lower Ordovician species *Colaptoconus quadraplicatus* occurs in half of all breccia samples (Table 12.2), and Lower Ordovician species occur at all four localities (Highway 13, Ash Grove Quarry, Coon Creek, and County Road SE 450). One of the youngest species, *Gnathodus texanus*, was recovered from nearly every sample of breccia and also is the most abundant species. This is also one of the most abundant species recovered from in situ conodont samples from the upper part of the Burlington-Keokuk and lowermost part of the Warsaw Formation in Springfield.

Especially noteworthy is the upward mixing of Lower Ordovician conodonts so that they occur with Middle Mississippian conodonts in an area where Mississippian and Pennsylvanian strata are exposed at the surface. This upward mixing is consistent with upward movement of Precambrian cataclastic granite in the Vista 1 core and with conodont samples containing rounded quartz pebbles that were probably derived from the basal Cambrian sandstone, which is identified as the Reagan Sandstone in this part of Missouri. Such upward movement is one criterion for recognizing a central uplift in the middle of an impact structure. Another criterion is the central uplift having a round shape, a feature that is shown in Figure 12.2.

Two important observations relate to the apparent absence of certain conodont taxa from the Weaubleau breccia. First, several taxa that are typical of the Osagean Burlington Limestone were not found, including species of *Bactrognathus* Branson and Mehl, 1941b; *Doliognathus* Branson and Mehl, 1941b; *Eotaphrus* Pierce and Langenheim, 1974; *Scaliognathus* Branson and Mehl, 1941b; and *Staurognathus* Branson and Mehl, 1941b. This situation is unusual considering that most of the crinoids and blastoids in the Weaubleau breccia are species typical of the Burlington Limestone. Lane et al. (2005) documented ranges of conodonts in the type area of the Osagean Series in Saint Clair County, slightly north of the Osage River (locality TA on Fig. 12.2 herein). Their three sections are only about 15 km northwest of the roadcuts of Weaubleau breccia along Highway 13. These authors did not find these genera in the Burlington Limestone, and they only recovered a modest, low-diversity fauna from the type Osagean. Only 19 of their 45 samples yielded conodonts. Thus, the probable reason that few typical Burlington conodonts were reworked into the breccia is that they were not present in the Burlington Limestone in this area before impact. Species that Lane et al. (2005) reported from the Burlington Limestone that also occur in the Weaubleau breccia are *Polygnathus communis* and *Pseudopolygnathus multistriatus* (Fig. 12.13). They assigned 24 m of strata to the Burlington Limestone and pointed out that Keyes (1893, p. 60) noted the absence of Keokuk-age strata along the bluffs of the Osage River in this part of Missouri. The absence of Keokuk-age strata in the Osagean type area was also noted by Kaiser (1950) and Thompson (1986).

The second observation regarding the absence of certain conodonts in the Weaubleau breccia is that no conodonts younger than the lower Meramecian *homopunctatus*–upper *texanus* Biozone were recovered, although two species typical of this biozone are present in the breccia, namely *Gnathodus texanus* and *Taphrognathus varians* (Fig. 12.13). *Cavusgnathus* Harris and Hollingsworth, 1933, occurs in the Salem Limestone and younger strata. The genus is diagnostic of the overlying Lower *Cavusgnathus* Biozone, but it has not been recovered from the Weaubleau breccia. Thompson recovered *Cavusgnathus* from the Salem and Saint Louis limestones (Fig. 12.3) in exposures along Horse Creek in northwestern Dade County (Fig. 12.1), about 60 km southwest of the Weaubleau structure. Also, Salem Limestone has been identified in the subsurface of Vernon and Bates counties (Thompson, 1986), so it is probable that strata containing *Cavusgnathus* were deposited near the Weaubleau structure. The absence of this genus from the Weaubleau breccia is consistent with an interpretation that formation of the breccia predates deposition of the Salem and Saint Louis limestones.

CONODONTS FROM WEAUBLEAU EGGS. Most of the chert nodules known as Weaubleau eggs have a clast of gray-green or brown clay mudstone at the center. Approximately 40 of these eggs from the roadcuts along Missouri Highway 13 and approximately 60 from the Ash Grove Quarry were collected and broken open, and the clay mudstone was removed from most (several had chert centers or were hollow). All of this mudstone—several kilograms—was processed so as to recover conodonts. Table 12.2 records that eggs from Highway 13 yielded three small *Siphonodella obsoleta* Hass, 1959 (Fig. 12.12.13), and the eggs from Ash Grove Quarry yielded one juvenile *Polygnathus* sp. These taxa are consistent with the clay mudstone being derived from shales and mudstones of the Lower Mississippian Northview Formation.

CORALS. Wayne Bamber (Geological Survey of Canada, Calgary; personal commun.) identified corals recovered from the Weaubleau breccia at the roadcuts along Highway 13. Solitary rugose corals include *Cyathaxonia?*

sp., *Sychnoelasma* sp., *Zaphrentites* sp., and a tabulate coral that probably is *Cladochonus* sp. Three of these taxa have long ranges, but *Sychnoelasma* Lang, Smith, and Thomas, 1940, is restricted almost entirely to the Osagean coral zone II of Sando and Bamber (1985, fig. 5; Fig. 12.13 herein). Some specimens could not be identified because of poor preservation. A single, silicified tabulate coral, probably a michelinid, was found at a small roadcut near the middle of the central uplift of the Weaubleau structure, along Saint Clair County Highway W (locality W on Fig. 12.2). This group of corals has a long range in the Mississippian. Although not as diagnostic, the ages of the corals are consistent with data from echinoderms and conodonts.

SHARK TOOTH. One shark tooth was found in the Weaubleau breccia at a small roadcut along Saint Clair County Road SE 450. The rock is coarse-grained resurge breccia with large crinoid columnals and chert clasts. The shark tooth is partly exposed on the broken edge of a lithologic sample and is associated with *Agaricocrinus planoconvexus*, *Platycrinites* sp., and a diverse conodont fauna (Table 12.2). The black tooth is flat and thin, of the type found in shell-crushing sharks. Such shark teeth commonly are found in southwest Missouri in sandstone of the basal Mississippian Bachelor Formation and rarely are found in the Burlington-Keokuk Limestone, so the presence of this tooth is consistent with other fossils found in the breccia.

Fossils from Chert in Graydon Conglomerate

Unconsolidated residual chert breccia, assigned to the Pennsylvnian Graydon Conglomerate, occurs above the Weaubleau breccia at roadcuts along Highway 13 (Figs. 12.4, 12.5). The chert clasts apparently remained after the dissolution of carbonate that was deposited above the Weaubleau breccia. Thus, identifying the formation from which this chert was derived will place an upper limit on the time of impact and breccia formation. We did not make extensive collections of fossils from this chert, but several brachiopods, bryozoa, and a single trilobite pygidium have been identified.

Alfred C. Spreng (University of Missouri, Rolla, written commun.) identified a diverse fauna of brachiopods preserved as molds in several chert clasts. The names and previously reported stratigraphic occurrences of these brachiopods are in Table 12.4, including *Camarotoechia mutata* (Hall, 1856); *Cleiothyridina hirsuta* (Hall, 1856); *Composita globosa* Weller, 1914; *Echinoconchus biseriatus* (Hall, 1856); *Linoproductus* Chao, 1927; *Rhipidomella* Oehlert, 1890; *Schizophoria* King, 1850; and *Spirifer bifurcatus* Hall, 1856. Additional specimens identified questionably to species or only to genus are not included in Table 12.4.

Spreng also identified several fenestrate and bifoliate bryozoa that are preserved silicified or as molds in chert clasts; their names and stratigraphic occurrences also are listed in Table 12.4. Ulrich (1890) named all of these bryozoa, including *Fenestella*. cf. *F. compressa*, *F. funicula*, *F. multispinosa*, *F. regalis*, *F. serratula*, *F. triserialis?*, *Hemitrypa* cf. *H. aspera*, *H. perstriata*, *H. proutana*, *Rhombopora attenuata*, and *Sulcoretepora nitida*.

One chert clast had a trilobite pygidium, which was identified by David Brezinski (Maryland Geological Survey, personal commun.) as *Sper-*

genaspis Brezinski, 1987. It is probably *S. mauvaisensis* (Hessler, 1965), which is known from the Keokuk and Warsaw formations, but it may be *S. salemi* Brezinski, 1987, which is known from the Salem and Saint Louis formations (Fig. 12.3).

Most of the fossils in chert clasts have been reported previously from the Keokuk, Warsaw, and Salem formations in the upper Mississippi River Valley, including Iowa and Missouri (Table 12.4). As noted in the discussion of these stratigraphic units, the contact between the Keokuk and Warsaw formations has changed over time, and it is difficult to know how Ulrich (1890), Weller (1914), Van Tuyl (1925), Moore (1928), and Branson (1944) identified that contact relative to its placement at the present. Consequently, there remains some uncertainty about the identity of the specific formation or formations from which this chert was derived. However, it is probable that at least some of the chert was derived from the Keokuk or Warsaw formations, and it is clear that these fossils are derived from a relatively thin stratigraphic interval in the upper Osagean or lower Meramecian (Fig. 12.3).

Age of Weaubleau Breccia

The presence of upper Osagean to lower Meramecian conodonts in the resurge breccia, and the absence of upper Meramecian species, suggests that meteorite impact occurred before deposition of late Meramecian strata. Brachiopods and bryozoans in the chert clasts above the Weaubleau breccia are of latest Osagean to earliest Meramecian age. They thereby constrain the upper age limit of the Weaubleau breccia to a time that is only slightly younger than the age of the youngest fossils in the breccia itself, rather than being from, e.g., the Upper Mississippian or Lower Pennsylvanian. Also, Dulin et al. (2004) and Dulin and Elmore (2006) studied the paleomagnetism of the Weaubleau breccia and documented a paleopole position that is consistent with nearby undeformed Burlington-Keokuk strata. This paleopole position plots on the Mississippian part of the polar wandering path for Laurentia, and they concluded that formation of the Weaubleau structure cannot be younger than Late Mississippian.

On the basis of echinoderms in the breccia, another argument supports the interpretation that the time of impact cannot be much younger than earliest Meramecian. The resurge breccia contains intact crinoids and blastoids that were derived from unconsolidated sediment of Burlington and Keokuk age, as discussed above. Meteorite impact resulted in resurge currents that reworked this loose sediment into the Weaubleau breccia. If the impact had occurred considerably later than the deposition of the youngest of these sediments, lithification of the sediment would have produced the typical hard crinoidal grainstone that is characteristic of Osagean strata in southwest Missouri, outside of the impact area.

If the time of impact and breccia formation was Chesterian or earliest Pennsylvanian, echinoderms in the Burlington-Keokuk sediments likely would have been locked into hard limestone, and echinoderms could not have been reworked into the breccia as intact fossils. Also, breccia exposures do not contain abundant large clasts of Burlington-Keokuk that could

	Reported Occurrences in Region			
Taxon	Mississippi River Valley (Weller, 1914)	Iowa (Van Tuyl, 1925)	Missouri (Moore, 1928)	Missouri (Branson, 1944)
Brachiopoda				
Camarotoechia mutata Hall, 1856	S	K, W	K, W	K, W
Camarotoechia mutata Hall, 1856	W, S, Steg	—	K, W	K, W
Composita globosa Weller, 1914	K	—	W	K, W
Echinoconchus biseriatus (Hall, 1856)	S	K, W, S	K, W	K, W, S
Linoproductus sp.	—	—	—	—
Rhipidomella sp.	—	—	—	—
Schizophoria sp.	—	—	—	—
Spirifer bifurcatus Hall, 1856	S	W, S	W	W, S
Bryozoa				
Fenestella cf. *compressa* Ulrich, 1890	K	K, W	K, W	K, W
Fenestella funicula Ulrich, 1890	K	W	W	K, W
Fenestella multispinosa Ulrich, 1890	K	K, W, S	K, W	K, W
Fenestella regalis Ulrich, 1890	K	—	W	K
Fenestella serratula Ulrich, 1890	K—Ch	K, W, S	K, W	K, W, S
Fenestella triserialis? Ulrich, 1890	K	K, W	K, W	K, W
Hemitrypa cf. *aspera* Ulrich, 1890	K	—	—	—
Hemitrypa perstriata Ulrich, 1890	K, W	K, W	—	K, W
Hemitrypa proutana Ulrich, 1890	K (rare W, St L)	K, W	W	—
Rhombopora attenuata Ulrich, 1890	upper K	K, W, S	K, W	K, W
Sulcoretepora nitida Ulrich, 1890	K	W	—	—

Note.—Identifications and reported occurrences provided by Albert S. Spreng (University of Missouri, Rolla). *Abbreviations.*—Ch, Chesterian Series; K, Keokuk Limestone; S, Salem Limestone; St L, Saint Louis Limestone; Ste G, Ste. Genevieve Limestone; W, Warsaw Formation.

have been weathered to produce the echinoderms. The most logical scenario is that meteorite impact and deposition of the Weaubleau breccia occurred soon after deposition of sediment containing the youngest fossils that occur in the breccia. Thus, a latest Osagean to early Meramecian time of impact is most likely probably during deposition of the *mehli*–lower *texanus* Biozone or the *homopunctatus*–upper *texanus* Biozone (Fig. 12.13). On the basis of current geochronologic data (Gradstein et al., 2004), the impact may have occurred about 335–340 million years ago.

Type Osagean Series

The proximity of the Weaubleau structure to the type area of the Osagean Series explains one of the peculiarities of that stratigraphic succession, which is that strata coeval with the Keokuk Limestone are absent along the Osage River bluffs near Osceola, Missouri (Thompson, 1986, p. 65). Lane et al. (2005) found that the youngest age-diagnostic conodonts in their sections in the type area of the Osagean Series (locality TA on Fig. 12.2 herein) are equivalent to those in the lower part of the Burlington Limestone. They considered that younger exposures, although lacking age-diagnostic faunas, are equivalent to the middle part of the Burlington Limestone. Thus, not only are strata coeval with the Keokuk Limestone missing along the Osage River bluffs, but strata coeval with the upper part of the Burlington Limestone (Fig. 12.3) are also missing. An obvious hypothesis to explain this situation is that strata coeval with the upper part of the Burlington and the Keokuk intervals were deposited near Osceola, but these strata were eroded away before Pennsylvanian deposition. An alternative hypothesis is that strata of that age were deposited, but resurge currents after the meteorite impact reworked the unconsolidated sediments into the Weaubleau breccia before they were lithified. The stratigraphic succession along the Osage River bluffs has the key to evaluating these two hypotheses.

A scenic viewpoint on Missouri Highway 82 is situated on the river bluffs above the confluence of the Sac and Osage rivers (locality OV on Fig. 12.2). Osagean limestones form cliffs that are overlain by a covered interval with Weaubleau eggs, and this interval is overlain by Pennsylvanian sandstone. Complete eggs and broken halves of Weaubleau eggs and slabs of the sandstone are incorporated into stone walls and stairs that descend from the highway to the scenic viewpoint. Loose Weaubleau eggs are common in regolith near Osceola, and they occur in the Weaubleau breccia at places such as the Highway 13 roadcuts (locality HS on Fig. 12.2). The presence of these distinctive chert nodules in regolith is a useful proxy for the original distribution of the Weaubleau breccia, even where the carbonate part of the breccia has been weathered away. Thus, the Weaubleau breccia likely was present along the Osage River above the mid-Osagean limestones and below the Pennsylvanian. It is not clear whether some of the actual breccia is still present in this area, or if all that remains is regolith containing Weaubleau eggs.

The stratigraphic succession in the Osage River bluffs is comparable to strata at the Highway 13 roadcuts where the Weaubleau breccia is exposed. However, the successions in the two areas differ in ways that complement each other. The Burlington-Keokuk Limestone is not present at the Highway 13 roadcuts, which is within the intensely deformed part of the Weaubleau structure. Presumably, all lithified Kinderhookian and Osagean strata were shattered during impact. Echinoderms and conodonts coeval with faunas that occur in the upper part of the Burlington and Keokuk limestones occur there in the resurge facies of the breccia. These fossils indicate that sediments of that age were not lithified and were reworked into the Weaubleau breccia. The Osage River is outside of the intensely deformed part of the Weaubleau structure. Only the lower part of the Burlington-Keokuk Limestone is present (Lane et al., 2005), and younger, unlithified Osagean sediments that presumably were present at the time of impact have been reworked into the Weaubleau breccia. There, the breccia apparently has been weathered away, and only the Weaubleau eggs are left behind to document its presence.

These data are not consistent with the first hypothesis listed above, i. e., strata coeval with the upper Burlington-Keokuk interval were deposited and eroded away before deposition of the Pennsylvanian. Weaubleau eggs occurring along the Osage River bluffs are consistent with the second hypothesis, which involves reworking unconsolidated upper Osagean sediments into the Weaubleau breccia.

Conclusions

Diagnostic criteria for conclusively identifying an anomalous geological feature as an impact structure include the presence of shatter cones or shocked quartz. Shatter cones have not been found at the Weaubleau structure, but both shatter cones and shocked quartz are present at the Decaturville and Crooked Creek structures (Fig. 12.1). Shocked quartz is common in the Weaubleau breccia, and Morrow and Evans (2007) reported the results of petrographic study of such quartz grains using a universal stage. Perhaps the presence of mixed-age fossils associated with polymict breccia now should be added to the list of diagnostic criteria for recognizing impact structures. Such reworked fossils are especially strong evidence if fossils from lower strata can be interpreted as having been moved upward so as to occur mixed with fossils from stratigraphic units that were at or near the sediment-water interface at the time of impact. The low CAI of conodonts from the Weaubleau breccia precludes breccia formation by a mechanism such as massive hydrothermal dissolution and subsequent collapse to form the breccia. Such a mechanism would produce downward mixing of younger fossils into older stratigraphic units, whereas the opposite (upward mixing) occurs in the Weaubleau breccia. Low CAI also precludes the long-discredited hypothesis of cryptovolcanic explosion, which has been proposed often but never once demonstrated for any feature on Earth. Volcanic activity would have probably produced a much higher CAI for conodonts derived from strata associated with such a feature.

No obvious extinctions resulted from the formation of the Weaubleau structure in late Osagean to early Meramecian time, and none should be expected because the structure is too small to have had widespread effects on the biosphere. If the Weaubleau, Decaturville, and Crooked Creek structures (and possibly other 38th parallel structures) comprise true serial impacts that occurred within minutes of each other in a shallow ocean, then the effects on the biosphere would have been greater. However, more precise timing of the impacts is needed so as to trace laterally the stratigraphic level of the Weaubleau breccia and to search for biological effects recorded at that level.

It is an accident of history that the Osagean Series was named for an area where the upper half of the Osagean is missing because of a geological process as unusual as a meteorite impact. Kaiser (1950), Thompson (1986), and Lane et al. (2005) documented that strata in the Osage River Valley are inadequate as a reference area for the Osagean Series, and a composite section of strata exposed along the Mississippi River Valley probably will continue to be used as a de facto standard. Knowing the degree of incompleteness of the type Osagean Series has set the stage for establishing a stratotype for the Osagean. Such a stratotype should be proposed after research by a diverse team of stratigraphers and paleontologists specializing in all relevant fossil groups, and the Mississippi River Valley may be the best place to designate such a stratotype.

Acknowledgments

Ash Grove Aggregates granted access to their quarry on Saint Clair County Road TT. W. Bamber, D. Brezinski, F. Gahn, S. Leslie, J. Over, J. Repetski, and A. Spreng identified fossils associated with the Weaubleau structure. M. Craig, director of the electron microscopy facility at Missouri State University, made the micrographs of conodonts. J. Ray evaluated the source of chert clasts in the Graydon Conglomerate. The Missouri Department of Transportation drilled several cores in and near the Weaubleau structure. Construction roadcuts along Highway 13 unexpectedly provided excellent exposures of the Weaubleau breccia. S. Marcus and C. Rexroad reviewed the chapter and made many helpful suggestions. R. Riis, while an undergraduate student at Missouri State University, processed and picked conodonts from random sample A at the Highway 13 roadcuts. B. Dattilo, L. Miller, and R. Vanwey donated crinoids they found at roadcuts of the Weaubleau breccia. B. Glass and R. Hays donated unusually shaped Weaubleau eggs for study.

References

Beveridge, T. R. 1951. Geology of the Weaubleau Creek area, Missouri. Missouri Geological Survey and Water Resources, 32, second series, 111 p.

Branson, E. B. 1944. The geology of Missouri. University of Missouri Studies, 19(3), 535 p.

Branson, E. B., and M. G. Mehl. 1933. Conodonts from the Jefferson City (Lower Ordovician) of Missouri. University of Missouri Studies, 8:53–64.

Branson, E. B., and M. G. Mehl. 1934a. Conodonts from the Grassy Creek Shale of Missouri. University of Missouri Studies, 8(3):171–259.

Branson, E. B., and M. G. Mehl. 1934b. Conodonts from the Bushberg Sandstone and equivalent formations in Missouri. University of Missouri Studies, 8(4):265–300.

Branson, E. B., and M. G. Mehl. 1938. Conodonts from the Lower Mississippian of Missouri. University of Missouri Studies, 13(4):128–148.

Branson, E. B., and M. G. Mehl. 1941a. Conodonts from the Keokuk Formation. Journal of the Science Laboratories, Denison University, 35:179–188.

Branson, E. B., and M. G. Mehl. 1941b. New and little known Carboniferous conodont genera. Journal of Paleontology, 15:97–106.

Brezinski, D. K. 1987. *Spergenaspis:* a new Carboniferous trilobite genus from North America. Annals of Carnegie Musuem, 56:245–251.

Chao, Y. T. 1927. Productidae of China. Part 1. Producti. China Geological Survey, Palaeontologica Sinica, series B, part 2, 192 p.

Cooper, C. L. 1939. Conodonts from a Bushberg-Hannibal horizon in Oklahoma. Journal of Paleontology, 13:379–422.

Davis, G. H., K. R. Evans, J. F. Miller, P. S. Mulvaney, and C. W. Rovey II. 2005. Weaubleau-Osceola, Decateurville, and Crooked Creek structures: new aspects of Missouri's 38th parallel structures, p. 13. *In* K. R. Evans, J. W. Horton Jr., M. F. Thompson, and J. E. Warme (eds.), SEPM Research Conference: The Sedimentary Record of Meteorite Impacts, Springfield, Missouri, May 21–23, Abstracts with Program.

Dulin, S. A., and R. D. Elmore. 2006. Paleomagnetism of the Decaturville impact breccias and the Weaubleau-Osceola structure, southwest Missouri: testing the serial impact hypothesis. Geological Society of America, Abstracts with Programs, 38(7):121.

Dulin, S. A., K. G. Gardner, R. D. Elmore, L. A. Totten, K. R. Evans, J. F. Miller, and C. W. Rovey II. 2004. Paleomagnetism of the Weaubleau-Osceola impact structure, SW Missouri. Eos, 85(7), Joint Assembly Supplement, abstract GP31A-07.

Earth Impact Data Base. 2006. Planetary and Space Science Centre, University of New Brunswick <http://www.unb.ca/passc/ImpactDatabase/>.

Epstein, A. G., J. B. Epstein, and L. D. Harris. 1977. Conodont color alteration—an index to organic metamorphism. U.S. Geological Survey Professional Paper, 995, 27 p.

Evans, K. R., K. L. Mickus, S. Fagerlin, J. Luczaj, E. Mantei, J. F. Miller, T. Moeglin, R. T. Pavlovski, and K. C. Thomson. 2003a. New dimensions of the Weaubleau structure: a possible meteorite impact site in southwestern Missouri. Geological Society of America, Abstracts with Programs, 35(2):51.

Evans, K. R., C. W. Rovey II, K. L. Mickus, J. F. Miller, T. Plymate, and K. C. Thomson. 2003b. Weaubleau-Osceola structure, Missouri: deformation, event stratification, and shock metamorphism of a mid-Carboniferous impact site. Third International Conference on Large Meteorite Impacts, Nördlingen, Germany, August 5–7, Lunar and Planetary Institute Contribution 1167, abstract 4111.

Furnish, W. M. 1938. Conodonts from the Prairie du Chien beds of the Upper Mississippi Valley. Journal of Paleontology, 12:318–340.

Gahn, F. J. 2002. Crinoid and blastoid biozonation and biodiversity in the Early Mississippian (Osagean) Burlington Limestone. Iowa Department of Natural Resources, Geological Survey Guidebook, 23:53–74.

Gentile, R. J., and T. L. Thompson. 2004. Paleozoic succession in Missouri, part 5: Pennsylvanian Subsystem. Missouri Department of Natural Resources, Geological Survey and Resource Assessment Division, Report of Investigations, 70(5). CD-ROM.

Gradstein, F. M., J. G. Ogg, A. G. Smith, et al. 2004. A Geologic Time Scale 2004. Cambridge University Press, Cambridge, Massachusetts, 589 p.

Gunnell, F. H. 1933. Conodonts and fish remains from the Cherokee, Kansas

City, and Wabaunsee Groups of Missouri and Kansas. Journal of Paleontology, 7:261–297.

Hall, J. 1856. Descriptions of new species of fossils from the Carboniferous limestone of Indiana and Illinois. Albany Institute Transactions, 4:1–36.

Hall, J. 1858. Report on the Geological Survey of the State of Iowa, embracing the results of investigations made during portions of the years 1855, 1856, and 1857. Iowa Geological Surveyl, 1(2):473–724.

Hall, J. 1860. Contributions to the palaeontology of Iowa: being descriptions of new species of Crinoidea and other fossils. Iowa Geological Survey, Geological Report of Iowa, Supplement, 1(2):1–94.

Hall, J. 1861. Descriptions of new species of Crinoidea and other fossils, from the Carboniferous rocks of the Mississippi Valley. Iowa Geological Survey Report of Investigations, Preliminary Notice, p. 1–19.

Harris, R. W., and R. V. Hollingsworth. 1933. New Pennsylvanian conodonts from Oklahoma. American Journal of Science, series 5, 25(147):193–204.

Hass, W. H. 1959. Conodonts from the Chappel Limestone of Texas. U.S. Geological Survey Professional Paper, 294-J:365–399.

Hendricks, H. E. 1954. The geology of the Steelville quadrangle, Missouri. Missouri Geological Survey and Water Resources, 36 (second series), 88 p.

Hessler, R. R. 1965. Lower Mississippian trilobites of the family Proetidae in the United States, part 2. Journal of Paleontology, 39:248–265.

Jackson, R. T. 1912. Phylogeny of the echini, with a revision of Paleozoic species. Boston Society of Natural History Memoir, 7, 491 p.

Kaiser, C. P. 1950. Stratigraphy of the lower Mississippian rocks in southwestern Missouri. American Association of Petroleum Geologists Bulletin, 35:2133–2175.

Kammer, T. W., and W. I. Ausich. 2006. The "Age of Crinoids": a Mississippian biodiversity spike coincident with widespread carbonate ramps. Palaios, 21:238–348.

Kammer, T. W., P. L. Brenckle, J. L. Carter, and W. I. Ausich. 1991. Redefinition of the Osagean-Meramecian boundary in the Mississippian stratotype region. Palaios, 5:414–431.

Keyes, C. R. 1893. The geological formations of Iowa. Iowa Geological Survey, 1:13–144.

King, W. 1850. A monograph of the Permian fossils of England. Palacontographical Society, Monograph 3, 258 p.

Lane, H. R., P. L. Brenckle, and J. F. Basemann. 2005. The type section of the Osagean Series (Mississippian Subsystem), west-central Missouri, USA. Bulletins of American Paleontology, 369:183–197.

Lane, H. R., C. A. Sandberg, and W. Ziegler. 1980. Taxonomy and phylogeny of some Lower Carboniferous conodonts and preliminary standard post-Siphonodella zonation. Geologica et Palaeontologica, 14:117–164.

Lane, N. G. 1958. The monobathrid camerate crinoid family: Batocrinidae. Unpublished Ph.D. dissertation, University of Kansas, 259 p.

Lang, W. D., Smith, S., and H. D. Thomas. 1940. Index of Palaeozoic coral genera. British Museum (Natural History), London, 231 p.

Laudon, L. R. 1937. Stratigraphy of northern extension of Burlington Limestone in Missouri and Iowa. American Association of Petroleum Geologists Bulletin, 21:1158–1167.

Lindströ, M., J. Ormö, E. Sturkell, and I. von Dalwigk. 2005. The Lockne Crater: revision and reassessment of structure and impact stratigraphy, p. 357–388. In C. Koeberl and H. Henkel (eds.), Impact Tectonics, Springer-Verlag. Berlin.

Macurda, D. B., Jr., and D. L. Meyer. 1983. Sea lilies and feather stars. American Scientist, 71:354–365.

Meek, F. B., AND A. H. Worthen. 1865. Descriptions of new species of Crinoidea, etc., from the Paleozoic rocks of Illinois and the adjoining states. Philadelphia Academy of Natural Sciences Proceedings, 1865:143–155.

Meek, F. B., and A. H. Worthen. 1866. Contributions to the palaeontology of Illinois and other western states. Philadelphia Academy of Natural Sciences Proceedings, 17:251–275.

Meek, F. B., and A. H. Worthen. 1868. Remarks on some types of Carboniferous Crinoidea with descriptions of new genera and species of the same, and of one echinoid. Philadelphia Academy of Natural Sciences Proceedings, 20:335–359.

Melosh, H. J. 1989. Impact cratering, a geological process. Oxford Monographs on Geology and Geophysics, 11, 245 p.

Miller, J. F., K. R. Evans, S. E. Bolyard, T. L. Thompson, G. H. Davis, W. I. Ausich, J. A. Waters, and R. L. Ethington. 2005. Mixed-age echinoderms, conodonts, and other fossils from a mid-Mississippian impact resurge breccia, St. Clair County, Missouri. Geological Society of America Abstracts with Programs, 37(7):62.

Miller, J. F., K. R. Evans, J. D. Loch, R. L. Ethington, J. H. Stitt, L. Holmer, and L. E. Popov. 2003. Stratigraphy of the Sauk III interval (Cambrian–Ordovician) in the Ibex area, western Millard County, Utah and Central Texas. Brigham Young University Geology Studies, 47:23–118, plus CD-ROM.

Miller, J. S. 1821. A natural history of the Crinoidea, or lily-shaped animals, with observation on the genera *Asteria, Euryale, Comatula,* and *Marsupites.* Bryan and Co., Bristol, 150 p.

Moore, R. C. 1928. Early Mississippian formations of Missouri. Missouri Bureau of Geology and Mines, second series, 21, 283 p.

Morrow, J. R., and K. R. Evans. 2007. Preliminary shocked-quartz petrography, upper Weaubleau breccia, Missouri, USA. 70th Annual Meeting of Meteoritical Society, August 12–13, Tucson, Arizona. Meteoritics and Planetary Science, 42(7):A112.

Offield, T. W., and H. A. Pohn. 1979. Geology of the Decaturville impact structure, Missouri. U.S. Geological Survey Professional Paper, 1042, 48 p.

Oehlert, D.-P. 1890. Note sur différents groupes établis dans le genere *Orthis* et en particulier sur *Rhipidomella* Oehlert (=*Rhipidomys* Oehlert, olim). Journal de Conchyliologie, series 3, 30:366–374.

Owen, D. D., and B. F. Shumard. 1850. Descriptions of fifteen new species of Crinoidea from the sub-Carboniferous limestone of Iowa, collected during the U.S. Geological Survey of Wisconsin, Iowa, and Minnesota in the years 1848–1849. Journal of the Philadelphia Academy of Natural Sciences, new series, 2(1):57–70.

Owen, D. D., and B. F. Shumard. 1852. Descriptions of seven new species of Crinoidea from the Subcarboniferous of Iowa and Illinois. Journal of the Philadelphia Academy of Natural Sciences, new series, 2(2):89–94.

Pierce, R. W., and R. L. Langenheim Jr. 1974. Platform conodonts of the Monte Cristo Group, Mississippian, Arrow Canyon Range, Clark County, Nevada. Journal of Paleontology, 48:149–169.

Rampino, M. R., and T. Volk. 1996. Multiple impact event in the Paleozoic: collision of a string of comets or asteroids? Geophysical Research Letters, 23:49–52.

Repetski, J. E., and R. L. Ethington. 1983. *Rossodus manitouensis* (Conodonta), a new Early Ordovician index fossil. Journal of Paleontology, 57:289–301.

Repetski, J. E., J. D. Loch, and R. L. Ethington. 1998. Conodonts and biostratigraphy of the Lower Ordovician Roubidoux Formation in and near the Ozark National Scenic Waterways, southeastern Missouri. National Park Service Geologic Resources Division Technical Report. NPS/NRGRD/GRDTR -98-01:109–115.

Repetski, J. E., J. D. Loch, and R. L. Ethington. 2000. A preliminary re-evaluation of the stratigraphy of the Roubidoux Formation of Missouri and correlative Lower Ordovician units in the southern Midcontinent. Oklahoma Geological Survey Circular, 31:103–106.

Rexroad, C., and C. Collinson. 1965. Conodonts from the Keokuk, Warsaw and Salem formations (Mississippian) of Illinois. Illinois Geological Survey Circular, 388, 26 p.

Roemer, C. F. 1854. *Dorycrinus*, ein neues Crinoidengeschlecht aus dem Kohlenkalke Noramerika's. Archiv für Naturgeschichte, Jahrgang 19, 1:207–218.

Roundy, P. V. 1926. Part II. The micro-fauna, p. 5–17. *In* P. V. Roundy, G. H. Girty, and M. I. Goldman (eds.), Mississippian Formations of San Saba County, Texas. U.S. Geological Survey Professional Paper, 146.

Sandberg, C. A. 2002. Mississippian. *In* Encyclopedia of Science and Technology (ninth edition). McGraw-Hill, New York, 11:254–258.

Sandberg, C. A., W. Ziegler, K. Leuteritz, and S. M. Brill. 1978. Phylogeny, speciation and zonation of *Siphonodella* (Conodonta), Upper Devonian and Lower Carboniferous. Newsletters on Stratigraphy, 7:102–120.

Sando, W., and W. Bamber. 1985. Coral zonation of the Mississippian System in the Western Interior Province of North America. U.S. Geological Survey Professional Paper, 1334, 61 p.

Shimer, H. W., and R. R. Shrock (eds.). 1944. Index Fossils of North America. John Wiley & Sons, New York, 837 p.

Shumard, B. F. 1855. Description of new species of organic remains. Missouri Geological Survey, Annual Reports, 1–2:185–208.

Snyder, F. G., and P. E. Gerdemann. 1965. Explosive igneous activity along an Illinois-Missouri-Kansas axis. American Journal of Science, 263:465–493.

Stockdell, R. B., K. Mickus, G. H. Davis, K. Evans, J. F. Miller, and C. Rovey. 2004. Bouger gravity and magnetic anomalies associated with the Weaubleau-Osceola impact structure, Missouri. Geological Society of America, Abstracts with Programs, 36(3), 18 p.

Thompson, T. L. 1967. Conodont zonation of lower Osagean rocks (Lower Mississippian) of southwestern Missouri. Missouri Geological Survey and Water Resources Report of Investigations, 39, 88 p.

Thompson, T. L. 1986. Paleozoic succession in Missouri, part 4, Mississippian System. Missouri Department of Natural Resources, Division of Geology and Land Survey Report of Investigations, 70(4), 182 p.

Thompson, T. L. 1991. Paleozoic succession in Missouri, part 2, Ordovician System. Missouri Department of Natural Resources, Division of Geology and Land Survey Report of Investigations, 70(2), 282 p.

Thompson, T. L. 1995. The stratigraphic succession in Missouri. Missouri Department of Natural Resources, Division of Geology and Land Survey, 40, second series, revised, 189 p.

Thompson, T. L., and L. D. Fellows. 1970. Stratigraphy and conodont biostratigraphy of Kinderhookian and Osagean rocks of southwestern Missouri and adjacent areas. Missouri Geological Survey and Water Resources Report of Investigations, 45, 263 p.

Ulrich, E. O. 1890. Paleozoic bryozoa. Illinois Geological Survey, 8:285–688.

Ulrich, E. O., and R. S. Bassler. 1926. A classification of the toothlike fossils, conodonts, with descriptions of American Devonian and Mississippian species. U.S. National Museum Proceedings, 68, article 12, 63 p.

van Tuyl, F. M. 1925. The stratigraphy of Mississippian formations of Iowa. Iowa Geological Survey, 30:33–349.

Voges, A. 1959. Conodonten aus dem Unterkarbon I und II (*Gattendorfia-* und *Pericyclus-*Stufe) des Sauerlandes. Paläontologische Zeitschrift, 33(4):266–314.

Wachsmuth, C., and F. Springer. 1881. Revision of the Palaeocrinoidea, part 2. Family Sphaeroidocrinidae, with the sub-families Platycrinidae, Rhodocrinidae, and Actinocrinidae. Academy of Natural Sciences of Philadelphia, Proceedings:175–411.

Wachsmuth, C., and F. Springer. 1897. The North American Crinoidea Camerata. Harvard College, Museum of Comparative Zoology Memoirs, 20, 21, 897 p.

Weller, S. 1914. The Mississippian Brachiopoda of the Mississippi Valley Basin. Illinois Geological Survey Monograph, 1, 598 p.

Youngquist, W., and A. K. Miller. 1949. Conodonts from the Late Mississippian Pella beds of south-central Iowa. Journal of Paleontology, 23:617–622.

Figure 13.1. *LeGrand (Iowa) fossil crinoids from the Mississippian (Kinderhookian). Each species is preserved in a different color. Each species is identified by a number pasted to the slab.* **1, 5, 7, 12, Platycrinites planus** *(Owen and Shumard, 1850);* **2, 22, Strimplecrinus inornatus** *(Wachsmuth and Springer, 1890);* **3, Aorocrinus radiatus** *(Wachsmuth and Springer in Miller, 1889);* **6, 19, Cribanocrinus watersianus** *(Wachsmuth and Springer in Miller, 1889);* **8, 9, Platycrinites symmetricus** *(Wachsmuth and Springer, 1888);* **11, 18, 24, Rhodocrinites kirbyi** *(Wachsmuth and Springer in Miller, 1889);* **21, Cusacrinus nodobrachiatus** *(Wachsmuth and Springer, 1890).*

CRINOID BIOMARKERS (BORDEN GROUP, MISSISSIPPIAN): IMPLICATIONS FOR PHYLOGENY

13

Christina E. O'Malley, William I. Ausich, and Yu-Ping Chin

In addition to the LeGrand Fauna, crinoids with species-specific coloration have been recorded from the Borden Formation in Indiana and are particularly well known from the Crawfordsville, Indiana, area (Lane, 1986).

Inorganic pigments such as iron oxides are the usual cause of coloration in fossils, but in the case of the LeGrand and Crawfordsville crinoids, species-specific coloration must have been produced by the organism itself because many individuals of different colors exist on a single bedding plane. Diagenetic and taphonomic causes can be ruled out because all fossils from the same horizon or even the same locality must experience the same depositional, diagenetic, tectonic, and taphonomic conditions. These fossils have a common history and differ only in species identity. Therefore, biomarker molecules (molecules produced biologically within the organism) responsible for this species-specific coloration can be found and extracted from these Mississippian crinoids (O'Malley, 2006).

Max Blumer, a petroleum geologist, did pioneering work in the study of organic molecules preserved in fossil crinoids from the European Alps (Blumer, 1960, 1962a, 1962b). Questions raised by his work include the following: What exactly caused the coloration? How did those molecules come to be trapped in the crinoids? How many different kinds of molecules are involved? Why does this only occur in Europe? Answering these questions has only begun in earnest recently as a result of advances in needed analytical techniques to assay compounds extracted from the Mesozoic European crinoids.

PREVIOUS WORK. In the early 1950s, Blumer began work on extracting molecules responsible for the blue and purple shades preserved in Jurassic crinoid fossils from Switzerland (Blumer, 1951, 1962a, 1962b; Thomas and Blumer, 1964). He named these molecules *fringelites*, after the village of Fringeli (Hess, 1972), where the spectacular fossils he studied were found. Blumer's work on the European Jurassic *Millericrinus* sp. focused on characterizing the largest of the organic molecules by using ultraviolet-visible (UV-Vis) absorbance spectroscopy.

Fringelites resemble hypericin compounds (a class of quinones) so closely that Blumer (personal commun.) indicated that fringelites may be structurally identical to hypericin (Fox, 1976; Falk et al., 1994). Fringelites are now understood to be polynuclear aromatic hydrocarbons with ex-

Introduction

Why is it that each of the forty species of crinoids found at LeGrand [Iowa] is colored in a characteristic way peculiar to that species alone? Are these colors indicative of the colors they wore in life? It is hoped that someday these mysteries, and others far more important, will be solved with clues that have not yet come to light, or have so far been overlooked.

Boyt (1962)

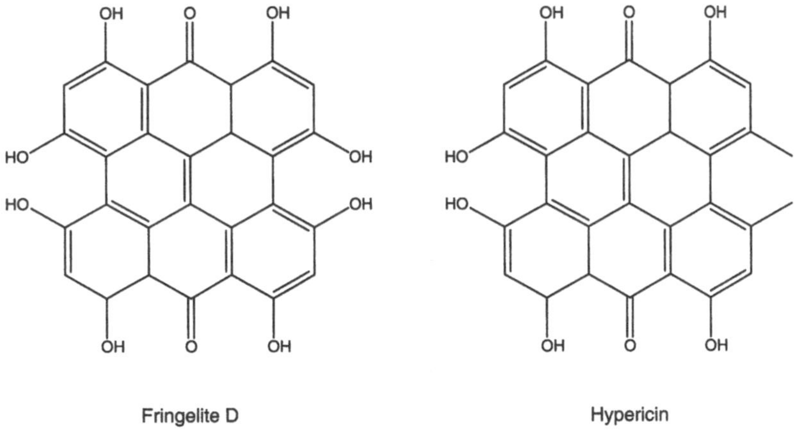

Figure 13.2. *Structure of fringelite D and hypericin. Fringelite D is remarkably similar to hypericin and is thought to have a similar function (Falk, 1999). A hypothesis for the origin of fringelites in crinoids is that living animals have hypericin in their tissues, and over time, two methyl groups are replaced by hydroxyl groups (Wolkenstein et al., 2006).*

Fringelite D

Hypericin

tended quinone functional groups and are similar to hypericin and its derivatives (Blumer, 1962b; Brockman, 1957). The ultraviolet and visible light spectra of the fringelites and their derivatives were first obtained by Blumer (1960, 1962a) and further described and characterized by Falk and Mayr (1995) and Falk et al. (1994) (Fig. 13.2). Additionally, the acid-base properties of fringelite D were characterized by their respective pK_a values in ground and excited states and compared with those of hypericin (Obermüller and Falk, 2001). The homo- and heteroassociation properties of fringelite D were found to be similar to those of hypericin.

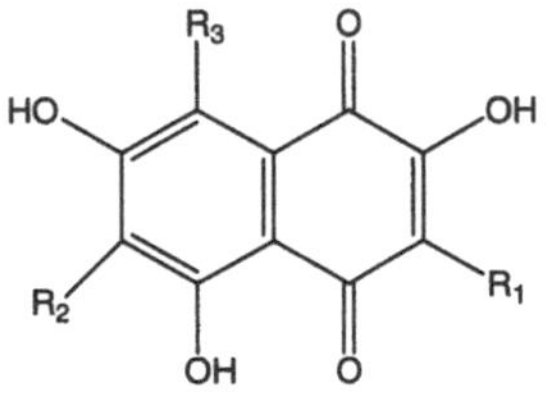

Figure 13.3. *Chemical structure of a selection of modern echinoderm pigments. These represent possible structures of fossil organic molecules found in Mississippian crinoids.*

Echinochrome A: $R_1 = R_3 = OH$, $R_2 = C_2H_5$
Spinochrome A: $R_1 = COCH_3$, $R_2 = H$, $R_3 = OH$
Spinochrome B: $R_1 = OH$, $R_2 = R_3 = H$
Spinochrome C: $R_1 = COCH_3$, $R_2 = R_3 = H$
Spinochrome D: $R_1 = R_3 = OH$, $R_2 = H$
Spinochrome E: $R_1 = R_2 = R_3 = OH$

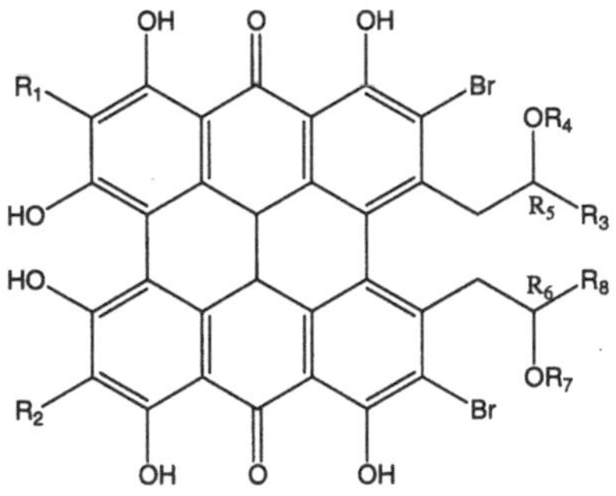

Gymnochrome A: $R_1 = R_2 = R_6 = Br$, $R_3 = R_8 = CH_3$, $R_4 = R_7 = H$, $R_5 = S$
Gymnochrome B:
   $R_1 = H$ and $R_2 = Br$, or $R_1 = Br$ and $R_2 = H$; $R_3 = R_8 = CH_3$, $R_4 = R_7 = H$, $R_5 = S$, $R_6 = H$ or Br
Gymnochrome C: $R_1 = R_2 = Br$, $R_3 = R_8 = CH_3$, $R_4 = H$, $R_5 = R_6 = S$, $R_7 = SO_3H$
Gymnochrome D: $R_1 = R_2 = Br$, $R_3 = R_8 = CH_2CH_2CH_3$, $R_4 = R_5 = R_6 = R_7 = (SO_3H$ or $H)$
Isogymnochrome D:
   $R_1 = R_2 = Br$, $R_3 = R_8 = CH_2CH_2CH_3$, $R_4 = R_7 = SO_3H$, $R_5 = SO_3H$ or H, $R_6 = SO_3H$ or H

Fringelites are structurally similar to the hydroxynaphthoquinones present in modern echinoderms, such as echinochromes and spinochromes (Dimelow, 1958) and gymnochromes (DeRiccardis et al., 1991) (Fig. 13.3). Hydroxyquinones are a group of pigments that occur in some plants and echinoderms, including the modern relatives of the fossil crinoids (Blumer, 1965). The UV-Vis absorption properties for these compounds have not been analyzed, and these represent a few of the more than 200 pigments that are known in modern echinoderms (Stonik and Elyakov, 1988).

The remarkable stability of these compounds is in part the result of the physiology of the echinoderm. Living echinoderms possess calcite ossicles, where each one is a single crystal in optical continuity with a high-Mg calcite composition. Echinoderm ossicles have stereomic microstructure, which is a three-dimensional reticulate pattern of calcite rods. When alive, the porosity within the stereom is filled with mesodermal soft tissue. During early diagenesis, the ossicle mineralogy transforms to low-Mg calcite, and the porosity within the stereom is occluded by a syntaxial crystal growth that preserves the original crystallinity of the ossicle. Fringelites may become incorporated into the calcite either during life or early diagenesis by forming lattices bonded by metals as chelates. An echinoderm fossil, with single-crystal, low-Mg calcite ossicles, is geochemically stable over geological time, and organic molecules encapsulated within this crystal are resistant to further diagenetic change and leaching (Falk and Mayr, 1997) under many circumstances.

It is unknown whether fringelites exist in fossil material in a free or bound state. Fringelites may derive their stability over geologic timescales by an interaction with other elements, i.e., transition metals or minerals present in the fossil. They may form metal complexes through coordination bonds at their *peri* regions or form salts by bonding at their *bay* sites. If these interactions occur, the stability of fringelites may ultimately be due to the formation of complex lattices of fringelites bonded together in crosslinked chains through the fossil (Fig. 13.4, Falk and Mayr, 1997). What is obvious is that fringelites are stable through geologic time.

Fringelites can exist in several forms (Fig. 13.2) (Thomas and Blumer, 1964), and the relative amount of each form of fringelite is determined by the oxidation-reduction (redox) status of its immediate environment (Blumer and Omenn, 1961). Inorganic minerals in fossil echinoderms with fringelite pigments were formed in a reducing environment and may have existed in a highly reduced form (Blumer, 1962b). In a reducing environment, fringelites would be expected to decompose to smaller hydrocarbons over time, and geological time certainly allows opportunity for this decomposition to occur. However, Thomas and Blumer (1964) showed that fringelite pigments are in significantly higher concentrations in fossil material than smaller hydrocarbon decomposition products, which suggests that reduction occurs slowly in an irreversible manner. Intermediates of this reaction are metastable to unstable and decompose readily (Thomas and Blumer, 1964). Fringelites are not present in the host rock at higher concentrations than observed within fossil crinoids. Thus, the presence of fringelite in country rock is because crinoids are a major component of the sediments in which they are preserved (Blumer, 1965).

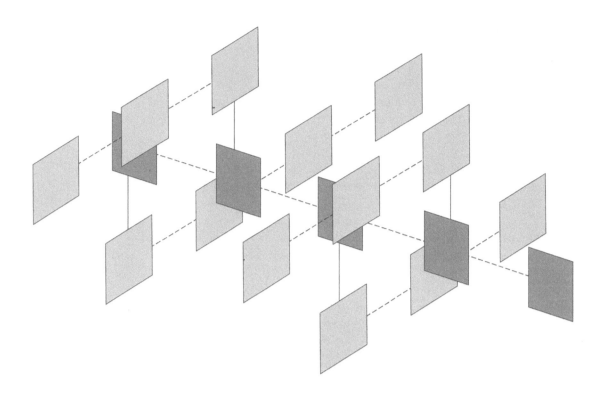

Figure 13.4. *Schematic structural aspects of the polymeric fringelite complex lattice in the mineral matrix of fossils. Squares denote individual Fringelite D molecules; dashed lines connecting them identify **bay** phenolate bonds to alkaline earth metal or transition metal ions; and solid lines denote links, so fringelites form chains by coordination to a transition metal ion (after Falk and Mayr, 1997).*

Blumer (1965) suggested that naturally occurring fringelites are the result of condensation and redox reactions between two molecules of napthaquinones. Napthaquinones have been confirmed in *Antedon bifida* Pennant, 1777 (a modern comatulid crinoid), by Dimelow (1958). The true origin of the preserved and extracted fringelites is still open to debate because they may be true chemical fossils or simply the products of diagenesis (Falk et al., 1994). Because other chemical fossils are known to exist and other modern molecules resemble the fringelite molecules (e.g., porphyrins, soprenoid hydrocarbons from phytol, and the algal pigment stentorian), these similar biomarkers also are likely to be chemical fossils that represent other phyla and classes of organisms (Falk et al., 1994).

Falk (1999) suggested that fringelites in fossil crinoids may have functioned as ingestion deterrents in the same way hypericin does in plants. We suspect that they may possibly play a role in the photosensory system of crinoids because hypericin is known to be a part of the phyotosynthetic pathway of plants, and ingestion of copious amounts of hypericin (such as are present in Saint John's wort [*Hypericum* sp.]) produces an oversensitivity to light in domesticated animals (usually sheep grazing on hillsides). Because many crinoids are sensitive to light stimulus (they have daily activity patterns and are known to prefer shade during the day; Boolootian, 1966), it is possible that their sensitivity to light is produced by photosensitive molecules in their skin--a "dermal light sense," as suggested by Boolootian.

A qualitative chemical analysis of organic molecules in fossil crinoids was carried out in order to reconstruct their phylogenetic relationships based on extractable organic molecules. The crinoids were analyzed by extracting these organic molecules with acetone and methanol and then analyzing them by UV-Vis absorption spectroscopy for the presence of fringelites.

Samples were analyzed from three sites. Samples from the Jurassic Liesberg Formation were used to duplicate the results of Blumer and Falk (Blumer, 1960, 1962a, 1962b; Falk et al., 1994). The two Mississippian sites in Indiana were used to test for consistency and differences within and between taxa. The Jurassic locality is within the Liesberg beds, Liesberg, Switzerland (Middle Oxfordian). The two Mississippian sites are from the late Osagean Edwardsville Formation, Borden Group, at the Boy Scout Camp (Edwardsville Formation, Borden Group, Middle Mississippian, late Osagean T7N, R1W, Sec. 8; SW ¼, SE ¼, NW ¼, Allen's Creek, Indiana Quadrangle, Bloomington, Indiana) and Etter Farm (Edwardsville Formation, Borden Group, Middle Mississippian, late Osagean T11N, R2E, sec. 8; NE ¼, NW ¼, NE ¼, Cope Indiana Quadrangle, Martinsville, Indiana). The two Mississippian localities are separated by approximately 20 miles and have nearly identical faunas. Their paleoenvironment has been interpreted as an interchannel mudstone delta platform (Ausich and Lane, 1980; Ausich, 1983; Powers and Ausich, 1990). The mudstone host rock effectively seals the crinoid ossicles from groundwater and allows little leaching of hydrocarbons from the ossicles into the country rock. Leaching is further inhibited by the low solubility of these compounds in water. Further, the mudstone allows the fossil material to be easily broken away from the host rock. The Edwardsville Formation rocks have never been buried deeply or heated significantly. Strata in Indiana have a conodont alteration index of 1.5 or less, except for a region in Putnam County that was presumably affected by hydrothermal alternation (Harris et al., 1990; C. Rexroad, personal commun., 2006). The area in Monroe and

This Study

Figure 13.5. *Plates and parts chosen for analysis by species and location. The designation "×2" indicates the number of duplicate samples when the number of samples is greater than 1. The unknown sample was a fragment of a radial plate, but it lacked the characters necessary for identification.*

	Platycrinites hemisphaericus	*Paradichocrinus planus*	*Gilbertsocrinus*	*Cyathocrinites* sp.	*Barycrinus* sp.	*Halysiocrinus tunicatus*	*Unknown*	*Liliocrinus* sp.
	Camerata			**Cladida**		**Disparida**		**Articulata**
Boy Scout Camp	Circlet Radial	Radial x2	Stem	Radial	-	Radial	Radial	-
Etter Farm	Radial	-	-	Radial x2	Stem	Radial	-	-
Basel, Switzerland	-	-	-	-	-	-	-	Stem

Systematic Paleontology of Collection

Phylum	Class	Subclass	Superorder	Order	Suborder	Superfamily	Family	Genus	Species	Locality
ECHINODERMATA (Wachsmuth and Springer, 1885)	CRINOIDEA (Miller, 1821)	CAMERATA (Wachsmuth and Springer, 1885)		DIPLOBATHRIDA (Moore and Laudon, 1943)	EUDIPLOBATHRINA (Ubaghs, 1953)	RHODOCRINITACEA (Roemer, 1855)	RHODOCRINITIDAE (Roemer, 1855)	GILBERTSOCRINUS (Phillips, 1836)	*Gilbertsocrinus* sp.	O
		CAMERATA (Wachsmuth and Springer, 1885)		MONOBATHRIDA (Moore and Laudon, 1943)	COMPSOCRININA (Ubaghs, 1978)	HEXACRINITACEA (Wachsmuth and Springer, 1885)	DICHOCRINIDAE (S.A. Miller, 1889)	PARADICHOCRINUS (Springer, 1926)	*Paradichocrinus planus* (Springer, 1926)	O
		CAMERATA (Wachsmuth and Springer, 1885)		MONOBATHRIDA (Moore and Laudon, 1943)	COMPSOCRININA (Ubaghs, 1978)	PLATYCRINITACEA (Austin and Austin, 1842)	PLATYCRINITIDAE (Austin and Austin, 1842)	PLATYCRINITES (Miller, 1821)	*Platycrinites hemisphaericus* (Meek and Worthen, 1865)	□
		DISPARIDA (Moore and Laudon, 1943)		CALCEOCRINIDA (Meek and Worthen, 1869)			CALCEOCRINIDAE (Meek and Worthen, 1869)	HALYSIOCRINUS (Ulrich, 1886)	*Halysiocrinus tunicatus* (Hall, 1860)	□ O
		CLADIDA (Moore and Laudon, 1943)	CYATHOCRININA (Bather, 1899)	CYATHOCRINIDA (Bassler, 1938)			CYATHOCRINITIDAE (Bassler, 1938)	CYATHOCRINITES (Miller, 1821)	*Cyathocrinites* sp.	□ O
		CLADIDA (Moore and Laudon, 1943)	CYATHOCRININA (Bather, 1899)	CYATHOCRINIDA (Bassler, 1938)			BARYCRINIDAE (Jaekel, 1918)	BARYCRINUS (Meek and Worthen, 1868)	*Barycrinus* sp.	□

O Boy Scout Camp
□ Etter Farm

Morgan counties studied here is far removed from the zone of alteration. The material examined in this study was collected by Ausich (1983), and additional material was collected by Powers for use in his master's thesis (Powers and Ausich, 1990). All collections are deposited in the Orton Geological Museum, The Ohio State University.

Crinoids sampled represent three subclasses to answer questions about the general occurrence of biomarkers within different crinoid clades (Fig. 13.5). Two genera (*Halysiocrinus* Hall, 1860, and *Cyathocrinites* Miller, 1821) were tested for consistency of the occurrence of the biomarkers within a genus. If there was a regional or environmental influence on the occurrence or concentrations of biomarkers in crinoids, that difference should be revealed by the duplicate samples.

Isolated radial plates were principally used for analysis because they are easily identified to species. Samples from other skeletal elements were analyzed to determine whether biomarker occurrence existed throughout the animal. The stem of *Gilbertsocrinus* sp. is distinct enough to be identified and was also analyzed, and an identifiable basal circlet of the calyx of *Platycrinites hemisphericus* (Meek and Worthen, 1865) was also tested (Fig. 13.5).

Specimens for analysis were selected to test for clade-specific biomarker distinctions at several taxonomic levels (Fig. 13.6). Representatives of three subclasses (Cladida Moore and Laudon, 1943; Camerata Wachsmuth and Springer, 1885; and Disparida Moore and Laudon, 1943) were compared to evaluate subclass level differences. Three genera (*Gilbertsocrinus* Phillips, 1836; *Paradichocrinus* Springer, 1926; and *Platycrinites* Miller, 1821) were tested to evaluate the differences of biomarkers in two orders of camerates (Diplobathrida Moore and Laudon, 1943, and Monobathrida Moore and Laudon, 1943). Two genera (*Paradichocrinus* and *Platycrinites*) representing different subfamilies (Hexacrinitacea Wachsmuth and Springer, 1885, and Platycrinitacea Austin and Austin, 1842) of the subclass Camerata were tested to evaluate the difference between biomarkers in two superfamilies. Two cladid genera from Etter Farm (*Cyathocrinites* and *Barycrinus* Wachsmuth and Springer, 1868) representing different families (Cyathocrinitidae Bassler, 1938, and Barycrinidae Jaekel, 1918) were compared to determine whether there is a difference in biomarkers at the family level.

Methodology

Initial color of the specimens is determined by comparison to a Munsel color chart. Pigments were extracted from crinoids with a series of organic solvents. UV-Vis absorbance spectroscopy was used to screen for fringelites and biomarker molecules. The benefit of UV-Vis absorbance spectroscopy is that the presence of organic pigments is easy to test, even with a small amount of material. Information gained by this method will enable identification of the fringelite compounds and provide a simple inexpensive screening method by which to search for the presence of fringelites in fossil extracts. These results were compared with other taxa to determine the similarity and differences among taxa at various taxonomic levels. The

Figure 13.6. *Systematic paleontology of the collection of samples chosen for study, listed by taxonomic groups.*

presence of biomarker molecules in various parts of crinoids was investigated to test the possibility of only performing destructive analysis on distal portions of particularly well-preserved fossil specimens.

CHEMICAL EXTRACTION TECHNIQUES. A relatively simple method was devised to extract the organic pigment molecules from the fossil crinoids based on the method of Thomas and Blumer as described in their study of fringelites (Thomas and Blumer, 1964).

Crinoid samples were ground to fine particle size to maximize surface area and speed up the reaction of calcite with a strong acid. The mortar and pestle was cleaned with methanol and dried between samples. Concentrated HCl was added by drops to the sample of ground crinoid. The reaction of calcite in the crinoids with HCl results in an organic residue (without calcite) solid in a slightly acidic solution of calcium chloride. The vial was then centrifuged for a few seconds to produce a loose pellet of sediment. The liquid was decanted, and extraction was continued to collect the organic compounds in the residue.

In order to separate the major organic fractions from the collected residue, a series of solvents was added, in order from a nonpolar to polar solvents. First, chloroform was added by drops, and then the sample was recentrifuged and the liquid decanted. Second, a 4:1 acetone-methanol solution was added by drops to the residue. This was covered and allowed to sit overnight. After 24 hours, the sample was centrifuged, and the solvent was removed by decanting (with dissolved pigments in solution) and labeled. Finally, neat methanol was added by drops to the remaining residue and processed in the same manner as the previous solvent.

Fringelites had previously been reported in the fraction collected in the 4:1 acetone-methanol solution, so that was the portion set aside to be analyzed in this study. Blumer (1962b) reported fossil fringelites, bianthrones, and related molecules as being soluble in methanol. Here, the acetone-methanol solution was the first methanol-containing solvent used, and by visual inspection, this solution extracted the organic molecules from the residue.

ANALYTICAL TECHNIQUES. UV-Vis light spectroscopy was used to search for the presence of fringelites in Mississippian and Jurassic crinoid samples. Fringelites are well described via UV-Vis spectroscopy (Blumer, 1965; Wolkenstein, 2005). The benefit of UV-Vis absorbance spectroscopy is that the presence of organic pigments is easy to test, even with a small amount of material. Information gained by this method enabled identification of the fringelite compounds and provided a simple, inexpensive screening method by which to search for the presence of fringelites in fossil extracts.

A Varian (Cary 3) UV-Vis spectrophotometer was used for the screening of samples for fringelites. A sample of stock solution of 4:1 acetone-methanol was used as a blank. Samples in 4:1 acetone-methanol were scanned at 200–700 nm. Data collected from the blank were subtracted from data collected from the samples in order to eliminate the effects of the solvent on the spectra for each sample.

Discriminant function analysis of hue, chroma, and saturation (determined by comparison to a Munsel color chart) (Fig. 13.7) indicate that on the basis of visual coloration alone, the analyzed crinoids sort into phylogenetic groups (subclasses) (Fig. 13.8). This simple analysis strongly supports what has been recognized in the field for some time (Boyt, 1962; Lane, 1986): the fossils of at least some different crinoid species have coloration that is more similar for more closely related species, and therefore chromophoric biomarkers may function as useful characters in phylogenetic analysis.

More detailed analyses were performed with UV-Vis spectroscopy of extracts produced from the simple extraction technique used in this study. Fringelites were extracted from a Jurassic *Liliocrinus munsterianus* Rollier, 1911, sample (from Basel, Switzerland) as indicated by noticeable peaks at 526 and 566 nm. These peaks were identified in the *L. munsterianus* sample, but they were not observed for any of the samples of Mississippian age (Figs. 13.9, 13.10), excluding fringelites from the possible organic molecules present in these rocks. This was confirmed by Wolkenstein (personal commun., 2007) when he analyzed a similar series of crinoid samples from the LeGrand, Iowa locality.

In the fraction of Mississippian fossil extracts in a solution of acetone and methanol, characteristic absorbance peaks are present for each species (Figs. 13.9, 13.11). The European Jurassic sample has absorbance peaks that were not observed in any other fossil sample; peaks are located at 441, 526, and 566 nm. Our peaks match those observed by Blumer for fringelite molecules extracted from European Jurassic fossil crinoids (Blumer, 1960, 1965). Other authors have reported peaks at 495, 530, and 571 nm (Wolkenstein, 2005) for related genera of fossil crinoids, and so are slightly different from the UV-Vis spectra collected for this study.

The Mississippian camerate genera shared common wavelengths at three peaks, but one differed. The cladids also have a degree of similarity in sharing two peaks, and each has two peaks not present in the other. The camerates had peak values that were not present in any other group (three peaks at 309, 321/322, and 360 nm). The disparids, cladids, and articulate all shared a peak at 324 nm, indicating that they are more closely related to one another than any is to the Camerata (Fig. 13.12). Absorbance peaks from samples of country rock only share two peaks with fossil sample extracts. One is the 211 nm, and the second is also present in *Barycrinus* and the Swiss country rock sample.

Further analysis of these samples should provide clues for the absence of fringelites in more ancient fossils relative to their younger European crinoids. Because of geographical and stratigraphical differences, the samples have undergone different diagenetic histories. The Mississippian crinoids are nearly twice as old as those from the Jurassic, but the Indiana Mississippian crinoids have been located in a stable portion of the North American craton with rocks having a conodont alteration index of 1.5 or less, as discussed above. In contrast, in Switzerland, there has been a significant amount of tectonism and metamorphism coincident with the formation of the Alps that may have altered the original biomarkers into the fringelites we observe today. Blumer (1965) suggested that fringelites may

Locality	Sample	Intact			Powdered		
		Hue	Value	Saturation	Hue	Value	Saturation
BSC	Country Rock	5Y	8	2	10YR	6	3
BSC	*Cyathocrinites* sp	5Y	7	1	10YR	8	2
BSC	*Gilbertsocrinus* sp.	10YR	7	2	10YR	7	2
BSC	*Halysiocrinus tunicatus*	5B	5	1	10YR	7	1
BSC	*Paradichocrinus planus* 1	5YR	5.5	1	5YR	8	1
BSC	*Paradichocrinus planus* 2	5Y	6.5	1	5Y	7	1
BSC	*Platycrinites hemisphaericus* (circ)	5B	4	1	5Y	8	2
BSC	*Platycrinites hemisphaericus* (rad)	5B	4	1	5RP	8	2
BSC	Unknown BSC	5B	5.5	1	5YR	8	2
EF	*Barycrinus* sp.	5Y	7	1	5YR	7	2
EF	Country Rock				5YR	7	2
EF	*Cyathocrinites* sp. 1	5Y	7	1	5YR	8	1
EF	*Cyathocrinites* sp. 2	5Y	6.5	1	5YR	8	1
EF	*Halysiocrinus tunicatus*	5Y	4.5	1	10YR	9	1
EF	*Platycrinites hemisphaericus*	5YR	5	1	5YR	8	1
SW	Country Rock				5GY	7	1
SW	*Liliocrinus* sp.	5RP	1.5	2	5P	4	2

Figure 13.7. *Hue, value, and saturation for fossil calcite and powdered calcite using the Munsel color chart.*

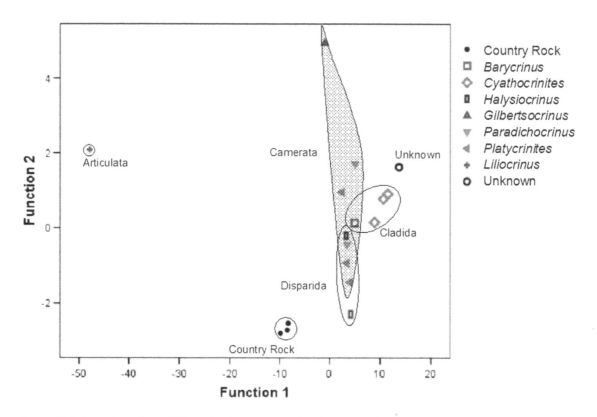

Figure 13.8. *Data based on visible color data measured with a Munsel color chart. Genera of fossil crinoids sort into morphologically determined clades. The extracts of the unknown sample appears to be closest to the extracts of the Disparida (Halysiocrinus) by this analysis.*

Subclass	Camerata		Disparida	Cladida		Articulata		
Genus	*Gilbertsocrinus*	*Platycrinites*	*Halysiocrinus*	*Barycrinus*	*Cyathocrinites*	*Liliocrinus*	Indiana Rock	Swiss Rock
211	X	X	X	X	X	X	X	X
309	X							
321, 322	X	X						
323							X	
324			X	X	X	X		
326								X
348				X				X
360	X	X			X			
366								X
368				X		X		
375			X					
399							X	
441						X		
445								X
526						X		
566						X		

(Wavelengths of Peak Absorbance)

Figure 13.9. *Characteristic UV-Vis absorption peaks for each genus. Peaks at 441, 526, and 566 nm indicate the presence of fringelites. Data for* **Paradichocrinus** *were not collected for wavelengths less than 400 nm and so are not included.*

Figure 13.10. *UV-Vis spectral data from Swiss country rock and* **Liliocrinus**. *Note the presence of peaks at 441, 526, and 566 nm, indicating the presence of fringelites.*

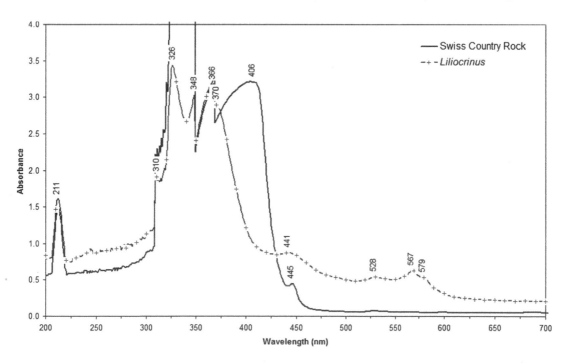

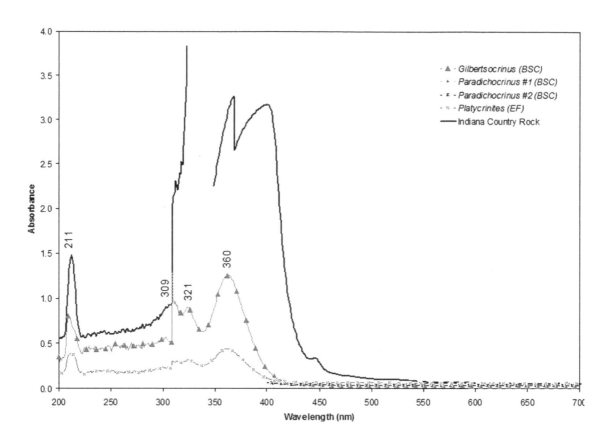

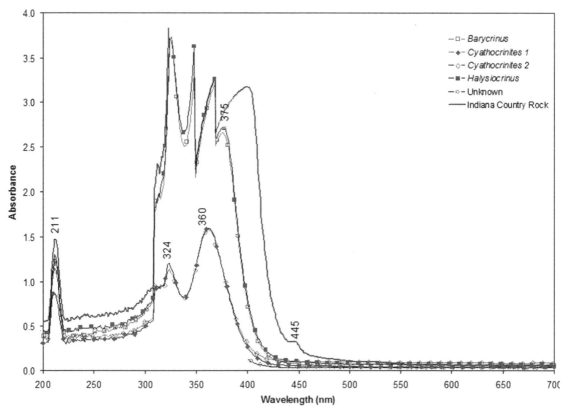

be the result of condensation reactions (a pathway that can be accelerated by heat and pressure) of anthraquinones, which exist in modern echinoderms. Our preliminary results suggest that these or similar and smaller quinones may be causing the coloration in the Indiana fossil extracts. These samples have not undergone the condensation processes to which the Swiss samples were subjected.

The *Liliocrinus* Rollier, 1911, stem was the only specimen to contain fringelite D (when UV-Vis data were used). This example was a test of different parts of fossil crinoid specimens meant to test the consistency of occurrence of fringelites in different parts of a crinoid. Because no molecules were identified positively from the Mississippian samples, the only evidence that can be used to test for consistency is the presence of similar peaks in UV-Vis data.

In the camerates, in addition to the general difference between *Gilbertsocrinus* and *Platycrinites*, *Gilbertsocrinus* was a sample of a stem, and *Platycrinites* samples were exclusively calyx plates. An additional peak was present in *Gilbertsocrinus*. It is most probable that this peak is the result of a slightly different molecule present in *Gilbertsocrinus*, but it is also possible that a difference between the chemical composition of the molecules in the stem and calyx is responsible.

In the cladids, there were two unique peaks in the *Barycrinus* sample (a columnal) and one unique peak in the *Cyathocrinites* samples (radial and circlet plates). Here again, one was a stem sample and one a calyx sample. There was no difference in the UV-Vis spectrum from the extracts from different calyx plates within a species.

The UV-Vis data indicate that it may be possible to identify a fossil crinoid specimen to subclass by a simple measurement. Unfortunately, this would require destructive testing of at least a fragment of the sample, but destruction of the calyx is not absolutely necessary. In this study, samples were taken both from the calyx (most genera) and the stem (*Liliocrinus*, *Barycrinus*, and *Gilbertsocrinus*), and all yielded usable results. Further investigation into this possibility will show whether this is a suitable substitution.

Figure 13.11. *UV-Vis spectral data from Mississippian rock and camerate fossil extracts in 4:1 acetone-methanol solution. No peaks appear between 500 and 600 nm, which would indicate the presence of fringelites.*

Figure 13.12. *UV-Vis spectral data from Mississippian rock for cladid and disparid (**Halysiocrinus**) fossil extracts in 4:1 acetone-methanol solution. No peaks appear between 500 and 600 nm, which would indicate the presence of fringelites.*

Conclusions

The relatedness among living organisms is determined by both morphological and genetic information. Both are subject to various processes that may confuse the primary phylogenetic signal, but having both types of information allows for comparison among these data. Among fossils, traditionally only morphology is used, with poor genetic information available only for a few extremely young fossils. The genes of organisms code for morphology, but they also code for specific proteins and other biogenic molecules, including pigments. Thus, a detailed understanding of ancient pigments may provide an independent proxy for phylogeny that can be used in conjunction with morphology.

Fossil echinoderms, specifically crinoids, possess chromophoric organic molecules--fringelites and other polycylic aromatic hydrocarbons--that are resistant to diagenetic leaching, are chemically stable over geologic time, and are species specific. Fringelites and fossil polynuclear aromatic hydrocarbons are ideal candidates to function as this proxy for phylogenetic reconstruction in echinoderm fossils. Organic biomarkers have not been previously used to determine phylogenetic relationships or to detect ancient biochemical processes because of the assumed destruction by diagenesis on such molecules. However, echinoderm skeletons are mineralogically distinct in that the stability of their fossilized remains allows for the preservation of small organic molecules as stable complexes (Falk and Mayr, 1997). This study has shown that chemical analysis of organic molecules in preserved crinoids can be used to group crinoid fossils by taxonomic groups supported by morphologic characters. Thus, this study suggests that considerable potential exists for developing a biomarker phylogeny for crinoids. Biomarkers may be an important tool for determining the identity of partially preserved organisms and for their relationship to one another. Additional work is needed to refine these techniques and to expand our knowledge about this group of molecules and their occurrence in fossil echinoderms.

Acknowledgments

D. Finkelstein and G.D. Sevastopulo significantly improved an earlier draft of this manuscript. This research was partially funded with a research grant to CEO from the Geological Society of America.

References

Ausich, W. I. 1983. Component concept for the study of paleocommunities with an example from the early Carboniferous of Southern Indiana (USA). Paleogeography, Paleoclimatology, Paleoecology, 44:251–282.

Ausich, W. I., and N. G. Lane. 1980. Field trip 2: platform communities and rocks of the Borden Siltstone Delta (Mississippian) along the south shore of Monroe Reservoir, Monroe County, Indiana, p. 36–67. In R. H. Shaver (ed.), Field Trips 1980 from the Indiana University Campus, Bloomington. Department of Geology, Indiana University; Bloomington, Indiana.

Austin, T., and T. Austin. 1842. XVIII.--Proposed arrangement of the Echinodermata, particularly as regards the Crinoidea, and a subdivision of the Class Adelostella (Echinidae). Annals and Magazine of Natural History, series 1, 10:106–113.

Bassler, R. S. 1938. Pelmatozoa Palaeozoica. In W. Quenstedt (ed.), Fossilium Catalogus, I: Animalia, Part 83. W. Junk's, Gravenhage, The Netherlands, 194 p.

Blumer, M. 1951. Fossile Kohlenwasserstoffe und Farbstoffe in Kalksteinen. Mikrochemie, 36/37:1048–1055.

Blumer, M. 1960. Pigments of a fossil echinoderm. Nature, 188:1100–1101.

Blumer, M. 1962a. The organic chemistry of a fossil--I: The structure of fringelite-pigments. Geochimica et Cosmochemica Acta, 26:225–227.

Blumer, M. 1962b. The organic chemistry of a fossil--II: Some rare polynuclear hydrocarbons. Geochimica et Cosmochemica Acta, 26:228–230.

Blumer, M. 1965. Organic pigments: their long-term fate. Science, 149:722–726.

Blumer, M., and G. S. Omenn. 1961. Fossil porphyrins: uncomplexed chlorins in a Triassic sediment. Geochimica et Cosmochemica Acta, 25:81–90.

Boolootian, R. A. (ed.). 1966. Physiology of Echinodermata. Interscience Publishers, New York. 822 p.

Boyt, R. 1962. Crinoid and Starfish Fossils from LeGrand, Iowa. Iowa State Department of History and Archives, LeGrand, 24 p.

Brockman, H. 1957. Photodynamisch wirksame Pflanzenfarbstoffe. Fortschritte der Chemie Organischer Naturstoffe, 14:141.

DEriccardis, F., M. Iorizzi, L. Minale, R. Riccio, B. Richer de Forges, and C. Debitus. 1991. The gymnochromes: novel marine brominated phenanthroperylenequinone pigments from the stalked crinoid *Gymnocrinus richeri*. Journal of Organic Chemistry, 56:6781–6787.

Dimelow, E. T. 1958. Some aspects of the biology of Antedon bifida with some references to *Neocomatella europa*. Unpublished Ph.D. dissertation, University of Reading.

Falk, H. 1999. From the photosensibilisator hypericin to the photoreceptor stentorin--the chemistry of the phenanthroperylene quinones. Angewandte Chemie International Edition, 38:3116–3136.

Falk, H., and E. Mayr. 1995. Synthesis and properties of fringelite-D (1,3,4,6,8,10,11,13-octahydroxy-phenantho[1,10,9,8,O,P,Q,R,A]perylene-7,14-dione). Monatshefte für Chemie, 126(6–7):699–710.

Falk, H., and E. Mayr. 1997. Concerning bay salt and perichelate formation of hydroxyphenanthoperlylene quinones (fringelites). Monatshefte für Chemie, 128(4):353–360.

Falk, H, E. Mayr, and A. E. Richter. 1994. Simple diffuse-reflectance UV-Vis spectroscopic determination of organic pigments (fringelites) in fossils. Mikrochim Acta, 117(1–2):1-5.

Fox, D. L. 1976. Animal Biochromes and Structural Colors. University of California Press, Berkeley, 433 p.

Hall, J. 1860. Observations upon a new genus of Crinoidea: *Cheirocrinus*, p. 121–124. *In* Appendix F. Contributions to Palaeontology, 1858 and 1859: Thirteenth Annual Report of the Regents of the University of the State of New York, on the Condition of the State Cabinet of Natural History, and the Historical and Antiquarian Collection Annexed thereto, State of New York in Senate Document, 89.

Harris, A. G., C. B. Rexroad, R. T. Lierman, and R. A. Askin. 1990. Evaluation of a CAI anomaly, Putnam County, Central Indiana, USA: possibility of a Mississippi Valley–type hydrothermal system. Courier-Forschungs Institut Senkenberg, 118:253–266.

Hess, H. 1972. The fringelites of the Jurassic Sea. CIBA-GEIGY, 2:14–17.

Jaekel, O. 1918. Phylogenie und System der Pelmatozoen. Paläontologische Zeitschrift, 3(1), 128 p.

Lane, N. G. 1986. The Fossil Crinoids of Montgomery County Indiana. Rocks and Minerals, 61:151–158.

Meek, F. B., and A. H. Worthen. 1865. Descriptions of new Crinoidea, etc. from the Carboniferous rocks of Illinois and some of the adjoining states. Proceedings of the Academy of Natural Sciences of Philadelphia, 17:155–166.

Miller, J. S. 1821. A natural history of the Crinoidea, or lily-shaped animals; with observations on the genera, *Asteria, Euryale, Comatula* and *Marsupites*. Bryan & Co., Bristol. 150 p.

Miller, S. A. 1889. North American geology and paleontology. Western Methodist Book Concern, Cincinnati, Ohio, 664 p.

Moore, R. C., and L. R. Laudon. 1943. Evolution and classification of Paleozoic crinoids. Geological Society of America Special Paper, 46, 151 p.

Obermüller, R. A., and H. Falk. 2001. Concerning the absorption and photochemical properties of an ω-4-dimethylaminobenzal hypericin derivative. Monatshefte für Chemie, 132:1519–1526.

O'Malley, C. E. 2006. Crinoid biomarkers (Borden Group, Mississippian): implications for phylogeny. Unpublished master's thesis, Ohio State University, Columbus, 75 p.

Owen, D. D., and B. F. Shumard. 1850. Descriptions of fifteen new species of Crinoidea from the Subcarboniferous limestone of Iowa, Wisconsin and Minnesota in the years 1848–1849. Journal of the Academy of Natural Sciences of Philadelphia, series 2, 2:57–70.

Pennant, T. 1777. The British Zoology (fourth edition, volume 4). Warrington, London.

Phillips, J. 1836. Illustrations of the geology of Yorkshire, or a description of the strata and organic remains, p. 203–208. *In* Part 2, the Mountain Limestone districts (2nd edition). John Murray, London.

Powers, B. G., and W. I. Ausich. 1990. Epizoan associations in a lower Mississippian Paleocommunity (Borden Group, Indiana, USA). Historical Biology, 4:245–265.

Rollier, L. 1911. Fossiles nouveaux ou peu connus. Memoires de la Societe Paleontologique Suisse, 37:1–7.

Springer, F. 1926. Unusual forms of fossil crinoids: Proceedings of the U.S. National Museum, 67(9), 137 p.

Stonik, V. A., and G. B. Elyakov. 1988. Secondary metabolites from echinoderms as chemotaxonomic markers. Bioorganic Marine Chemistry, 2:43–86.

Thomas, D. W., and M. Blumer. 1964. The organic chemistry of a fossil--III. The hydrocarbons and their geochemistry. Geochimica et Cosmochemica Acta, 28:1467–1477.

Wachsmuth, C., and F. Springer. 1880–1886. Revision of the Palaeocrinoidea: Proceedings of the Academy of Natural Sciences of Philadelphia. Part I. The families Ichthyocrinidae and Cyathocrinidae, (1880):226–378. Part II. Family Sphaeroidocrinidae, with the sub-families Platycrinidae, Rhodocrinidae, and Actinocrinidae, (1881):177–411. Part III, Section 1. Discussion of the classification and relations of the brachiate crinoids, and conclusion of the generic descriptions, (1885):225–364. Part III, Section 2. Discussion of the classification and relations of the brachiate crinoids, and conclusion of the generic descriptions, (1886):64–226.

Wachsmuth, C., and F. Springer. 1888. The summit plates in blastoids, crinoids, and cystids, and their morphological relations. American Geologist, 1:61.

Wachsmuth, C., and F. Springer. 1890. New species of crinoids and blastoids from the Kinderhook Group of the Lower Carboniferous rocks at LeGrand, Iowa, and a new genus from the Niagaran Group of Western Tennessee. Illinois Geological Survey, Palaeontology of Illinois, 8(2):155–206.

Wolkenstein, K. 2005. Phenanthroperylenchinon-Pigmente in fossilen Crinoiden: Charakterisierung, Vorkommen und Diagenese. Unpublished Ph. D. dissertation, Ruprecht-Karls-Universität, Heidelberg, Germany, 113 p.

Wolkenstein, K., J. H. Gross, F. Heinz, H. F. Schöler. 2006. Preservation of hypericin and related polycyclic quinone pigments in fossil crinoids. Proceedings of the Royal Society B, 273:451–456.

PART 5. ECHINODERM
FAUNAL STUDIES

INTRODUCTION TO PART 5

William I. Ausich and Gary D. Webster

On the Origin of Species (Darwin, 1859) shocked conventional thought and changed how humankind views its place on Earth. If this theory has any validity, there must be some proof in the fossil record. The industrial revolution needed raw materials. Consequently, continents had to be explored for their geological resources, and new methods to explore for these resources needed to be developed. Even competition to fill museums with display specimens fueled the late nineteenth and early twentieth century era of paleontologic exploration. Understanding the occurrence of fossils was (and remains) an integral part of understanding the temporal, geographic, and paleoenvironmental distribution of the geological record; and the late nineteenth and early twentieth century was an era of discovery for paleontology. This era was primarily concentrated in regions or on continents with large universities, and much was learned. The importance of paleontology during this time is highlighted by the fact that James Hall, the great nineteenth-century invertebrate paleontologist, was the first president of the American Association for the Advancement of Science.

Eventually this quest for new information waned for undoubtedly many reasons; the paleontological record has historically been one with a strong geographic and monographic biases. There was a bias for regions or continents with large universities and for the fossil groups that had received extensive study.

An important and continuing aspect of the paleobiology movement has been to document Phanerozoic trends in biodiversity: When were there times of exceptional extinctions and adaptive radiations? Did standing diversity vary through time? Are there latitudinal variations in biodiversity patterns? Data compilations that allow a Phanerozoic-scale perspective are the Sepkoski Compendium (2003) and the Paleobiology Database (<http://paleodb.org/cgi-bin/bridge.pl>). These compilations, and others like the *Treatise on Invertebrate Paleontology*, represent monumental works and are vital for understanding large-scale problems. However, serendipitously, such compilations also highlight what we do not know and establish a new research agenda. Faunal compilations and *Treatise* summaries make evident taxa that are in dire need of revision; and they point out temporal, geographic, and paleoenvironmental gaps in our knowledge. Modern phylogenetically based systematics are essential in order for faunal databases to have maximum effectiveness for interpretation, and temporal trends are more meaningful once historical monographic biases are eliminated. Neither of these tasks is easy to accomplish. We refer to the collection of new echinoderm faunas that expand our geographic or temporal coverage as the frontier paleontology of the twenty-first century.

Part 5 includes an account and summary of completing the poorly known early diversification of the Crinoidea (Sprinkle, Guensburg, and Gahn, Chapter 14). Donovan et al. (Chapter 15) bring comprehensive order to previously known but poorly understood faunas from the Silurian crinoids of England that are essential for an understanding of the base of the Middle Paleozoic crinoid evolutionary fauna.

Echinoderm faunas of Puerto Rico help to complete the faunal distribution of the Caribbean region (Vélez-Juarbe and Santos, Chapter 17). Complete faunal data are necessary for a thorough understanding of the climate system of the Cenozoic. Waters et al. (Chapter 16) summarize the pelmatozoan echinoderms that are known from China. This research team led by Gary Lane is responsible for our knowledge of the Devonian of China. From the dinosaur faunas and egg clusters discovered in the 1920s to the feathered reptiles and Cambrian Lagerstätten reported in recent years, China has revealed numerous fine fossil occurrences. Biogeographically, China is complex, and completing our knowledge of its faunal associations is essential for understanding true global patterns in the evolutionary history of life on Earth.

References

Darwin, C. 1859. *On the origin of species by means of natural selection, or preservation of favoured races in the struggle for life.* Murray, London., 490 p.

Sepkoski, J. J., Jr. 2002. *A compendium of fossil marine animal genera.* Bulletins of American Paleontology, 363, 560 p.

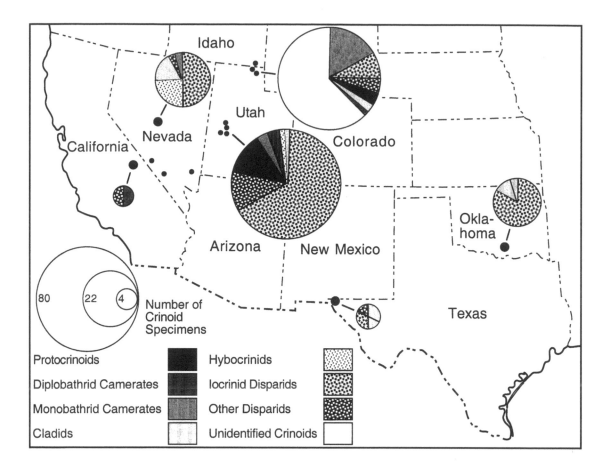

Figure 14.1. *Geographic distribution of the six Early Ordovician crinoid faunas in the western United States showing the localities where crinoids have been collected (large and small black dots) and the number of crinoid specimens known (size of the circle). The Garden City crinoid fauna from northern Utah and southern Idaho include many unidentified crinoids because recently collected specimens mostly buried in slabs have not yet been prepared for study and identification.*

OVERVIEW OF EARLY ORDOVICIAN CRINOID DIVERSITY FROM THE WESTERN AND SOUTHWESTERN UNITED STATES

14

James Sprinkle, Thomas E. Guensburg, and Forest J. Gahn

Introduction

The western and southwestern United States have produced the largest collection of Early Ordovician crinoids, both in terms of number of specimens and number of taxa, of any region in the world. Most of these crinoid specimens have been collected since 1989, although a few older specimens date back to the 1960s and 1970s in different areas. Several older crinoids collected by Lehi Hintze from the Ibex area of western Utah were described by Gary Lane (1970) in one of the first papers on Early and Middle Ordovician crinoids from western North America. Crinoids are now known from six faunas in the western and southwestern United States, including two large faunas each with 86–97 partial to complete crinoid specimens, two medium-size faunas each with 18–22 specimens, and two small faunas each with four to six specimens (Fig. 14.1). In most of these faunas, the crinoids are accompanied by numerous blastozoans (primarily glyptocystitid rhombiferans and eocrinoids), edrioasteroids, mitrate and rare cornute stylophorans, and rare asteroids and ophiuroids. Most of these crinoid faunas have been extensively collected during the past 17 years, and we are in the process of describing five of these faunas. This is a progress report on that work.

The Lower Ordovician sections that we have studied in the western and southwestern United States had been divided into lettered trilobite (and a few brachiopod) assemblage zones A–J by Ross (1951), who studied the Garden City Limestone in northern Utah and southern Idaho. These lettered zones were also used by Hintze (1953) working in the Ibex area of western Utah (and many later authors working in other areas), who established named trilobite zones mostly or entirely equivalent to these lettered zones (for example, Ross's zone D is the same as Hintze's *Leiostegium-Kainella* zone). This shelly fossil zonation for the Early and Middle Ordovician (Hintze, 1973; Ross et al., 1997, fig. 10), along with parallel systems of biozones based on conodonts and graptolites (see Webby et al., 2004), have been widely used for correlation and relative age dating. Because the Ross lettered shelly fossil zones have been so widely used and their order is easy to understand, we continue their use here to designate parts of our sections (Fig. 14.2).

The faunas identified here are regionally defined crinoid associations for various parts of the Early Ordovician. They have been listed below and

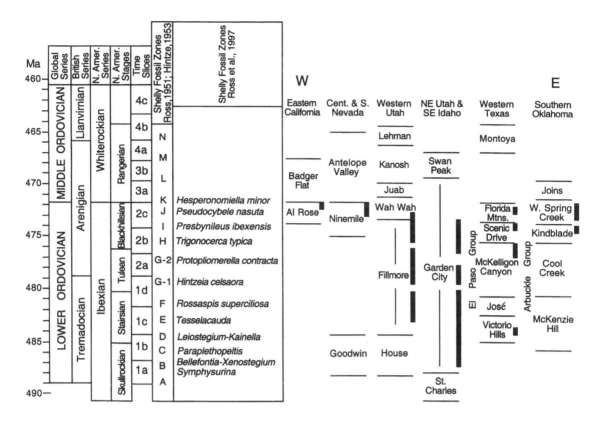

The figure shows a stratigraphic chart with columns (left to right): Ma timescale (460–490), Global Series (LOWER ORDOVICIAN / MIDDLE ORDOVICIAN), British Series (Tremadocian, Arenigian, Llanvirnian), N. Amer. Series (Ibexian, Whiterockian), N. Amer. Stages (Skullrockian, Stairsian, Tulean, Blackhillsian, Rangerian), Time Slices (1a–4c), Shelly Fossil Zones Ross, 1951; Hintze, 1953 (A–N), Shelly Fossil Zones Ross et al., 1997.

Shelly Fossil Zones (Ross et al., 1997) listing:
Hesperonomiella minor — K — J
Pseudocybele nasuta — J
Presbynileus ibexensis — I
Trigonocerca typica — H
Protopliomerella contracta — G-2
Hintzeia celsaora — G-1
Rossaspis superciliosa — F
Tesselacauda — E
Leiostegium-Kainella — D
Paraplethopeltis — C
Bellefontia-Xenostegium — B
Symphysurina — A

Sections (W to E): Eastern California (Al Rose, Badger Flat), Cent. & S. Nevada (Antelope Valley, Ninemile, Goodwin), Western Utah (Lehman, Kanosh, Juab, Wah Wah, Fillmore, House), NE Utah & SE Idaho (Swan Peak, Garden City, St. Charles), Western Texas / El Paso Group (Montoya, Florida Mtns., Scenic Drive, McKelligon Canyon, José, Victorio Hills), Southern Oklahoma / Arbuckle Group (Joins, W. Spring Creek, Kindblade, Cool Creek, McKenzie Hill).

Figure 14.2. *Stratigraphic distribution of the six Early Ordovician stratigraphic sections that have produced crinoid faunas arrayed from west (left) to east (right) showing the Ross (1951) lettered shelly fossil zones (A–N), the equivalent Ross-Hintze shelly fossil zones from Ross et al. (1997), and the intervals within formations (marked by black bars) from which echinoderms have been collected. Timescale at left with dates and time slices from Webby et al. (2004).*

described in order of crinoid specimen abundance. For each fauna, we have described 1) the lithology, thickness, and distribution of this faunal unit, 2) the crinoid collecting history, and 3) a summary of the crinoid fauna and stratigraphic distribution of its members. Because most of the Early Ordovician crinoids from these faunas are undescribed, we generally provide only a coarse taxonomic summary for each fauna. However, we have made a careful, conservative, species-level assessment of currently known Early Ordovician crinoids from the western and southwestern United States from the available collections. Crinoid species diversity in these faunas will probably increase to more than the numbers reported here as fieldwork continues and as previously collected material is prepared and studied.

The two large Early Ordovician crinoid faunas are from the Fillmore Formation and Wah Wah Limestone of western Utah, with about 97 specimens in at least 20 species (Sprinkle and Guensburg, 1995, 1997; Guensburg and Sprinkle, 2003), and from the Garden City Formation of northern Utah and southern Idaho, with about 86 specimens in at least 18 species, most of these recently discovered since 2003 (Fig. 14.1). The two medium-size Early Ordovician crinoid faunas are from the Ninemile Shale in central and southern Nevada, with 22 specimens in at least seven species (Sprinkle and Guensburg, 1995), and from the Arbuckle Group in southern Oklahoma, with 18 specimens in three species (Sprinkle and McKinzie, 1996). The two small Early Ordovician crinoid faunas are from the El Paso Group of West Texas, with six specimens in four species (Sprinkle and

Wahlman, 1994; Sprinkle and McKinzie, 1996), and from the Al Rose Formation of eastern California, with four specimens in two named genera and species (Strimple and McGinnis, 1972; Ausich, 1986). We have collected crinoids from all but the last fauna.

Each of these crinoid faunas occurs in a slightly different depositional setting, but all six of them are located in the western to southwestern, passive shelf margin of Early Ordovician Laurentia. The farthest west localities (central Nevada and California) represent deeper-water, outer-shelf, mud-dominated environments, in contrast to the localities farther east and southeast, which are mostly shallow-water, inner-shelf, mixed carbonate and clastic environments (Fillmore Formation and Wah Wah Limestone) or carbonate environments (Garden City Formation, El Paso Group, and Arbuckle Group). The Arbuckle Group was formed on the northeastern flank of a more rapidly subsiding aulacogen cutting into the southern Laurentia continental margin (Ham, 1969), and the sediments deposited there were similar to other shelf areas, but thicker. All of the shallow-water units have strong storm influence (Dattilo, 1993; Sprinkle and Guensburg, 1995), with numerous flat-pebble conglomerate beds, megaripples in grainstones, and sponge-algal mounds with scalloped and eroded top surfaces. Hardgrounds and lithified mound surfaces are common in these shallow-water lithologies, and the crinoids and several other echinoderm groups lived attached to these hard substrates until they were either knocked over by storm currents and rapidly buried in place, or torn loose by storm currents and rapidly buried nearby. Nearly all of the complete crinoid specimens that we have recovered appear to have been preserved in this manner. In contrast, the crinoids that lived, died, and fell apart on the seafloor during quiet times between major storms are now represented only by disarticulated echinoderm fragments in muddy facies or in winnowed grainstones (Sprinkle and Guensburg, 1995).

FILLMORE FORMATION AND WAH WAH LIMESTONE IN WESTERN UTAH. These formations are part of a thick, mixed carbonate and clastic, Lower Ordovician section exposed in the Ibex area of western Utah (Hintze, 1951, 1953, 1973; Dattilo, 1988, 1993). The Fillmore Formation is about 550 m (1800 feet) thick and consists of six informal members (Hintze, 1973) that are more resistant limestone-dominated units alternating with less resistant shale and siltstone-dominated units. The Wah Wah Limestone is much thinner, about 75 m, and contains the base of the Whiterockian (Lower–Middle Ordovician boundary) about three-quarters of the way above the base (Fig. 14.2). Flat-pebble conglomerate beds are common in both formations, and several megarippled grainstones occur near the contact between these two formations. Many of these bed surfaces have developed into hardgrounds marked by discoloration, attached echinoderm and sponge holdfasts, and rare borings (Dattilo, 1988, 1993). Sponge-algal mound layers occur at several levels in both formations, and some mound intervals are as much as 3 m thick and persistent throughout the Ibex area. On the basis of these distinctive lithologies, evidence for numerous depo-

Large Crinoid Faunas

sitional cycles within the middle part of the Fillmore Formation, and the shelly fauna and abundant ichnofauna, the Fillmore Formation and Wah Wah Limestone have been interpreted as shallow-water, storm-dominated, inner- to middle-shelf deposits (Hintze, 1973, 1979; Dattilo, 1988, 1993).

The Fillmore Formation has produced sporadic crinoid specimens for many years, dating back at least to the 1960s. Lehi Hintze collected a few crinoids from the Lower and Middle Ordovician parts of the Ibex section in the 1960s that were described by Gary Lane (1970). Most of these crinoids were from the lower Middle Ordovician Kanosh Shale and Lehman Formation; only a few partial specimens and fragments were from Lower Ordovician units, including *Hybocrinus* sp. A, which is a distinctive undescribed disparid. Another slab with two crinoids collected in the early 1970s by Steven Church, then a graduate student at Brigham Young University, was also sent to Lane and later described by Kelly and Ausich (1978, 1979) as the eustenocrinid disparid *Pogonipocrinus*. These crinoids came from a widespread mound horizon (Church, 1974) in the middle part of the Fillmore Formation (zone G-2), now known as Church's Reef. The first crinoid from the lower part of the Fillmore Formation (zone E), a large, complete, well-preserved iocrinid disparid, was found in the spring of 1979 in Skull Rock Pass by Jerry Mansfield, then a student at Weber State College, and given to his professor, Richard Moyle (Guensburg, personal commun.). In 1986–1987, Ben Dattilo, then a Brigham Young University master's student working on the depositional environments and cyclicity of the middle part of the Fillmore Formation, collected a small echinoderm fauna, including a few crinoids, near the Pyramid Section (Hintze, 1973) just west of Skull Rock Pass. Dattilo discovered that most of the complete Fillmore Formation echinoderms were attached to hardgrounds developed on flat-pebble conglomerates and sponge-algal mounds and had been smothered and preserved on or just above these surfaces by high-energy storm events (Dattilo, 1988, 1993). When Sprinkle contacted Dattilo about borrowing this collection in November 1988, Dattilo told us how these specimens occurred and sent us his collection to study.

Guensburg and Sprinkle began joint fieldwork in the Fillmore Formation in June 1989. We used Dattilo collecting technique to gradually accumulate a large crinoid-dominated echinoderm fauna associated with hardgrounds in several parts of the section (Guensburg and Sprinkle, 1990, 1992, 1994, 2001, 2003; Sprinkle and Guensburg, 1993, 1995, 1997). We also discovered a second echinoderm association, including mostly rhombiferans, mitrates, and a few eocrinoids, associated with soft substrates, and we extended both of these echinoderm communities into the overlying lower Wah Wah Limestone, of latest Early Ordovician age. Colin Sumrall, then a University of Texas graduate student, contributed many important specimens in 1991, 1993–1994, and 1997. Beginning in 1994, we were assisted by several commercial collectors, such as Terry Abbott, Susan Abbott, and Ed Cole of Delta, Utah, and more recently by students and amateur collectors, such as Seth Finnegan from the University of California, Riverside, and Marcus Donivan of Salt Lake City, who gave us echinoderm specimens found while searching for trilobites or other fossils or went out in the field with us to new collecting localities. Many echinoderm specimens were found in 1997–1998, including some of the best crinoids, but echinoderms have continued to be found up through 2006. In 17 summers of collecting (typically 3–6 days each summer), we have accumulated 97 crinoids out of an echinoderm fauna of at least 213 specimens. Many echinoderms were preserved in highly indurated matrix and required up to several hundred hours to prepare by hand; others were found in softer and more easily removed matrix.

The 97 crinoid specimens from the Fillmore Formation and Wah Wah Limestone (Fig. 14.1) belong to at least 20 species. These include two protocrinoids (based on 11 specimens), three monobathrid camerates (three specimens), and a reteocrinid diplobathrid camerate (three specimens) that have been named and described (Guensburg and Sprinkle, 2003). A conical cladid specimen and a probable specimen of *Hybocrinus* Billings, 1857, are being described by Guensburg and Sprinkle, along with similar taxa from the Ninemile Shale in Nevada. At least nine iocrinid species (based on about 67 specimens) and three other disparids, including Lane's *Hybocrinus* sp. A (five specimens), *Pogonipocrinus* (two specimens), and one other disparid (four specimens), await description or revision.

Protocrinoids clearly dominate the lower part of the Fillmore Formation (zones D–F), and these occur with a few iocrinids (or other disparids) and the oldest known monocyclic camerate. Iocrinids dominate the middle part of the Fillmore Formation (zones G-1 to G-2), and they occur with other disparids, the oldest known diplobathrid camerate, and the oldest known hybocrinid. Iocrinids are also common in the upper Fillmore Formation and lower Wah Wah Limestone (zones H–J), along with several disparids, monobathrids, and a cladid. Stratigraphic ranges for about half of these crinoids are plotted in Sprinkle and Guensburg (1997, plate 1, chart C).

GARDEN CITY FORMATION IN NORTHEASTERN UTAH AND SOUTHEASTERN IDAHO. The Garden City Formation occurs stratigraphically above the Saint Charles Formation and below the Swan Peak Formation and is approximately 370–550 m (1200–1800 feet) thick (Ross, 1951). The basal Garden City Formation is thought to be early Ibexian in age (zone B; see Taylor and Landing, 1982), and the uppermost beds are early Whiterockian in age (zone L; Fig. 14.2). The formation has been divided lithologically into two unnamed members (Ross, 1949), a lower member consisting mostly of thin- to medium-bedded, laminated, carbonate mudstone and flat-pebble conglomerate beds, and an upper member consisting of skeletal grainstone beds with abundant black chert. Near the top of the lower member, flat-pebble conglomerates become particularly abundant, and these are commonly deposited in lenticular channels that scour underlying beds and grade laterally into mudstone facies. The lower member of the Garden City Formation is generally poorly fossiliferous, consisting primarily of fragmentary trilobites, echinoderms, and graptolites; but it also contains numerous stromatolitic and thrombolitic mound horizons. These mounds are generally much smaller and less conspicuous (termed *miniherms* by Toomey and Nitecki, 1979, p. 146) than those of the Fillmore Formation and Wah Wah Limestone, but as in those units, most of the articulated echinoderms that we have found are associated with these mounds.

Crinoids were first reported from the Garden City Formation in the Logan Quadrangle in the late 1940s (Williams, 1948), yet to date, no named species have been described from this formation. Ross (1951, 1968) listed blastozoan plates from the uppermost Garden City Formation of Middle Ordovician (Whiterockian) age in northeastern Utah that were described by Sprinkle (1973). In 1979, Ed Landing, then with the U.S. Geological Survey, found several complete echinoderm specimens (mostly rhombiferans) in the lowermost beds of the Garden City Formation in the Bear River Range in southeastern Idaho that were sent to Sprinkle for identification. About 25 additional echinoderm specimens from this locality were added to the collection in August 1981, during a Cambrian Symposium field trip (see Landing and Taylor, 1981). Sprinkle returned to this locality and nearby areas in 1984–1986 for a few days each summer and doubled the number of echinoderms recovered to about 53, mostly glyptocystitid rhombiferans. In 1989, a poorly preserved large conical crinoid (a protocrinoid or cladid) was found on the flanks of a nearby sinkhole; this is the oldest Ordovician crinoid (upper zone A–lower zone B) that has ever been found. In 2000, Guensburg discovered (during preparation) that this crinoid specimen had floor plates within the brachials of its atomous arms (figured by Guensburg and Sprinkle, 2001). We also recently discovered that three other badly eroded small "eocrinoids" collected in the mid-1980s were in fact poorly preserved crinoids, perhaps the same taxon as the sinkhole specimen. About 54 echinoderm specimens, including these four crinoids, are now known from the lowermost Garden City Formation.

In fall 2003, Jake and Nick Skabelund of Logan and Cove, Utah, found two float slabs with a small complete iocrinid, several small eocrinoids, and a large rhombiferan at a Garden City locality in Logan Canyon (see Williams, 1958). In the spring of 2004, Paul Jamison of Logan, Utah, led a Utah Friends of Paleontology (UFOP) field trip to this new locality, and three more articulated crinoids were collected: Jamison discovered a new iocrinid, and Glade Gunther of Brigham City, Utah, found a slab with two exceptionally preserved camerates. In late July 2004, Gahn and Sprinkle visited these collectors and looked for additional echinoderms at these new and old localities. After finding a few additional crinoids and other echinoderms in Logan Canyon and at Landing's original Bear River Range locality, Gahn, Sprinkle, J. Skabelund, and Jamison discovered an almost completely exposed middle Garden City section on a mountaintop with several small sponge-algal mound layers covered by dozens of crinoids and other echinoderms lying between or attached to the mounds. At least 42 echinoderms were discovered the first day, a record for echinoderms at a Lower Ordovician section in the western United States. Gahn, Jamison, and J. Skabelund then spent the next week measuring this new section, looking for additional echinoderms (nearly 100 were eventually located), and collecting the best 25 specimens. The newly collected Garden City specimens were taken to the Smithsonian Institution, and Gahn spent the winter and spring months of 2004–2005 preparing the best crinoid specimens with visits from both Guensburg and Sprinkle to help clean, identify, and study these new Garden City echinoderms.

In mid-July 2005, this group of collectors, along with Guensburg and Dan Blake, University of Illinois, revisited the new Garden City locality and collected more than 40 echinoderm specimens, some left from the previous summer and others newly discovered. Most of the group then hiked into a more remote locality in the Bear River Range, where the Garden City Formation was well exposed, and collected about 35 additional echinoderms (including many kirkocystid mitrates) over several days. These new echinoderms went back to the Smithsonian Institution with Gahn, and many are being prepared for study. A few additional crinoids were collected in 2006 from localities in Logan and Green Canyons. About 137 new echinoderms, with 82 crinoids, have been recovered in 2003–2006, bringing the total Garden City echinoderm fauna to about 191 specimens, including 86 crinoids.

Many of the crinoids from the Garden City Formation are still unprepared and cannot yet be identified, but 30 are well enough preserved and exposed to tentatively recognize 18 species. The crinoids that can be identified include protocrinoids (three specimens, one species), camerates (13 specimens, seven species), cladids (two specimens, one species), disparids (five specimens, four species), and five additional species (seven specimens) that are now unassigned to a crinoid group but clearly represent distinct species. These include four probable disparids and either a camerate or a protocrinoid (Fig. 14.1).

Camerates are more abundant and diverse in the Garden City fauna than any of the other western faunas. Monobathrids appear to be most prevalent, some of which show affinity to Fillmore–Wah Wah genera such as *Habrotecrinus* and *Adelphicrinus* (both Guensburg and Sprinkle, 2003). However, others are quite different from camerates described in other Early Ordovician crinoid faunas and instead have affinities with the Xenocrinacea, Reteocrinidae, and the enigmatic Cleiocrinidae. Disparids, especially iocrinids, are at least as diverse as the camerates and probably more abundant, given the probable taxonomic affinity of most of the 55 unassigned specimens. At least one cladid is present, a dendrocrinid similar to *Compagicrinus* Jobson and Paul (1979) from Greenland, and *Elpasocrinus* Sprinkle and Wahlman (1994) from west Texas.

On the basis of position in the formation and associated trilobites, about 35% of the echinoderms are from the lowermost part of the Garden City Formation (zones B–F), which has numerous rhombiferans and cornutes, several protocrinoids and eocrinoids, and a possible cladid. Another 30% of the echinoderms are from the lower part of the middle Garden City Formation (zone G-1), including many kirkocystid mitrates, iocrinids and other disparids, camerates, and a starfish. The other 35% of the echinoderms are from the upper part of the middle Garden City Formation (zones G-2 to I), including iocrinids, other disparids, camerates, and cladids, along with rhombiferans, eocrinoids, and a few edrioasteroids, starfish, and cornutes.

Medium-Size Crinoid Faunas

NINEMILE SHALE IN CENTRAL AND SOUTHERN NEVADA. The Ninemile Shale in central Nevada is a relatively thin (43–118 m) but usually poorly exposed, greenish, platy shale containing thin, nodular, micritic limestone beds scattered throughout, a few thicker, highly burrowed, grainstone beds, and thin, tan-brown, calcareous siltstone beds up to 1 cm thick that become more common toward the top. The Ninemile Shale is usually overlain by the thick, cliff-forming Antelope Valley Limestone that protects the underlying Ninemile Shale but also partly covers it in limestone float (and younger fossils). At some sections, the shale beds are crumpled, baked, and cut by numerous joints and calcite veins that make fossil collecting difficult or impossible. However, at two nearby repeated sections in Whiterock Canyon, southwest of Eureka, central Nevada, the exposed Ninemile Shale is better preserved and more fossiliferous, even though the lower Front Section (WR-1) is thrust over a Devonian orange siltstone and black chert unit in the Woodruff Formation (Ross and Ethington, 1991). Most of the echinoderms collected from the Ninemile Shale have come from the lower and middle parts of these two adjacent sections, especially near a distinctive 20–30-cm burrowed grainstone bed that is only a few meters above the thrust fault in the Front Section. This bed may also be present near the exposed base of the Ninemile Shale in the nearby Narrows Section (WR-2). Graptolites in the shales and conodonts in the nodular limestone beds indicate a zone J (latest Ibexian) age for these two Whiterock Canyon sections (Mitchell, *in* Ross and Ethington, 1991). The Ninemile Shale in central Nevada lacks shallow-water indicators, such as sponge-algal mounds, flat-pebble conglomerates, hardgrounds, or erosional surfaces, and has been interpreted as a deeper water, outer shelf to upper slope, fine clastic deposit (Ross et al., 1989). Only a few echinoderms and only one crinoid have been found at three other Ninemile Shale sections in southern Nevada (Fig. 14.1).

A few echinoderms (several rhombiferans, a rhipidocystid, and a mitrate stylophoran) had previously been found in the Ninemile Shale in central and southern Nevada during the past 30 years. In 1985, Noel Eberz of San Diego State University found a rhipidocystid theca and stem on a slab while on a field trip to Whiterock Canyon led by Reuben J. Ross Jr. This rhipidocystid was sent to Sprinkle, who then visited Whiterock Canyon in August 1986 on another field trip led by Ross, and discovered two partial rhombiferans, a small mitrate stylophoran, and many plates and stems of rhombiferans and other echinoderms. Intensive collecting for echinoderms and other fossils began in June 1989 by Sprinkle with University of Texas graduate students Ronald Johns (1989–1990; now at Austin Community College) and Colin Sumrall (1991, 1993; now at University of Tennessee). We gradually accumulated a large fauna dominated by glyptocystitid rhombiferans but containing many other groups, totaling about 225 partial or complete echinoderm specimens (Sprinkle, 1990; Sprinkle and Guensburg, 1995).

Crinoids are not common in the Ninemile Shale, but since 1989, we have collected 22 crinoid specimens, about 10% of the echinoderm fauna, belonging to at least seven species. These include two or more species of iocrinid disparids (five specimens and six arm sets with long anal tubes on slabs), *Hybocrinus* Billings, 1857 (five specimens), one homocrinid disparid specimen, a cup-shaped cladid with atomous arms resembling *Porocrinus* Billings, 1857, but lacking goniospires (three specimens), one thin conical cladid specimen, and one large but poorly preserved camerate specimen with partial tegmen and scattered arms (Fig. 14.1). All of these appear to be from zone J (latest Early Ordovician; Fig. 14.1) and are still undescribed.

UPPER ARBUCKLE GROUP IN SOUTHERN OKLAHOMA. The Arbuckle Group is a thick sequence of Lower Ordovician carbonates exposed in the uplifted cores of the Arbuckle and Wichita Mountains in southern and southwestern Oklahoma (Ham, 1969). The Arbuckle Group section is at least 1600 m (5300 feet) thick and consists of five formations. Most of the carbonates are thin- to medium-bedded mudstones to wackestones, and typically only the more resistant lithologies are exposed because of moderate rainfall and vegetation cover. Several sponge-algal mound units are present, especially in the middle and upper parts of the section (Toomey and Nitecki, 1979), and metazoan diversity is higher in these more massive boundstone units that are based on slabbed and polished sections. Unfortunately, most fossils weathering from these lithologies have been coarsely silicified, and parts of the Arbuckle Group carbonates have also been dolomitized, further hindering fossil collecting. Only the top two units, the Kindblade and the West Spring Creek formations, which are about 910 m (3000 feet) thick, have produced any complete echinoderms (Fig. 14.2; Sprinkle and McKinzie, 1996). The best section for fossil collecting is the long series of roadcuts along Interstate 35 in the central Arbuckle Mountains. These roadcuts, constructed in the 1960s, expose both the resistant carbonates and the thin shaly interbeds that separate them.

Only a few crinoids have been collected from the upper Arbuckle Group in southern Oklahoma (Sprinkle and McKinzie, 1996). In 1983, Barry Webby found a fairly large crinoid calyx on a limestone slab from the West Spring Creek Formation in an Interstate 35 roadcut while on a field trip led by Robert O. Fay, Oklahoma Geological Survey, who forwarded this specimen to Sprinkle in 1984. In June 1993, Mark McKinzie, an amateur fossil collector from Dallas, Texas, collected four small iocrinid disparids on limestone slabs from a shaly interval in the middle West Spring Creek Formation on Interstate 35. McKinzie and Sprinkle returned to this locality in July 1993 and collected 11 additional iocrinids. In June 1994, McKinzie found another productive layer in the underlying Kindblade Formation that produced two larger, abraded, or covered crinoids in a flat-pebble conglomerate bed. When excavated from between two near-vertical pebbles, the better-preserved crinoid was found to be a cladid resembling *Cupulocrinus* d'Orbigny (1849).

The Arbuckle Group echinoderm fauna now includes 18 crinoid specimens in three species (Fig. 14.1). One small iocrinid disparid species with a long anal tube (15 specimens; see reconstruction in McKinzie, 1995) occurs in the West Spring Creek Formation (probably zones I–J), one cladid species comes from the upper Kindblade Formation (two specimens; probably zone G-2), and the single larger merocrinid-like cladid?

specimen also occurs in the lower West Spring Creek Formation (probably zone H) (Fig. 14.2). The low diversity and lack of other echinoderm groups probably result from the limited exposures and the absence of echinoderms from the few mound horizons that have been examined.

Small Crinoid Faunas

EL PASO GROUP IN WEST TEXAS. The El Paso Group consists of thin-bedded to massive carbonates about 400 m (1312 feet) thick in west Texas. Most of the known echinoderm specimens have come from several well-exposed sections in the southern Franklin Mountains just north of downtown El Paso, Texas. The history of echinoderm collecting in the El Paso Group from the 1960s to 1991 has been summarized in Sprinkle and Wahlman (1994, p. 324–325). Since then, Mark McKinzie, visiting from Dallas, found a large, distinctive, but badly crushed rhombiferan below Scenic Drive in December 1992 and wrote a preliminary description of it (McKinzie, 1993) before sending it to Sprinkle. In May 1995, McKinzie found three additional echinoderms, including two small iocrinids, at the same locality. In December 1995, Sprinkle and McKinzie collected from these old and new localities, finding a group of 11 small rhombiferan thecae on a large limestone float block, two small mitrate stylophorans, and a heavily plated disparid crinoid. These new finds led to a joint abstract on this growing echinoderm fauna (Sprinkle and McKinzie, 1996). In 2000, Kevin Durney, an amateur collector then living in El Paso, found a second, more complete slab specimen of the unusual crinoid found in the 1960s from the upper part of the El Paso Group. These two crinoid specimens are to be described together in a multiauthored paper in preparation.

The El Paso Group has now produced six crinoid specimens belonging to four taxa (Fig. 1), 26% of the total echinoderm fauna of 23 specimens. These include the unusual pseudomonocyclic "cladid" (the two Cincinnati and Durney specimens), the cladid *Elpasocrinus* (Sprinkle and Wahlman, 1994; one specimen), a small iocrinid disparid (two specimens), and a heavily plated and armed disparid (one specimen). Three crinoids have come from the McKelligon Canyon Limestone (middle part of the El Paso Group, zone G-2), and three from the Scenic Drive and Florida Mountains formations (upper part of the El Paso Group, zones I–J) (Fig. 14.2).

AL ROSE FORMATION IN EASTERN CALIFORNIA. This formation ranges up to 122 m of orange to reddish-brown shale and siltstone with thin interbeds of chert and black, nodular limestone (Ross, 1966). It is commonly faulted at the base, folded internally, and cut by small granitic intrusions. Trilobites from limestone and shale beds throughout the formation and graptolites from shale beds near the top are the only common fossils, and they indicate a late Early Ordovician age ranging from trilobite zones H–J (Ross, 1966). Lack of shallow-water indicators, such as sponge-algal mounds or storm deposits, and the abundance of graptolites indicate deeper-water, offshore shelf, depositional environments.

Philip Walker collected two crinoids on limestone slabs from the top of the Al Rose Formation (zone J) near Independence, California, in the

1960s that were obtained in approximately 1969 by Michael McGinnis, then working in Los Angeles. These two crinoids were subsequently described as the monobathrid camerate *Proexenocrinus inyoensis* Strimple and McGinnis (1972). The camerate calyces on these original slabs were small and somewhat abraded, so that some plate sutures were difficult to see. Ausich (1986) redescribed *Proexenocrinus* as a rhodocrinid diplobathrid on the basis of barely visible infrabasals in the stem cavity of the paratype. He also pointed out that two additional tiny, stemmed, disparid crinoids, which he named *Inyocrinus strimplei* Ausich (1986), were present on the holotype slab. These two diplobathrid camerates and two disparids from zone J are the only complete echinoderms that have been found in the Al Rose Formation (Figs. 14.1, 14.2), but not much echinoderm collecting has been done in this deformed unit.

Recent discoveries of much larger crinoid faunas in the Early Ordovician of the western and southwestern United States have greatly increased the diversity, morphological variability, and stratigraphic range of the earliest crinoids. These faunas indicate that the diversification of crinoids during the Great Ordovician Biodiversification Event (Sprinkle and Guensburg, 2004) started earlier than previously thought, almost at the beginning of the Early Ordovician. Most Paleozoic crinoid subclasses and orders (except for flexibles) were present in these faunas by the end of the Early Ordovician, although crinoid diversity only increased slowly until the early Late Ordovician when crinoid diversity and abundance greatly expanded (Sprinkle and Guensburg, 2003). High diversity and disparity among Late Ordovician crinoids have long been evidence of an underrepresented Early Ordovician record.

Guensburg and Sprinkle (2003) recently described two genera of protocrinoids from western Utah, arguably the most primitive crinoids ever found. The presence of dozens to hundreds of extra plates in the midcup of protocrinoids implies that the direction of crinoid cup evolution was from complex and irregular to simple and organized. This major transition from complex to simple crinoid cups appears to have taken place relatively quickly near the beginning of the Ordovician (zones B–E), perhaps within only a few million years.

Several crinoid subclasses now extend nearly to the base of the Ordovician, establishing an early timing for acquisition of traditional crinoid traits. A new set of characters for phylogenetic analysis is being developed by using the greatly expanded Early Ordovician crinoid record. Initial analysis indicates that characters considered fundamental for defining high-level taxa, such as number of plate circlets in the cup, are not of great phylogenetic significance early in crinoid history, but other characters, particularly posterior cup morphology, remain robust nearly to the tree base. Character distribution supports previously identified protocrinoids as the basal crinoid taxon.

The discovery of protocrinoids makes a rhombiferan (or even an advanced blastozoan) origin for crinoids (Ausich, 1997) highly unlikely be-

Implications of these Crinoid Faunas

cause rhombiferans already had a well-organized theca with relatively few plate circlets, plus an advanced holomeric stem and specialized respiratory rhombs, by this time in the Early Ordovician. The many morphologic differences also make rhombiferans a poor outgroup for polarizing characters in early crinoid phylogenetic analysis (Ausich, 1997; Guensburg and Sprinkle, 1997; Sumrall, 1998).

Although cup plating was rapidly simplified, these Early Ordovician crinoids retained relatively primitive arms, stems, and holdfasts. Many early crinoid arm features would be considered simple (uniserial arms, isotomous branching or no branching, lack of pinnules), whereas stems were more complex (irregularly plated distal stems and conical holdfasts, pentameric proximal stems). No crinoids with pinnules or biserial arms have been found in these Early Ordovician faunas, and only one Fillmore crinoid has more advanced arm branching, indicating that arms remained relatively primitive and simple until later in the Ordovician. Most crinoid stems remained pentameric with homeomorphic or simple heteromorphic (nodal and internodal) patterns, and distal stems and holdfasts were mostly multiplated cementing types for attachment to hard substrates. These were again primitive but more complex than the holomeric stems and simplified but more varied holdfasts in later crinoids (Brett, 1981).

Although protocrinoids dominated the earliest Ordovician crinoid faunas (zones B–E), disparids became the most abundant and diverse crinoids by the middle of the Early Ordovician (zone G-1; Sprinkle and Guensburg, 2004), followed in abundance by various groups of monobathrid camerates. This pattern may be partly facies controlled because iocrinid disparids seem to be most common in mixed clastic and carbonate facies, such as the Fillmore Formation, the Ninemile Shale, and shaly parts of the West Spring Creek Formation. In contrast, monobathrid camerates appear to be more common in pure carbonates, such as the lower part of the Garden City Formation.

Finally, many of the monocyclic camerates that we have recently found in the middle Garden City Formation appear strikingly similar to dicyclic camerates known from other crinoid faunas later in the Ordovician. Although two-circlet versus three-circlet cup designs are a convenient and easy-to-use morphologic character to separate major groups of later crinoids (Moore and Laudon, 1943; Lane and Webster, 1980), this dichotomy may not work so well in the primitive camerate crinoids now being collected in the Early Ordovician.

Conclusions

1. We have collected and are now describing five large, medium, and small crinoid faunas from the Early Ordovician of the western and southwestern United States that total (with an additional sixth small fauna) more than 233 specimens and represent as many as 54 different species. These are the largest and most diverse collections of Early Ordovician crinoids that have ever been assembled from a single geographic region.

2. These faunas include early representatives of most Paleozoic clades of crinoids (except for flexibles) that were present by the end of the Early Ordovician.

They indicate that Laurentia was clearly a center for the major initial diversification in Ordovician crinoids.

3. The change from complex and irregular cup plating (protocrinoids) to simple and highly organized cup plating (disparids, hybocrinids) appears to have occurred rapidly near the beginning of the Ordovician (zones B–E), perhaps within only a few million years.

4. Although cup plating was rapidly simplified, these Early Ordovician crinoids retained relatively primitive arms, stems, and holdfasts. No crinoids with pinnules or biserial arms have been found in these Early Ordovician faunas, and few examples of advanced arm branching are known. Nearly all crinoid stems remained pentameric, sometimes grading distally into multiplated regions, or into terminal holdfasts that were multiplated cementing types for attachment to hard substrates.

5. Although protocrinoids dominated the earliest Ordovician crinoid faunas (zones B–E), disparids became the most abundant and diverse crinoids by the middle of the Early Ordovician (zone G-1), followed by various groups of monobathrid camerates. Several of the monobathrids from the Garden City Formation show a strong resemblance to dicyclic camerates from other occurrences later in the Ordovician.

6. Crinoids average about 34% (range, 10%–100%) of the specimens in these six Early Ordovician echinoderm faunas. Other echinoderm groups, such as glyptocystitid rhombiferans and eocrinoids in Laurentia and diploporans and stylophorans in northern Gondwana, were more abundant and diverse than crinoids during parts of the Early and Middle Ordovician. This pattern persisted until the early Late Ordovician, when crinoids underwent a major expansion, increasing their dominance over other echinoderm groups in most environments and regions from then on through the end of the Paleozoic.

7. Recent discovery of a large new crinoid fauna in the middle Garden City Formation supports previous observations that most early crinoids attached to hard substrates (mounds, hardgrounds), and that complete, articulated, crinoid specimens were preserved only if rapidly buried alive by storms or sediment slumps (Gahn et al., 2006).

Acknowledgments

We thank all the mentioned academic, amateur, and commercial fossil collectors who either contributed echinoderm specimens or helped us in the field during the past 17 years of work on this project. Also, R. Ethington, University of Missouri, Columbia, and A. Aase, Fossil Butte National Monument, Wyoming, identified conodont and trilobite collections, respectively, to help date our echinoderm faunas. Reviewers C. Sumrall, University of Tennessee, Knoxville, and J. Miller, Missouri State University, Springfield, made many helpful suggestions. Funding for fieldwork, travel, specimen preparation and study, and publication expenses were provided by National Science Foundation grants BSR-8906568 (1989–1991, to JS) and EAR-9304253 (1993–1994, to JS, TEG), by a PRF-ACS grant to M. Wilson, College of Wooster (1989–1990, to TEG), by CRDF Cooperative research grant RG1-242 (1997–1998, to S. V. Rozhnov, JS, and T. E. Guensburg), by the University of Texas at Austin (JS), by Southern Illinois University, Edwardsville, and Rock Valley College, Rockford, Illinois (TEG), and by the Smithsonian Institution, Washington, D.C., and Brigham Young University, Idaho, Rexburg (FJG).

References

Ausich, W. I. 1986. The crinoids of the Al Rose Formation (Early Ordovician, Inyo County, California, United States of America). Alcheringa, 10:217–224.

Ausich, W. I. 1997. Calyx plate homologies and early evolutionary history of the Crinoidea, p. 289–304. *In* J. A. Waters and C. G. Maples (eds.), Geobiology of Echinoderms. Paleontology Society Papers, 3.

Billings, E. 1857. New species of fossils from Silurian rocks of Canada. Canada Geological Survey, Report of Progress 1853–1856, Report for the Year 1856:247–345.

Brett, C. E. 1981. Terminology and functional morphology of attachment structures in pelmatozoan echinoderms. Lethaia, 14:343–370.

Church, S. B. 1974. Lower Ordovician patch reefs in western Utah. Brigham Young University Geology Studies, 21(3):41–62.

Dattilo, B. F. 1988. Depositional environments of the Fillmore Formation (Lower Ordovician) of western Utah. Unpublished master's thesis, Brigham Young University, Provo, Utah, 94 p.

Dattilo, B. F. 1993. The Lower Ordovician Fillmore Formation of western Utah: storm-dominated sedimentation on a passive margin. Brigham Young University Geology Studies, 39(1):71–100.

D'Orbigny, A. D. 1849. Cours élémentaire de paléontologie et géologie stratigraphiques, 1. Victor Masson, Paris, 299 p.

Gahn, F. J., J. Sprinkle, and T. E. Guensburg. 2006. Garden City of echinoderms: a new Early Ordovician Lagerstätte from Idaho and Utah. Geological Society of America Abstracts with Programs, 38(7):383.

Guensburg, T. E., and J. Sprinkle. 1990. Early Ordovician crinoid-dominated echinoderm fauna from the Fillmore Formation of western Utah. Geological Society of America Abstracts with Programs, 22(7):A220.

Guensburg, T. E., and J. Sprinkle. 1992. Rise of echinoderms in the Paleozoic evolutionary fauna: significance of paleoenvironmental controls. Geology, 20:407–410.

Guensburg, T. E., and J. Sprinkle. 1994. Revised phylogeny and functional interpretation of the Edrioasteroidea based on new taxa from the Early and Middle Ordovician of western Utah. Fieldiana (Geology), new series, 29:1–43.

Guensburg, T. E., and J. Sprinkle. 1997. Rhombiferans are not the ancestors of crinoids. Geological Society of America Abstracts with Programs, 29(7): A-341.

Guensburg, T. E., and J. Sprinkle. 2001. Earliest crinoids: new evidence for the origin of the dominant Paleozoic echinoderms. Geology, 29:131–134.

Guensburg, T. E., and J. Sprinkle. 2003. The oldest known crinoids (Early Ordovician, Utah) and a new crinoid plate homology system. Bulletins of American Paleontology, 364:1–43.

Ham, W. E. 1969. Regional Geology of the Arbuckle Mountains, Oklahoma. Oklahoma Geological Survey Guidebook 17, 51 p.

Hintze, L. F. 1951. Lower Ordovician detailed stratigraphic sections for western Utah. Utah Geological and Mineralogical Survey Bulletin, 39, 99 p.

Hintze, L. F. 1953. Lower Ordovician trilobites from western Utah and eastern Nevada. Utah Geological and Mineralogical Survey Bulletin, 48, 249 p.

Hintze, L. F. 1973. Lower and Middle Ordovician stratigraphic sections in the Ibex area, Millard County, Utah. Brigham Young University Geology Studies, 20(4):3–36.

Hintze, L. F. 1979. Preliminary zonation of Lower Ordovician of western Utah by various taxa. Brigham Young University Geology Studies, 26(2):13–19.

Jobson, L., and C. R. C. Paul. 1979. *Compagicrinus fenestratus*, a new Lower

Ordovician inadunate crinoid from North Greenland. Rapport Grönlands Geologiske Undersøgelse, 91:71–81.

Kelly, S. M., and W. I. Ausich. 1978. A new Lower Ordovician (Middle Canadian) disparid crinoid from Utah. Journal of Paleontology, 52:916–920.

Kelly, S. M., and W. I. Ausich. 1979. A new name for the Lower Ordovician crinoid *Pogocrinus* Kelly and Ausich. Journal of Paleontology, 53:1433.

Landing, E., and M. E. Taylor. 1981. Upper St. Charles and lower Garden City formations at Franklin Basin, Bear River Range, Idaho, p. 165–168. *In* M. E. Taylor and A. R. Palmer (eds.), Cambrian Stratigraphy and Paleontology of the Great Basin and Vicinity, Western United States. Second International Symposium on the Cambrian System, Guidebook for Field Trip, 1.

Lane, N. G. 1970. Lower and Middle Ordovician crinoids from west-central Utah. Brigham Young University Geology Studies, 17(1):3–17.

Lane, N. G., and G. D. Webster. 1980. Crinoidea, p. 144–157. *In* T. W. Broadhead and J. A. Waters (eds.), Echinoderms, Notes for a Short Course. University of Tennessee Department of Geological Sciences, Studies in Geology, 3.

McKinzie, M. G. 1993. First reported occurrence of macrocystellid eocrinoid from the Lower Ordovician of El Paso, Texas. MAPS Digest (EXPO XV edition), 16(4):91–96.

McKinzie, M. G. 1995. Significance of an Early Ordovician crinoid discovery in relation to other recent occurrences from El Paso, Texas. MAPS Digest (EXPO XVII edition), 18(4):20–27.

Moore, R. C., and L. R. Landon. 1943. Evolution and classification of Paleozoic crinoids. Geological Society of America Special Paper, 46, 153 p.

Ross, D. C. 1966. Stratigraphy of some Paleozoic formations in the Independence Quadrangle, Inyo County, California. U.S. Geological Survey Professional Paper, 396, 64 p.

Ross, R. J., Jr. 1949. Stratigraphy and trilobite faunas of the Garden City Formation, northeastern Utah. American Journal of Science, 247:472–491.

Ross, R. J., Jr. 1951. Stratigraphy of the Garden City Formation in northeastern Utah, and its trilobite faunas. Yale Peabody Museum Bulletin, 6, 161 p.

Ross, R. J., Jr. 1968. Brachiopods from the upper part of the Garden City Formation (Ordovician), north-central Utah. U.S. Geological Survey Professional Paper, 593-H, 13 p.

Ross, R. J., Jr., and R. L. Ethington. 1991. Stratotype of Ordovician Whiterock Series (with an appendix on graptolite correlation of the topmost Ibexian by Charles E. Mitchell). Palaios, 6:156–173.

Ross, R. J., Jr., L. F. Hintze, R. L. Ethington, J. F. Miller, M. E. Taylor, and J. E. Repetski. 1997. The Ibexian, lowermost series in the North American Ordovician, p. 1–50. *In* M. E. Taylor (ed.), Early Paleozoic Biochronology of the Great Basin, western United States. U.S. Geological Survey Professional Paper, 1579.

Ross, R. J., Jr., N. P. James, L. F. Hintze, and F. G. Poole. 1989. Architecture and evolution of a Whiterockian (early Middle Ordovician) carbonate platform, Basin Ranges of western USA, p. 167–185. *In* P. D. Crevallo, J. L. Wilson, J. F. Sarg, and J. F. Read (eds.), Controls on Carbonate Platform and Basin Development. Society of Economic Paleontology and Mineralogy, Special Publication, 44.

Sprinkle, J. 1973. Morphology and evolution of blastozoan echinoderms. Museum of Comparative Zoology, Harvard University, Special Publication, 283 p.

Sprinkle, J. 1990. New echinoderm fauna from the Ninemile Shale (Lower Ordovician) of central and southern Nevada. Geological Society of America Abstracts with Programs, 22(7):A219.

Sprinkle, J., and T. E. Guensburg. 1993. Appendix D—echinoderm biostratigraphy, p. 61–63. *In* R. J. Ross Jr., L. F. Hintze, R. L. Ethington, J. F. Miller, M. E. Taylor, and J. E. Repetski, The Ibexian Series (Lower Ordovician): a replacement for "Canadian Series" in North American stratigraphy. U.S. Geological Survey Open-File Report, 93-598.

Sprinkle, J., and T. E. Guensburg. 1995. Origin of echinoderms in the Paleozoic evolutionary fauna: the role of substrates. Palaios, 10:437–453.

Sprinkle, J., and T. E. Guensburg. 1997. Appendix D4—echinoderm biostratigraphy, p. 49–50 and pl. 1, chart C. *In* R. J. Ross Jr., L. F. Hintze, R. L. Ethington, J. F. Miller, M. E. Taylor, and J. E. Repetski, The Ibexian, lowermost series in the North American Ordovician. *In* M. E. Taylor (ed.), Early Paleozoic Biochronology of the Great Basin, western United States. U.S. Geological Survey Professional Paper, 1579.

Sprinkle, J., and T. E. Guensburg. 2003. Major expansion of echinoderms in the early Late Ordovician (Mohawkian, middle Caradoc) and its possible causes, p. 327–332. *In* G. L. Albanesi, M. S. Beresi, and S. H. Peralta (eds.), Ordovician from the Andes. Instituto Superior de Correlación Geológica (INSUGEO), Serie Correlación Geológica, 17.

Sprinkle, J., and T. E. Guensburg. 2004. Crinozoan, blastozoan, echinozoan, asterozoan, and homalozoan echinoderms, p. 266–280. *In* B. D. Webby, M. L. Droser, F. Paris, and I. Percival (eds.), The Great Ordovician Biodiversification Event. Columbia University Press, New York.

Sprinkle, J., and M. G. McKinzie. 1996. Early Ordovician echinoderms from southern Oklahoma and west Texas. Geological Society of America Abstracts with Programs, 28(1):64.

Sprinkle, J., and G. P. Wahlman. 1994. New echinoderms from the Early Ordovician of west Texas. Journal of Paleontology, 68:324–338.

Strimple, H. L., and M. R. McGinnis. 1972. A new camerate crinoid from the Al Rose Formation, Lower Ordovician of California. Journal of Paleontology, 46:72–74.

Sumrall, C. D. 1998. A phylogenetic test of the glyptocystitid rhombiferan origin of crinoids. Geological Society of America Abstracts with Programs, 30(7): A-94.

Taylor, M. E., and E. Landing. 1982. Biostratigraphy of the Cambrian–Ordovician transition in the Bear River Range, Utah and Idaho, western United States, p. 181–191. *In* M. G. Bassett and W. T. Dean (eds.), The Cambrian–Ordovician Boundary: Sections, Fossil Distributions, and Correlations. National Museum of Wales, Geological Series, 3.

Toomey, D. F., and M. H. Nitecki. 1979. Organic buildups in the Lower Ordovician (Canadian) of Texas and Oklahoma. Fieldiana (Geology), new series, 2, 181 p.

Webby, B. D., R. A. Cooper, S. M. Bergström, and F. Paris. 2004. Stratigraphic framework and time slices, p. 41–47. *In* B. D. Webby, M. L. Droser, F. Paris, and I. Percival (eds.), The Great Ordovician Biodiversification Event. Columbia University Press, New York.

Williams, J. S. 1948. Geology of the Paleozoic rocks, Logan Quadrangle, Utah. Geological Society of America Bulletin, 59:1121–1164.

Williams, J. S. 1958. Geologic Atlas of Utah. Cache County. Utah Geological and Mineralogical Survey Bulletin, 64, 98 p.

Figure 15.1. *Typical British Llandovery crinoids, all from the North Esk Inlier, Pentland Hills, near Edinburgh, Scotland (see Brower, 1975; images after Donovan et al., 2007).* **1, 2, Macrostylocrinus silurocirrifer** *Brower, 1975; **1**, NMS 1897.32.292, paratype, natural mold of crown; **2**, NMS 1897.32.293, paratype, pluricolumnal bearing elongate radices. **3**, **Dimerocrinites pentlandicus** Brower, 1975, NMS 1885.26.78h, holotype, crown and proxistele. **4, Myelodactylus parvispinifer** (Brower, 1975), NMS 1897.32.285, holotype, pluricolumnal lacking crown, but showing the double recurvature of the proxistele (near top). **5, Pisocrinus campana** Miller, 1891, NMS 1970.43L, crown and column. **6, Dendrocrinus extensidiscus** Brower, 1975, NMS 1897.32.291, paratype, natural mold of crown and proxistele. All are latex casts of natural molds unless stated otherwise.*

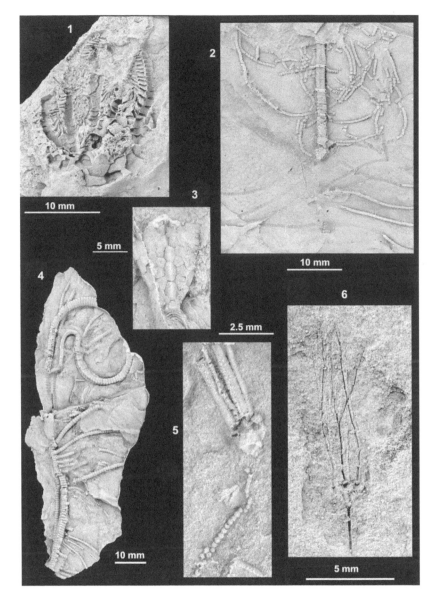

EVER SINCE RAMSBOTTOM: SILURIAN CRINOIDS OF THE BRITISH ISLES SINCE 1954

15

Stephen K. Donovan, David N. Lewis,
Rosanne E. Widdison, and Fiona E. Fearnhead

After the Lower Carboniferous, the greatest diversity of well-preserved crinoids in the Paleozoic of the British Isles is undoubtedly in the Silurian; yet the only comprehensive study of this fauna remains unpublished. W. H. C. "Bill" Ramsbottom (1926–2004), better known for his work on Carboniferous stratigraphy (Saunders, 2005; Holliday and Saunders, 2006), studied the systematics and biodiversity of Ordovician and Silurian crinoids of the British Isles for his Ph.D. (1953). His monograph of Ordovician crinoids appeared in 1961, yet only a fraction of his Silurian work was published as separate papers (Ramsbottom, 1950, 1951a, 1951b, 1952, 1954, 1958). The lack of a single monographic reference or comprehensive field guide to the Silurian crinoids is the most obvious gap in the literature of British and Irish Paleozoic echinoderms, emphasizing the importance of the monographic treatment of Ramsbottom's doctoral research.

British Paleozoic crinoids have been well served by monographic studies, most particularly those of the Ordovician (Ramsbottom, 1961) and Carboniferous (Wright, 1950–1960). Devonian and Permian marine deposits are of limited areal extent, because these parts of the succession are dominated by the widespread Old and New Red Sandstones, respectively, and related nonmarine–marginal marine deposits. Thus, crinoids are relatively rare from these intervals (see Lane et al., 2001; Donovan et al., 1986, for the most recent accounts of British Devonian and Permian crinoids, respectively). The Silurian is the odd unit out. Horizons bearing well-preserved crinoids are well known from the Llandovery, Wenlock, and Ludlow. However, crinoid studies on these are scattered through the literature and known only to the specialist. The purpose of the present chapter is not to correct this omission per se; rather, it is to introduce a wider audience to the full array of British Silurian crinoids and to highlight the importance of Ramsbottom's doctoral research. We provide a systematic summary of all the British Silurian Crinoidea that is based on the best available information (Tables 15.1–15.3; Fig. 15.1–15.3); we particularly want to emphasize the advances in knowledge that have occurred in the more than 50 years since Ramsbottom's (1953) thesis, which marked the beginning of the modern knowledge base of these echinoderms.

Except where explicitly stated, the abundant record of disarticulated columnals from the British Silurian is not considered. Specimens illus-

Introduction

In 1961 you published a monograph on British Ordovician Crinoidea, but since this is the standard work on the subject there seems no point in holding it against you.

Moore (1973, p. 562, at the time that Ramsbottom received the John Phillips Medal from the Yorkshire Geological Society)

trated herein are deposited in the Natural History Museum, London (BMNH); National Museums of Scotland, Edinburgh (NMS); and the British Geological Survey, Keyworth (BGS GSM). Other important collections are in the Dudley Museum, the Ludlow Museum, the Lapworth Museum of the University of Birmingham, the Sedgwick Museum of the University of Cambridge, the Hunterian Museum of the University of Glasgow, and the Oxford University Museum of Natural History. Although detailed locality and stratigraphic data are not provided in the present account, this may be gleaned from Cocks et al. (1971, 1992) and other references herein.

Notes on Ramsbottom (1953)

Bill Ramsbottom's Ph.D. thesis was a crucial contribution to echinoderm studies, documenting all of the crinoids then known from the lower Paleozoic of the British Isles, including descriptions of many new taxa. That such an important work was not published in its entirety is a poor comment on the Ph.D. system in Britain (echoing Halsted, 1988). The systematic account of the Ordovician crinoids was published as a volume essentially derived straight from the Ph.D. (Ramsbottom, 1961; see Donovan, 1992, p. 152, for a discussion of a "lost" Ramsbottom taxon). The importance of this work is easily demonstrated by examination of the contents. Until that time, there were only five nominal crinoid species recorded from the British Ordovician: *Ramseyocrinus cambriensis* (Hicks, 1873); *Merocrinus salopiae* Bather, 1896; *Balacrinus basalis* (M'Coy, 1851); *Diabolocrinus globularis* (Nicholson and Etheridge, 1880); and *Trochocrinites laevis* Portlock, 1843. To this meager total, Ramsbottom added 17 nominal species—a 340% increase in species diversity—including the first published account of crinoids from the Rawtheyan (Ashgill) Lady Burn Starfish Bed, an immediately pre–mass extinction echinoderm Lagerstätte (an earlier account of this deposit by James Wright was not published; Donovan, 1992, p. 152). Thus, Ramsbottom (1961) provided the basis for subsequent research in the British Isles; more than 70 species are now described from this interval (not including the many columnal morphotaxa; Donovan, 1986–1995; Donovan and Wright, 1995), making it one of the best documented areas for Ordovician crinoids globally. Yet none of this research was by Ramsbottom, who by this time had become a distinguished upper Paleozoic stratigrapher with the Institute of Geological Sciences (later British Geological Survey). It is his many contributions in this field for which he is best known internationally (e.g., Moore, 1973; Holliday and Saunders, 2006).

In contrast, the echinoderm Lagerstätte of the Much Wenlock Limestone Formation at the Wren's Nest, Dudley, West Midlands, has been well known since the nineteenth century, but new systematic studies have been few during the past 100 years (Widdison, 2001a, p. 10–12). Ramsbottom's Ph.D. dissertation was the first monograph to treat the British Wenlock crinoids and included descriptions of some new species. More importantly, Ramsbottom provided the first systematic account of the British Llandovery crinoids, which were essentially unknown at this time; this original account was referenced by Sevastopulo et al. (1989) in a review of the global distribution of echinoderms in the Silurian. Apart from the Telychian crinoids of the North Esk Inlier, Pentland Hills, south of Edinburgh, which were described by Brower (1975), most of our knowledge of British Llandovery crinoids derives from Ramsbottom's original account that has only recently being supplemented by new discoveries (Donovan and Lewis, 2005; Fearnhead and Donovan, 2005). Publication of Ramsbottom's Llandovery taxa has had to be undertaken by other authors (e.g., Donovan, 1993). More happily, he did publish an account of the Ludlow taxa that he recognized from the Lake District and the Leintwardine Starfish Bed of Shropshire (Ramsbottom, 1958), although a projected paper on *Scyphocrinites* from the British Isles never appeared (Donovan et al., 2007).

British Llandovery Crinoids

This account is adapted from Donovan and Harper (2003, p. 93–95). The Llandovery has a depauperate record of fossil echinoderms globally (Paul, 1982; Eckert and Brett, 2001). Table 15.1 lists the known occurrences of crinoids from the British Llandovery. The oldest pelmatozoan fauna from the British Silurian is from Haverfordwest, Dyfed, southwest Wales (Donovan, 1993, 1994), from either the top of the Haverford Mudstone Formation or the bottom of the Gasworks Sandstone Formation (or, less probably, from both). Cocks and Price (1975, table 1) and Cocks et al. (1971, fig. 2) placed the boundary between these two units in the *cyphus* graptolite Biozone, placing the fauna high in the lower Llandovery (Rhuddanian), although this has subsequently been revised to suggest that these units may lie low in the Middle Silurian *triangulatus* graptolite Biozone (Aeronian; Cocks et al., 1992, fig. 4). Although the higher strata of the Haverford Mudstone Formation have yielded Rhuddanian assemblages, the faunas of the Gasworks Sandstone Formation are stratigraphically undiagnostic. Thus, the boundary between these two units may be in the uppermost Rhuddanian, lowermost Aeronian or could represent the Rhuddanian–Aeronian boundary. Whatever their correlation, the crinoid fauna from Haverfordwest is the oldest and one of the most diverse described from the British Llandovery (Table 15.1). Although some taxa have been left in open nomenclature, at the familial level, this crinoid fauna is comparable to others from the Ashgill and younger Silurian. New specimens of cladids and camerates were recently collected by one of us (FEF) (Fearnhead and Donovan, 2005).

A slightly younger fauna from the Newlands Formation of the Craighead Inlier, near Girvan, Strathclyde, southwest Scotland, is dominated by distinctive columnals and pluricolumnals of the rhodocrinitid? *Floricolumnus* (col.) *girvanensis* Donovan and Clark, 1992. Similar columnals also occur in the Haverfordwest fauna and the bead bed of the Brassfield Formation (Aeronian) of Ohio (Ausich, 1996, fig. 17–5.11). Ramsbottom (1953, p. 250, pl. 12, fig. 6) described a poorly preserved crinoid cup from the Newlands Formation as *Dimerocrinites* sp. nov. He compared this specimen to *Dimerocrinites? vagans* Foerste, 1919, from the Brassfield Formation of Ohio of comparable age (=periechocrinid *incertae sedis* [sic] of Ausich, 1987). Recently, Ramsbottom's specimen has been redescribed as *Ptychocrinus mullochillensis* Fearnhead and Donovan, 2007.

Taxon	Locality*
DISPARIDA	
Calceocrinus turnbulli Donovan, 1993	a
Calceocrinus? sp. *in* Donovan (1993)	a
Pisocrinus campana S. A. Miller, 1891	h
Myelodactylus hibernicus Donovan and Sevastopulo, 1989	d
Myelodactylus parvispinifer (Brower, 1975)	h
Myelodactylus penkillensis Donovan and Sevastopulo, 1989	c
Myelodactylus convolutus Hall, 1852	g
CLADIDA	
Petalocrinus bifidus Bather MS *in* Fearnhead and Donovan (2007)	i
Dendrocrinus extensidiscus Brower, 1975	h
Dendrocrinus? gasworksensis Donovan, 1993	a
Segmentocolumnus (col.) *clarksoni* Donovan and Harper, 2003	e
SAGENOCRINIDA	
Cryptanisocrinus kilbridensis Donovan et al., 1992	e
Clidochirus? sp. *in* Donovan and Lewis (2005)	g
MONOBATHRIDA	
Macrostylocrinus silurocirrifer Brower, 1975	h
Macrostylocrinus? sp. indet.	a
Clematocrinus ramsbottomi Fearnhead and Donovan, 2007	f
DIPLOBATHRIDA	
Chaosocrinus ornatus Donovan, 1993	a
Dimerocrinites pentlandicus Brower, 1975	h
Ptychocrinus mullochillensis Fearnhead and Donovan, 2007	b
Ptychocrinus longibrachialis Brower, 1975	h
DIPLOBATHRIDA?	
Floricolumnus (col.) *girvanensis* Donovan and Clark, 1992	b
Floricolumnus (col.) cf. *girvanensis* Donovan and Clark, 1992	a
INCERTI ORDINIS	
Crinoid indet. sp. A *in* Donovan (1993)	a
Crinoid indet. sp. B *in* Donovan (1993)	a

Note.—Revised after Donovan and Harper (2003, table 1, fig. 4). All occurrences are Telychian (upper Llandovery) unless stated otherwise.

* Key to localities: a, top Haverford Mudstone Formation–basal Gasworks Sandstone Formation, Haverfordwest (Donovan, 1993, 1994) (high Rhuddanian or low Aeronian); b, Newlands Formation, Craighead Inlier, Girvan (Ramsbottom, 1953; Donovan and Clark, 1992; Fearnhead and Donovan, 2007) (Aeronian); c, Penkill Formation, Girvan (Donovan and Sevastopulo, 1989); d, Kilbride Formation, Lettershanbally, Co. Galway (Donovan and Sevastopulo, 1989); e, upper Finny School member, Kilbride Formation, Kilbride Peninsula, Co. Mayo (Donovan et al., 1992; Donovan and Harper, 2003); f, Damery Beds, Tortworth Inlier (Ramsbottom, 1953; Fearnhead and Donovan, 2007); g, Devil's Dingle, Buildwas, Shropshire (Donovan and Lewis, 2005); h, Reservoir Formation, North Esk Inlier (Brower, 1975); i, *Petalocrinus* Limestone, Woolhope Inlier (Ramsbottom, 1953; Fearnhead and Donovan, 2007).

The crinoid fauna of the North Esk Inlier, Pentland Hills, southwest of Edinburgh, was described by Brower (1975) (Fig. 15.1). Robertson (1989, p. 128) noted that well-preserved echinoderms become common toward the top of the Reservoir Formation and that crinoids come from two starfish beds in this part of the succession. Robertson (1989, p. 138) correlated the Reservoir Formation, along with the overlying Deerhope Formation and the lower member of the Wether Law Linn Formation, with the *crenulata* graptolite Biozone. Thus, the starfish beds appear to be in approximately the middle of the *crenulata* graptolite Biozone. The crinoid fauna is perhaps more typically Silurian than that from Haverfordwest, including *Myelodactylus* Hall, 1852, and common *Pisocrinus* de Koninck, 1858 (Robertson, 1989).

Other Llandovery occurrences are of limited diversity. Three have yielded members of the unusual disparid genus *Myelodactylus*. *Myelodactylus penkillensis* Donovan and Sevastopulo, 1989, was described on the basis of distinctive pluricolumnals from the Telychian Penkill Formation in the main Silurian outcrop at Girvan, southwest Scotland. In contrast, *Myelodactylus hibernicus* Donovan and Sevastopulo, 1989, is based on an external mold without counterpart of an essentially complete crinoid from the Kilbride Formation. It is probable that this specimen comes from the lower half of this formation, on the basis of the sandstone matrix (E. N. Doyle, personal commun.). The upper part of the same formation has yielded the external mold, once again without counterpart, of the sagenocrinid flexible *Cryptanisocrinus kilbridensis* Donovan et al., 1992, and the common columnal morphotaxon *Segmentocolumnus* (col.) *clarksoni* Donovan and Harper, 2003. Donovan and Lewis (2005) recorded *Myelodactylus convolutus* Hall, 1852, and the sagenocrinid *Clidochirus?* sp. from Devil's Dingle, Buildwas, Shropshire.

Further occurrences have yet to be published. In his Ph.D. dissertation, Ramsbottom (1953) described a new species of *Haplocrinus* Jaekel, 1895, from the Damery Beds of the Tortworth Inlier, north of Bristol. This species has been described as *Clematocrinus ramsbottomi* Fearnhead in Fearnhead and Donovan, 2007. Of potentially greater stratigraphic significance is the occurrence of distinctive *Petalocrinus bifidus* Fearnhead and Donovan, 2007, in a distinct marker band near the top of the Llandovery (Aldridge et al., 2002, fig. 37) in the Woolhope Inlier of the Welsh borderlands (Pocock, 1930) (Fig. 15.2.3).

However, despite the growing number of localities yielding complete Llandovery crinoids from the British Isles, only two of nine have produced even a moderate diversity of taxa, arbitrarily defined for this chapter as five or more identifiable taxa (Haverfordwest and the North Esk Inier). Further, seven of the nine occurrences are upper Llandovery (Telychian) (Table 15.1; Donovan and Harper, 2003, fig. 4), indicating that the interval (Rhuddanian and Aeronian) must still be considered a stratigraphic gap after the relative abundance of Ashgill crinoids from the British Isles.

Table 15.1. *Crinoids from the Llandovery of the British Isles, excluding undescribed fragmentary specimens.*

British Wenlock Crinoids

The diverse crinoids of the Much Wenlock Limestone Formation of the Silurian inlier at the Wren's Nest, Dudley, were described mainly during the nineteenth and early twentieth centuries, yet there is no comprehen-

Figure 15.2. *Typical British Silurian crinoids (Wenlock unless stated otherwise).* **1, Periechocrinus costatus** *Austin and Austin, 1843, BMNH 57249.* **2, Eucalyptocrinites decorus** *(Phillips, 1839), BMNH E1456.* **3, Petalocrinus bifidus** *Bather MS in Fearnhead and Donovan, 2007, BMNH E70060, Llandovery,* **Petalocrinus** *Limestone.* **4, Cicerocrinus elegans** *Sollas, 1900, BMNH 26071, Ludlow, Leintwardine Formation.* **5, Crotalocrinites verucosus** *(Schlotheim, 1820), BMNH 40204.* **6, Lecanocrinus baccus** *(Salter, 1873), BMNH 57162.* **7, Meristocrinus minor** *Springer, 1920, BMNH E5721.* **8, Cyathocrinites monile** *Salter, 1873, BMNH E1450.* **9, Temnocrinus tuberculatus** *(Miller, 1821), BMNH 57468.* **10, Ichthyocrinus pyriformis** *(Phillips, 1839), BMNH E5598.* **11, Calceocrinus serialis** *(Salter, 1859), BMNH E57427.* **12, Calpiocrinus intermedius** *Springer, 1920, BMNH 57470.* **13, Homalocrinus nanus** *(Salter, 1873), BMNH E1422.*

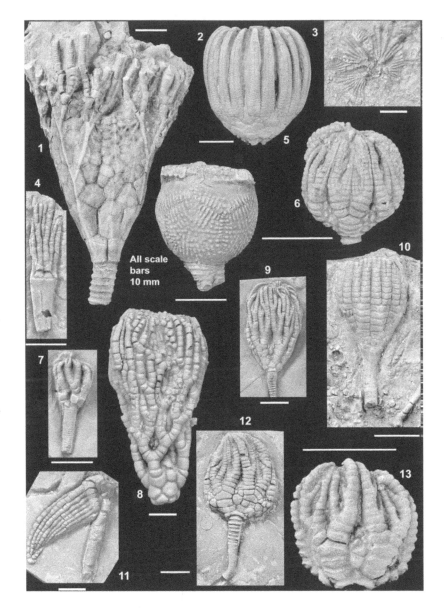

sive guide to this fauna. It is a superficially strange, but easily explained, feature of the British Silurian Crinoidea that the most diverse crinoid fauna (Table 15.2) comprising the best-preserved specimens (Fig. 15.2.1, 15.2.2, 15.2.5–15.2.13) has received relatively little attention during the past 100 years. Notable exceptions include two theses (Ramsbottom, 1953; Widdison, 2001a) and a few published papers. However, active discovery of new crinoids at Wren's Nest essentially ended when active quarrying of the limestone at this site ceased during the 1920s (Hardie, 1971; Cutler et al., 1990; Ausich et al., 1999, p. 43). Much of what has been published subsequently has focused on analysis of existing museum specimens.

Ramsbottom published a number of systematic papers on the British Wenlock Crinoidea (1950, 1951a, 1951b, 1952, 1954). Brower (1976) redescribed *Promelocrinus* Jaekel, 1902, from Dudley and used this to analyze its ontogeny (see also Cowen, 1981). Rozhnov (1981), Sevastopulo et al. (1989, p. 266), and Donovan and Sevastopulo (1989) revised the pisocrinids and myelodactylids, respectively. Other published papers have been more concerned with paleobiological aspects of the fauna (see below).

Crinoids are poorly known from the Ludlow of the British Isles (Table 15.3). A paper by Ramsbottom (1958), derived from his Ph.D. thesis, is the standard reference to the crinoids of this interval, in which he described eight taxa from two areas, Leintwardine in the type area of the Ludlow, in Herefordshire (Fig. 15.2.4), and the Lake District, northwest England. To these few taxa can be added *"Scyphocrinites" pulcher* (M'Coy, 1851) from North Wales, the Lake District and elsewhere (Donovan et al., personal

British Ludlow Crinoids

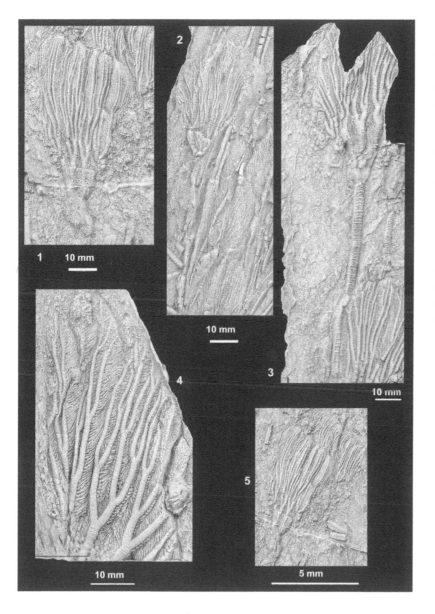

Figure 15.3. *"Scypho-crinites" pulcher* (M'Coy, 1851) from the Lower Ludlow of the Lake District, England (Donovan et al., personal data); *1, 4, 5,* BGS GSM 89997; *1, 4,* detail of free arms of another specimen; *5,* crown and crinoid debris. *2,* BMNH 40284, detail of slab preserving multiple crowns and associated proxisteles. *3,* BGS GSM 73997, well-preserved specimen retaining proximal part of long(?) column. All are latex casts of natural molds.

DISPARIDA	*Homalocrinus nanus* (Salter, 1873)
Myelodactylus fletcheri (Salter, 1873)	*Calpiocrinus intermedius* Springer, 1920
Myelodactylus ammonis (Bather, 1893)	*Temnocrinus tuberculatus* (Miller, 1821)
Myelodactylus sp. A *in* Donovan and Sevastopulo (1989)	*Icthyocrinus intermedius* Angelin, 1878
Pisocrinus pilula de Koninck, 1858	*Icthyocrinus phillipsianus* Springer, 1920
Chirocrinus fletcheri (Salter, 1873)	*Icthyocrinus pyriformis* (Phillips, 1839)
Synchirocrinus pugil (Bather, 1893)	*Lecanocrinus bacchus* (Salter, 1873)
Synchirocrinus inclinus (Ramsbottom, 1952)	*Hormocrinus anglicus* Springer, 1920
Synchirocrinus gradatus (Salter, 1873)	*Pycnosaccus bucephalus* (Bather, 1890a)
Calceocrinus serialis (Austin MS in Salter, 1859)	MONOBATHRIDA
Calceocrinus anglicus (Springer, 1926)	*Promelocrinus anglicus* Jaekel, 1902
CLADIDA	*Promelocrinus* sp. *in* Brower, 1976
Cyathocrinites monile Salter, 1873	*"Actinocrinus" wynnei* Baily, 1860
Cyathocrinites vallatus Bather, 1892b	*"Mariacrinus" flabellatus* Salter,
Cyathocrinites sp. nov. *in* Ramsbottom (1953)	*Periechocrinus costatus* Austin and Austin, 1843
Gissocrinus arthriticus (Phillips, 1839)	*Periechocrinus limonium* Salter, 1873
Gissocrinus capillaris (Phillips, 1839)	*Periechocrinus simplex* Salter, 1873
Gissocrinus cyrili Bouska, 1944	*Periechocrinus* sp. nov. *in* Ramsbottom
Gissocrinus goniodactylus (Phillips, 1839)	*Carpocrinus simplex* (Phillips, 1839)
Gissocrinus scoparius (Salter, 1873)	*Carpocrinus* sp. nov. A *in* Widdison (2001a)
Gissocrinus squamiferus (Salter, 1873)	*Carpocrinus* sp. nov. B *in* Widdison (2001a)
Crotalocrinites pulcher (Hisinger, 1840)	*Desmidocrinus pentadactylus* Angelin, 1878
Crotalocrinites verucosus (Schlotheim, 1820)	*Eucalyptocrinites decorus* (Phillips, 1839)
Crotalocrinites sp. nov. *in* Widdison (2001a)	*Eucalyptocrinites granulatus* (Lewis, 1847)
"Enallocrinus" punctatus (Hisinger, 1828) gen. nov. *in* Widdison (2001a)	*Eucalyptocrinites* sp. nov. *in* Widdison (2001a)
Mastigocrinus arboreus (Salter, 1873)	*Calliocrinus* cf. *beyrichianus* (Angelin, 1878)
Mastigocrinus quinquelobus (Bather, 1892a)	*Calliocrinus* sp. nov. *in* Ramsbottom (1953) and Widdison (2001a)
Bathericrinus ramosus (Bather, 1891b)	*Clonocrinus polydactylus* (M'Coy, 1849)
Dictenocrinus decadactylus (Salter, 1873)	*Macrostylocrinus granulatus* (Salter, 1873) nomen nudum [=*M. anglicus* Jaekel, 1918?]
Dictenocrinus pinnulatus (Bather, 1891b)	*Marsupiocrinites coelatus* Phillips, 1839
Dendrocrinus sp. nov. *in* Ramsbottom (1953)	*Clematocrinus retiarius* (Phillips, 1839)
Thenarocrinus callipygus Bather, 1890b	*Cordylocrinus pecten* (Salter, 1873) nomen nudum
Thenarocrinus gracilis Bather, 1891a	*Cordylocrinus* sp. nov. *in* Ramsbottom (1953)
TAXOCRINIDA	DIPLOBATHRIDA
Gnorimocrinus sp. nov. *in* Ramsbottom (1953)	*Lyriocrinus britannicus* Ramsbottom, 1950
Meristocrinus minor Springer, 1920	*Dimerocrinites decadactylus* Phillips, 1839
Meristocrinus nodulosus (Salter, 1873) nomen nudum *in* Ramsbottom (1953)	*Dimerocrinites speciosus* (Angelin, 1878)
SAGENOCRINIDA	*Dimerocrinites icosidactylus* Phillips, 1839
Sagenocrinites expansus (Phillips, 1839)	*Dimerocrinites uniformis* (Salter, 1873)

Note.—Excludes undescribed fragmentary specimens. Compiled principally from Ramsbottom (1952, 1953), Brower (1976), and Widdison (2001a), with generic revisions after Brower (1982) and Webster (2003). This list does not differentiate between specimens from the Wren's Nest, Dudley, and other Wenlock localities and horizons.

data) (Fig. 15.3). Yet there are surely more crinoids to be recognized from this interval. Disarticulated columnals and pluricolumnals showing a moderate morphological diversity are locally abundant, for example, in the type succession of the Ludlow Anticline; columnals occur in most units within this area (Siveter et al., 1989, fig. 30). Indeed, a mural in the Ludlow Museum illustrates numerous crinoids in the benthos of the Bringewood Group, yet none has been described from these strata. The fragmentary crinoid remains of this interval are awaiting systematic study.

British Pridoli(?) Crinoids

Ramsbottom (1953, p. 201–203, pl. 29, figs. 1–3, text-fig. 12A–C) described poorly preserved loboliths of *Scyphocrinites* Zenker, 1833, from Cornwall, which he considered to be "probably Silurian" (p. 201). These specimens, which are held in the BGS GSM collection, are incorrectly labeled as Wenlock on the basis of a lobolith misidentified as the rhombiferan cystoid *Pseudocrinites magnificus* Forbes, 1848 (=*Pseudocrinites bifasciatus* Pearce, 1843; Paul, 1967, p. 321, 322). Rather, the loboliths of *Scyphocrinites* indicate that these specimens are either Pridolian or lowermost Lochkovian (Haude, 1992). This remains the only, and admittedly tentative, identification of a Pridoli crinoid from the British succession.

Paleobiology

There are few studies of the more paleobiological aspects of the Silurian crinoids of the British Isles. Crinoids, particularly from the Wenlock of the Wren's Nest, Dudley, have provided data for the analysis of the paleobiogeography of Silurian crinoids (e.g., Holland, 1971; Sevastopulo et al.,

DISPARIDA
    Parapisocrinus cf. *sphaericus* (Rowley, 1904)
    Cicerocrinus elegans Sollas, 1900
    Cicerocrinus anglicus (Jaekel, 1900)
CLADIDA
    Gissocrinus ludensis (Sollas, MS) Ramsbottom, 1958
    Mastigocrinus bravoniensis Ramsbottom, 1958
TAXOCRINIDA
    Eutaxocrinus maccoyanus (Salter, 1873)
SAGENOCRINIDA
    Meristocrinus orbignyi (M'Coy, 1850)
MONOBATHRIDA
    Clematocrinus quinquepennis (Salter, MS) (Ramsbottom, 1958)
    "Scyphocrinites" pulcher (M'Coy, 1851)
    Scyphocrinites sp. *in* Ramsbottom (1953)

Note.—Data based mainly on Ramsbottom (1953, 1958), Webster (2003), and Donovan et al. (personal commun.), excluding undescribed fragmentary specimens.

Taxon	Llandovery	Wenlock	Ludlow	Pridoli
Disparida	7	10	3	0
Cladida	4	21	2	0
Taxocrinida	0	3	1	0
Sagenocrinida	2	10	1	0
Monobathrida	3	23	2	1?
Diplobathrida	4 + 2?	5	0	0
Incertae ordinis	2	0	0	0
Totals	24	72	9	1?

Table 15.4. *Summary of data in Tables 15.1–15.3 showing the numbers of crinoid species recognized from each major stratigraphic interval of the Silurian of the British Isles.*

1989). More general studies of the paleoautecology of Silurian pelmatozoans have again relied on our knowledge of Wenlock taxa (e.g., Cowen, 1981), most notably the wide-ranging review of Brett (1984), although Donovan and Sevastopulo (1989) considered all British Llandovery and Wenlock taxa in their discussion of the functional morphology of the myelodactylids (see also Donovan, 2006).

The only detailed study of the paleosynecology of British Silurian crinoids was by Watkins and Hurst (1977), who used slabs from the Wren's Nest in museums, particularly those that preserve multiple specimens, to determine relationships within the community of such a skeletal-rich carbonate environment. However, interactions between British Silurian crinoids and other organisms remain poorly known and have generally only been, at best, illustrated (e.g., Donovan and Harper, 2003, fig. 2C). An exception is the study by Widdison (2001b) of host specificity of some of the organisms that produced embedment structures in Wenlock crinoids.

Conclusions

Table 15.4 summarizes the distribution data discussed herein. Most groups have a peak in diversity in the Wenlock, particularly the cladids and monobathrids, but the disparids maintain a moderate diversity throughout. The total richness of approximately 100 species is undoubtedly a minimum estimate, which will grow as new sites are found, collected, and described. Although the largest number of species is from the Wenlock, mainly from the echinoderm Lagerstätte of the Much Wenlock Limestone Formation at the Wren's Nest, Dudley, it seems likely that future discoveries will be in the Llandovery and Ludlow, which remain understudied.

We hope that this chapter gives something of a flavor of Ramsbottom's (1953) unpublished Ph.D. thesis, his related publications, and their important position in influencing studies on British lower Paleozoic crinoids since the early 1950s. We rejoice in Ramsbottom's many contributions to paleontology and biostratigraphy, most particularly his monograph of the British Ordovician crinoids (Ramsbottom, 1961), yet we remain sad that a companion volume on the Silurian never appeared. We consider such a publication highly desirable. It remains a long-term goal of the authors to build on the legacy of Ramsbottom's original investigation. Until a more comprehensive study can be published, we present this chapter as a celebration of the British Silurian Crinoidea and Bill Ramsbottom's Silurian research, in which we provide the most comprehensive listing of known taxa published to date, and, we hope, a starting point for future studies.

Acknowledgments

The loan of specimens from the North Esk Inlier, Pentland Hills (Fig. 15.1), was kindly arranged by L. Anderson (NMS). We thank P. Crabb of the Photographic Unit, BMNH, for providing the images. The constructive comments of our reviewers, W. I. Ausich (The Ohio State University, Columbus) and C. E. Brett (University of Cincinnati), are gratefully acknowledged.

References

Aldridge, R. J., M. G. Bassett, E. N. K. Clarkson, C. Downie, C. H. Holland, P. D. Lane, R. B. Rickards, C. T. Scrutton, and C. Taylor. 2002. Telychian rocks in the British Isles, p. 43–72. *In* C. H. Holland and M. G. Bassett (eds.), Telychian Rocks of the British Isles and China. National Museums and Galleries of Wales, Cardiff.

Angelin, N. P. 1878. Iconographia Crinoideorum in Stratis Sueciae Siluricus fossilium. Samson and Wallin, Holmiae, 62 p.

Ausich, W. I. 1987. Brassfield Compsocrinina (Lower Silurian crinoids) from Ohio. Journal of Paleontology, 61:552–562.

Ausich, W. I. 1996. Phylum Echinodermata, p. 242–261. *In* R. M. Feldmann and M. Hackathorn (eds.), Fossils of Ohio. Ohio Department of Natural Resources Division of Geological Survey Bulletin, 70.

Ausich, W. I., S. K. Donovan, H. Hess, and M. J. Simms. 1999. Fossil occurrence, p. 41–49. *In* H. Hess, W. I. Ausich, C. E. Brett, and M. J. Simms, Fossil Crinoids. Cambridge University Press, Cambridge.

Austin, T., and T. Austin JR. 1843. Description of several new genera and species of Crinoidea. Annals and Magazine of Natural History, series 1, 11:195–207.

Baily, W. H. 1860. Palaeontological Notes, p. 10–14. *In* Explanations to Accompany Sheet 145 of the Maps of the Geological Survey of Ireland. Memoirs of the Geological Survey of Ireland.

Bather, F. A. 1890a. British fossil crinoids II—the classification of the Inadunata Fistulata. Annals and Magazine of Natural History, series 6, 5:310–334, 373–388, 485–486.

Bather, F. A. 1890b. British fossil crinoids III, *Thenarocrinus callipygus* gen. et sp. nov., Wenlock Limestone. Annals and Magazine of Natural History, series 6, 6:222–235.

Bather, F. A. 1891a. British fossil crinoids IV, *Thenarocrinus gracilis* sp. nov., Wenlock Limestone, and a note on *T. callipygus*. Annals and Magazine of Natural History, series 6, 7:35–40.

Bather, F. A. 1891b. British fossil crinoids V, *Botryocrinus*, Wenlock Limestone. Annals and Magazine of Natural History, series 6, 7:389–413.

Bather, F. A. 1892a. British fossil crinoids VI, *Botryocrinus quinquelobus*, sp. nov., Wenlock Limestone; and note on *Botryocrinus pinnulatus*. Annals and Magazine of Natural History, series 6, 9:189–194.

Bather, F. A. 1892b. British fossil crinoids VIII, *Cyathocrinus: C. acinotubus* Angelin and *C. vallatus* sp. nov. Annals and Magazine of Natural History, series 6, 9:202–227.

Bather, F. A. 1893. The Crinoidea of Gotland, Part 1, the Crinoidea Inadunata. Svenska Vetenskaps-Akademien, Handlingar (ny följd), 25(2), 182 p.

Bather, F. A. 1896. *Merocrinus salopiae*, n. sp., and another crinoid, from the Middle Ordovician of west Shropshire. Geological Magazine, 33:71–75.

Bouska, J. 1944. Rod *Gissocrinus* Angelin a jeho druhy v ceském siluru. Rozpravy II. Trídy Ceské Akademie, 53(44):1–12.

Brett, C. E. 1984. Autecology of Silurian pelmatozoan echinoderms, p. 87–120. *In* M. G. Bassett and J. D. Lawson (eds.), Autecology of Silurian Organisms. Special Papers in Palaeontology, 32.

Brower, J. C. 1975. Silurian crinoids from the Pentland Hills, Scotland. Palaeontology, 18:631–656.

Brower, J. C. 1976. *Promelocrinus* from the Wenlock at Dudley. Palaeontology, 19:651–680.

Brower, J. C. 1982. Phylogeny of primitive calceocrinids, p. 90–110. *In* J. Sprinkle (ed.), Echinoderm Faunas from the Bromide Formation (Middle Ordovician) of Oklahoma. University of Kansas Paleontological Contributions, Monograph, 1.

Cocks, L. R. M., and D. Price. 1975. The biostratigraphy of the Upper Ordovician and Lower Silurian of south-west Dyfed, with comments on the *Hirnantia* fauna. Palaeontology, 18:703–724.

Cocks, L. R. M., C. H. Holland, and R. B. Rickards. 1992. A revised correlation of the Silurian rocks in the British Isles. Geological Society Special Report, 21:1–32.

Cocks, L. R. M., C. H. Holland, R. B. Rickards, and I. Strachan. 1971. A correlation of Silurian rocks in the British Isles. Journal of the Geological Society, London, 21:103–136.

Cowen, R. 1981. Crinoid arms and banana plantations: an economic harvesting analogy. Paleobiology, 7:332–343.

Cutler, A., P. G. Oliver, and C. G. R. Reid. 1990. Wren's Nest National Nature Reserve: Geological Handbook and Field Guide. Dudley Leisure Services Department, Dudley, 29 p.

De Koninck, L. G. 1858. Sur quelques Crinoides palaeozoiques nouveaux de l'Angleterre. Bulletin de l'Academie Royale des Sciences, des Lettres et des Beaux-Arts de Belgique, series 2, 4:93–108.

Donovan, S. K. 1986–1995. Pelmatozoan columnals from the Ordovician of the British Isles (in three parts). Monograph of the Palaeontographical Society, 138 (568):1–68 [1986]; 142 (580):69–114 [1989]; 149 (597):115–193 [1995].

Donovan, S. K. 1992. New cladid crinoids from the Late Ordovician of Girvan, Scotland. Palaeontology, 35:149–158.

Donovan, S. K. 1993. A Rhuddanian (Silurian, lower Llandovery) pelmatozoan fauna from south-west Wales. Geological Journal, 28:1–19.

Donovan, S. K. 1994. The Late Ordovician extinction of the crinoids in Britain. National Geographic Research and Exploration, 10:72–79.

Donovan, S. K. 2006. Comment: crinoid anchoring strategies for soft bottom dwelling (Seilacher and MacClintock, 2005). Palaios, 21:406–408.

Donovan, S. K., and N. D. L. Clark. 1992. An unusual crinoid columnal morphospecies from the Llandovery of Scotland and Wales. Palaeontology, 35:27–35.

Donovan, S. K., and D. A. T. Harper. 2003. Llandovery Crinoidea of the British Isles, including description of a new species from the Kilbride Formation (Telychian) of western Ireland. Geological Journal, 38:85–97.

Donovan, S. K., and D. N. Lewis. 2005. Upper Llandovery (Telychian) crinoids (Echinodermata) of Devil's Dingle, Buildwas, Shropshire. Geological Journal, 40:343–350.

Donovan, S. K., and G. D. Sevastopulo. 1989. Myelodactylid crinoids from the Silurian of the British Isles. Palaeontology, 32:689–710.

Donovan, S. K., and A. D. Wright. 1995. Pelmatozoan (crinoid?) columnals from the Hirnantian (Ordovician, Ashgill) of Keisley, Cumbria, UK. Proceedings of the Yorkshire Geological Society, 50:229-238.

Donovan, S. K., E. N. Doyle, and D. A. T. Harper. 1992. A flexible crinoid from the Llandovery (Silurian) of western Ireland. Journal of Paleontology, 66:262–266.

Donovan, S. K., F. E. Fearnhead, and D. N. Lewis. 2007. Crinoids. In E. N. K. Clarkson, D. A. T. Harper,, and C. Taylor (eds.), Silurian Fossils of the Pentland Hills, Scotland. Palaeontological Association Field Guides to Fossils, 11:172–180.

Donovan, S. K., N. T. J. Hollingworth, and C. J. Veltkamp. 1986. The British Permian crinoid "Cyathocrinites" ramosus (Schlotheim). Palaeontology, 29:809–825.

Eckert, J. D., and C. E. Brett. 2001. Early Silurian (Llandovery) crinoids from the Lower Clinton Group, western New York State. Bulletins of American Paleontology, 360:1–88.

Fearnhead, F. E., and S. K. Donovan. 2005. Tales of the unpublished: "new" crinoids from the Llandovery of Britain. Geological Society of America Abstracts with Programs, 37(7):368

Fearnhead, F. E., and S. K. Donovan. 2007. New crinoids (Echinodermata) from the Llandovery (Lower Silurian) of the British Isles. Palaeontology, 50:905–915.

Foerste, A. F. 1919. Echinodermata of the Brassfield (Silurian) Formation of Ohio. Bulletin of the Science Laboratory, Denison University, 19:3–32.

Forbes, E. 1848. On the Cystideae of the Silurian rocks of the British Isles. Memoirs of the Geological Survey of the United Kingdom, 2(2):483–538.

Hall, J. 1852. Palaeontology of New York, volume 2, containing descriptions of the organic remains of the lower Middle Division of the New-York system. D. Appleton, Albany, New York, 362 p.

Halsted, L. B. 1988. The thesis that won't go away. Nature, 331:497–498.

Hardie, W. G. 1971. Wren's Nest Hill, Dudley, p. 7–9. In W. G. Hardie, G. M. Bennison, P. A. Garrett, J. D. Lawson, and F. W. Shotton, The Area around Birmingham (second revised edition). Geologists' Association Guide, 1.

Haude, R. 1992. Scyphocrinoiden, die Bojen-Seelilien im Hohen Silur-Tiefen Devon. Palaeontographica, A222:141–187.

Hicks, H. 1873. On the Tremadoc rocks in the neighbourhood of St. David's, South Wales, and their fossil contents. Quarterly Journal of the Geological Society of London, 29:39–52.

Hisinger, W. 1828. Anteckningar I Physik och Geognosie und resor uti Sverige och Norrige. volume 4. Sjunde Haftet, Stockholm, 260 p.

Hisinger, W. 1840. Anteckinger I Physik och Geognosie under resor uti Sverige och Norrige. Sjunde Haflet, Uppsala, p. 15, 45, 64, 65, 72.

Holland, C. H. 1971. Silurian faunal provinces? p. 61–76. In F. A. Middlemiss, P. F. Rawson, and G. Newall (eds.), Faunal Provinces in Time and Space. Geological Journal Special Issue, 4. Seel House Press, Liverpool.

Holliday, D. W., and W. B. Saunders. 2006. W. H. C. (Bill) Ramsbottom (1926–2004). Proceedings of the Yorkshire Geological Society, 56:55–56.

Jaekel, O. 1895. Beiträge zur Kenntniss der palaeozoischen Crinoiden Deutschlands. Paläontologisches Abhandlungen, 7(3):1–116.

Jaekel, O. 1900. Über einen neuen Pentacrinoideen-Typus aus dem Obsersilur. Deutsche Geologische Gesellscaft, Zeitschrift, 52:480–487.

Jaekel, O. 1902. Über verschiedene Wege phylogenetischer Entwicklung. 5th Verhandlungen der International Zoological-Congress Berlin, 1901:1058–1117.

Jaekel, O. 1918. Phylogenie und System der Pelmatozoen. Paläeontologisches Zeitschrift, 3(1), 128 p.

Lane, N. G., C. G. Maples, and J. A. Waters. 2001. Revision of Late Devonian (Famennian) and some Early Carboniferous (Tournaisian) crinoids and blastoids from the type Devonian area of north Devon. Palaeontology, 44:1043–1080.

Lewis, W. A. 1847. On a new species of *Hypanthocrinites* from the Wenlock Shale of Walsall. London Geological Journal, 1847:99–100.

M'Coy, F. 1849. On some new Palaeozoic Echinodermata. Annals and Magazine of Natural History, series 2, 3:244–254.

M'Coy, F. 1850. On some new genera and species of Silurian Radiata in the collection of the University of Cambridge. Annals and Magazine of Natural History, series 2, 6:270–290.

M'Coy, F. 1851. A systematic description of the British palaeozoic fossils in the geological museum of the University of Cambridge. Cambridge University Press, Cambridge, 184 p.

Miller, J. S. 1821. A Natural History of the Crinoidea or Lily-Shaped Animals, with observation on the genera *Asteria, Euryale, Comatula* and *Marsupites*. Bryan and Co., Bristol, 150 p.

Miller, S. A. 1891. The structure, classification, and arrangement of American Palaeozoic crinoids into families. Indiana Department of Geology and Natural History, 16th Annual Report (for 1889):302–326.

Moore, L. R. 1973. Presentation of the John Phillips Medal by the president to W. H. C. Ramsbottom, Ph.D. Proceedings of the Yorkshire Geological Society, 39:562–563.

Nicholson, H. A., and R. Etheridge Jr. 1880. A monograph of the Silurian fossils of the Girvan District in Ayrshire, with special reference to those contained in the "Gray collection," 1 (III). William Blackwood, Edinburgh, Echinoderms, 237–334.

Paul, C. R. C. 1967. The British Silurian cystoids. Bulletin of the British Museum (Natural History), Geology, 13:299–355.

Paul, C. R. C. 1982. The adequacy of the fossil record, p. 75–117. *In* K. A. Joysey and A. E. Friday (eds.), Problems of Phylogenetic Reconstruction. Academic Press, London.

Pearce, J. C. 1843. On an entirely new form of encrinite from the Dudley Limestone. Annals and Magazine of Natural History, 12:160.

Phillips, J. 1839. Organic remains, p. 670–675. *In* R. I. Murchison, The Silurian System, Part 2. John Murray, London.

Pocock, R. W. 1930. The *Petalocrinus* Limestone horizon at Woolhope (Herefordshire). Quarterly Journal of the Geological Society of London, 86:50–63.

Portlock, J. E. 1843. Report on the Geology of the County of Londonderry and of parts of Tyrone and Fermanagh. Dublin and London, 784 p.

Ramsbottom, W. H. C. 1950. A new species of *Lyriocrinus* from the Wenlock Limestone. Annals and Magazine of Natural History, series 12, 3:651–656.

Ramsbottom, W. H. C. 1951a. Two species of *Gissocrinus* from the Wenlock Limestone. Annals and Magazine of Natural History, series 12, 4:490–497.

Ramsbottom, W. H. C. 1951b. The type species of *Periechocrinites* Austin and Austin. Annals and Magazine of Natural History, series 12, 4:1040–1043.

Ramsbottom, W. H. C. 1952. Calceocrinidae from the Wenlock Limestone of Dudley. Bulletin of the Geological Survey of Great Britain, 4:33–48.

Ramsbottom, W. H. C. 1953. The British Lower Palaeozoic Crinoidea. Unpublished Ph.D. thesis, University of London, 290 p.

Ramsbottom, W. H. C. 1954. *Periechocrinus* versus *Periechocrinites*. Annals and Magazine of Natural History, series 12, 7:687–688.

Ramsbottom, W. H. C. 1958. British Upper Silurian crinoids from the Ludlovian. Palaeontology, 1:106–115.

Ramsbottom, W. H. C. 1961. The British Ordovician Crinoidea. Monograph of the Palaeontographical Society, 114(492):1–37.

Robertson, G. 1989. A palaeoenvironmental interpretation of the Silurian rocks in the Pentland Hills, near Edinburgh, Scotland. Transactions of the Royal Society of Edinburgh, Earth Sciences, 80:127–141.

Rowley, R. R. 1904. The Echinodermata of the Missouri Silurian and a new brachiopod. American Geologist, 34:269–282.

Rozhnov, S. V. 1981. Morskie lilie nadsemeistva Pisocrinacea. Akademiya Nauk SSSR, Paleontologicheskii Institut Trudy, 192, 127 p.

Salter, J. W. 1859. Appendix A, 531–552. In R. I. Murchison, Siluria (third edition). John Murray, London.

Salter, J. W. 1873. A catalogue of the collection of Cambrian and Silurian fossils, contained in the geological museum of the University of Cambridge. Cambridge University Press, Cambridge, 204 p.

Saunders, W. B. 2005. Reminiscences of W. H. C. Ramsbottom (1926–2004). Newsletter on Carboniferous Stratigraphy, 23:47–48.

Schlotheim, F. F. Von. 1820. Die Petrefactenkunde auf ihrem jetzigen Standpunkte durch die Beschreibung einer Sammlung versteinerter und fossiler überretse des Theirund Pflanzenreichs der Vorwelt erläutert. Beckersche Buchhandlung, Duchy of Saxe-Coburg-Gotha, Gotha, 437 p.

Sevastopulo, G. D., W. I. Ausich, and C. Franzén-Bengston. 1989. Echinoderms, p. 264–267. In C. H. Holland and M. G. Bassett (eds.), A Global Standard for the Silurian System. National Museum of Wales, Cardiff.

Siveter, D. J., R. M. Owens, and A. T. Thomas. 1989. Silurian Field Excursions: A Geotraverse across Wales and the Welsh Borderland. National Museum of Wales, Cardiff, 133 p.

Sollas, W. J. 1900. On two new genera and species of Crinoidea. Quarterly Journal of the Geological Society of London, 56:264–272.

Springer, F. 1920. The Crinoidea Flexibilia. Smithsonian Institution Publications, 2501, 486 p.

Springer, F. 1926. American Silurian crinoids. Smithsonian Institution Publications, 2871, 239 p.

Watkins, R., and J. M. Hurst. 1977. Community relations of Silurian crinoids at Dudley, England. Paleobiology, 3:207–217.

Webster, G. D. 2003. Bibliography and index of Paleozoic crinoids, coronates, and hemistreptocrinoids 1758–1999. Geological Society of America Special Paper, 363. <http://crinoid.gsajournals.org/crinoidmod>.

Widdison, R. E. 2001a. The Palaeobiology of Crinoids from the Much Wenlock Limestone Formation. Unpublished Ph.D. thesis, University of Birmingham, 270 p.

Widdison, R. E. 2001b. Symbiosis in crinoids from the Wenlock of Britain, p. 139–143. In M. Barker (ed.), Echinoderms 2000: Proceedings of the Tenth International Conference, Dunedin, 31 January–4 February 2000. Balkema, Lisse.

Wright, J. 1950–1960. A monograph of the British Carboniferous Crinoidea. Monograph of the Palaeontographical Society volume 1(1):1–24 [1950]; (2):25–46 [1951a]; (3):47–102 [1951b]; (4):103–148 [1952]; (5):149–190 [1954]; 2(1):191–254 [1955a]; (2):255–272 [1955b]; (3):273–306 [1956]; (4):307–328 [1958]; (5):329–347 [1960].

Zenker, J. C. 1833. Beiträge zur Naturgeschichte der Urwelt. Organische Reste (Petrefacten) aus der Altenburger Braunkohlen-formation, dem Blankenburger Quadersandstein, jenaischen bunten Sandstein und böhmischen Übergangsgebirge. Friedrich Mauke, Jena, Germany, 67 p.

Figure 16.1. *Schematic map of China showing the major Paleozoic plates and accretion zones. Junggar and Northeast China are Laurasian in origin. The other plates are Gondwanan.*

OVERVIEW OF PALEOZOIC STEMMED ECHINODERMS FROM CHINA

16

Johnny A. Waters, Sara A. Marcus, Christopher G. Maples, N. Gary Lane, Hongfei Hou, Zhouting Liao, Jinxing Wang, and Lujun Liu

Stemmed echinoderms were integral parts of benthic marine communities for much of the Paleozoic and have been used in studies of community evolution through time (e.g., Ausich and Bottjer, 1986). Studies of paleobiogeography have used stemmed echinoderms to show possible faunal realms and oceanic connections during the Paleozoic, and recent work has focused on the evolution of clades within Paleozoic stemmed echinoderms (e.g., Simms, 1994; Ausich, 1996, 1998). For these types of studies, it is imperative to have a taxonomically accurate, global database of stemmed echinoderms in order to have a complete picture of community evolution, paleobiogeography, and clade evolution through time. At present, however, stemmed-echinoderm communities are known in detail only from Europe and North America, are known to a lesser extent in Australia, Indonesia, and Africa, and largely are unknown from Asia and South America (Lane and Sevastopulo, 1987, 1990; Webster, 1973, 1977, 1986, 1988, 1990, 1998), although some notable exceptions have been published in recent years (Haude, 1995; Lane et al., 1996, 1997; Jell, 1999; Jell and Jell, 1999; Jell and Theron, 1999; Webster and Jell, 1999a, 1999b; Waters et al., 2003a; Webster et al., 2005). Nonetheless, interpretations of Paleozoic echinoderms do not fully take into account data from major areas of available Paleozoic outcrop belt—clearly a situation that can and should be addressed. Recent reports of fossils from the Early Cambrian of Chengjiang, China, as the most primitive echinoderms yet discovered (Shu et al., 2004; but see Smith, 2004, for an alternative view) underscore the need for increased emphasis on study of the Asian outcrop belts.

Our purpose in this chapter is to discuss the current state of knowledge of Paleozoic stemmed-echinoderm faunas from China in order to fill the large temporal and spatial gap in the echinoderm fossil record from part of the Asian continent. Because this paper is meant to be an overview, we discuss work in progress in China. We also review faunal studies from the published literature, some of which may be difficult to obtain in the United States. Taxa based only on stems are not included in our review because we have serious doubts about taxonomic consistency and uniformity of stem taxa in interbasinal and larger-scale studies, as well as doubts about the meaning of stem taxa as indicators of real taxonomic diversity and phylogenetic relationships.

Tien (1926) published one of the earliest reviews of echinoderms known from Asia. In addition to providing the first published descriptions

Introduction

of Chinese crinoids, Tien (1926) and Tien and Mu (1955) also listed known echinoderm occurrences from China and compared these echinoderms with other known faunas. Later papers on Asian echinoderm occurrences are listed in Bassler and Moody (1943) and Webster (1973, 1977, 1986, 1988, 1990, 2003), which provide taxonomic compilations of echinoderms with brief notes on geographic occurrences. Our discussion of Asian echinoderms listed in these sources is not intended to be exhaustive, and we have made no attempt to discuss every paper compiled by Webster describing isolated occurrences of stemmed echinoderms in China.

Unfortunately, seemingly well-documented faunas reported from the Himalayas (Gupta and Webster, 1976, and other citations with V. J. Gupta as coauthor) cannot be trusted, given the aspersions cast on the validity of papers authored or coauthored by V. J. Gupta as noted by Talent (1989), Ahluwalia (1989), and Webster (1991). Therefore, these reports were excluded from our analysis, further reducing the already slim database of known echinoderms from the Asian continent. As noted above, we also have not included studies of occurrences of crinoid stems, which are ubiquitous in certain facies but may not convey taxonomic information at a generic level.

Despite the large number of paleontologists in China, few Chinese scientists specialize in the study of Paleozoic stemmed echinoderms, and few collections of Chinese echinoderms exist in museums. Consequently, reports of Paleozoic Chinese echinoderms have been relatively rare and surely underestimate their actual diversity.

Tectonic Overview of China

China is a mosaic of plate fragments, microcontinents, and accretionary wedges. China contains parts or all of numerous substantial continental blocks, including Tarim, North China (=Sino-Korea), South China, Junggar, Kazakhstania, Angara (=Siberia), Indochina, and Sibumasu (=Shan Thai, Malaya), as well as smaller continental fragments and accretionary complexes that have coalesced since the middle Paleozoic (McKerrow and Scotese, 1990; Nie et al., 1990; Sengör et al., 1993; Van der Voo, 1993; Metcalfe, 1996; Torsvik and Cocks, 2004) (Fig. 16.1). China also contains the geologic history of the convergence between terranes having origins to the north of the Paleo-Tethys (Junggar, Kazakhstania, Angara) with those terranes (Tarim, North China, South China, Indochina, and Sibumasu) that rifted away from Gondwana in the middle Paleozoic. Echinoderm fossils in this diverse collage generally occur in areas of platform and shelf sedimentation on the margins of larger continental blocks such as North China, South China, and Sibumasu. However, echinoderms also have been collected from sedimentary accretionary-wedge facies on what were active plate margins between Kazakhstania–northern Junggar and southern Junggar-Tarim. Because Chinese terranes were situated on both the northern and southern margins of the Paleo-Tethys, they have complex temporal, spatial, and climatic relationships (Fig. 16.2). Many echinoderms tend to be highly endemic; thus, their geographic distribution within faunas can play extremely important roles in delineating paleogeographic

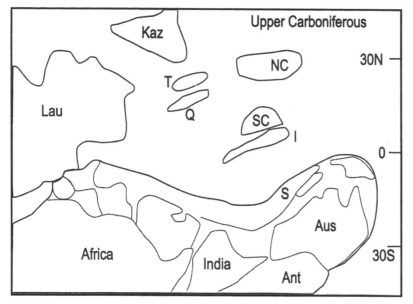

Figure 16.2. *Series of paleogeographic maps modified from Metcalfe (1996) and Torsvik and Cocks (2004) showing the position of the Chinese plates during the Ordovician, Late Devonian and Upper Carboniferous. Plates with significant echinoderm faunas are marked with an asterisk. Abbreviations: Ant, Antarctica; Aus, Australia; Kaz, Kazakhstania; I, Indochina; NC, North China; Q, Qiangtang; S, Sibumasu; SC, South China; and T, Tarim.*

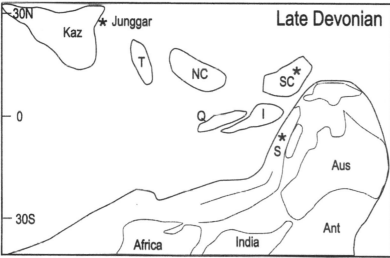

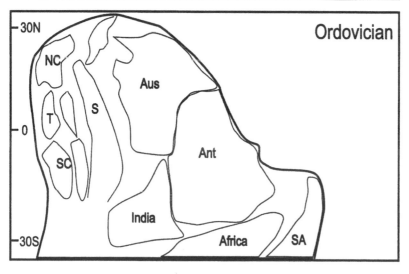

positions of these terranes through time. Consequently, we discuss the Chinese echinoderm faunas in the contexts of the terranes in which they are occur in order to facilitate paleobiogeographical comparisons with contemporaneous faunas around the world.

Junggar Terrane Faunas

Carroll et al. (1990) have suggested that Kazakhstania sutured to Siberia (=Angara), forming a cratonic area in what is now northern Xinjiang, during the Late Devonian to Early Carboniferous. Rocks of these ages from the Junggar Terrane represent near-arc, basin-fill sediments that developed on top of older Paleozoic seafloor that was trapped during suturing of the Junggar Terrane to the Kazakhstania-Siberia cratonic block. The near-arc, basin-fill facies from Junggar likely represent the eastern extension of similar island-arc assemblages in southern Kazakhstania. Paleomagnetic data placed the southern margin of the Kazakhstania-Junggar Terrane at 40° north latitude in the Late Devonian (McKerrow and Scotese, 1990; Li et al., 1991), although Torsvik and Cocks (2004) place the terranes in northern tropical latitudes. After Kazakhstania-Siberia-Junggar coalesced, the Tien Shan Ocean (Palaeo-Tethys of Nie et al., 1990; Junggar Ocean of Carroll et al., 1990; part of the Turkestan Ocean of Sengör et al., 1993) developed between Kazakhstania-Siberia-Junggar and the Tarim continental block. Tarim then sutured with Kazakhstania-Siberia-Junggar in latest Permian time, forming the Tien Shan suture zone.

HONGGULELENG FORMATION. We have collected an extensive, diverse Late Devonian echinoderm fauna from the Hongguleleng Formation, Junggar Basin, Xinjiang-Uygur Autonomous Region (Fig. 16.2). The Hongguleleng Formation is part of a Late Devonian accretionary wedge that formed along the southern margin of Kazakhstania-Junggar. The fauna from the Hongguleleng Formation is characterized by benthic marine invertebrate fossils and is dominated by brachiopods, corals, trilobites, crinoids, blastoids, and bryozoans (Liao and Cai, 1987; Sartenaer and Xu, 1989; Soto and Lin, 2000). Although comprehensive taxonomic studies of the faunal elements are ongoing, the total diversity of the Hongguleleng will likely exceed 100 genera.

Monographic taxonomic studies of the echinoderm collections made from the Hongguleleng Formation (Lane et al., 1997; Waters et al., 2003a) indicate that crinoids (Table 16.1) consist of one new family, 21 genera (four of which are new), and 32 species (29 of which are new). The faunas also contain eleven genera of blastoids (nine of which are new) and 13 species (all of which are new). Most of the crinoid genera are reported from Asia for the first time. The faunas are significant because they are the most abundant and diverse Famennian echinoderms known and because they suggest that rebound from Late Devonian extinction events happened soon after the events. In addition, several of the taxa imply terrane proximity in areas where these data are otherwise sparse.

Agathocrinus Schevchenko, 1967, is a member of the Family Parahexacrinidae, which is a family of specialized Devonian crinoids previously considered endemic to Kazakhstania (Lane et al., 1997). Its presence in this fauna provides some of the strongest evidence yet of the close biogeographic ties between these two areas during the Late Devonian. *Eutaxocrinus* Springer, 1906, is a flexible crinoid that is about equally common in rocks of Devonian and Early Carboniferous time and is also about equally common, in terms of species, in Europe and in North America. Lane et al. (1997) first reported *Eutaxocrinus* from Asia and showed the affinity between this fauna and those of the same ages previously reported from North America and Europe.

Conodonts identified by Zhao and Wang (1990) indicate an early to middle Famennian age (*Palmatolepis crepida* Biozone to *P. marginifera* Biozone) in the lower part of the Hongguleleng Formation, ranging as high as the latest Famennian (*Siphonodella praesulcata* Biozone) in the upper part of the Formation. Coral faunas reported by Soto and Lin (1997, 2000) from the Hongguleleng Formation have strong affinities with Famennian corals from Poland. The echinoderm faunas that we have collected have strong affinity with Famennian, Etroeungtian (Strunian or latest Famennian), and Early Missisippian European and North American faunas, which supports the interpretation of a Famennian age for the Hongguleleng Formation, as determined by Zhao and Wang (1990) and Soto and Lin (2000).

The affinities of this echinoderm fauna are important because the latest Famennian age assigned to the upper portion of the Hongguleleng (where the echinoderms were collected) has been questioned by Xia (1996, 1997), who suggested that the entire Hongguleleng Formation is actually late Frasnian to early Famennian in age. We do not accept Xia's (1996, 1997) assignment of late *P. rhenana* to middle *P. crepida* Biozone for the entirety of the Hongguleleng Formation for several reasons. First, we are uncertain where some of Xia's (1996, 1997) conodont samples were actually collected. This area is not straightforward structurally, and in the Boulongour area, the Hongguleleng Formation is deformed into a recumbent syncline. One limb of the syncline has a stratigraphically normal sequence, and one limb has a stratigraphically reversed sequence (Hou et al., 1993). It is apparent that Xia (1997) recognized that his stratigraphic section was reversed. However, he makes references to additional collections made from the other limb of the syncline, but there is no measured section for these samples. We also are unsure as to exactly where in the section these other samples were collected, or on what basis these samples were correlated with Xia's (1997) measured section.

Final determination of the age of the Hongguleleng Formation will depend on recollection of the type locality within a better-defined stratigraphic and temporal framework. A working group headed by John Talent from Australia recollected the sections of the Hongguleleng Formation in 2005 with this objective in mind. Although samples are still being processed for conodonts, fieldwork at the type section of the Hongguleleng Formation did call into question the assignment of a late Famennian age for the top of the section.

Determination of the age of the Hongguleleng Formation is critical for a variety of reasons. If the formation spans the Frasnian–Famennian

Table 16.1. (on following pages) *Paleozoic echinoderms found in China by stratigraphic position and plate.*

Era	Junggar	South China	North China	Sibumasu
Permian	—	Unidentified blastoid	—	—
	—	*Deltoblastus*	—	—
	—	—	—	—
Pennsylvanian	*Platycrinites*	*Amphipsalidocrinus*	*Erisocrinus*	*Texacrinus*
	Actinocrinites	*Kallimorphocrinus*	*Sinocrinus*	—
	Sinocrinus	*Neolageniocrinus*	*Mathericrinus*	—
	Graphiocrinus	*Dichostreblocrinus*	*Delocrinus*	—
	Cromyocrinus	*Litocrinus*	*Platycrinites*	
	Petschoracrinus	*Passalocrinus*	—	—
	Paracatillocrinus	*Allagecrinus*	—	—
	Paragaricocrinoid	Dichocrinoids	—	—
	Hexacrinid	Poteriocrines	—	—
	Acrocrinoid	Spiraculate blastoid	—	—
	—	Codiacrinoids	—	—
	—	Flexible crinoids	—	—
Mississippian	—	*Guilinocrinus*	*Mesoblastus*	*Mesoblastus*
	—	—	—	*Cribanocrinus*
	—	—	—	*Rhodocrinites*
	—	—	—	*Gilbertsocrinus*
	—	—	—	*Yunnanocrinus*
	—	—	—	*Amphoracrinus*
	—	—	—	*Pimlicocrinus*
	—	—	—	*Stomiocrinus*
	—	—	—	*Platycrinites*
	—	—	—	*Brahmacrinus*
	—	—	—	*Synbathocrinus*
	—	—	—	*Cyathocrinites*
	—	—	—	*Parabarycrinus*
	—	—	—	*Bollandocrinus*
	—	—	—	*Exaetocrinus*
	—	—	—	*Texacrinus*
	—	—	—	*Sinorbitremites*
Devonian	*Junggaroblastus*	*Devonoblastus*	—	*Shidianocrinus*
	Orophocrinus	—	—	*Melocrinites*
	Emuhablastus	—	—	*Changninocrinus*
	Tripoblastus	—	—	*Haplocrinites*
	Uyguroblastus	—	—	*Ovalocrinus*
	Sinopetaloblastus	—	—	*Quasicydonocrinus*
	Breimeriblastus	—	—	*Megaradialocrinus*
	Conoblastus	—	—	*Petalocrinus*
	Hyperoblastus	—	—	*Cupressocrinites*
	Houiblastus	—	—	*Breimeriblastus*

Era	Junggar	South China	North China	Sibumasu
	Xinjiangoblastus	—	—	Eutaxocrinus
	Taxocrinus	—	—	Parascyphocrinites
	Euonychocrinus	—	—	Scyphocrinites
	Forbesiocrinus	—	—	—
	Athabascocrinus	—	—	—
	Hexacrinites	—	—	—
	Agathocrinus	—	—	—
	Abactinocrinus	—	—	—
	Actinocrinites	—	—	—
	Chinacrinus	—	—	—
	Deltacrinus	—	—	—
	Euspirocrinidae indet.	—	—	—
	Parisocrinus	—	—	—
	Bridgerocrinus	—	—	—
	Logocrinus	—	—	—
	Julieticrinus	—	—	—
	Cosmetocrinus	—	—	—
	Sostronocrinus	—	—	—
	Amadeusicrinus	—	—	—
	Grabauicrinus	—	—	—
	?Graphiocrinus	—	—	—
	Holcocrinus	—	—	—
Silurian	—	Caelocrinus	Dazhucrinus	Pisocrinus
	—	Dazhucrinus	—	Spirocrinus
	—	Petalocrinus	—	Scyphocrinites
	—	Sinopetalocrinus	—	Carolicrinus
	—	Spirocrinus	—	Marhoumacrinus
	—	Pisocrinus	—	—
	—	Neoarchaeocrinus	—	—
Ordovician	—	—	—	Caryocrinites
	—	—	—	Paracaryocrinites
	—	—	—	Heliocrinites
	—	—	—	Echinosphaerities
	—	—	—	Sphaeronites
	—	—	—	Aristocystites
	—	—	—	Sinocystis
	—	—	—	Ovacystis
Cambrian	—	Sinoeocrinus	—	—

Note.—Data complied from a variety of sources contained in Webster (2003), the online crinoid bibliography. Taxonomic assignments have not been reevaluated beyond that contained in Webster or other printed references. Major localities and faunas are discussed in the text.

boundary, as suggested by Xia (1997), then the Frasnian-Famennian extinction event is contained within the formation. There is little evidence observed on the outcrop of a major disruption of sedimentation associated with this event. If the majority of the formation is early Famennian in age, rather than late Famennian as previously thought, then the observed taxonomic and morphologic patterns of echinoderm diversification happened even more rapidly after the extinction event than previously interpreted.

QIJIAGOU FORMATION. During fieldwork in Xinjiang-Uygur Autonomous Region, China, we visited exposures of the Qijiagou Formation (Moscovian; Upper Carboniferous) exposed in the Taoshigo Valley near the village of Daheyan, north and west of Turpan (Lane et al., 1996). This locality, on the southern flanks of the Bogda Shan at the northwestern margin of the Turpan Basin, was also visited as part of Excursion 4 of the Eleventh International Congress of Carboniferous Stratigraphy and Geology held in 1987 (Liao et al., 1987).

The stemmed-echinoderm fauna from the Qijiagou Formation seems to have its closest affinities with Moscovian-age crinoids described from Russia by Yakovlev and Ivanov (1956) (Lane et al., 1996). It is dominated by species of the camerate crinoid *Platycrinites* Miller, 1821. Other camerate crinoids in the fauna include a paragaricocrinid, *Actinocrinites* Miller, 1821, a hexacrinitid, and an acrocrinoid. Previously, the only other occurrence of acrocrinoids outside North America was the occurrence of *Springeracrocrinus* Moore and Strimple, 1969, from Moscovian-age rocks of Russia and a late Namurian or early Westphalian occurrence from Queensland (Webster and Jell, 1999a). In addition to the camerates, several advanced cladids are present that are more typical of many European and North American Late Carboniferous crinoid faunas. These advanced cladids include an erisocrinoid, possibly *Sinocrinus* Tien, 1926; *Graphiocrinus*, de Koninck and Le Hon, 1854; and *Cromyocrinus* Trautschold, 1867. *Petschoracrinus* Yakovlev, 1928, an agassizocrinoid represented by partly fused infrabasal cones, is present, as well as a single radial plate with an angustary facet that may represent a cyathocrinoid. *Paracatillocrinus* Wanner, 1916, a catillocrinoid, and an unknown genus of troosticrinid blastoid also occur in this fauna.

The Qijiagou Formation consists of carbonates, porphyritic diorites, and tuffaceous conglomeratic sandstones (Liao et al., 1987). The crinoid fauna is preserved in a series of graded conglomeratic sandstones that overlie a carbonate mound and contain clasts up to 3 m in dimension. Lane et al. (1996) postulated that the crinoids may have been swept off of the carbonate mound and deposited as debris-flow bed load on a paleoslope with the volcanic debris. The dominant clasts found with the crinoids are porphyritic andesite pebbles and cobbles ranging in size from 2 to 20 cm, suggesting that they formed in an actively eroding island-arc complex. This is consistent with interpretations that the Tien Shan Ocean, which lay between Kazakhstania-Siberia-Junggar and Tarim during Late Carboniferous time (Fig. 16.2), was bounded by emergent volcanic arcs. The Bogda Shan, which defines part of the northern margin of this paleo-ocean, consisted of active emergent arcs and associated sedimentary basin fills (Liao et al., 1987). The arcs contributed thousands of meters of volcanics and volcaniclastic debris during deposition of the Qijiagou Formation and other Late Carboniferous units in this area.

South China Terrane

The South China terrane consists of Precambrian basement with Phanerozoic sediments that have been deformed by multiple subsequent orogenies (Van der Voo, 1993). The South China block originated on the northern margin of Gondwana (Metcalfe, 1996) but had rifted away by the end of the Devonian. Nie (1994) concluded that the Devonian to Triassic carbonate platform on the South China block recorded initial rifting from Gondwana and passive-margin sedimentation associated with continental separation. Available data suggest that the South China block was equatorial during the Carboniferous and Permian but was neither permanently attached to Gondwana during the Paleozoic nor to Pangaea during the late Paleozoic to early Mesozoic (Metcalfe, 1996). Torsvik and Cocks (2004) offered alternate interpretations of the detailed placement of South China during the Paleozoic, but their reconstructions are consistent with a broad equatorial placement of the terrane during the late Paleozoic. Our faunas support the interpretation of warm equatorial climates during the late Paleozoic.

KAILI FORMATION. Parsley and Zhao (2002, 2006) and Zhao and Parsley (2003) reported on abundant, well-preserved populations of eocrinoids from the Kaili Formation (basal Middle Cambrian) in Guizhou Province. The Kaili Formation contains a Lagerstätte with significant soft-bodied preservation that is unique among similar Middle Cambrian biotas in having abundant gogiid eocrinoid echinoderms. A single species of *Sinoeocrinus* Zhao et al. (1994) has been identified by Parsley and Zhao (2002) as an unidentified edrioasteroid.

SILURIAN, FORMATION UNDESIGNATED. A Silurian crinoid fauna, with the genera *Caelocrinus* Xu, 1962; *Dazhucrinus* Mu and Wu, 1974; *Petalocrinus* Weller and Davidson, 1896; *Sinopetalocrinus* Mu and Lin, 1987; *Spirocrinus* Mu and Wu, 1974; *Pisocrinus* de Koninck, 1858; and *Neoarchaeocrinus* Strimple and Watkins, 1955, has been described from the Llandoverian formations in Guizhou and Sichuan by Mu (1950), Mu and Wu (1974), and Mu and Lin (1987). We have not visited these localities and have no further information about these faunas.

LUOCHENG FORMATION. We have collected a diverse echinoderm fauna from two localities in the upper part of the Luocheng Formation (late Early Carboniferous) near Xinxu, Guangxi Province. The Luocheng Formation is approximately 50 to 100 m of gray-black mudstones to siltstones, with siderite occurring in nodules and layers throughout the section. The Luocheng Formation is exposed in a series of roadcuts, quarries, and natural outcrops along an approximately 8-km-long, east-west transect in a tectonically complex area. There are virtually no fossils found in the lower portion of the Luocheng Formation. The only fossils in the basal 50 m of the formation were the trace fossil *Planolites* Nicholson, 1873, and one specimen of the brachiopod *Crurithyrus* George, 1931. Sedimentary structures observed include laminations and rare bedding-plane unidirectional ripple marks and bidirectional (oscillating) ripple marks. The exposures are folded, tilted, and are in fault contact with overlying strata in some portions of the outcrop area, but are in stratigraphic continuity in other areas.

Contained within the upper portion of Luocheng Formation is a single, fossil-rich interval that contains the following exceptionally well-preserved, diminutive faunal assemblage (in decreasing order of abundance): microgastropods (approximately 12 genera), bivalves (one species), crinoids, ostracodes, foraminiferans, conularids, brachiopods, blastoids, echinoids, bryozoans, cephalopods, rostroconchs, vertebrate remains, and corals. The Xinxu 1 locality also preserves a hardground, which is a synsedimentary lithified carbonate seafloor often marked by the presence of borers and encrusters. The hardground at Xinxu is represented by dense micritic and sideritic clasts that measure 5 to 10 cm in the long axis and are 1 to 2 cm thick and 2 to 10 cm in the shorter axis. These clasts are bored on both sides by a *Trypanites*-like trace maker, and they are encrusted by trepostome bryozoans, crinoid holdfasts, and corals that are usually found on both sides of the clasts.

A second measured section (Xinxu 2) of the Luocheng Formation, where the fossiliferous zone is well exposed, is located approximately 8 km to the west of Xinxu 1, Brick Pit section (discussed above). Xinxu 2 has an approximately 1-m-thick fossiliferous layer exposed in the upper portion of the Luocheng Formation. This fossiliferous zone is divided into two subzones: a coral subzone and a trilobite subzone. The lower fossiliferous subzone at Xinxu 2, the coral subzone, is dominated by the coral *Cladochonus* M'Coy, 1847, which is from an in situ mat. The coral subzone also contains large (> 4-cm-diameter stem), in situ crinoid holdfasts, and nonechinoderm faunas, including orthocone cephalopods, gastropods, bivalves, brachiopods, and trilobites. Echinoderms include microcrinoids, small disparid crinoids from the Family Allagecrinidae, small cladid cyathocrinids, and fragments of cladid poteriocrinids, flexible and camerate crinoids, blastoids, ophiuroids, and echinoids (Table 16.1).

The Luocheng Formation was interpreted to have been deposited in a brackish-water, nearshore setting (Kuang et al., 1987), but one having strong marine influences for periods of time to establish such diverse benthic marine assemblages (Marcus, 1993).

TANGBAGOU FORMATION. Chi (1943) described the blastoid *Mesoblastus* Etheridge and Carpenter, 1886, from Dushan, Guizhou Province, from the Kolaoho Series or the lower part of the Aikuan Group (Lower Carboniferous). Although Chi's specimens were assigned Geological Survey of China collection numbers, we were unable to ascertain these specimens' whereabouts. However, we do have specimens that were collected by Wang Shubei, Chengdu Institute of Geology and Minerals. These specimens come from the Qilinzhai Reservoir section, 7 km west of Dushan City, Guizhou Province. This locality is near, but not the same as, Chi's (1943) locality. We visited this locality in May 1993, and despite the help of several enterprising young Chinese students who were on their school lunch break, we were able to find only one additional specimen.

At the Qilinzhai Reservoir locality, blastoids occur in unit 2 of the Tangbagou Formation, Carboniferous (Tournaisian), *Pseudouralinia* Coral Biozone. The Tangbagou Formation, which overlies the Gelaohe Formation and underlies the Xiangbai Formation, is divided into 10 beds at this locality. Bed 1 of the Tangbagou Formation consists of approximately 12 m of weakly dolomitized micrite, forming the base of the section. Bed 2 overlies Bed 1, is approximately 50 m thick, and consists of micritic limestones and interbedded shales containing many fossils, including brachiopods, corals, bivalves, trilobites, and echinoderms. This part of the Tangbagou Formation at the Qilinzhai Reservoir section was previously placed in the Kolaoho Formation, but it is now included in the Tangbagou Formation because of the occurrence of the coral *Pseudouralinia* Poty and Xu, 1996. Our specimen, those collected by Wang Shubei, and the specimens collected by Chi probably came from similar stratigraphic horizons from localities only a few kilometers apart. The specimens do not belong to the genus *Mesoblastus*, but represent a new genus of the family Schizoblastidae Etheridge and Carpenter, 1886, which will be described elsewhere.

LONGHUIXIAN, FORMATION UNKNOWN. Luo (1984) reported an occurrence of *Deltoblastus* Fay, 1961, from the Middle Permian of Longhuixian, Hunan (Fig. 16.2), south-central China. Three blastoid specimens were collected from a dark limestone at this locality. We have not visited this site and attempts to locate Luo's collection have been unsuccessful. However, on the basis of our evaluation of the figures in Luo (1984), we have no reason to doubt the generic assignment of the specimens to *Deltoblastus*. If these specimens do belong to *Deltoblastus*, they represent one of the few occurrences of the genus outside Timor, where it is the dominant blastoid in the fauna. Yakovlev and Faas (1938) reported specimens of *Schizoblastus* cf. *permicus*, now *Deltoblastus permicus*, from the Sosio Limestone in Sicily. Partial specimens of blastoids questionably assigned to *Deltoblastus* have been described from Sicily by Gupta and Webster (1976). More recently, Webster and Sevastopulo (2001) have reported *Deltoblastus* from Oman. A previous report of *Deltoblastus* from India (Gupta and Webster, 1976, 1980), as well as the Sicily report, has been rejected by Webster (1991).

CHANGHSING FORMATION. Chen et al. (2004) described ophiuroids and figured a blastoid from Permian–Triassic boundary beds from Zhejiang Province from sections located approximately 40 km from the Permo–Triassic Global Stratotype section at Meishan. The blastoid was collected in the upper meter of the Changhsing Formation, which is uppermost Permian (Changhsingian) in age. Although we have neither visited these sections nor examined the blastoid material, the occurrence (if confirmed) is significant in that it represents the youngest known occurrence of a blastoid and extends the range of the class to the Permo–Triassic boundary.

Sino-Korea to Siberian Trend (North China Terrane)

Like the South China block, the North China block also consists of Phanerozoic sediments overlying Precambrian basement and deformed by many orogenic events (Van der Voo, 1993). Plots of pre-Permian paleolatitudes for North China, as well as lithofacies studies and other available data, support an equatorial climate for North China in the late Paleozoic (Van der Voo, 1993; Metcalfe, 1996; Torsvik and Cocks, 2004). Hou and Boucot (1990) noted the presence of a Hercynian fold belt between the Sino-Korea and Siberian plates that extends from Balkhsh, through Junggar, Mongolia, and Greater and Lesser Khingan to the Okhtsk Sea. Hou and Boucot (1990) also noted that the Heitai Formation in Mishan County is situated at the eastern end of the Hercynian fold belt and may be a part of the Burrya-Jiamusi Terrane. Nie et al. (1990) placed the North China Terrane in the tropical Cathaysian floral province in an ever-wet climatic setting during the Pennsylvanian and Early Permian.

HEITAI FORMATION. Mu (1955) described a new species of the blastoid *Devonoblastus* Reimann, 1935, from the Heitai Formation, Mishan County, Heilongjiang Province, in northeastern China. The blastoids were collected from siliceous, sandy limestone and calcareous sandstone with a total thickness of 32.9 m in the lower part of the Heitai Formation now referred to the Xiaheitai Formation (Wang et al., 1986). On the basis of the analysis of the conodonts and ostracods, Wang et al. (1986) assigned an early Eifelian age to these units. We have plaster casts of Mu's types and two additional specimens collected by Su Yangzheng from Taipinglu, Baoqing County. On the basis of these specimens, Mu's species will be removed from *Devonoblastus* and placed into a new genus in a study to be published later.

CHIHLI (QIJI) TAIYUAN SERIES. Tien (1926) described a well-preserved crinoid fauna from the Upper Carboniferous Taiyuan Series at the Lin Cheng (Lincheng) coalfield near Chihli, (Qiji) Hopeh (Hebei) Province, northern China. More recently, Huang and Chen (1987) reported the age of the Taiyuan Series as Pennsylvanian to Early Permian on the basis of fusulinids, with species of *Triticites* Girty, 1904, found in the lower beds and species of *Pseudoschwagerina* Dunbar and Skinner, 1936, found in the upper beds. Tien (1926) noted that crinoids were collected from three limestone beds within the Taiyuan Series, along with numerous well-preserved specimens of corals, brachiopods, bivalves, gastropods, and cephalopods. Unfortunately, the stratigraphic descriptions given in Tien (1926) are not precise enough to assess whether the crinoids were collected from Pennsylvanian or Lower Permian beds.

Crinoids described by Tien (1926) include *Sinocrinus* (a new genus); *Eupachycrinus* Meek and Worthen, 1865, species reassigned to *Mathericrinus* by Webster (1981); *Graphiocrinus* de Koninck and Le Hon, 1854, species reassigned to *Delocrinus* by Bassler and Moody (1943); and *Platycrinites* (based on stems only). After Tien's (1926) report, *Sinocrinus* subsequently has been reported from Spain and the United States, and the only other known occurrence of *Mathericrinus* is from the Pennsylvanian of the United States. *Delocrinus* has a much wider temporal and geographic distribution and is known from the Middle Pennsylvanian to Upper Permian, with reported occurrences from the United States, Russia, Timor, Japan, and Bolivia. The echinoderm fauna from the Taiyuan Formation clearly has paleobiogeographic affinities with Pennsylvanian echinoderms from the United States and warrants further investigation and recollection to establish these ties firmly.

The Sibumasu Terrane derives its name from Thailand (Siam), Burma, Malaysia, and Sumatra, although the northern end of the terrane extends into western Yunnan Province. Yunnan Province, which is complex tectonically, can be divided roughly into western and eastern tectonic regions by a suture zone (=Langcangjiang Fault Zone) that trends along the eastern margin of the Gaoligong Shan and the western margin of the Nujiang Valley. Western Yunnan is divided further into three tectonic zones, from west to east: the Tengchong Block, the Baoshan Block, and the Changning-Menglian Belt (Jin, 1997). These western tectonic zones have Gondwanan affinities in contrast to eastern Yunnan, which is composed of two blocks that have Cathaysian affinities (Jin, 1997). Van der Voo (1993) and Metcalfe (1996) illustrated a Devonian paleolatitudinal position for the Yunnan portion of Sibumasu at more than 40° south latitude, with a rapid northward drift into the tropics during the Permian (Fig. 16.2). All echinoderm localities we visited are on the Baoshan block and are in Devonian and Mississippian strata.

BAOSHAN, ORDOVICIAN UNDIFFERENTIATED. Paleozoic echinoderms from Yunnan Province have been known for many years (Reed, 1906, 1908, 1917; Jahnke and Shi, 1989; Haude, 1992), but the only major monographic study of the faunas from Yunnan published recently is by Chen and Yao (1993). The echinoderm fossils described and discussed in Chen and Yao (1993) were collected from Ordovician through Mississippian strata in Shidian and Baoshan counties, Yunnan Province, by the authors and students of the Geological Department of Kunming Institute of Technology during fieldwork in 1984–1985. Types from Chen and Yao's 1993 monograph are now housed in the Nanjing Institute for Geology and Paleontology.

In the 1993 monograph, Chen and Yao described a collection of more than 1000 individuals of cystoids that are referred to eight genera (one new) and 14 species (three new) from Ordovician strata in Shidian County. Genera include *Caryocrinites* Say, 1825; *Paracaryocrinites* Chen and Yao, 1993; *Heliocrinites* Eichwald, 1840; *Echinosphaerites* Walhlenberg, 1818; *Sphaeronites* Hisinger, 1828; *Aristocystites* Barrande, 1887; and *Sinocystis* Reed, 1917. We did not visit the Ordovician localities and are not in a position to comment on the taxonomic assignment of the cystoids.

BAOSHAN, SILURIAN, UNDIFFERENTIATED. Reed (1906, 1908), Jahnke and Shi (1989), Haude (1992), and Chen and Yao (1990) all have described species of *Scyphocrinites* Zenker, 1833, from the Upper Silurian "*Camerocrinus* beds" in Shidian County, western Yunnan. We did not visit the Silurian localities and are not in a position to comment further on the occurrences.

BAOSHAN, DEVONIAN, HEYUANZHAI FORMATION. Chen et al. (1983) described *Parascyphocrinites* from the upper portion of the Heyuanzhai Formation. We visited Professor Chen in Kunming during May 1993 to examine his collections and visit localities in western Yunnan. One of the localities visited was a quarry near the village of Youwang in western Yunnan that exposed the Heyuanzhai Formation, a Givetian to Frasnian dolomite, where reasonably common echinoderms, mostly small crinoids, can be collected from weathered surfaces of the dolomite. Chen and Yao (1993) reported the following genera from this locality: *Shidianocrinus* Chen, Yao, and Yu, 1985; *Melocrinites* Goldfuss, 1831; *Changninocrinus* Chen, Yao, and Yu, 1985; *Haplocrinites* Steininger, 1837; *Ovalocrinus* Chen and Yao, 1993; *Quasicydonocrinus* Chen and Yao, 1993; *Megaradialocrinus* Chen and Yao, 1993; *Petalocrinus* Weller and Davidson, 1896; and *Cupressocrinites* Goldfuss, 1831.

The stratigraphic sequence exposed at the quarry is approximately 5 m of dolomitic packstone to wackestone, which is overlain by approximately 0.33 m of conglomerate with phosphatic pebbles at the Frasnian–Tournaisian boundary (the Famennian is absent at this locality). Chen and Yao (1993) reported the blastoid *Cryptoschisma* Etheridge and Carpenter, 1886, from this locality, but Waters et al. (2003b) reassigned the specimens to *Breimeriblastus* Waters et al. (2003a), a Devonian hyperoblastid genus.

BAOSHAN, CARBONIFEROUS, PUMEQIAN FORMATION. Approximately 1000 well-preserved specimens of Carboniferous crinoids belonging to 28 genera and 64 species were also collected and described by Chen and Yao (1993) from Baoshan County. Most of these crinoid fossils are first occurrences in China, including five new genera and 34 new species (Table 16.1). We visited localities near the villages of Lutu and Pumeqian and collected specimens of Chen and Yao's (1993) echinoderm fauna from the Pumeqian Formation, which is Tournaisian to Viséan in age, and was mainly limestone. Although localities from which most of the 1000+ crinoid specimens in Chen's collections originally were collected are no longer are accessible because of development of a major sinkhole in the immediate vicinity of the original localities, we were able to collect crinoids at other nearby exposures of the Pumeqian Formation. The crinoid fauna is dominated by rhodocrinoids and platycrinoids, and our general perception of the fauna is that it is similar to Waulsortian echinoderm faunas from North Yorkshire but does contain some new taxa, e.g., *Yunnanocrinus* Chen and Yao (1993). Blastoids from this locality were described by Chen and Yao (1993) as *Pentremites* Say, 1825. However the specimens were reassigned to *Sinorbitremites* Waters et al., 2003b, a genus of orbitremitid blastoids. Clearly, more knowledge of echinoderms from this area will aid in determining paleobiogeographic affinities of the Baoshan block during the Paleozoic.

Tarim Terrane

In the summer of 2000, our working group completed a 7000-km reconnaissance trip in the Tarim terrane along the route of Urumqi to Korla, Tazhong, Minfeng, Hotan, Kashi, Aksu, Kuqu, Xinyuan, Yining, Jinghe, and back to Urumqi. In conducting the preparatory work for this trip and on the actual trip itself, we visited about 20 Late Devonian–Early Permian sections located in the middle Tianshan Mountains, the western Tianshan Mountains, the southwestern portion of the Tarim Basin, and the northwestern Kunlun Mountains. The faunas at these localities were mainly Gondwanan (South facies type) and commonly abundant and diverse, but fossil echinoderms were scarce. Echinoderm collecting in the Tarim terrane is hampered by poor access; variable weather conditions, such as sandstorms and heavy floods, which make transportation difficult; and the location of key sections in politically sensitive border regions. Many sections we visited exhibited low-grade metamorphism and significant brittle deformation associated with mountain building, which further decreased the likelihood of collecting echinoderm calyxes.

Although systematic Paleozoic echinoderm studies in China are just beginning, existing collections and our field experiences suggest that future studies will continue to yield diverse and significant faunas in many of the tectonic terranes. As the Chinese paleontological community becomes increasingly aware of the potential of echinoderms for biostratigraphic and biogeographic studies, we hope that study of these important fossils in China will increase significantly. Existing fossil data will have to be reconciled with the paleolatitudinal models derived from the expanding paleomagnetic database. For example, biogeographic affinities of the Junggar echinoderms with Western Europe agree with the interpretation of the Junggar Terrane amalgamating with Kazakhstania in the Early Devonian. The presence of tropical echinoderm faunas at presumed high latitudes is problematic but is consistent with data from other paleontological studies in various Asian terranes that suggest tropical paleolatitudes, whereas the paleomagnetic data indicate higher latitudes (Van der Voo, 1993). The echinoderm faunas from the South China and North China blocks are consistent with equatorial paleolatitudes for these blocks, but the large new faunas from Yunnan and their apparent affinities with European faunas need to be reconciled with paleomagnetic data that place Sibumasu in the high southern latitudes.

Conclusions

This work was supported in part by grants to JAW (EAR9404729; EAR9117673), NGL (EAR9406116; EAR9117472), and CGM (EAR9405207; EAR9117505) from the Geology and Paleontology Program of the National Science Foundation and from the National Geographic Society, and SAM from the American Association of Petroleum Geologists. We are grateful to these organizations for their support of our work. We also thank our many colleagues in China who graciously allowed us access to specimens, aided us in the field, showed us echinoderm localities, and helped us in innumerable ways. Reviews by J. Sprinkle and R. Parsley improved the chapter significantly.

Acknowledgments

Ahluwalia, A. D. 1989. The peripatetic fossils: part 3. Upper Palaeozoic in Lahul-Spiti. Nature, 431:13–15.
Ausich, W. I. 1996. Crinoid plate circlet homologies. Journal of Paleontology, 70:955–964.
Ausich, W. I. 1998. Early phylogeny and subclass division of the Crinoidea (Phylum Echinodermata). Journal of Paleontology, 72:499–510.

References

Ausich, W. I., and D. J. Bottjer. 1986. Phanerozoic development of tiering in soft substrata suspension-feeding communities. Paleobiology, 12:400–420.

Barrande, J. 1887. Classe des Echinodermes. Ordre des Cystidees. Ouvrage post-hume de feu Joachim Barrande, publie par le Doct. W. Waagen. *In* Systeme Silurien du Centre de la Boheme, Part I. Recherches Paleontologiques, continuation editee par le Musee Boheme, 7(1), 233 p.

Bassler, R. S., and M. W. Moody. 1943. Bibliographic and faunal index of Paleozoic Pelmatozoan echinoderms. Geological Society of America Special Paper, 45, 732 p.

Carroll, A. R., Y. Liang, S. A. Graham, X. Xiao, M. S. Hendrix, J. Chu, and C. L. McKnight. 1990. Junggar Basin, northwest China: trapped Late Paleozoic ocean. Tectonophysics, 181:1–14.

Chen, Z.-Q., G.-R. Shi, and K. Kaiho. 2004. New ophiuroids from the Permian/Triassic boundary beds of South China. Palaeontology, 47:1301–1312.

Chen, Z.-T., And J.-H. Yao. 1990. First discovery of *Scyphocrinites* in China. Acta Palaeontologica Sinica, 29:228–236.

Chen, Z.-T., and J.-H. Yao. 1993. Palaeozoic echinoderm fossils of western Yunnan, China. Geological Publishing House, Beijing, 102 pp. (In Chinese with English summary)

Chen, Z.-T., J.-H. Yao, and G.-R. Yu. 1983. A new genus *Parascyphocrinites* from the Middle Devonian in west Yunnan. Yunnan Geology, 2:337–340.

Chen, Z.-T., J.-H. Yao, and G.-R. Yu. 1985. Two new genera of Middle Devonian crinoids from western Yunnan. Acta Palaeontologica Sinica, 24:548–552.

Chi, S.-Y. 1943. A Lower Carboniferous blastoid from the Tushan District Kueichou. Bulletin of the Geological Society of China, 23:111–113.

De Koninck, L. G. 1858. Notice sur un nouveau genre de crinoides du terrain carbonifère de l'Angleterre. Academy Royal Belgique, Mémoire, 28(3, Supplement):208–217.

De Koninck, L. G., and H. Le Hon. 1854. Recherches sur les crinoids du terrain carbonifere de la Belgique. Academie Royal de Belgique Memoir, 28(3), 217 p.

Dunbar, C. O., and J. W. Skinner. 1936. *Schwagerina* versus *Pseudoschwagerina* and *Paraschwagerina*. Journal of Paleontology, 10:83–91.

Eichwald, E. von. 1840. Uber das silurische Schichtensystem in Esthland. Zeitschrift fur Natur- und Heilkunde der konigliche Medicinisch-chirurgische Akademie St. Petersburg, 1, 210 p.

Etheridge, R., Jr., and P. H. Carpenter. 1886. Catalogue of the Blastoidea in the Geological Department of the British Museum (Natural History), with an account of the morphology and systematic position of the group, and a revision of the genera and species. British Museum Catalog, London, 322 p.

Fay, R. O. 1961. *Deltoblastus*, a new Permian blastoid genus from Timor. Oklahoma Geology Notes, 21(2):36–40.

George, T. N. 1931. *Ambocoelia* Hall and certain similar British Sprifieridae. Geological Society of London Quarterly Journal, 87:30–61.

Girty, G. H. 1904. *Triticites*, a new genus of Carboniferous foraminifers. American Journal of Science, series 4, 17:234–240.

Goldfuss, G. A. 1831 [1826–1844]. Petrefacta Germaniae, tam ea, Quae in Museo Universitatis Regiae Borussicae Fridericiae Wilhelmiae Rhenanea, serventur, quam alia quaecunque in Museis Hoeninghusiano Muensteriano aliisque, extant, iconibus et descriiptionns illustrata.—Abbildungen und Beschreibungen der Petrefacten Deutschlands und der Angränzende Länder, unter Mitwirkung des Hern Grafen Georg zu Münster, herausgegeben von August Goldfuss, 1(1826–1833), Divisio secunda. Radiariorum reliquiae: 115–221.

Gupta, V. J., and G. D. Webster. 1976. *Deltoblastus batheri* from the Kashmir Himalaya. Rivista Italiana di Paleontogia e Stratigrafia, 82:279–284.

Gupta, V. J., and G. D. Webster. 1980. *Deltoblastus*, paleontologic data for plate tectonic relationship of India and Timor. Journal Geological Society of India, 21:362–364.

Haude, R. 1992. Scyphocrinoiden, die Bojen-Seelilien im hohen Silur-tiefen Devon: Paleontographica Abteilung A. Palaeozoologie und Stratigraphie, 222:141–187.

Haude, R. 1995. Lower Devonian echinoderms from the Precordillera (Argentina). Neues Jahrbuch für Geologie und Paläontologie Abhandlungen, 197:37–86.

Hisinger, W. 1828. Antekningar i physik ock geognosie under resor uti Sverige ock Norrige, 4, 260 p.

Hou, H.-F., and A. J. Boucot. 1990. The Balkhash-Mongolia-Okhotsk region of the Old World Realm (Devonian), p. 297–303. *In* W. S. McKerrow and C. R. Scotese (eds.), Paleozoic Palaeogeography and Biogeography. Geological Society Memoir, 12.

Hou, H.-F., N. G. Lane, J. A. Waters, and C. G. Maples. 1993 [1994]. Discovery of a new Famennian echinoderm fauna from the Hongguleleng Formation of Xinjiang, with redefinition of the formation. Stratigraphy and Paleontology of China, 2:1–18.

Huang, J., and B.-W. Chen. 1987. The evolution of the Tethys in China and adjacent regions. Geological Publishing House, Beijing, 109 p.

Jahnke, H., and Y. Shi. 1989. The Silurian–Devonian boundary strata and the Early Devonian of the Shidian-Baoshan area (W. Yunnan, China). Courier Forschungsinstitut Senckenberg, 110:137–193.

Jell, P. A. 1999. Silurian and Devonian crinoids from central Victoria. Memoirs of the Queensland Museum, 43, 114 p.

Jell, P. A., and J. S. Jell. 1999. Crinoids, a blastoid and a cyclocystoid from the Upper Devonian reef complex of the Canning Basin, Western Australia. Memoirs of the Queensland Museum, 43:201–236.

Jell, P. A., and J. N. Theron. 1999. Early Devonian echinoderms from South Africa. Memoirs of the Queensland Museum, 43:115–199.

Jin, X. 1997. Sedimentary and Paleogeographic Significance of Permo-Carboniferous Sequences in Western Yunnan, China. Geologisches Institut der Universitaet zu Koeln Sonderveroeffentlichungen, 99, 136 p.

Kuang, G., B. Yin, and W. Wei. 1987. Carboniferous carbonate sequences in Guangxi: 11th International Congress of Carboniferous Stratigraphy and Geology Guidebook Excursion 6. Printing House of the Juangsu Academy of Agricultural Sciences. Nanjing, China, 28 p.

Lane, N. G., and G. D. Sevastopulo. 1987. Stratigraphic distribution of Mississippian camerate crinoid genera from North America and Western Europe. Courier Memorial Volume, Senckenbergiana Lethaea, 68:17.

Lane, N. G., and G. D. Sevastopulo. 1990. Biogeography of Lower Carboniferous crinoids, p. 333–338. *In* W. S. McKerrow and C. R. Scotese (eds.), Paleozoic Palaeogeography and Biogeography. Geological Society Memoir, 12.

Lane, N. G., J. A. Waters, and C. G. Maples. 1997. Echinoderm faunas of the Hongguleleng Formation, Late Devonian (Famennian), Xinjiang-Uygur Autonomous Region, China. Paleontological Society Memoir, 47, 43 p.

Lane, N. G., J. A. Waters, C. G. Maples, S. A. Marcus, and Z.-T. Liao. 1996. A camerate-dominated late Carboniferous (Moscovian) crinoid fauna from volcanic conglomerates, Northwestern People's Republic of China. Journal of Paleontology, 70:117–128.

Li, Y.-P., R. Sharps, M. McWilliams, Y. Li, Q. Li, and W. Zhang. 1991. Late Paleozoic paleomagnetic results form the Junggar block, northwestern China. Journal of Geophysical Research, 96:16047–16060.

Liao, W.-H., and T.-C. Cai. 1987. Sequence of Devonian rugose coral assemblages from northern Xinjiang. Acta Palaeontologica Sinica, 26:689–707.

Liao, Z., L. Lu, N. Jiang, F. Xia, F. Sun, and Y. Zhou. 1987. Carboniferous and Permian in the western part of the East Mts. Tianshan, Guidebook for Excursion 4. 11th International Congress of Carboniferous Stratigraphy and Geology. Beijing, China, 39 p.

Luo, J. 1984. The first record of occurrence of Genus *Deltoblastus* from China, p. 100–104. *In* Selected Papers for Memorial of Professor Yoh Senshing, Working on Geological Science and Education for Sixty Years. Geological Publishing House, Beijing.

Marcus, S. A. 1993. A marine fossiliferous interval from a mid-Carboniferous, organic-rich shale, People's Republic of China. Geological Society of America Abstracts with Program, 26:A-104.

McKerrow, W. S., and C. R. Scotese. 1990. Palaeozoic palaeogeography and biogeography. Geological Society, London, Memoir 12, 435 p.

M'Coy, F. 1847. On the fossil botany and zoology of the rocks associated with the coal of Australia. Annals and Magazine of Natural History, series 1, 20:145–147, 226–236, 298–312.

Meek, F. B., and A. H. Worthen. 1865. Descriptions of new species of Crinoidea, etc. from the Paleozoic rocks of Illinois and some of the adjoining states. Proceedings of the Academy of Natural Sciences of Philadelphia, 17:143–155.

Metcalfe, I. 1996. Pre-Cretaceous evolution of SE Asian terranes, p. 97–122. *In* R. Hall and D. Blundell (eds), Tectonic Evolution of Southeast Asia. Geological Society Special Publication, 106.

Miller, J. S. 1821. A natural history of the Crinoidea, or lily-shaped animals; with observations on the genera, *Asteria, Euryale, Comatula* and *Marsupites*. Bryan & Co., Bristol, England, 150 p.

Moore, R. C., and H. L. Strimple. 1969. Explosive evolutionary differentiation of unique group of Mississippian–Pennsylvanian camerate crinoids (Acrocrinidae). University of Kansas Paleontological Contributions Paper, 39, 44 p.

Mu, A.-T. 1950. *Petalocrinus* from the Shihniulan Limestone of Wuchuan. Bulletin of the Geological Society of China, 29:93–96.

Mu, A.-T. 1955. A Devonian blastoid from Kirin. Acta Palaeontologica Sinica, 3:131–134.

Mu, A.-T., and C.-H. Lin. 1987. Petalocrinidae from the Silurian of Shiqian District, Guizhou. Bulletin of the Nanjing Institute Geology and Palaeontology Academica Sinica, 12:1–22.

Mu, A.-T., and Y.-R. Wu. 1974. Crinoid sections. *In* A Handbook of the Stratigraphy and Paleontology in Southwest China. Nanking Institute of Geology and Paleontology Academia Sinica, Beijing, 455 p.

Nicholson, H. A. 1873. Contributions to the study of the errant Annelides of the older Paleozoic rocks. Royal Society London Proceedings, 21:288–290.

Nie, S.-Y. 1994. Devonian rifting of South China from Gondwana—a case study. Twelfth Australian Geological Convention, Perth. 1994. Geological Society of Australia, Abstracts, 37:319.

Nie, S.-Y., D. B. Rowley, and A. M. Ziegler. 1990. Constraints on the locations of Asian microcontinents in Palaeo-Tethys during the Late Palaeozoic, p. 397–409. *In* W. S. McKerrow and C. R. Scotese (eds.), Paleozoic Palaeogeography and Biogeography. Geological Society Memoir, 12.

Parsley, R. L., and Y.-L. Zhao. 2002. Eocrinoids of the Middle Cambrian Kaili biota, Taijiang County, Guizhou, China. Geological Society of America Abstracts with Program, 34(6):81.

Parsley, R. L., and Y.-L. Zhao. 2006. Long stalked eocrinoids in the basal Middle Cambrian Kaili Biota, Taijiang County, Guizhou Province, China. Journal of Paleontology, 80:1058–1071.

Poty, E., and S.-C. Xu. 1996. Rugosa from the Devonian–Carboniferous transition in Hunan, China. Mémoires de L'Institut Géologique de L'Université de Louvain, 36:89–139.

Reed, F. R. C. 1906. The Lower Palaeozoic fossils of the Northern Shan States, Burma. With a section of Ordovician cystidea by F. A. Bather [q.v.]. Palaeontologica Indica, series 2, Memoir 3, 158 p.

Reed, F. R. C. 1908. The Devonian faunas of the Northern Shan states. India Geological Survey, Memoirs, 2(5):1–157.

Reed, F. R. C. 1917. Ordovician and Silurian fossils from Yunnan. Geological Survey of India, Palaeontologia India, new series, Memoir, 6, 20 p.

Reimann, I. G. 1935. New species and some new occurrences of Middle Devonian blastoids. Bulletin of the Buffalo Society of Natural Sciences, 17(1):43–45.

Sartenaer, P., and H.-K. Xu. 1989. The upper Famennian rhynchonellid genus *Planovatirostrum* Sartenaer, 1970 from Africa, China, Europe and the USSR. Bulletin de l'Institut Royal des Sciences Naturelles de Belgique, Sciences de la Terre, 59:37–48.

Say, T. 1825. On two genera and several species of Crinoidea. Journal of the Academy of Natural Sciences Philadelphia, series 1, 4(2):289–296.

Schevchenko, T. W. 1967. Rannedevonskie morskie lilii semeistva Parahexacrinidae fam. nov. Servschanskogo hrebta [Early Devonian crinoids of the Parahexacrinidae fam. nov. from the Zeravshan Ridge]. Paleontologicheskii Zhurnal, 3:76–88.

Sengör, A. M. C., B. A. Natal'in, and V. S. Burtman. 1993. Evolution of the Altaid tectonic collage and Palaeozoic crustal growth in Eurasia. Nature, 364:299–307.

Shu, D.-G., S. Conway Morris, J. Han, Z.-F. Zhang, and J.-N. Liu. 2004. Ancestral echinoderms from the Chengjiang deposits of China. Nature, 430:422–428.

Simms, M. J. 1994. Re-interpretation of thecal plate homology and phylogeny in the class Crinoidea. Lethaia, 26:303–312.

Smith, A. B. 2004. Echinoderm roots. Nature, 430:411–412.

Soto, F., and B. Lin. 1997. Biostratigraphic and biogeographic affinities of Famennian rugose corals in the Dzungar-Hinggan Basin (Northern China). Coral Research Bulletin, 5.239–246.

Soto, F., and B. Lin. 2000. Corales rugosos de la Formación Hongguleleng (Fameniense) en el N de Sinkiang (NO de China). Geobios, 33:527–541.

Springer, F. 1906. Discovery of the disk of *Onychocrinus* and further remarks on the Crinoidea Flexibilia. Journal of Geology, 14:467–523.

Steininger, J. 1837. Notes from meeting in which 2 new genera, with their types were published. Bulletin de la Société Géologique de France, 8:230–232.

Strimple, H. L., and W. T. Watkins. 1955. I. Three new genera. *In* New Ordovician echinoderms. Journal of the Washington Academy of Science, 45:347–353.

Talent, J. A. 1989. The case of the peripatetic fossils. Nature, 338:613–615.

Tien, C.-C. 1926. Crinoids from the Taiyuan Series of North China. Palaeontologia Sinica, series B, Geological Survey of China, 5, 58 p.

Tien, C.-C., and A.-T. Mu. 1955. Echinodermata, p. 83–95. *In* Y. Z. Sun (ed.),

Index Fossils of China, Invertebrate, volume 1, 83–95, pl. 45–51. Publishing House of Geology, Beijing.

Torsvik, T. H., and L. R. M. Cocks. 2004. Earth geography from 400 to 250 Ma: a palaeomagnetic, faunal and facies review. Journal of the Geological Society, London, 161:555–572.

Trautschold, H. 1867. Einige crinoideen und andere Thierreste des Jungeren Bergkalks im Gouvernment Moskau. Bulletin de la Societie Imperial Naturalistes de Moscou, 15(3–4):1–49.

van der Voo, R. 1993. Paleomagnetism of the Atlantic, Tethys, and Iapetus oceans. Cambridge University Press, Cambridge, 411 p.

Walhlenberg, G. G. 1818. Om svensks jordena bildning. Svea, Tidskrift för Vetenskap, Konst, 1, 77 p.

Wang, C.-Y., C.-G. Shi, and G.-S. Qu. 1986. Conodonts and ostracods from the Devonian Heitai Formation of Mishan County, Heilongjiang Province. Acta Micropalaeontologica Sinica, 3:205–214

Wanner, J. 1916. Die Permischen echinodermen von Timor, I. Teil. Palaontologie von Timor, 11, 329 p.

Waters, J. A., N. G. Lane, C. G. Maples, S. A. Marcus, Z. Liao, and L. Liu. 2003a. A quadrupling of Famennian Pelmatozoan diversity: new Late Devonian blastoids and crinoids from the Late Devonian (Famennian) of Northwest China. Journal of Paleontology, 77:922–948.

Waters, J. A., Z. T. Liao, L. Liu, N. G. Lane, C. G. Maples, and S. A. Marcus. 2003b. Redescription of Devonian and Carboniferous blastoids from western Yunnan, China. Journal of Paleontology, 77:933–936.

Webster, G. D. 1973. Bibliography and index of Paleozoic crinoids, 1942–1968. Geological Society of America Memoir, 137, 341 p.

Webster, G. D. 1977. Bibliography and index of Paleozoic crinoids, 1969–1973. Geological Society of America Microform Publication, 8, 235 p.

Webster, G. D. 1981. New crinoids from the Naco Formation (Middle Pennsylvanian) of Arizona and a revision of the family Cromyocrinidae. Journal of Paleontology, 55:1176–1199.

Webster, G. D. 1986. Bibliography and index of Paleozoic crinoids, 1974–1980. Geological Society of America Microform Publication, 16, 405 p.

Webster, G. D. 1988. Bibliography and index of Paleozoic crinoids, 1981–1985. Geological Society of America Microform Publication, 18, 235 p.

Webster, G. D. 1990. Index of Paleozoic crinoids and coronate echinoderms, publications through 1990. Geological Society of America Microform Publication, 1305 p.

Webster, G. D. 1991. An evaluation of the V. J. Gupta echinoderm papers, 1971–1989. Journal of Paleontology, 65:1006–1008.

Webster, G. D. 1998. Palaeobiogeography of Tethys Permian crinoids. Strzelecki International Symposium on Permian of Eastern Tethys: biostratigraphy, palaeogeography and resources. Proceedings of the Royal Society of Victoria, 110(1–2):289–308.

Webster, G. D. 2003. Bibliography and index of Paleozoic crinoids, coronates, and hemistreptocrinoids, 1758–1999. Geological Society of America, Special Paper 363. <http://crinoid.gsajournals.org/crinoidmod/>.

Webster, G. D., and P. A. Jell. 1999a. New Carboniferous crinoids from eastern Australia. Memoirs of the Queensland Museum, 43:237–277.

Webster, G. D., and P. A. Jell. 1999b. New Permian crinoids from Australia. Memoirs of the Queensland Museum, 43:279–339.

Webster, G. D., and G. D. Sevastopulo. 2001. Early Permian blastoids and crinoids from northeastern Oman. International Conference, Geology of Oman, Abstracts Volume, p. 87.

Webster, G. D., R. T. Becker, and C. G. Maples. 2005. Biostratigraphy, paleoecology and taxonomy of Devonian (Emsian and Famennian) crinoids from southeastern Morocco. Journal of Paleontology, 79:1052–1071.

Weller, S., and A. D. Davidson. 1896. *Petalocrinus mirabilis* (n. sp.) and a new American fauna. Journal of Geology, 4:166–173.

Xia, F.-S. 1996. New knowledge on the age of Hongguleleng Formation in northwestern margin of Junggar Basin, Northern Xinjiang. Acta Micropalaeontologica Sinica, 13:277–287.

Xia, F.-S. 1997. Marine microfaunas (bryozoans, conodonts and microvertebrate remains) from the Frasnian—Famennian interval in Northwestern Junggar Basin of Xinjiang in China. Beiträge zur Paläontologie, 22:91–207.

Xu, I.-W. 1962. *Caelocrinus*—Noavyi rod morskikh lilii iz srednesiluriiskogo otdela v provintsii sychuan [*Caelocrinus*, new genus of crinoid of Middle Silurian age from the province of Si-Tuhouan-I]: Acta Palaeontologica Sinica, 10:45–54.

Yakovlev, N. N. 1928. Dva novikh roda morskikh lilii (Poteriocrinidae) iz verkhnepaleozoiskikh otlozhenii Pechorekogo kraya [Two new genera of crinoids (Poteriocrinidae) of the Upper Paleozoic]. Trudy Akademii Nauk SSSR, Geologicheshogo Muzeya, 3:1–8, pl. 1–2.

Yakovlev, N. N., and A. Faas. 1938. Nuovi echinodermi permiani di Sicilia. Paleontographia Italica, 38:116–126.

Yakovlev, N. N., and A. P. Ivanov. 1956. Morskie lilii i blasatoidei kamennougolnykh i permskikh otlozhenii SSSR. Vsesoiuznyi nauchno-Issledsvatelskii Geologicheskii Institut, Trudy, new series 11, 142 p.

Zenker, J. C. 1833. Beiträge zur Naturgeschichte der Urwelt. Organische Reste (Petrefacten) aus der Altenburger Braunkohlen-formation, dem Blankenburger Quadersandstein, jenaischen bunten Sandstein und böhmischen Übergangsgebirge. Jena, Friedrich Mauke, (ed.), 67 p.

Zhao, Y. L., and R. L. Parsley. 2003. Short stemmed eocrinoids of the Middle Cambrian Kaili biota, Guizhow Province, China. Geological Society of America Abstracts with Program, 35(6):106.

Zhao, Z.-T., and C. Wang. 1990. Age of the Hongguleleng Formation in the Junggar Basin of Xinjiang. Journal of Stratigraphy, 14:145–146.

Zhao, Y.-L., Y.-Z. Huang, and X.-Y. Gong. 1994. Echinoderm fossils of Kaili Fauna from Taijiang, Guizhou. Acta Palaeontologica Sinica, 33(3):305–324.

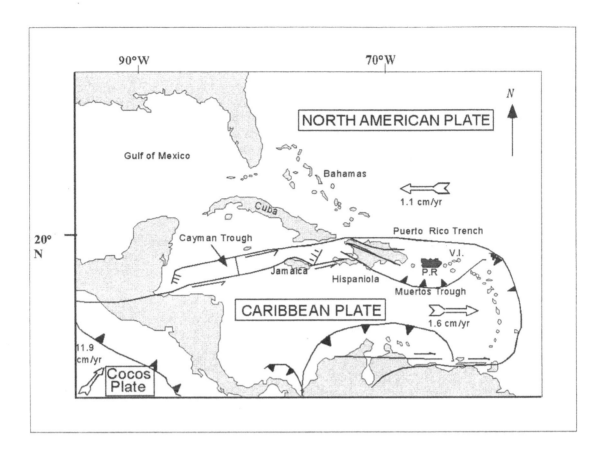

Figure 17.1. *Map of the Caribbean region and its major geologic features and plate boundaries. Modified from Mann et al. (1991).*

FOSSIL ECHINODERMATA FROM PUERTO RICO

17

Jorge Vélez-Juarbe and Hernán Santos

Fossil Echinodermata from Puerto Rico, the easternmost Greater Antilles (Fig. 17.1), have been reported in the literature (Jackson, 1922; Gordon, 1963; Cutress, 1980). However, these reports only mention the occurrence of several species of sea urchins and sand dollars (Class Echinoidea), mostly from Tertiary rocks, and a few species from the Cretaceous. A study of fossil echinoderms from Puerto Rico that includes both Cretaceous and Tertiary echinoids species and that also includes Crinoidea and Stelleroidea will help in the understanding of this group in Puerto Rico. This is the first attempt to make a comprehensive study of the Echinodermata from the Cretaceous, Tertiary, and Quaternary of Puerto Rico. Paleoecological interpretations of the occurrence of these fossils are discussed.

Methodology

Fieldwork was completed by searching for outcrops of Cretaceous, middle Tertiary, and Quaternary rocks, mostly along the western half of the island, where recent road development has made fresh exposures available. Prospecting in limestones of Paleocene to Eocene age within volcaniclastic-dominated formations was not successful because outcrops are largely covered, unlike the Oligocene and Miocene localities. Some fossils had been collected throughout the years and were already deposited at the Paleontological Collection, Department of Geology, University of Puerto Rico (UPRMP). A revision of the available literature that included fossil Echinodermata from Puerto Rico was complete, and we include previous reports with our work in Tables 17.4 to 17.10. Identification of the fossils was done by comparison with literature from other areas of the Caribbean (Jackson, 1922; Sánchez-Roig, 1926, 1949; Gordon, 1963; Kier and Grant, 1965; Kier, 1975, 1984, 1992; Cutress, 1980; Poddubiuk, 1987; Dixon et al., 1994; Donovan, 1994, 1996, 2000; Donovan et al. 1994, 2005; Donovan and Collins, 1997; Donovan and Harper, 2000; Mooi and Peterson, 2000).

The Echinodermata found during the course of this study can be used for paleoenvironmental interpretation of the deposits where they are found. These interpretations can be used for comparison with interpretations obtained from studying the sediments and other fossils. For this, we also use the echinoderms that have been reported in previous works where possible. Previous reports that were not certain of the formation where the fossils were collected are not used (i.e., some of Gordon's [1963] lower Ponce Limestone localities could now be considered as part of a new formation). Also, some formations yielded echinoids from different facies within the same locality (i.e., new formation locality NF-1 and San Sebastián Formation locality SS-2), but each locality will be discussed separately to avoid confusion about its paleoenvironment.

Geological Setting

During the Cretaceous and the early part of the Tertiary, the island of Puerto Rico was part of an island-arc system and was divided into three main volcanic provinces: Southwestern Igneous Province, Central Igneous Province, and Northeastern Igneous Province (Jolly et al., 1998; Schellekens, 1998). During the Late Cretaceous though the Eocene, these igneous provinces were united into the block forming the island of Puerto Rico and separated by two main fault zones (Jolly et al., 1998; Schellekens, 1998). Rocks younger than Eocene are not separated by these fault zones, and they even cover parts of the fault zones in some areas (Jolly et al., 1998). Here, we briefly discuss the age and environment of the formations where fossil Echinodermata were collected, which range in age from Cretaceous (Fig. 17.2), middle Tertiary, and Quaternary age (Fig. 17.3), and we list the corresponding localities (Fig. 17.4).

Cretaceous

RÍO MATÓN LIMESTONE MEMBER OF THE ROBLES FORMATION. Located in the Central Igneous Province (Jolly et al., 1998; Schellekens, 1998), the Río Matón Member was described by Berryhill and Glover (1960) as a limestone unit within the sandstone and siltstone dominated Robles Formation. The age of the Río Matón Limestone Member had been previously dated as late Albian (Jolly et al., 1998), but new data based on the occurrence of the rudist *Coalcomana ramosa* (Boehm, 1898) place the formation in the early Albian (Skelton, personal commun., 2004). On the basis of fossil content, this member represents a shallow marine environment with abundant nerineid gastropods, chondrodont bivalves, and rudists. One locality, RM-1, within this formation yielded fossil echinoderms. At this locality, fossil echinoderms consist of fragmentary radioles. Other fossils such as gastropods and bivalves are heavily recrystallized.

PARGUERA LIMESTONE. This limestone unit is located in the Southwestern Igneous Province (Jolly et al., 1998; Schellekens, 1998). Almy (1965) divided the Parguera Limestone into three members: the lower Bahia Fosforecente Member, the middle Punta Papayo Member, and the upper Isla Magueyes Member. This formation represents slope to basin environments (Almy, 1965). On the basis of the rudist fauna, the Bahia Fosforecente Member belongs to the *Durania curasavica* Biozone of Rojas et al. (1995) of ?Santonian age, and the Punta Papayo and Isla Magueyes members belong to the *Barrettia monilifera* Biozone of Campanian age. Echinoderm fossils were collected from localities PL-1, which consist of units of the Punta Papayo Member, and from PL-2 from the Bahia Fosforecente Member. Echinoids from locality PL-1 are recrystallized; at locality PL-2, fossils consist of various types of trace fossils.

YAUCO FORMATION. Located in the Southwestern Igneous Province (Jolly et al., 1998; Schellekens, 1998), the Yauco Formation consists mostly

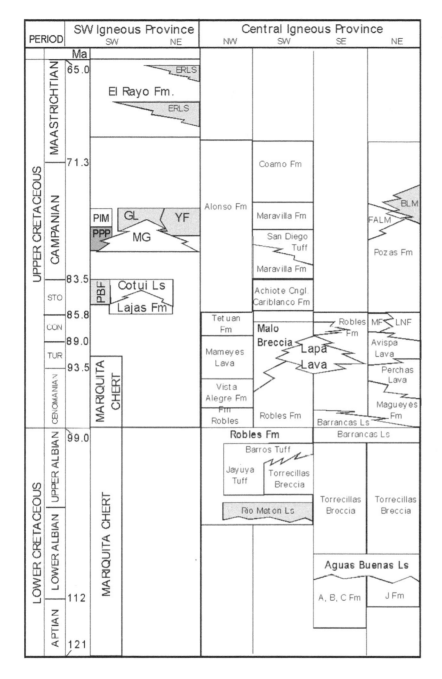

Figure 17.2. *Simplified stratigraphic relationships of Cretaceous formations in the Southwestern and Central igneous provinces of Puerto Rico (modified from Jolly et al., 1998). Units shaded in light gray are units where fossil Echinodermata were found in this study; units where fossil Echinodermata were collected both by this and previous works are shaded in dark gray. Abbreviations: BLM, Botijas Limestone Member of the Robles Formation; Cngl, conglomerate; ERLS, El Rayo Formation limestone lenses; FALM, Flor de Alba Limestone Member of the Pozas Formation; Fm, formation; GL, Guaniquilla Limestone; LNF, Los Negros Formation; Ls, limestone; MG, Monte Grande Formation; MF, Maricaboa Formation; NE, northeast; NW, northwest; PBF, Parguera Limestone, Bahia Fosforecente Member; PIM, Parguera Limestone, Isla Magueyes Member; PPP, Parguera Limestone, Punta Papayo Member; SE, southeast; SW, southwest; YF, Yauco Formation.*

of beds of mudstone, siltstone, and sandstone (Volckmann, 1984). The Yauco Formation in the area of study is early to middle Campanian age on the basis of its macro- and microinvertebrate fauna, and represents basinal environments. Only one locality, YF-1, yielded fossil echinoderms. At locality YF-1, the fossils consist mostly of disarticulated crinoid stems and rare inoceramid bivalves, the latter preserved as internal molds.

BOTIJAS LIMESTONE MEMBER OF THE POZAS FORMATION. The Pozas Formation is exposed in the Central Igneous Province of Puerto Rico (Jolly et al., 1998; Schellekens, 1998) between the towns of Jayuya and Coamo in south-central Puerto Rico. The formation consists of volcanic breccia, tuff, and volcanic sandstone interbedded with ash-flow tuff, lava, and limestone lenses, one of which is the Botijas Limestone Member, which consists of massive to thick-bedded limestone and calcarenite (Krushensky and Schellekens, 2001). The Botijas Limestone represents a shallow marine environment and contains abundant rudists from the middle Campanian Stage (Mitchell, personal commun., 2005). Two fossil echinoderms where found in locality BL-1. The echinoderm tests from this locality are recrystallized, and bivalves and gastropods are mostly preserved as internal molds, with rare recrystallized shells occurring.

GUANIQUILLA LIMESTONE. This formation is exposed in the Southwestern Igneous Province of Puerto Rico (Jolly et al., 1998; Schellekens, 1998). Originally described as part of the San Germán Limestone by Mitchell (1922), these units were later split into several formations on the basis of stratigraphy and paleontological data. Later, Santos (1999) assigned these rocks as a separate formation on the basis of the stratigraphy and the paleontological content. The age of the Guaniquilla Limestone is middle to late Campanian and represents shelf environments (Santos, 1999). The specimens from this locality consist of complete tests of irregular echinoids.

EL RAYO FORMATION. The El Rayo Formation is exposed between the towns of Lajas and Sabana Grande in the Southwestern Igneous Province (Jolly et al., 1998; Schellekens, 1998). The formation was first described by Slodowski (1956) as feldspathic olivine basalt porphyry with interbedded lenses of volcanic conglomerate and dark massive limestone. The limestones later were described by Santos (1999) as a carbonate platform/ramp developed during a relative increase in sea level that overlies the basalt porphyries and conglomerates. The age of El Rayo Limestone lenses is early late Maastrichtian on the basis of the occurrences of the ammonoid *Eubaculites labyrinthicus* (Morton, 1834) (Martínez-Colón, 2003) and the rudists *Parastroma guitarti* (Palmer, 1933) and *Titanosarcolites giganteus* (Whitfield, 1897) (Santos, 1999). Two localities from the inner platform environment, ER-E and ER-W, yielded fossil Echinodermata. The specimens from locality ER-W consist of complete tests of irregular echinoids and fragmentary radioles of regular echinoids. Gastropods at this locality generally consist of internal molds, whereas bivalves are found with the shell and as internal molds. At locality ER-E, they consist of incomplete tests of regular echinoids; these were the only fossils collected at this locality.

Middle Tertiary and Quaternary

JUANA DÍAZ FORMATION AND NEW FORMATION. The Juana Díaz Formation and an undescribed new formation both exposed in southern Puerto Rico (Monroe, 1980; Frost et al., 1983). All terrigenous beds, limestones, and chalk, which uncomformably overlie Cretaceous- to Eocene-age rocks and occur below the unconformity with the Ponce Limestone, were assigned to the Juana Díaz Formation (Monroe, 1973). Moussa and Seiglie (1970) and Monroe (1980) informally divided the Juana Díaz Formation into an upper carbonate unit and a lower clastic unit. Frost et al. (1983)

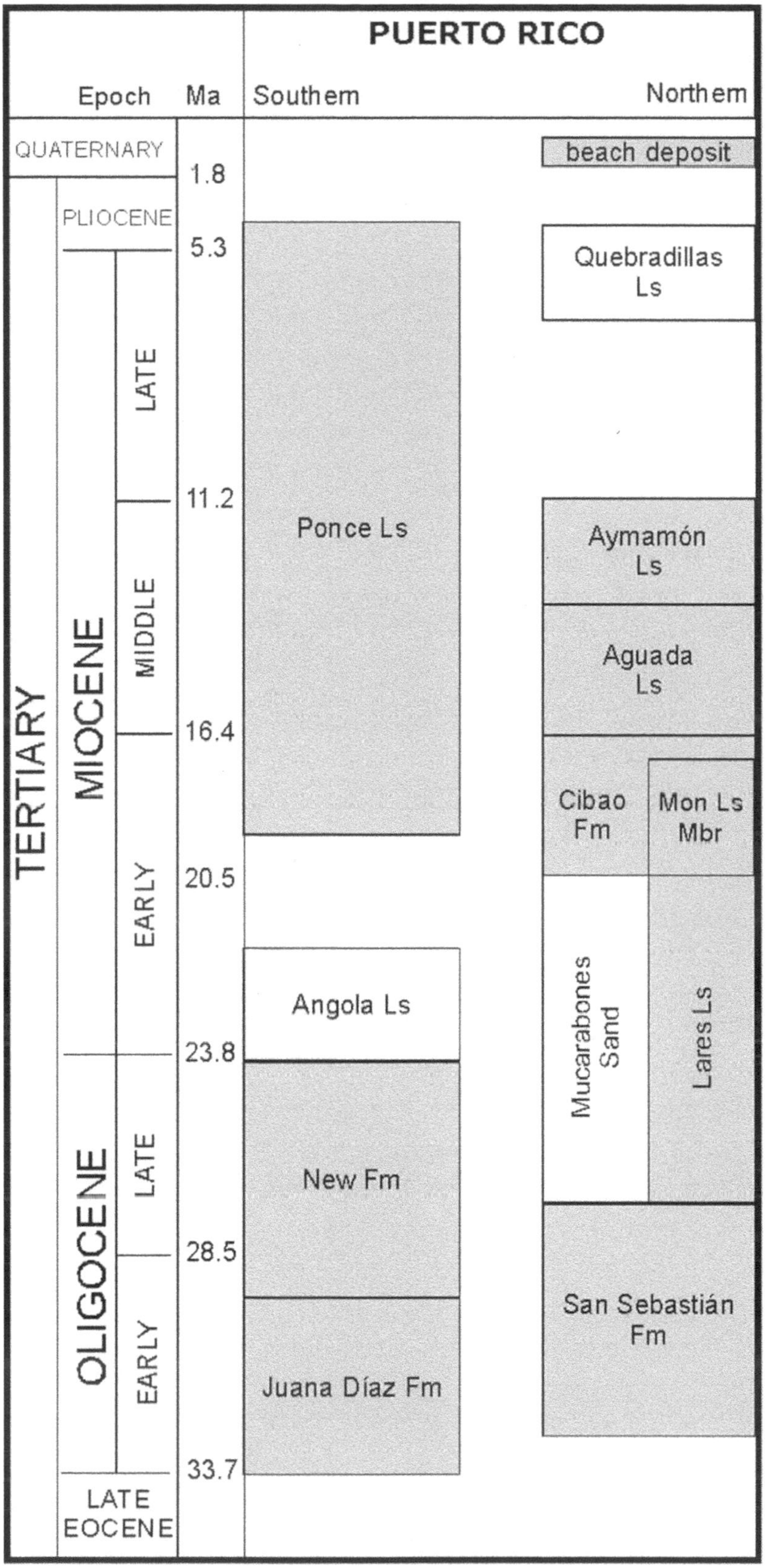

Figure 17.3. *Middle Tertiary and Quaternary stratigraphic units of Puerto Rico. The units shaded in gray are units where echinoids were collected during this and previous works. Abbreviations: Fm, formation; Ls, limestone; Mon Ls Mbr, Montebello Limestone Member.*

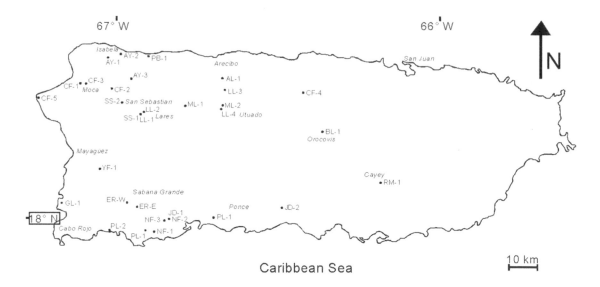

Figure 17.4. *Map of Puerto Rico showing the localities where fossil Echinodermata were collected for this study.*

proposed that the Juana Díaz be restricted to the lower clastic member and that the upper carbonate member be assigned as a new formation, yet to be named. The age of the Juana Díaz Formation, as proposed by Frost et al. (1983), is early Oligocene (Moussa and Seiglie, 1970; Frost et al., 1983; MacPhee and Iturralde-Vinent, 1995). The new formation is late Oligocene in age (Frost et al., 1983). The Juana Díaz Formation represents terrestrial to shallow marine, lagoonal environments; and the new formation represents open-shelf environments (Frost et al., 1983; MacPhee and Iturralde-Vinent, 1995). A total of two localities within the Juana Díaz Formation, JD-1 and JD-2, and three from the new formation, NF-1, NF-2, and NF-3, yielded fossil echinoderms. Preservation of the fossils collected consists of the following: JD-1, complete test of regular echinoids and disarticulated starfish ossicles; JD-2, fragmentary test of an irregular echinoid; NF-1, complete irregular echinoid tests and articulated ophiuroids; NF-2, isolated radioles of regular echinoids; and NF-3, a complete regular echinoid test.

SAN SEBASTIÁN FORMATION. The San Sebastián Formation is exposed in Northern Puerto Rico and consists predominantly of clay and sandy clay, with marl and soft limestone near the top of the formation (Monroe, 1980). On the basis of foraminifera, the age of the formation is late Oligocene as designated by Seiglie and Moussa (1984). The depositional environment of this formation is that of nearshore, lagoonal environments (Monroe, 1980). Echinoderms were collected from two localities, SS-1 and SS-2. The echinoids from SS-1 are well preserved and consist of complete test of irregular echinoids and isolated starfish ossicles. At SS-2, complete irregular echinoid tests were found, but one of them, from a mudstone unit, has dorsoventral compression that obscures some of its features.

LARES LIMESTONE. Exposed in northern Puerto Rico, the name was first used by Berkey (1915) as the Lares Shale; later, Hubbard (1923) redefined the formation and raised it to the Lares Formation. Zapp et al. (1948) used the name Lares Limestone and included two large regional bioherms in the formation. Nelson and Monroe (1966) restricted the formation to limestone beds below the Cibao Formation and excluded the bioherms once again. The upper parts of the bioherms of Zapp et al. (1948) were assigned to the Cibao Formation by Monroe (1980). The age of the Lares limestone was designated as late Oligocene to early Miocene by Seiglie and Moussa (1984) on the basis of foraminifera. The Lares Limestone represents backreef complexes and coral bioherm environments (Seiglie and Moussa, 1984). From this formation, four localities, LL-1, LL-2, LL-3, and LL-4, yielded echinoderms. At locality LL-1, complete test and fragmentary radioles of regular echinoids were found; isolated starfish and brittle star ossicles were also found. Localities LL-2, LL-3, and LL-4 yielded complete tests of irregular echinoids.

CIBAO FORMATION. Exposed in northern Puerto Rico, this formation was first described by Hubbard (1923) as the Cibao Limestone. Later, Zapp et al. (1948) used the name Cibao Marl for deposits of clastic materials between the Lares Limestone and the Aguada Limestone. Monroe (1962) redefined the Cibao Formation as consisting of several members of calcareous clay, limestone, sand, and gravel. The age of these deposits was determined as late early Miocene on the basis of its foraminiferal assemblage (Seiglie and Moussa, 1984). Echinoderms where collected from four localities in this formation, CF-1, CF-2, CF-3, CF-4, and CF-5. Three of the localities, CF-1, CF-4, and CF-5, yielded complete tests of irregular echinoid; localities CF-2 and CF-3 yielded isolated plates and radioles of regular echinoids.

The Montebello Limestone Member of the Cibao Formation was defined by Nelson and Monroe (1966) as weakly indurated, medium to coarse calcarenite composed largely of foraminifers and fragments of molluscan shells. Large beds of oysters are present above the Lares Limestone at most localities, marking the boundary between the Lares Limestone and the Montebello Limestone Member (Monroe, 1980). The Montebello Limestone represents reef complex environments. This member is highly fossiliferous at some localities where corals, bivalves, gastropods, and echinoderm test fragments are common. Localities within this formation are ML-1 and ML-2. At locality ML-1, complete tests of irregular echinoids were found, which differed from ML-2, where only isolated radioles of regular echinoids and disarticulated starfish ossicles were found.

PONCE LIMESTONE. Exposed in southern Puerto Rico, the name was first used by Berkey (1915), but he probably was referring to the limestone that is now assigned to the new formation of Frost et al. (1983). The limestone was divided by Hubbard (1923) into the lower and upper Ponce Limestones; later, Monroe (1973) restricted the name to the upper limestone. The foraminifera of the Ponce Limestone were studied by Seiglie and Bermúez (1969), and on the basis of the occurrence of *Globigerinoides ruber* (d'Orbigney, 1839) and *Globorotalia ungulate* Bermúez, 1960, they assigned a middle to late Miocene age to the formation. This formation represents reef environments (Monroe, 1980). This formation had only one locality, PO-1, which yielded fossil echinoderms in the form of complete tests of irregular echinoids.

AGUADA (LOS PUERTOS) LIMESTONE. The name was originally used by Zapp et al. (1948) for limestone exposures in the town of Aguada in northern Puerto Rico. Later, Monroe (1968) assigned these units to the Cibao Formation and the mapped rocks northeast of Aguada as the Aguada Limestone. Hubbard's (1923) type locality for the Los Puertos Limestone is the same as for the base of the Aguada Limestone, but at the top of the type unit, he included a limestone that is now assigned to the Aymamon Limestone (Monroe, 1980). Because Zapp et al. (1948) considered the exposures in the town of Aguada as typical, and because they were uncertain of the top of the Los Puertos Limestone, they abandoned the name "Lost Puertos" and replaced it with "Aguada Limestone" (Monroe, 1980). This formation represents reef complex environments. The age of the Aguada (Los Puertos) Limestone is late early Miocene to early middle Miocene (Ramirez, 2001) on the basis of its foraminiferal content. Fossils collected at locality AL-1 consist of disarticulated starfish ossicles; gastropods collected at this locality had heavily recrystallized shells.

AYMAMÓN LIMESTONE. Exposed in northern Puerto Rico, the lower beds of this formation represent backreef lagoonal environments, the middle units represent coral reef deposits, and the upper units represent forereef environments (Seiglie, 1978; Seiglie and Moussa, 1984). The age of this formation is middle Miocene on the basis of the foraminiferal content (Seiglie and Moussa, 1984). AY-1, AY-2, and AY-3 are the localities yielding echinoids. Both complete and incomplete irregular echinoid tests were collected at locality AY-1; at locality AY-2, an incomplete test of an irregular echinoid was collected; a heavily recrystallized, but still complete, test of an irregular echinoid was collected at locality AY-3.

ANCIENT BEACH DEPOSITS. Exposed in northern Puerto Rico, these deposits are mainly composed of fine- to coarse-grained carbonate sand representing high-energy intertidal marine environments. The age of this deposit is Pleistocene on the basis of isotopes obtained from corals (Taggart and Joyce, 1990). One locality within this formation yielded echinoderms, PB-1. Fossils at this locality consist mostly of irregular echinoderm tests, and gastropod shell fragments are also abundant and well preserved.

Results

The fossil Echinodermata from Puerto Rico are in a total of 14 formations ranging in age from Early Cretaceous to Quaternary. Cretaceous echinoderms were collected from a total of six formations outcropping in eight localities. A faunal list of Cretaceous species (Table 17.1) shows a total of 11 species, of which only one genus, *Hemiaster* Agassiz *in* Agassiz and Desor, 1847, was previously reported (Jackson, 1922). Of these 11 echinoids, four are from the El Rayo Formation, three in the Parguera Limestone; two are from the Botijas Limestone; and one species is in each of the remaining formations reported in this study. The only echinoid ichnofossil found during the course of this study was from a locality of the Bahia Fosforecente Member of the Parguera Limestone, and it is assigned to the genus *Scolicia* sp. This locality is also known for yielding other ichnofossils (M. Ruiz-Yantín, personal commun.).

Species	Locality	Formation	Specimen
Tylocidaris sp.	RM-1	Río Matón Limestone	UPRMP-2836
Cidaroida sp. A	ER-E	El Rayo Formation	UPRMP-2760
Cidaroida sp. B	ER-W	El Rayo Formation	UPRMP-2747
Codiopsis sp.	PL-1	Parguera Limestone	UPRMP-2685
Petalobrissus (*Petalobrissus*) sp.	BL-1	Botijas Limestone	UPRMP-2831
Hemiaster sp.	BL-1	Botijas Limestone	UPRMP-2833; UPRMP-2834
Hemiaster sp.	ER-W	El Rayo Formation	UPRMP-2671; UPRMP-2768
Proraster dalli	ER-W	El Rayo Formation	UPRMP-863
Nucleolitid	GL-1	Guaniquilla Limestone	UPRMP-2735
Crinoidea	YF-1	Yauco Formation	UPRMP-2688
Scolicia sp.	PL-2	BF-Parguera Limestone	NA

Note.—Refer to Figure 17.4 for key localities.
Abbreviation.—NA, not available.

There are a total of 13 species in the Oligocene rocks from a total of 11 localities (Table 17.2). From the Juana Díaz Formation, three species are present; five species occur in the new formation, four species were recovered from the San Sebastián Formation, and seven species were found in the Lares Limestone. The Miocene yielded a total of 15 species from a total of 13 localities. In the Cibao Formation, there are six species present, three of these in the Montebello Limestone Member; one from the Aguada Limestone; four from the Aymamón Limestone; and two from the Ponce Limestone. One echinoderm species was recovered from the Quaternary beach deposits (Table 17.3).

Table 17.1. *List of Cretaceous Echinodermata from Puerto Rico.*

Discussion

The fossil echinoids found in this study expand the fauna known from earlier works. Our main attempt was to include the different groups encompassing the Echinodermata. Even though we tried to collect as many specimens as we could, it was also important to use previous works to capture total diversity (Jackson, 1922; Gordon, 1963; Cutress, 1980). We identified eight Cretaceous genera from Puerto Rico, whereas only one genus, *Hemiaster*, was previously described (Jackson, 1922). Also, this is the first report of the echinoid ichnofossil, *Scolicia* sp., from the Cretaceous of Puerto Rico, as well as the first report of crinoids, also from the Cretaceous. The complex Cretaceous stratigraphy of the Central Igneous Province (Fig. 17.2) will undoubtedly yield more fossil Echinodermata in the future.

The Tertiary Echinodermata comprise mostly regular and irregular sea urchins and sand dollars, but we also report the occurrence of starfish (goniasterids) and brittle stars (ophiuroids), which had not been reported previously from Puerto Rico. The Tertiary sea urchins were described extensively by Gordon (1963), who concentrated mostly on irregular echinoids; and the cidaroids were later studied by Cutress (1980). We report *Clypeaster* cf. *C. oxybaphon* Jackson, 1922, as occurring in the Lares Lime-

Species	Locality	Formation	Specimen
Phyllacanthus peloria	LL-1	Lares Limestone	UPRMP-2751
Prionocidaris spinidentatus	NF-2	New Formation	UPRMP-2749
Echinometra sp.	NF-3	New Formation	UPRMP-2740
Gagaria sp.	JD-1	Juana Díaz Formation	UPRMP-2764
Echinometra sp.	LL-1	Lares Limestone	UPRMP-3035
Echinolampas lycopersicus	SS-1	San Sebastián Formation	UPRMP-2745
	LL-2	Lares Limestone	UPRMP-2736; UPRMP-2744
	LL-3	Lares Limestone	UPRMP-2771
Antillaster elegans	NF-1	New Formation	UPRMP-2955; UPRMP-2956
Agassizia clevei	LL-2	Lares Limestone	UPRMP-2765
Schizaster sp.	JD-2	Juana Díaz Formation	UPRMP-2769
	SS-2	San Sebastián Formation	UPRMP-2753
Clypeaster oxybaphon	NF-1	New Formation	UPRMP-2742; UPRMP-2828; UPRMP-2829
Clypeaster cf. *C. oxybaphon*	LL-4	Lares Limestone	UPRMP-2755
Echinarachnius sebastiani	SS-2	San Sebastián Formation	UPRMP-2770
Goniasteridae	JD-1	Juana Díaz Formation	UPRMP-2953
	SS-1	San Sebastián Formation	Uncat.
	LL-1	Lares Limestone	UPRMP-2762
Ophiuroidea	NF-1	New Formation	UPRMP-2767
	LL-1	Lares Limestone	UPRMP-3037

Note.—Refer to Figure 17.4 for key localities.

Table 17.2. *List of Oligocene Echinodermata from Puerto Rico.*

stone, previously not known to occur in this formation (Jackson, 1922; Gordon, 1963). The occurrence of *Prionocidaris spinidentatus* (Palmer *in* Sánchez-Roig, 1949) in the new formation and Cibao Formation represents the first record of this echinoid in these formations. Its occurrence in the new formation increases its age range. Previously known only from the Miocene (Cutress, 1980), it now ranges from late Oligocene to the middle Miocene. Fossils of the genus *Schizaster* Agassiz, 1836, occurring in the Juana Díaz and San Sebastián formations and in the Aymamón Limestone represent the first report of this genus in these formations. Some of the clypeasteroids that have not been reported previously are *Clypeaster* sp. 1, characterized by its unique test morphology, from the Cibao Formation, and *Clypeaster rosaceus* (Linnaeus, 1758) from Pleistocene beach deposits. The addition of these clypeasteroids to our fauna portrays the high diversity of this genus in the region. The genus *Encope* Agassiz, 1841, is, herein also reported for the first time as occurring in the Aymamón Limestone, adding one more genus to the ones already known to occur in the formation (Zapp et al., 1948; Gordon, 1963). *Phyllacanthus peloria* (Jackson, 1922), previously reported from the San Sebastián Formation and Lares Limestone

Species	Locality	Formation	Specimen
Phyllacanthus peloria	ML-2	Montebello Limestone	UPRMP-2752
Prionocidaris katherinae	CF-2	Cibao Formation	UPRMP-2758
Prionocidaris spinidentatus	CF-3	Cibao Formation	UPRMP-2748
Echinolampas aldrichi	CF-4	Cibao Formation	UPRMP-2750
Echinolampas semiorbis	ML-1	Montebello Limestone	UPRMP-2743
Eupatagus sp.	AY-1	Aymamón Limestone	UPRMP-2737
Paraster sp.	CF-1	Cibao Formation	UPRMP-2739
Schizaster sp.	AY-1	Aymamón Limestone	UPRMP-2734
Clypeaster concavus	CF-1	Cibao Formation	UPRMP-2741
Clypeaster cubensis	PO-1	Ponce Limestone	UPRMP-2957
	AY-3	Aymamón Limestone	UPRMP-2738
Clypeaster rosaceus	PB-1	Ancient Beach Deposits	UPRMP-2754
Clypeaster sp. 1	CF-5	Cibao Formation	UPRMP 2766
Clypeaster sp. 2	PO-1	Ponce Limestone	UPRMP-2912
Encope cf. *E. latus*	AY-2	Aymamón Limestone	UPRMP-2756
Goniasteridae	ML-2	Montebello Limestone	UPRMP-2930
	AL-1	Aguada Limestone	UPRMP-2761

Note.—Refer to Figure 17.4 for key localities.

Table 17.3. *List of Miocene and Quaternary Echinodermata from Puerto Rico.*

(Jackson, 1922; Cutress, 1980), is herein reported for the first time from the Montebello Limestone Member of the Cibao Formation, increasing its spatial distribution from late Oligocene to upper early Miocene.

We also present here the first report of fossil Asterozoa from Puerto Rico. These consist of starfish and brittle stars. Brittle stars have been reported previously from the Cenozoic of Trinidad and Jamaica (Dixon et al., 1994, Donovan, 2001); Puerto Rico is thus the second area of the Caribbean where these have been reported. Previous workers in Puerto Rico have largely ignored the starfish ossicles: there are no reports of their occurrence, even though they are fairly common in Oligocene to Miocene rocks. This is possibly because of the incompleteness of their remains, which makes identification difficult. Starfishes have been reported previously from the Cenozoic of other localities in the Caribbean (Donovan, 2001).

One group of fossils that should receive more attention are the clypeasteroids. Our *Clypeaster* Lamarck, 1801, fauna consists of a total of six species: one from the Oligocene, four from the Miocene, and one from the Quaternary. This demonstrates that Puerto Rico was not an exception in having a diverse *Clypeaster* fauna, at least during the Miocene, like other localities in the Caribbean region (Donovan, 2001).

Other clypeasteroids also show an interesting pattern. The occurrence of the clypeasteroid *Echinarachnius sebastiani* Jackson, 1922, presents an interesting biogeographical history. This genus is restricted today to high latitudes in the Pacific and northwestern Atlantic (Seilacher, 1979; Ghiold and Hoffman, 1986). The occurrence of this genus could be an indicator of a colder climate during the Oligocene in Puerto Rico and the Carib-

Figure 17.5. *Codiopsis sp.*, UPRMP-2685. *Dorsolateral view of specimen, still partially surrounded by matrix, from the Punta Papayo Member of the Parguera Limestone, locality PL-1, southwestern Puerto Rico. Scale bar = 10 mm.*

bean. Studies that used other fossils from this period (i.e., Graham and Jarzen, 1969) also indicate that the region was colder than it is today.

Cretaceous Paleoenvironments

RÍO MATÓN LIMESTONE MEMBER OF THE ROBLES FORMATION.. In this formation, only one regular echinoid was collected. The fossil consists of an isolated radiole assigned to the cidaroid genus *Tylocidaris* sp. This genus, like other short-spined cidaroids, probably preferred hard substrates (i.e., reef, sea grass beds) and can be an indicator of such an environment (Fell, 1966). That the fossil consists of an isolated radiole suggests that it was transported from its original depositional environment or that burial rate was low as in a hard substrate environment. Other fossils in the locality also indicate the occurrence of hard substrates, such as chondrodont bivalves in growth position.

PARGUERA LIMESTONE, PUNTA PAPAYO MEMBER.In this member, two echinoids are found, *Codiopsis* sp. and *Hemiaster berkeyi* Jackson, 1922. The occurrence of *Codiopsis* sp. (Fig. 17.5), a shallow-water form (Fell and Dawson, 1966), in this slope deposit indicates that the fossil was transported from shallower waters. *Hemiaster berkeyi*, like other spatangoids, possibly lived buried in mud or silt, which were present in the deposits where it was found.

PARGUERA LIMESTONE, BAHIA FOSFORECENTE MEMBER. The only echinoid fossil in this member consists of trace fossils assigned to *Scolicia* sp. (Fig. 17.6). This trace fossil is made by spatangoid echinoids. The fossils are found in mudstone units deposited in slope environments. The occurrence of this trace fossil and other trace fossils at this locality (such as *Zoophycus* Massalongo, 1855, and *Thalassinoides* Ehrenberg, 1944) indicates that this was a well-oxygenated bottom.

Figure 17.6. *Echinoid trace fossil **Scolicia** sp. from the Bahia Fosforecente Member of the Parguera Limestone, locality PL-2, southwestern Puerto Rico. Pencil is about 53 mm long.*

YAUCO FORMATION. Echinoderm fossils in this formation consists of stalked crinoid stems. The sedimentary rocks where these fossils were collected consist mostly of mudrock and siltstones deposited in a basinal environment. The occurrence of stalked crinoids in this formation agrees with the interpretation provided by the sedimentology, as these were eliminated from shallow-water environments by the Middle Cretaceous (Aronson and Blake, 1997).

BOTIJAS LIMESTONE MEMBER OF THE POZAS FORMATION. The spatangoid *Hemiaster* sp. and the cassiduloid *Petalobrissus* sp. were found in this formation. Both species were collected from mudstone deposited in shallow marine environments, possibly in a lagoonal setting. Species of *Hemiaster* probably lived buried in mud (Kier, 1984); the occurrence of callianassid crabs supports this interpretion about the locality.

GUANIQUILLA LIMESTONE. The only echinoid from this formation is a nucleolitid. The depositional environment in this formation is shallow marine. The fossil has a highly domed test morphology (Fig. 17.7), indicating that this echinoid probably lived partially buried in the sediment in forereef environments. This interpretation is supported by its similarity with *Antillaster* Lambert, 1909, which is believed to have lived in a similar habitat as recent species of *Paleopneustes* Agassiz, 1836 (Kier, 1984).

EL RAYO FORMATION. A total of four echinoderms are known from this formation (Table 17.1). The cidaroids indicate a hard substrate, needed for

Figure 17.7. *Lateral view of nucleolitid, UPRMP-2735, from the Guaniquilla Limestone, locality GL-1, southwestern Puerto Rico. Scale bar = 10 mm.*

the epifaunal life mode of these echinoids. The other echinoids, the spatangoids *Hemiaster* sp. and *Proraster dalli* (Clark, 1893), lived buried in mud in lagoonal environments. At locality ER-W, the cidaroid remains consist of fragmentary radioles. This indicates that these were transported into soft, low-energy settings where the spatangoids were living. The other cidaroids, found at locality ER-W, consist of articulated test fragments in a grainstone; these were probably transported into a high-energy nearshore environment.

Middle Tertiary and Quaternary Paleoenvironments

JUANA DÍAZ FORMATION. In this formation, five echinoderms were found (Table 17.4). The regular echinoids *Gagaria* sp. and *Cidaris bermudezi* Cutress, 1980, were echinoids with an epifaunal life mode, needing hard substrates, such as reefs or sea grass prairies growing in hard substrates. The spatangoid *Schizaster* sp. had an infaunal life mode, living in mud, like modern species of this genus. The modern species *Schizaster doederleini*

Table 17.4. *List of fossil Echinodermata from the Juana Diaz Formation.*

Species	Source
Cidaris bermudezi	Cutress (1980)
Gagaria sp.	This work
Schizaster sp.	This work
Clypeaster oxybaphon	Gordon (1963)
Goniasteridae	This work

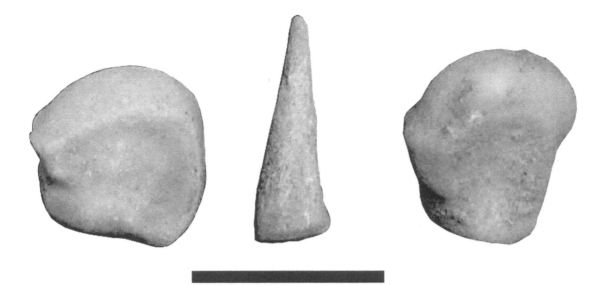

Figure 17.8. *Goniasterid ossicles, UPRMP-2953, from the Juana Diaz Formation, locality JD-1, southwestern Puerto Rico. Scale bar = 10 mm.*

(Chester, 1972) lives buried from 20 to 100 mm below the sediment surface (Kier, 1984).

Clypeasteroids are represented by *Clypeaster oxybaphon*, an Oligocene species with a generally flat test outline. This morphology is also exhibited by the modern species *Clypeaster subdepressus* Gray, 1825, which not only resembles sand dollars in shape but also in behavior (Seilacher, 1979). We believe that this was also the case with *Clypeaster oxybaphon*, living shallowly buried in nearshore, high-energy environments.

Starfish, also found in this formation, consist of isolated ossicles, therefore making identification difficult. We tentatively assign these to the Goniasteridae (Fig. 17.8). These are mobile benthos that could tolerate changes in salinity; therefore, their fossils are expected in a variety of depositional environments. Disarticulated asteroids (Fig. 17.8) indicate slow burial in high-energy depositional environments, an interpretation supported by the sandy facies where they were found.

NEW FORMATION. A total of five Echinodermata are known in this formation (Table 17.5). At locality NF-1, we collected three echinoids, each from different sedimentary facies. *Clypeaster oxybaphon* was from a well-sorted sandstone, which represents the nearshore environments preferred

Species	Source
Prionocidaris spinidentatus	This work
Echinometra sp.	This work
Antillaster elegans	Jackson (1922)*; this work
Clypeaster oxybaphon	Gordon (1963)†; this work
Ophiuroidea	This work

Table 17.5. *List of fossil Echinodermata from the new formation.*

* Reported as *Eupatagus elegans*, assigned to *Antillaster* by Kier (1984).
† Reported from the "Lower Ponce Limestone," the locality is now known to be part of the new formation.

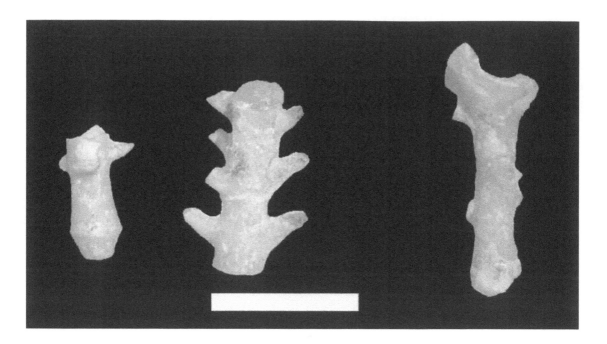

Figure 17.9. *Radioles of* **Prionocidaris spiniden-** **tatus**, *UPRMP-2749, from the New Formation, locality NF-2, southwestern Puerto Rico. Scale bar = 10 mm.*

by this echinoid, as discussed above. The spatangoid *Antillaster elegans* (Jackson, 1922) was found in what is interpreted as forereef deposits (Frost et al., 1983). According to Kier (1984), this echinoid lived on top of the substrate in deep tropical waters. Brittle stars (Ophiuroidea) were found articulated in fine-grained sandstone facies that represent beach deposits. The fact that these fossils are still articulated indicates rapid burial. Brittle stars have a cryptic life mode, but they can be found in a variety of environments within the shallow marine environments. The cidaroid *Prionocidaris spinidentatus* (Fig. 17.9) was collected from locality NF-2. *P. spinidentatus* had relatively long spines with coarse thorns; this adaptation could have aided in camouflage in areas with low turbulence, such as deep lagoonal environments (Poddubiuk, 1987). The regular echinoid *Echinometra* sp. was found in locality NF-3. This echinoid needed a hard substrate for its epifaunal life mode, like modern species of this genus. A modern example of the life mode of *Echinometra* Gray, 1825, can be found in the species *E. viridis* Agassiz, 1863, and *E. lucunter* (Linneaus, 1758). Both modern species are restricted to water depths of less than 3 m (Kier, 1975). This makes *Echinometra* useful for determination of water depth (if found in situ), at least for the Tertiary of Puerto Rico.

SAN SEBASTIÁN FORMATION. Eight echinoids are now known from this formation (Table 17.6). Because this formation contains a variety of facies within the same outcrop, only those echinoids collected during our research will be discussed thoroughly. At locality SS-1, we collected the cassiduloid *Echinolampas lycopersicus* Guppy, 1866, and goniasterid ossicles from well-sorted sandstone facies. *E. lycopersicus* is restricted to this environment, where they fed on large sediment particles (Poddubiuk, 1987). Goniasterids live in a variety of environments, so their occurrence in this locality is not surprising. Locality SS-2 yielded echinoids from two

Species	Source
Phyllacanthus peloria	Jackson (1922)
Echinometra lucunter	Gordon (1963)
Tylechinus sp.	Gordon (1963)
Echinolampas lycopersicus	Jackson (1922); this work
Schizaster sp.	This work
Echinarachnius sebastiani	Jackson (1922); this work
Scutella sp.	Hubbard (1923)
Goniasteridae	This work

Table 17.6. *List of fossil Echinodermata from the San Sebastian Formation.*

very different facies. The first was *Echinarachnius sebastiani* collected from well-sorted sandstone units. *E. sebastiani* probably lived shallowly buried in high-energy, nearshore sands like *Echinarachnius parma* (Lamarck, 1816), a modern species (Seilacher, 1979). The other echinoid collected in this locality, the spatangoid *Schizaster* sp., was collected from a stratigraphically higher mudstone unit. This echinoid is a large *Schizaster* sp. similar in size to *Schizaster* sp. from the Aymamón Limestone (see below) and *Schizaster munozi* Sánchez-Roig, 1949, from Cuba. Another similarity between *Schizaster* sp. from the San Sebastián Formation and *S. munozi* is the possibility that the former was also wedge shaped with deep petals, like the latter (Kier, 1984; Poddubiuk, 1987). However, dorsoventral deformation of the San Sebastián specimen obscures this feature. If the San Sebastián *Schizaster* is indeed similar to *S. munozi*, then it too could have lived buried in fine-grained sediment (Poddubiuk, 1987), much like the mudstone unit where it was found.

LARES LIMESTONE. A total of nine echinoids are known from this formation (Table 17.7). From locality LL-1, we collected four Echinodermata from the same facies. *Phyllacanthus peloria* and *Echinometra* sp. are echinoids that live in shallow lagoonal environments where hardgrounds are available (Kier, 1975; Poddubiuk, 1987). Goniasterids and ophiuroids live in a variety of environments and thus are poor paleoenvironmental indica-

Species	Source
Cidaris bermudezi	Jackson (1922)*; Cutress (1980)
Phyllacanthus peloria	Jackson (1922); Cutress (1980); this work
Echinometra sp.	This work
Echinolampas lycopersicus	Gordon (1963); this work
Echinolampas aldrichi	Gordon (1963)
Agassizia clevei	Gordon (1963); this work
Clypeaster cf. *C. oxybaphon*	This work
Goniasteridae	This work
Ophiuroidea	This work

Table 17.7. *List of echinoderm fossils from the Lares Limestone.*

* Reported as *Cidaris* sp. A, assigned to the species by Cutress (1980).

Figure 17.10. *Dorsal view of **Echinolampas lycopersicus**, UPRMP-2736, from the Lares Limestone, locality LL-2, northern Puerto Rico. Scale bar = 10 mm.*

tors, as discussed above. The echinoids collected at localities LL-2 (*Echinolampas lycopersicus* [Fig. 17.10] and *Agassizia clevei* Cotteau, 1875), LL-3 (*E. lycopersicus*), LL-4 (*Clypeaster* cf. *C. oxybaphon*) were found in grainstone facies representing outer bank deposits where these echinoids lived buried in the sand at different depths (Poddubiuk, 1987).

CIBAO FORMATION. Fifteen fossil Echinodermata occur in this formation and its members (Table 17.8). The spatangoid *Paraster* sp. and the clypeasteroid *Clypeaster concavus* Cotteau, 1875, were collected at locality CF-1 from surface deposits; nonetheless, the occurrence of *C. concavus* is indicative of sandy bottoms within shallow marine environments (Poddubiuk, 1987). Localities CF-2 and CF-3 yielded the cidaroids *Prionocidaris katherinae* Cutress, 1980, and *Prionocidaris spinidentatus*, respectively. These two species of *Prionocidaris* have similar radioles (with coarse spines), which in the case of *P. katherinae* could also have served a camouflage function. The cassiduloid *Echinolampas aldrichi* Twitchell, 1915 (Fig. 17.11) was collected at locality CF-4. Like other species of *Echinolampas* (Poddubiuk, 1987), this species probably fed by sieving coarse sediments and lived shallowly buried. The difference in size between the different species reported herein (Figs. 17.10–17.12) could be related to the depth at

Species	Source
Cidaris bermudezi	Gordon (1963)*
Phyllacanthus peloria	This work
Prionocidaris cojimarensis	Gordon (1963)†; Cutress (1980)
Prionocidaris katherinae	Cutress (1980); this work
Prionocidaris spinidentatus	This work
Echinolampas aldrichi	This work
Echinolampas lycopersicus	Gordon (1963)
Echinolampas semiorbis	Gordon (1963); this work
Agassizia clevei	Gordon (1963)
Paraster loveni	Gordon (1963)
Paraster sp.	This work
Clypeaster concavus	Gordon (1963); this work
Clypeaster oxybaphon	Gordon (1963)
Clypeaster sp. 1	This work
Goniasteridae	This work

Table 17.8. *List of fossil Echinodermata from the Cibao Formation.*

* Reported as *Cidaris* sp. assigned to species by Cutress (1980).
† Originally reported as *Phyllacanthus* sp., emended by Cutress (1980).

which they lived. Species of *E. lycopersicus*, having the smaller diameter, lived buried completely, whereas *E. semiorbis* Guppy, 1866 (Fig. 17.12), with the largest diameter, lived shallowly buried (Poddubiuk, 1987). *Echinolampas aldrichi*, with a diameter more or less between the other species, could have been a shallow burrower. The last locality, CF-5, yielded an interesting species of clypeasteroid echinoid *Clypeaster* sp. 1. It is characterized by being highly domed dorsally and flattened ventrally (Fig. 17.13). This test morphology indicates that this species lived on top of the seafloor, possibly as an active grazer in sea grass prairies much like *Clypeaster altus* Lamarck, 1816, or other high-domed echinoids like *Antillaster elegans* (Kier, 1984; Poddubiuk, 1987).

Figure 17.11. *Dorsal and ventral views of* **Echinolampas aldrichi,** *UPRMP-2750, from the Cibao Formation, locality CF-4, northern Puerto Rico. Scale bar = 10 mm.*

Figure 17.12. *Dorsal view of **Echinolampas semi-orbis**, UPRMP-2743, from the Montebello Limestone Member of the Cibao Formation, locality ML-1, northern Puerto Rico. Scale bar = 10 mm.*

Two localities within the Montebello Limestone Member of the Cibao Formation each yielded one echinoderm. At locality ML-1, we found the third species of *Echinolampas* reported in this work. This species, *E. semiorbis*, is characterized by its larger size and highly domed test, as discussed above. This echinoid fed on organic-rich sediments near sea grass bafflestones (Poddubiuk, 1987). Goniasterid ossicles were collected from locality ML-2. As already discussed, starfish inhabit a variety of habitats within the marine realm and thus are not useful indicators of paleoenvironments.

AGUADA (LOS PUERTOS) LIMESTONE. A total of four echinoderms are found in this formation (Table 17.9). We will only discuss the occurrence of goniasterid ossicles in locality AL-1. Again, these offer little information on paleoenvironment of the locality. The occurrence of goniasterids in this locality provides the youngest record of starfish as fossils in Puerto Rico. Future discoveries will, we hope, yield starfish fossils from younger sediments, as in other regions of the Caribbean (Donovan, 2001).

AYMAMÓN LIMESTONE. Five fossil Echinodermata are in this formation (Table 17.10). Echinoids collected in locality AY-1 include the spatangoids *Eupatagus* sp. and *Schizaster* sp. collected from well-sorted grainstones. Echinoids of the genus *Eupatagus* Agassiz *in* Agassiz and Desor, 1847, were shallow burrowers, living in sand (Kier, 1984). These sands

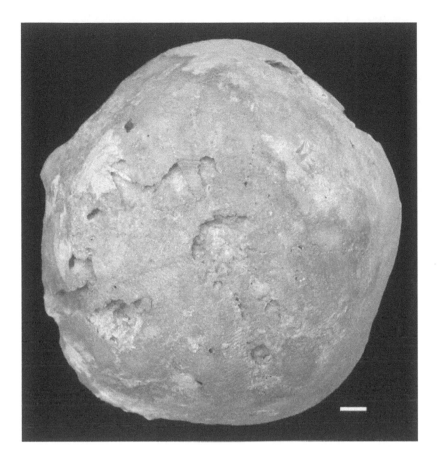

Figure 17.13. *Dorsal view of **Clypeaster** sp. 1, UP-RMP-2766, from the Cibao Formation, locality CF-5, western Puerto Rico. Scale bar = 10 mm.*

seem to have been deposited in deep waters within the open shelf because of the occurrence of teeth of *Carcharodon megalodon* (Agassiz, 1835) in this deposit (Nieves-Rivera et al., 2003). *Schizaster* sp. had an infaunal life mode, sharing similar test morphology with *Schizaster munozi* from Cuba. We assume that life mode of *Schizaster* sp. was similar to S. *munozi*, living buried in fine-grained sediment (Poddubiuk, 1987). The specimen collected in this locality was transported from its preferred habitat to the grainstone facies. Evidence for this is indicated by the growth of barnacles on its ventral surface. Echinoids from locality AY-2 include the sand dollar *Encope* sp. This echinoid probably lived shallowly buried in the sand, like the modern species *Encope michelini* Agassiz, 1841 (Kier and Grant, 1965). At locality AY-3, we collected *Clypeaster cubensis* Cotteau, 1875, which shares similar test morphology with the modern species *Clypeaster rosaceus*. The latter lives on top of the sediment, partially buried, or in sea-grass

Species	Source
Prionocidaris cojimarensis	Cutress (1980)
Agassizia clevei	Gordon (1963)
Clypeaster oxybaphon	Gordon (1963)
Goniasteridae	This work

Table 17.9. *List of fossil Echinodermata from Aguada (Los Puertos) Limestone.*

Species	Source
Eupatagus cubensis	Zapp et al. (1948)*
Eupatagus sp.	This work
Schizaster sp.	This work
Clypeaster cubensis	Gordon (1963); this work
Encope sp.	This work

* Originally described as *E. depressus*, synonymized with *E. cubensis* in Kier (1984).

prairies more than 5 m deep (Kier, 1975). We believe that *C. cubensis* had a similar life mode.

PONCE LIMESTONE. At least six echinoderms are in this formation (Table 17.11). Only two clypeasteroids were collected from locality PO-1: *Clypeaster cubensis* and *Clypeaster* sp. 2 (which could be a juvenile of *C. concavus*). As discussed above, these echinoids most probably had the same life modes as *C. rosaceus* and *C. subdepressus*, respectively. The fossils were collected from grainstones deposited in a backreef setting.

Species	Source
Prionocidaris spinidentatus	Cutress (1980)
Tripneustes sp.	Gordon (1963)
Psamechinus sp.	Gordon (1963)
Echinolampas lycopersicus	Jackson (1922)
Clypeaster concavus	Zapp et al. (1948)
Clypeaster cubensis	Gordon (1963); this work
Clypeaster sp. 2	This work

ANCIENT BEACH DEPOSITS. Only one echinoid is known from these Pleistocene deposits in Puerto Rico, *Clypeaster rosaceus*. The fossils were collected from well-sorted, medium-grained sandstone facies where it possibly lived partially buried like modern species (Kier, 1975). The deposit at this locality represents a nearshore sand environment now exposed as a result of sea-level changes.

Conclusions

Fossil Echinodermata from Puerto Rico were studied and used for paleoenvironmental interpretations. Changes in the spatial distribution of some of the species were made, as well as new reports for the island. The Echinodermata reported herein were found in a total of 15 formations. These formations rage in age from Cretaceous to Quaternary. Comparisons with other published works on fossil echinoderms from Puerto Rico showed that diversity is greater than thought previously, especially in the Cretaceous. The occurrence of starfish (Goniasteridae) and brittle stars

(Ophiuroidea), reported here from Tertiary-age rocks, are the first records of these groups of fossils in Puerto Rico. The occurrence of some of the echinoderms in this study expands the understanding of the diversity of these organisms in the region. The echinoid pattern of distribution through time proposed by Donovan (2001) for other islands in the Caribbean can also be applied to the Puerto Rican echinoid fauna. Paleoenvironmental interpretations were provided for the localities where we found echinoids. These interpretations were applied to each locality and the units where the fossils were found.

This research, as well as more research in the near future, will further increase our knowledge of the echinoderm biodiversity known from Puerto Rico, especially for the Cretaceous, which will be useful for obtaining a better understanding of the paleobiology of this group in the Caribbean region.

Acknowledgments

JVJ extends his gratitude to S. Mitchell (UWI) for helping in the identification of some of the echinoid species. We are thankful to C. L. Schneider and J. Nebelsick, whose reviews greatly improved the chapter. We very much appreciate the opportunity to participate in this project given to us by W. I. Ausich and G. D. Webster. Part of this research was performed during an undergraduate research project with a grant from the PR-LSAMP undergraduate research stipend provided to JVJ, and he expresses his gratitude. The GSA and the GSA Sedimentary Geology Division are recognized for providing travel grants to the senior author to attend and present preliminary results at the 2005 GSA annual meeting. Facilities for preparation and depositing of specimens were available at the Geology Museum at the Department of Geology, University of Puerto Rico.

References

Agassiz, A. 1863. List of the echinoderms sent to different institutions in exchange for other specimens, with annotations. Museum of Comparative Zoology, 1:17–28.

Agassiz, L. 1833–1843. Recherches sur les poisons fossils. Volume 3. Neuchatel, 390 p.

Agassiz, L. 1836. Prodrome d'un monographie des radiaires ou échinodermes. Société d'histoire naturelle de Neuchatel mémoire, 1:168–199.

Agassiz, L. 1841. Monographies d'echinodermes vivans et fossils, 2. Solothium, Neuchatel, 151 p.

Agassiz, L., and E. Desor. 1847. Catalogue raisonné des familles, des genres et des especes de la classe des echinoderms, 167 p.

Almy, C. C., Jr. 1965. Parguera Limestone, Upper Cretaceous, Mayaguez Group, Southwest Puerto Rico. Unpublished Ph.D. dissertation, Rice University, Houston, 203 p.

Aronson, R. B., and D. B. Blake. 1997. Evolutionary paleoecology of dense ophiuroid populations. Paleontological Society Papers, 3:107–119.

Berkey, C. P. 1915. Geological reconnaissance of Porto Rico. Annals of the New York Academy of Science, 70 p.

Bermúez, P. J. 1960. Consideraciones sobre los sedimentos del Mioceno medio al reciente de las costas central y oriente de Venezuela. Minas e Hidrocar-buros, Boletin de la Sociedad Venezoana de Géologos, 7:333–412.

Berryhill, H. L., and L. Glover. 1960. Geology of the Cayey quadrangle, Puerto Rico. U.S. Geologic Survey Miscellaneous Geologic Investigations, Map I-319, scale 1:20,000, 1 sheet.

Boehm, G. 1898. Ueber Caprinidenkalke aus Mexico. Zeitschrift der deutschen geologischen Gesellschaft, 50:323–332.

Chester, R. H. 1972. A new *Paraster* (Echinoidea: Spatangoida) from the Carib-bean. Bulletin of Marine Science, 22:10–25.

Clark, W. B. 1893. The Mesozoic Echinodermata of the United States. U.S. Geological Survey Bulletin, 97, 207 p.

Cotteau, G. H. 1875. Description des échinides tertiaires des iles St. Barthélemy et Anguilla. Kongl. svenska vetenskapaakademien Handlingar, n. F., 13:1–47.

Cutress, B. M. 1980. Cretaceous and Tertiary Cidaroida (Echinodermata, Echi-noidea) of the Caribbean area. Bulletins of American Paleontology, 77, 221 p.

Dixon, H. L., S. K. Donovan, And C. J. Veltkamp. 1994. Crinoid and ophiuroid ossicles from the Oligocene of Jamaica. Caribbean Journal of Science, 30:143–145.

Donovan, S. K. 1994. Some fossil echinoids (Echinodermata) from the Ceno-zoic of Jamaica, Cuba and Guadaloupe. Caribbean Journal of Science, 30:164–170.

Donovan, S. K. 1996. A regular echinoid from the Walderston Formation (lower Oligocene) of Jamaica. Caribbean Journal of Science, 32:78–82.

Donovan, S. K. 2000. A fore-reef echinoid fauna from the Pleistocene of Barba-dos. Caribbean Journal of Science, 36:314–320.

Donovan, S. K. 2001. Evolution of Caribbean echinoderms during the Ceno-zoic: moving towards a complete picture using all of the fossils. Palaeogeog-raphy, Palaeoecology, Palaeoclimatology, 166:177–192.

Donovan, S. K., and J. S. H. Collins. 1997. Unique preservation of an *Echinome-tra* Gray (Echinodermata: Echinoidea) in the Pleistocene of Jamaica. Ca-ribbean Journal of Science, 33:123–124.

Donovan, S. K., and D. A. T. Harper. 2000. The irregular echinoid *Brissus* Gray from the Tertiary of Jamaica. Caribbean Journal of Science, 36:332–335.

Donovan, S. K., H. L. Dixon, D. T. J. Littlewood, C. V. Milsom, and J. C. Nor-man. 1994. The clypeasteroid echinoid *Encope homala* Arnold and Clark, 1934, in the Cenozoic of Jamaica. Caribbean Journal of Science, 30:171–180.

Donovan, S. K., D. N. Lewis, and P. Davis. 2005. Fossil echinoids from Belize. Caribbean Journal of Science, 41:323–328.

Ehrenberg, K. 1944. Ergänzende Bemerkungen zu den seinerzeit aus dem Miozän von Burgschleinitz beschriebenen Gangtrernen und Bauten deka-poder Krebse. Paläontologische Zeitschrift, 23:354–359.

Fell, H. B. 1966. Cidaroids, p. U312–U339. *In* R. C. Moore (eds.), Treatise on Invertebrate Paleontology, pt. U. Echinodermata 3. Geological Society of America and University of Kansas Press, Lawrence.

Fell, H. B., and D. L. Dawson. 1966. Echinacea, p. U367-U437. *In* R. C. Moore (ed.), Treatise on Invertebrate Paleontology. Pt. U. Echinodermata 3. Geo-logical Society of America and University of Kansas Press, Lawrence.

Frost, S. H., J. L. Harbour, M. J. Realini, and P. M. Harris. 1983. Oligocene reef tract development, southwestern Puerto Rico. Sedimenta, 9:1–144.

Ghiold, J., and A. Hoffman. 1986. Biogeography and biogeographic history of Clypeasteroid echinoids. Journal of Biogeography, 13:183–206.

Gordon, W. A. 1963. Middle Tertiary Echinoids of Puerto Rico. Journal of Paleontology, 37:628–642.

Graham, A., and D. M. Jarzen. 1969. Studies in neotropical paleobotany. I. The Oligocene communities of Puerto Rico. Annals of the Missouri Botanical Garden, 56:308–357.

Gray, J. E. 1825. An attempt to divide the Echinida or sea eggs, into natural families. Annals of Philosophy, 10:423–431.

Guppy, R. J. L. 1866. On Tertiary echinoderms from the West Indies. Geolological Society of London, Quarterly Journal, 22:297–301.

Hubbard, B. 1923. The geology of the Lares District, Porto Rico. New York Academy of Science Survey of Porto Rico and the Virgin Islands, 2, 115 p.

Jackson, R. T. 1922. Fossil echini of the West Indies. Carnegie Institution of Washington Publication, 306, 103 p.

Jolly, W. T., E. G. Lidiak, J. H. Schellekens, and H. Santos. 1998. Volcanism, tectonics, and stratigraphic correlations in Puerto Rico, p. 1–34. In E. G. Lidiak, and D. A. Larue (eds.), Tectonics and Geochemistry of the Northeastern Caribbean. Geological Society of America Special Paper, 322.

Kier, P. M. 1975. The echinoids of Carrie Bow Cay, Belize. Smithsonian Contributions to Zoology, 206, 45 p.

Kier, P. M. 1984. Fossil spatangoid echinoids of Cuba. Smithsonian Contributions to Paleobiology, 55, 339 p.

Kier, P. M. 1992. Neogene Paleontology in the Northern Dominican Republic 13. The Class Echinoidea (Echinodermata). Bulletins of American Paleontology, 102:13–35.

Kier, P. M., and R. E. Grant. 1965. Echinoid distribution and habits, Key Largo Coral Reef Preserve, Florida. Smithsonian Miscellaneous Collections, 149, 68 p.

Krushensky, R. H., and J. H. Schellenkens. 2001. Geology of Puerto Rico, p. 24–36. In W. J. Bawiec (comp.), Geology, geochemistry, geophysics, mineral occurrences and mineral resource assessment of the Commonwealth of Puerto Rico. U.S. Geological Survey Open-File Report, 98-38, CD-ROM.

Lamarck, J. B. P. A. De M. De. 1801. Systéme des animaux sans vertebres précéde du discourse d'ouverture de Cours de Zoologie, donné dans le Muséum National d'Histoire Naturelle, l'an 8. Paris, 432 p.

Lamarck, J. B. P. A. De M. De. 1816. Histoire naturelle des animaux sans vertebres, présentant les caracteres, genéralaux et particuliers de ces animaux, leur distribution, leur classes, leurs familles, leur genres et la citation synonymique des principales especes qui s'y rapportent (first edition, volume 3), 586 p.

Lambert, J. M. 1907–1909. Description des Echinides fossils des terrains miocéniques Sardaigne. Mémoire de la Société Paléontologie Suisse, 34–35:1–142.

Linnaeus, C. 1758. Systema Naturae per Regna tria naturae, secundum Classes, Ordines, Genera, Species, cum characteribus differentis, synonymis, locis (tenth revised edition), 824 p.

MacPhee, R. D. E., and M. A. Iturralde-Vinent. 1995. Origin of the Greater Antillean land mammal fauna, 1: new Tertiary fossils from Cuba and Puerto Rico. American Museum Novitates, 3141:1–30.

Mann, P., G. Drapper, and J. F. Lewis. 1991. An overview of the geologic and tectonic development of Hispaniola, p. 1–28. In P. Mann, G. Drapper, and J. F. Lewis (eds.), Geologic and tectonic development of the North American-Caribbean plate boundary in Hispaniola. Geological Society of America Special Paper, 262.

Martínez-Colón, M. 2003. Geologic and tectonic history of the Eastern section of the Sabana Grande Quadrangle. Unpublished Ph.D. dissertation, University of Puerto Rico, Mayagüez, 109 p.

Massalongo, A. 1855. Monografía delle nereidi fossili del M. Bolca. G. Antonelli, Verona, 35 p.

Mitchell, G. J. 1922. Geology of the Ponce District, Porto Rico. New York Academy of Sciences, Scientific Survey of Puerto Rico and the Virgin Islands, 1:229–300.

Monroe, W. H. 1962. Geology of the Manati Quadrangle. U.S. Geological Survey Miscellaneous Investigation Series, Map I-0334, scale 1:20,000.

Monroe, W. H. 1968. The Aguada Limestone of northwestern Puerto Rico. Geological Survey Bulletin, 1274-G.

Monroe, W. H. 1973. Geologic map of the Bayamon quadrangle, Puerto Rico. U.S. Geological Survey Miscellaneous Geological Investigation, Map I-565, scale 1:20,000.

Monroe, W. H. 1980. Geology of the middle Tertiary formations of Puerto Rico. U.S. Geological Survey Professional Paper, 953, 93 p.

Mooi, R., and D. Peterson. 2000. A new species of *Leodia* (Clypeasteroida: Echinoidea) from the Neogene of Venezuela and its importance in the phylogeny of Mellitid sand dollars. Journal of Paleontology, 74:1083–1092.

Morton, S. G. 1834. Synopsis of the organic remains of the Cretaceous group. American Journal of Science, 17:1–88.

Moussa, M. T., and G. A. Seiglie. 1970. Revision of mid-Tertiary stratigraphy of southwestern Puerto Rico. American Association of Petroleum Geologist Bulletin, 54:1887–1898.

Nelson, A. R., and W. H. Monroe. 1966. Geology of the Florida Quadrangle, Puerto Rico. U.S. Geological Survey Bulletin B, 1221C:C1–C22.

Nieves-Rivera, A. M., M. Ruiz-Yantin, and M. D. Gottfried. 2003. New record of the lamnid shark *Carcharodon megalodon* from the Middle Miocene of Puerto Rico. Caribbean Journal of Science, 39:223–227.

D'Orbigney, A., 1839. Voyage dans l'Amerique Meridionale. T. 5, pt. 5, Foraminiféres, Strasbourg, 85 p.

Palmer, R. H. 1933. Nuevos rudistas de Cuba. Revista de Agricultura, Comercio y Trabajo, 14:95–125.

Poddubiuk, R. H. 1987. Sedimentology, echinoid palaeoecology and palaeobiogeography of Oligo–Miocene eastern Caribbean Limestones. Ph. D. dissertation, Royal Holloway and Bedford College, University of London, 419 p.

Ramirez, W. R. 2001. The Lares Limestone and the Montebello Member of the Cibao Formation along Highway PR10. *In* Ramirez W. R. (ed.), Field Trip Guide, 2001: A Karst Odyssey. 19th Symposium of Caribbean Geology, University of Puerto Rico, Mayaguez.

Rojas, R., M. A. Iturralde-Vinent, and P. W. Skelton. 1995. Stratigraphy, composition and age of Cuban rudist-bearing deposits, p. 272–291. *In* G. Alencaster, and B. E. Buitron-Sanchez, (eds.), The Third International Conference on Rudistae. Revista Mexicana de Ciencias Geologicas, 12(2).

Sánchez-Roig, M. 1926. Contribucíon a la paleontología Cubana. Los equinodermos fósiles de Cuba. Boletín Minas, 10, 179 p.

Sánchez-Roig, M. 1949. Los equinodermos fósiles de Cuba. Paleontología Cubana, 1, 330 p.

Santos, H. 1999. Stratigraphy and depositional history of the Upper Cretaceous strata in the Cabo Rojo–San German structural block, southwestern Puerto Rico. Unpublished Ph.D. dissertation, University of Colorado, Boulder, 185 p.

Schellekens, J. H. 1998. Geochemical evolution and tectonic history of Puerto

Rico, p. 35–66. *In* E. G. Lidiak, and D. A. Larue (eds.), Tectonics and Geochemistry of the Northeastern Caribbean. Geological Society of America Special Paper, 322.

Seiglie, G. A. 1978. Comments on the Miocene–Pliocene boundary in the Caribbean region. Annales du Centre Universitaire de Savoie, Sciences Naturelles, 3:71–82.

Seiglie, G. A., and P. J. Bermúez. 1969. Informe preliminar sobre los foraminiferos del terciario del sur de Puerto Rico. Parte I. Caribbean Journal of Science, 9:67–80.

Seiglie, G. A., and M. T. Moussa. 1984. Late Oligocene–Pliocene transgressive-regressive cycles of sedimentation in northwestern Puerto Rico. Memoir American Association of Petroleum Geologist, 36:89–95.

Seilacher, A. 1979. Constructional morphology of sand dollars. Paleobiology, 5:191–221.

Slodowski, T. R. 1956. Ecology of the Yauco area, Puerto Rico. Unpublished Ph. D. dissertation, Princeton University, Princeton, 177 p.

Taggart, B. E., and J. Joyce. 1990. Radiometrically dated marine terraces on Northwestern Puerto Rico, p. 248–258. *In* D. K. Larue, and G. Draper (eds.), Transactions of the 12th Caribbean Geological Conference. Miami Geological Society.

Twitchell, M. W. 1915. *In* W. B. Clark and M. W. Twitchell (eds.), The Mesozoic and Cenozoic Echinodermata of the United States. U.S. Geological Survey Monograph, 54, 341 p.

Volckmann, R. P. 1984. Upper Cretaceous stratigraphy of southwestern Puerto Rico: a revision. *In* Stratigraphic notes, 1983: contributions to stratigraphy. Geological Survey Bulletin, 1537A:A73–A83.

Whitfield, R. P. 1897. Description of species of Rudistae from the Cretaceous rocks of Jamaica, W.I., collected and presented by Mr. F. C. Nicholas. Bulletin of the American Museum of Natural History, 9:185–196.

Zapp, A. D., H. R. Bergquist, and C. R. Thomas. 1948. Tertiary geology of the coastal plains of Puerto Rico. U.S. Geological Survey Oil and Gas Investigations Map, OM-0085:52–54.

APPENDICES

Appendix 5.1.

*Locality, size, and specimen number for **Uintacrinus** slabs.*

Institution	Catalog No.	State	County	Township	Range	Section	Site Information	Length (m)	Width (m)	n
AMNH	46600	KS	Logan	14S	32W	27	Springer loc. 1	1.63	0.97	65
AMNH	38378	KS	Logan	15S	34W	8	—	—	—	—
U AR	—	KS	Gove	13S	27W	SW13	S of Quinter	—	—	—
U Chicago	—	KS	—	—	—	—	—	2.13	1.22	—
CMC	50684	KS	Logan	14S	32W	SW SW NW26	—	—	—	—
CMC	50685	CO	Mesa	2S	2E	NE NW SW22	—	—	—	—
CMC	50686	CO	Mesa	2S	2E	NE NW SW22	—	—	—	—
CMC	50687	CO	Mesa	2S	2E	—	—	—	—	—
CMC	50688	CO	Mesa	2S	2E	—	—	—	—	—
DMNHS	6000	CO	Mesa	2S	2E	SE NE SW6	—	0.83	0.74	27
FFM	—	KS	Logan	15S	32W	NW6	—	—	—	26
MCZ	—	KS	Logan	14S	32W	27	Springer loc. 1	2.44	1.22	125
U IA	3-1567	KS	Logan	—	—	—	Elkader	~1.0	~1.0	44
U KS	—	KS	Logan	—	—	—	Elkader	1.80	1.20	—
U KS	—	KS	Logan	—	—	—	Springer loc. 2	—	—	—
SMNK	—	KS	Gove	—	—	—	—	0.25	0.13	6
SMNK	—	CO	Mesa	—	—	—	—	—	—	—
Keystone G	—	KS	Logan	15S	32W	NW SW8	—	0.99	0.60	25
LACM	—	KS	Logan	14S	32W	NE22	—	1.14	0.81	20?

Institution	Catalog No.	State	County	Township	Range	Section	Site Information	Length (m)	Width (m)	n
UMMP	14755	KS	—	—	—	—	—	~1.0	~1.0	25
NHM, London	—	KS	Logan	—	—	—	Springer loc. 1	—	—	—
NHM, Paris	—	KS	—	—	—	—	—	—	—	—
NSM	16845	KS	—	—	—	—	—	1.20	1.13	100
U OK	—	KS	—	—	—	—	—	1.22	0.76	—
SMNH	SMF XXIII 174	KS	Gove	—	—	—	near Quinter	1.33	1.10	68
SI, NMNH	S6933	KS	Logan	14S	32W	27	Springer loc. 1	1.52	1.52	154
SI, NMNH	S6934	KS	Logan	—	—	—	20 Mi. W of Elkader	1.67	1.58	187
SI, NMNH	351878	KS	—	—	—	—	—	—	—	—
SM	IP-1	KS	Logan	15S	32W	NW6	—	2.22	1.08	240
SM	14063	KS	Logan	14S	32W	NE22	Springer loc. 2?	—	—	—
U Tübingen	—	KS	—	—	—	—	—	—	—	—
U WI	—	KS	Logan	14S	33W	NW SW9	—	1.83	1.52	—
YPM	24758	KS	Logan	14S	32W	Near Elkader	Springer loc. 1	2.00	1.38	226

Note.—AMNH, American Museum of Natural History, New York; CMC, Cincinnati Museum Center, Cincinnati, Ohio; CO, Colorado; DMNHS, Denver Museum of Natural History and Science, Denver, Colorado; FFM, Fick Fossil Museum, Oakley, Kansas; Keystone G, Keystone Gallery, Oakley, Kansas; KS, Kansas; LACM, Natural History Museum of Los Angeles County, California; MCZ, Museum of Comparative Zoology, Harvard University, Cambridge, Massachusetts; NHM, London, Natural History Museum, London, UK; NHM, Paris, Natural History Museum, Paris, France; NSM, University of Nebraska, State Museum, Lincoln, Nebraska; SI, NMNH, Smithsonian Institution, National Museum of Natural History, Washington, D.C.; SM, Fort Hays State University, Sternberg Museum of Natural History, Fort Hays, Kansas; SMNH, Senckenberg Museum of Natural History, Frankfurt, Germany; SMNK, Staatliches Museum für Naturkunde, Karlsruhe, Germany; U AR, University of Arkansas, Fayetteville, Arkansas; U Chicago, University of Chicago, Chicago, Illinois; U IA, University of Iowa, Iowa City, Iowa; U KS, University of Kansas, Museum of Natural History, Lawrence, Kansas; U OK, University of Oklahoma, Norman, Oklahoma; U Tübingen, Geologisches-Paläontologisches Institut, Tübingen Univerity, Germany; U WI, University of Wisconsin, Geological Museum, Madison, Wisconsin; UMMP, University of Michigan, Museum of Paleontology, Ann Arbor, Michigan; YPM, Yale University, Peabody Museum of Natural History, New Haven, Connecticut.

Appendix 5.2.

Collection information
*for **Uintacrinus** slabs.*

Institution	Catalog No.	Collector	Date	References	Comments
AMNH	46600	F. Springer	1898	Hovey (1902)	F. Springer, donor
AMNH	38378	"EM73"	—	—	—
U AR	—	—	—	Miller et al. (1957)	—
U Chicago	—	—	—	Meyer and Milsom (2001)	On display
CMC	50684	C. Bonner, D. Meyer	1997	—	—
CMC	50685	D. Meyer	1997	—	Fragments
CMC	50686	D. Meyer	1998	—	—
CMC	50687	D. Meyer	1999	—	Twin 1
CMC	50688	D. Meyer	1999	—	Baculite
DMNHS	6000	W. Hawes	1990	Meyer and Milsom (2001)	On display
FFM	—	G. Sternberg	1925	—	On display; part of SM slab
MCZ	—	F. Springer	1898	Springer (1901)	On display; part of 6 × 15 m lens
U IA	3-1567	—	—	—	On display
U KS	—	—	1891	Williston (1897)	On display
U KS	—	—	1898	Springer (1901)	IP collections
SMNK	—	G. Sternberg?	—	—	On display
SMNK	—	W. Hawes	1989	—	—
Keystone G	—	C. Bonner	—	—	On display
LACM	—	C. Bonner	—	—	On display
UMMP	14755	—	—	—	—
NHM, London	—	F. Springer	1898	Springer (1901)	—
NHM, Paris	—	—	—	—	On display
NSM	16845	—	—	—	On display
U OK					
SMNH	SMF XXIII 174	G. Sternberg	1907	Struve (1957)	On display
SI, NMNH	S6933	F. Springer	1898	Springer (1901), Schuchert (1904)	Part of 6 × 15 m lens
	S6934			Bassler (1909), Rogers (1991)	
SI, NMNH	S6934	—	—	—	Baculite mold

Institution	Catalog No.	Collector	Date	References	Comments
SM	IP-1	—	—	Miller et al. (1957)	On display
SM	14063	M. Rolfs	1968	—	—
U. Tübingen	—	G. Sternberg	—	—	3 slabs on display
U WI	—	J. Skulan, P. Druckenmiller	1992	Meyer and Milsom (2001)	On display
YPM	24758	H. Martin	1898	Beecher (1900), Springer (1901)	—

Note.—Abbreviations as in Appendix 5.1.

Appendix 7.1.

Additional references not cited in text but used for data compilation in figures.

Bassler, R. S. 1935. The classification of the Edrioasteroidea. Smithsonian Miscellaneous Collections, 93:1–11.

Bassler, R. S. 1936. New species of American Edrioasteroidea. Smithsonian Miscellaneous Collections, 95:1–33.

Bell, B. M. 1975. Ontogeny and systematics of *Timeischytes casteri*, n. sp.: An enigmatic Devonian edrioasteroid. Bulletins of American Paleontology, 67:33–56.

Bell, B. M. 1976. A study of North American Edrioasteroidea. New York State Museum Memoir, 21:446 p.

Berg-Madsen, V. 1986. Middle Cambrian cystoid (sensu lato) stem columnals from Bornholm, Denmark. Lethaia, 19:67–80.

Bockelie, J. F. 1981. The Middle Ordovician of the Oslo region, Norway; 30, The eocrinoid genera *Cryptocrinites*, *Rhipidocystis*, and *Bockia*. Norsk Geologisk Tidsskrift, 61:123–147.

Bockelie, J. F., and C. R. C. Paul. 1983. *Cyathotheca suecica* and its bearing on the evolution of the Edrioasteroidea. Lethaia, 16:257–264.

Brett, C. E., W. D. Liddell, and K. L. Derstler. 1983. Late Cambrian hard substrate communities from Montana/Wyoming: The oldest known hardground encrusters. Lethaia, 16:281–289.

Dean, J., and A. B. Smith. 1998. Palaeobiology of the primitive Ordovician pelmatozoan echinoderm *Cardiocystites*. Palaeontology, 41:1183–1194.

Durham, J. W. 1967. Notes on the Helicoplacoidea and early echinoderms. Journal of Paleontology, 41:97–102.

Durham, J. W. 1968. Lepidocystoids, p. 631–637. In R.C. Moore (ed.), Treatise on Invertebrate Paleontology. Pt. S. Echinodermata 1. Geological Society of America and Kansas University Press, Lawrence.

Durham, J. W. 1978. A Lower Cambrian eocrinoid. Journal of Paleontology, 52:195–199.

Durham, J. W., and K. E. Caster. 1963. Helicoplacoidea: a new class of echinoderms. Science, 140:820–822.

Fatka, O., and V. Kordule. 1984. *Acanthocystites* Barrande, 1887 (Eocrinoidea) from the Jince Formation (Middle Cambrian) of the Barrandian area. Bulletin of the Geological Survey of Prague, 59:299–302.

Guensburg, T. E. 1983. Echinoderms of the Middle Ordovician Lebanon Limestone, central Tennessee. Bulletins of American Paleontology, 86:5–100.

Guensburg, T. E. 1988. Systematics, functional morphology, and life modes of Late Ordovician edrioasteroids, Orchard Creek Shale, southern Illinois. Journal of Paleontology, 62:110–126.

Guensburg, T. E., and J. Sprinkle. 1994. Revised phylogeny and functional interpretation of the Edrioasteroidea based on new taxa from the Early and Middle Ordovician of western Utah. Fieldiana. Geology, 29:1–43.

Harker, P. 1953. A new edrioasteroid from the Carboniferous of Alberta. Journal of Paleontology, 27:288–289.

Harker, P., and R. D. Hutchinson. 1953. A new occurrence and redescription of *Gogia prolifica* Walcott. Journal of Paleontology, 27:285–287.

Holloway, D. J., and P. A. Jell, P. A., 1983. Silurian and Devonian edrioasteroids from Australia. Journal of Paleontology, 57:1001–1016.

Jell, P. A., C. F. Burrett, and M. R. Banks. 1985. Cambrian and Ordovician echinoderms from eastern Australia. Alcheringa, 9:183–208.

Kammer, T. W., E. C. Tissue, and M. A. Wilson. 1987. *Neoisorophusella*, a new edrioasteroid genus from the Upper Mississippian of the eastern United States. Journal of Paleontology, 61:1033–1043.

Lewis, R. D., J. Sprinkle, J. B. Bailey, J. Moffit, and R. L. Parsley. 1987. *Man-*

dalacystis, a new rhipidocystid eocrinoid from the Whiterockian Stage (Ordovician) in Oklahoma and Nevada. Journal of Paleontology, 61:1222–1235.

Parsley, R. L. 1975. Systematics and functional morphology of *Columbocystis*, a Middle Ordovician "cystidean" (Echinodermata) of uncertain affinities. Bulletins of American Paleontology, 67:349–360.

Parsley, R. L., and Y. L. Zhao. 2002. Eocrinoids of the Middle Cambrian Kaili biota, Taijiang County, Guizhou, China. Geological Society of America Annual Meeting, Abstracts with Programs, 34(6):81.

Parsley, R. L., and Y. L. Zhao. 2004. Functional morphology of brachioles in gogiid and other Early and Middle Cambrian eocrinoids, p. 479–484. In T. Heinzeller and J. Nebelsik (eds.), Echinoderms: München. Taylor & Francis, London.

Parsley, R. L., and Y. L. Zhao. 2005. Ontogeny and functional morphology of basal Middle Cambrian gogiid eocrinoids in the Kaili biota, Guizhou Province, China. Geological Society of America Annual Meeting, Abstracts with Programs, 37(7):63.

Paul, C. R. C., and A. B. Smith. 1984. The early radiation and phylogeny of echinoderms. Biological Reviews, 59:443–481.

Plas, V., and R. J. Prokop. 1979. *Argodiscus rarus* sp. n. (Edrioasteroidea) from the Sarka Formation (Llanvirn) of Bohemia. Bulletin of the Geological Survey of Prague, 54:41–43

Regnéll, G. 1967. Edrioasteroids, p. 136–173. In R. C. Moore (ed.), Treatise on Invertebrate Paleontology. Pt. S. Echinodermata 1. Geological Society of America and Kansas University Press, Lawrence.

Robison, R. A. 1965. Middle Cambrian eocrinoids from western North America. Journal of Paleontology, 39:355–364.

Robison, R. A. 1991. Middle Cambrian biotic diversity: Examples from four Utah Lagerstätten, p. 77–93. In A. M. Simonetta, and S. Conway Morris (eds.), The Early Evolution of Metazoa and the Significance of Problematic Taxa. Cambridge University Press, Cambridge.

Rozhnov, S. V., and P. V. Fedorov. 2001. A new cryptocrinid genus (Eocrinoidea, Echinodermata) from the bioherm-related facies of the Volkhov Stage (late Arenigian, Ordovician), Leningrad region. Paleontological Journal, 35:606–613.

Schuchert, C. 1919. A Lower Cambrian edrioasteroid *Stromatocystites walcotti*. Smithsonian Miscellaneous Collections, 70:1–7.

Simms, M. J., A. S. Gale, P. Gilliland, E. P. F. Rose and G. D. Sevastopulo. 1993. Echinodermata, p. 491–528. In M. J. Benton (ed.), The Fossil Record 2. Chapman and Hall, London.

Smith, A. B. 1980. *Floridiscus girvanensis*, a new edrioasteroid (Echinodermata) from the Ordovician of Ayrshire. Scottish Journal of Geology, 16:275–279.

Smith, A. B. 1983. British Carboniferous Edrioasteroidea (Echinodermata). Bulletin of the British Museum of Natural History (Geology), 37:113–138.

Smith, A. B. 1984. Classification of the Echinodermata. Palaeontology, 27:431–459.

Smith, A. B. 1985. Cambrian eleutherozoan echinoderms and the early diversification of edrioasteroids. Palaeontology, 28:715–756.

Smith, A. B. 1988. Patterns of diversification and extinction in early Palaeozoic echinoderms. Palaeontology, 33:799–828.

Smith, A. B., and M. A. Arbizu. 1987. Inverse larval development in a Devonian edrioasteroid from Spain and the phylogeny of Agelacrinitinae. Lethaia, 20:49–62.

Sprinkle, J. 1973. New occurrence of the Ordovician eocrinoid *Lingulocystis* from Bolivia, South America.

Sprinkle, J. 1992. Radiation of Echinodermata, 375–398. In J. H. Lipps, and P. H. Signor (eds.), Origin and Early Evolution of the Metazoa. Plenum Press, New York.

Sprinkle, J., 1995. Origin of echinoderms in the Paleozoic evolutionary fauna: The role of substrates. Palaios, 10:437–453.

Sprinkle, J., and B. M. Bell. 1978. Paedomorphosis in edrioasteroid echinoderms. Paleobiology, 4:82–88.

Sprinkle, J., and G. P. Wahlman. 1994. New echinoderms from the Early Ordovician of west Texas. Journal of Paleontology, 68:324–338.

Strimple, H. L. 1977. Discovery of a Middle Devonian edrioasteroid colony. Earth Science, 30:11–14.

Sumrall, C. D. 1992. *Spiraclavus nacoensis*, a new species of clavate agelacrinitid edrioasteroid from central Arizona. Journal of Paleontology, 66:90–98.

Sumrall, C. D. 1994. Thecal designs in isorophinid edrioasteroids. Lethaia, 26:289- 302.

Sumrall, C. D. 2000. The biological implications of an edrioasteroid attached to a pleurocystitid rhombiferan. Journal of Paleontology, 74:67–71.

Sumrall, C. D. 2001. Paleoecology and taphonomy of two new edrioasteroids from a Mississippian hardground in Kentucky. Journal of Paleontology, 75:136–146.

Sumrall, C. D., and A. L. Bowsher. 1996. *Giganticlavus*, a new genus of Pennsylvanian edrioasteroid from North America. Journal of Paleontology, 70:986–993.

Sumrall, C. D., and R. L. Parsley. 2003. Morphology and biomechanical implications of isolated discocystinid plates (Edrioasteroidea, Echinodermata) from the Carboniferous of North America. Palaeontology, 46:113–138.

Sumrall, C. D., J. Sprinkle, and T. E. Guensburg. 1997. Systematics and paleoecology of Late Cambrian echinoderms from the western United States. Journal of Paleontology, 71:1091–1109.

Sumrall, C. D., J. Garbisch, and J. P. Pope. 2000. The systematics of postibullinid edrioasteroids. Journal of Paleontology, 74:72–83.

Ubaghs, G. 1967. Eocrinoidea, p. 445–495. In R. C. Moore, R.C. (ed.), Treatise on Invertebrate Paleontology. Pt. S. Echinodermata 1. Geological Society of America and Kansas University Press, Lawrence.

Ubaghs, G. 1975. Early Paleozoic echinoderms. Annual Review of Earth and Planetary Sciences, 3:79–97.

Ubaghs, G. 1987. Échinodermes nouveaux du Cambrien moyen de la Montagne Noire (France). Annales de Paléontologie, 73:1–27.

Ubaghs, G., and R. A. Robison. 1985. A new homoiostelean and a new eocrinoid from the Middle Cambrian of Utah. University of Kansas Paleontological Contributions Paper, 115:1–24.

Ubaghs, G., and D. Vizcaïno. 1990. A new eocrinoid from the Lower Cambrian of Spain. Palaeontology, 33:249–256.

Waddington, J. B. 1980. A soft substrate community with edrioasteroids, from the Veralum Formation (Middle Ordovician) at Gamebridge, Ontario. Canadian. Journal of Earth Science, 17:674–679.

Zhao, Y. L., and R. L. Parsley. 2003. Short-stemmed eocrinoids of the Middle Cambrian Kaili biota, Guizhou Province, China. Geological Society of America Annual Meeting, Abstracts with Programs, 35(6):106.

A. Calyx shape
    0: Low bowl, 1: Medium globe, 2: High cone, 3: Medium cone, 4: Low cone, 5: Medium bowl, 6: Very high bowl.
B. Basal concavity
    0: No, 1: Yes.
C. Calyx lobation
    0: No, 1: Yes
D. Calyx plate thickness
    0: Thin, 1: Thick
E. Median ray ridges
    0: Yes, 1: No
F. Basal plate height
    0: Low, 1: High
G. Radial plate dimensions
    0: H > W, 1: H~W, 2:H > W
H. Orientation of radial plates
    0: Horizontal, 1: Vertical
I. Radial plates unequal in size
    0: No, 1: Yes
J. Radial circlet separated or partially separated in all rays.
    0: No, 1: Yes
K. First primibrachial (fixed) dimensions
    0: H > W, 1: H~W, 2: W > H
L. Primaxil shape
    0: 5-sided, 1: 6-sided, 2: 7-sided
M. Primaxil
    0: Second primibrachial, 1: First primibrachial
N. Fixed brachials isotomously branched
    0: Yes, 1: No
O. Fixed rays symmetrical
    0: Yes, 1: No
P. Distal-most fixed brachitaxis
    0: Tertibrachitaxis, 1:Quartibrachitaxis, 2: Secundibrachitaxis
Q. Relative number of regular interray plates
    0: Few, 1: Numerous
R. Regular interrays with plating in biseries
    0: No, 1: Yes
S. Regular interrays depressed
    0: No, 1: Yes
T. CD interray proximal plating
    0: 1-2, 1: 1-3
U. Anitaxial ridge
    0: Yes, 1: No
V. CD interray width in relation to regular interray width
    0: Wider than, 1: Very wider than
W. Tegmen height [not used in the PAUP analyses]
    0: Low, 1: High

Appendix 8.1.

Character and character states for genus diagnosis among the Periechocrinidae.

X. Tegmen strength [not used in the PAUP analyses]
    0: Incompetent, 1: Robust
Y. Tegmen depressed interradially [not used in the PAUP analyses]
    0: No, 1: Yes
Z. Ambulacrals and orals differentiated on tegmen [not used in the PAUP analyses]
    0: No, 1: Yes
AA. Well-developed anal tube [not used in the PAUP analyses]
    0: Yes, 1: No
AB. Anus position [not used in the PAUP analyses]
    0: Eccentric, 1: Subcentral, 2: Central
AC. Free arms branch [not used in the PAUP analyses]
    0: Yes, 1: No
AD. Brachial type
    0: Uniserial, 1: Biserial
AE. Arms composed of pinnulate ramules
    0: No, 1: Yes
AF. Gonioporoids on tegmen
    0: No, 1: Yes.

Appendix 8.2.

Character states for periechocrinids.

Genus	A	B	C	D	E	F	G	H	I	J	K	L	M	N	O
Acacocrinus	0	0	0	0	0	0	0	1	0	0	0	0, 1	0	0	0
Aryballocrinus	5	0	0	0	0, 1	0	0	1	0	0	0	0	0	0	0
Athabascacrinus	4	1	1	1	1	0	0	0	0	0	2	0	0, 1	0	0
Beyrichocrinus	4	1	0	1	1	0	0	0	1	1	2	0	0	0	0
Clitheroecrinus	0	1	0	1	1	0	0	0	0	0	2	0	0	0	0
Corocrinus	3	0	1	1	0	1	2	1	1	0	1	2	0	0	0
Gennaeocrinus	0	0	1	0	0	0	1	0	0	0	1	0, 1, 2	0	1	0
Ibanocrinus	0	0	0	0	0	?	0	0	0	0	0	0	0	0	0
Lenneocrinus	2	0	1	0	0	0	2	1	0	0	0	2	0	0	0
Mandelacrinus	4	0	1	1	0	1	2	1	0	0	2	0	0	0	0
Mcoycrinus	0	1	0	1	1	0	0	0	0	0	1	0	0	0	0
Megistocrinus	0, 4	1	0	1	1	0	0	0	0	0	1	0, 1, 2	0	0	0
Periechocrinus	2	0	0	0	0, 1	1	2	1	0	0	0	2	0, 1	0	0
Pithocrinus	0	0	0	1	1	0	0	1	0	0	2	0	0, 1	0	1
Pradocrinus	6	0	1	0	0	1	2	1	0	0	0	2	0	0	0
Pyxidocrinus	2	0	1	0	1	0	1	1	0	0	1	0	0	0	0
Stamnocrinus	2	0	1	1	1	1	1	1	0	0	2	2	0	0	0
Stiptocrinus	2	0	1	1	1	1	2	1	0	0	1	0	0	0	0
Thamnocrinus	2	0	1	1	0	1	2	1	0	0	1	2	0	1	1
Tirocrinus	1	0	1	0	1	0	0	0	0	0	0	0	0	0	0
Glyptocrinus	3	0	0	0	0	0	0	1	0	0	0	1	0	0	0
Periglyptocrinus	5	0	0	0	0	0	1	1	0	0	1	0,1	0	0	0
Pycnocrinus	1	0	0	0	0	0	1	1	0	0	1	1	2	0	0

Note.—Appendix 8.1 provides an explanation of characters and character states.

P	Q	R	S	T	U	V	W	X	Y	Z	AA	AB	AC	AD	AE	AF
2	0	0	0	0	1	0	?	?	?	?	?	?	?	0	0	?
2	0	0	0	1	0, 1	1	0	0	?	?	1	0	0	1	0	?
0, 2	1	0	0	1	1	0	0	1	1	1	1	0	?	?	?	0
0	1	0	0	0	1	0	0	1	0	0	1	1	?	?	?	0
0, 2	0	0	0	0	1	0	?	?	?	?	?	?	?	?	?	?
2	0	0	1	1	0	0	0	0	?	?	?	?	0	0, 1	0	?
0, 2	1	0	0	1	0	1	0	1	1	1	1	0	0, 1	1	0	0
0	1	0	1	1	0	0	?	?	?	?	?	?	?	?	?	?
2	0	1	1	1	0	0	?	?	?	?	?	?	0	1	0	?
2	0	0	1	1	0	0	?	?	?	?	?	?	1	1	0	0
0	1	0	0	1	1	0	1	1	0	0	0	2	?	?	?	0
0, 2	1	0	0	1	1	1	0	1	1	1	1	0	0	1	0	0
0, 1	1	1	0	1	0	0	0	0	0, 1	0	0	0, 1	0	1	0	0
0, 1, 2	0	1	0	1	1	0	1	1	0	0	0	1	0	1	0	0
2	0	1	0	1	0	1	0	1	1	1	1	0	1	1	0	0
0	0	1	0	1	1	0	1	1	1	1	1	0	?	?	?	1
0	1	0	0	1	1	0	0	1	0	0	0	0	?	?	?	0
2	1	0	1	1	1	0	0	1	0	0	1	0	?	0	?	0
0	1	1	0	1	1	0	?	?	1	?	?	?	0	0	1	?
0	0	1	0	0	?	?	?	?	?	?	?	?	1	0	0	?
0	1	1	0	1	0	0	0	0	1	0	0	?	0	0	0	0
0	1	1	0	1	0	0	?	?	?	?	?	?	0	1	0	0
2	1	1	0	1	0	0	0	0	1	?	0	?	1	0	0	0

Appendix 9.1.

Characters, character states, and character matrix for 43 genera of Osagean–late Meramecian pinnulate cladid genera from eastern North America.

Genus	Characters								
	1	2	3	4	5	6	7	8	9
Abrotocrinus	2	1	3	3	1	1	0	1	(12)
Acylocrinus	2	1	1	3	1	0	1	1	1
Adinocrinus	3	1	3	3	1	1	0	0	1
Aphelecrinus	1	1	3	3	1	0	0	1	1
Armenocrinus	1	1	1	3	4	0	0	1	1
Ascetocrinus	1	1	3	2	1	0	0	1	2
Aulocrinus	2	0	3	3	1	0	1	1	2
Blothrocrinus	1	1	3	?	2	1	0	1	2
Bollandocrinus	1	1	3	3	2	0	1	0	1
Bursacrinus	1	1	1	3	1	0	0	0	1
Cercidocrinus	1	1	3	3	1	1	0	0	1
Coeliocrinus	2	1	3	3	1	0	0	1	1
Corythocrinus	1	1	1	3	4	0	0	1	1
Cosmetocrinus	1	1	3	3	1	0	0	1	1
Cromyocrinus	2	1	3	3	0	0	0	0	1
Culmicrinus	1	1	(23)	3	4	1	0	1	1
Cydrocrinus	1	1	3	2	2	0	0	1	1
Decadocrinus	2	1	3	3	2	1	1	1	(12)
Dinotocrinus	2	1	3	3	1	0	0	1	1
Eratocrinus	3	1	(23)	3	1	1	0	0	1
Gilmocrinus	1	1	3	?	0	0	0	1	1
Graphiocrinus	3	1	1	3	1	0	1	0	1
Histocrinus	1	1	3	2	2	1	1	1	1
Holcocrinus	1	1	1	3	1	0	1	1	2
Hylodecrinus	2	0	3	3	2	0	0	1	2
Hypselocrinus	1	1	3	3	1	1	1	1	1
Lanecrinus	2	1	3	3	1	0	1	1	1
Lebetocrinus	1	0	1	3	1	0	0	1	1
Lekocrinus	3	1	3	3	1	1	0	1	1
Linocrinus	3	0	3	3	1	1	0	1	1
Nactocrinus	1	1	1	3	1	0	1	0	1

Genus	Characters								
	1	2	3	4	5	6	7	8	9
Ophiurocrinus	1	1	3	3	0	0	0	0	1
Pachylocrinus	2	1	3	3	2	0	0	1	2
Parascytalocrinus	2	1	3	3	1	1	1	1	1
Pelecocrinus	1	1	3	1	1	1	0	1	1
Poteriocrinites	1	1	3	1	4	0	0	0	1
Sarocrinus	3	1	3	3	2	1	0	0	1
Scytalocrinus	1	1	3	3	1	0	1	1	1
Springericrinus	1	0	(123)	1	1	0	0	1	1
Stinocrinus	2	0	3	3	2	1	0	0	2
Tropiocrinus	3	1	1	3	1	1	1	1	(12)
Ulrichicrinus	1	1	3	3	1	0	0	1	1

Note.—These nine characters are based on high-frequency characters in genera with only one anal plate. This character matrix was used for the calculations of binomial probabilities (Tables 10.2 and 10.3) and is not a thorough character matrix for cladistic analysis.

1. Cup height: 1 = high, infrabasals clearly visible in side view; 2 = medium, infrabasals not readily visible in side view; 3 = low, radials dominate side view.
2. Cup plate surface: 0 = ornamented with plications, ridges, nodes, etc.; 1 = smooth.
3. Number of anals in cup: 1, 2, or 3.
4. Radial facets: 1 = angustary, 2 = peneplenary, 3 = plenary.
5. Primibrachial axillary: 1, 2, 3, 4 (> 3), 0 = atomous.
6. A-ray arm: 0 = same as other rays; 1 = different from other rays, either atomous or with extra primibrachials.
7. Number of arms: 0 = > 10; 1 = 9 or 10 arms.
8. Brachial shape: 0 = rectangular; 1 = cuneate or subcuneate brachials.
9. Column: 1 = round, 2 = pentagonal or subpentagonal.

Appendix 10.1

Listing of the type species of cyathocrinids

Cyathocrinida	Age	Uf	Bf	Tf	Rtlr	SCu	CvPl	Rm	R/A	Arl	Pin
Barycrinidae											
Barycrinus spurius (Hall, 1858); Kammer and Ausich, 1996	M Miss	—	X	—	X	—	—	X	—	—	—
?Meniscocrinus magnitubus Kammer and Ausich, 1996	M Miss	—	—	—	X	—	—	—	—	X	—
Parabarycrinus complanatus Chen and Yao, 1993	E Miss	X	—	—	—	—	—	—	—	—	—
Pellecrinus hexadactylus (Lyon and Cassadey, 1860)	M Miss	—	—	—	X	—	—	—	X	—	—
Cupressocrinitidae											
Cupressocrinites crassus Goldfuss, 1831	M Dev	—	—	X	X	—	—	—	—	—	X
Abbreviatocrinites abbreviatus (Goldfuss, 1839); Bohaty, 2005	M Dev	—	—	X	—	—	—	—	—	—	—
Rhopalocrinus gracilis (Schultze, 1866)	M Dev	—	X	—	X	—	—	—	—	—	X
Robustocrinites scaber (Schultze, 1866); Bohaty, 2005	M Dev	—	—	X	—	—	—	—	—	—	—
Cyathocrinitidae											
Cyathocrinites planus Miller, 1821	E Miss	—	X	—	X	—	X	—	—	—	—
Anarchocrinus rossicus Jaekel, 1918	L Ord	—	—	—	X	—	?	—	—	—	—
Ceratocrinus exornatus Wanner, 1937	M Perm	—	—	—	—	X	X	—	—	—	—
Conicocyathocrinites wilsoni (Springer, 1926)	M Sil	—	—	—	X	—	?	—	—	—	—
Gissocrinus typus (Bather, 1893)	L Sil	—	—	—	X	—	X	—	—	—	—
Levicyathocrinites acinotubus (Angelin, 1878); Bather, 1893	M Sil	—	X	—	—	—	—	—	—	—	—
Occiducrinus australis Webster, 1990	E Perm	—	X	—	X	—	?	—	—	—	—

Cyathocrinida	Age	Uf	Bf	Tf	Rtlr	SCu	CvPl	Rm	R/A	Arl	Pin
Rugosocyathocrinites pauli Frest, 1977	M Sil	X	—	—	—	—	—	—	—	—	—
Euspirocrinidae											
Euspirocrinus spiralis Angelin, 1878; Bather, 1893	L Sil	X	—	—	X	—	?	—	—	—	—
Ampheristocrinus typus Hall, 1879	M Sil	—	—	—	X	—	?	—	—	—	—
Anaglyptocrinus willinki Webster and Jell, 1999	E Perm	—	X	—	X	—	X	—	—	—	—
Caelocrinus stellifer Xu, 1962	E Sil	—	—	—	—	—	—	—	—	—	—
Eoparisocrinus siluricus (Springer, 1926); Ausich, 1986	E Sil	X	—	—	X	—	?	—	—	—	—
Kooptoonocrinus nutti Jell and Holloway, 1983	L Sil	—	—	—	X	—	X	—	—	—	—
Kopriacrinus mckellari Webster and Jell, 1999	M Penn	—	X	—	X	—	?	—	—	—	—
Monaldicrinus johni Jell and Theron, 1999	E Dev	—	X	—	X	—	—	X	—	—	—
Necopinocrinus tycherus Webster and Jell, 1999	E Perm	—	X	—	—	X	?	—	—	—	—
Neerkolocrinus typus Webster and Jell, 1999	M Penn	—	X	—	X	—	—	X	—	—	—
Parisocrinus perplexus (Wachsmuth and Springer, 1869)	M Miss	—	—	—	X	—	?	—	—	—	—
Vasocrinus valens Lyon, 1857	M Dev	—	X	—	X	—	X	—	—	—	—
Zygotocrinus fragilis Kirk, 1943	M Miss	—	—	—	X	—	?	—	—	—	—
Gasterocomidae											
Gasterocoma antiqua Goldfuss, 1839	M Dev	X	—	—	—	—	—	—	—	—	—
Arachnocrinus bulbosus (Hall, 1862); Springer, 1911	M Dev	X	—	—	X	—	?	—	—	—	—
Kopficrinus pustuliferus Goldring, 1954	E Dev	X	—	—	—	—	—	—	—	—	—
Mictocrinus robustus Goldring, 1923	M Dev	X	—	—	X	—	?	—	—	—	—

Cyathocrinida	Age	Uf	Bf	Tf	Rtlr	SCu	CvPl	Rm	R/A	Arl	Pin
Myrtillocrinus elongatus Sandberger and Sandberger, 1856	M Dev	X	—	—	—	—	—	—	—	—	—
Nanocrinus paradoxus Müller, 1856	M Dev	X	—	—	X	—	X	—	—	—	—
Schultzicrinus typus Springer, 1911	M Dev	—	—	—	X	—	—	—	—	—	—
Scoliocrinus eremita Jaekel, 1895	M Dev	X	—	—	—	—	X	—	—	—	—
Lecythocrinidae											
Lecythocrinus eifelianus Müller, 1858	M Dev	—	—	—	X	—	?	—	—	—	—
Ancyrocrinus bulbosus (Hall, 1852); McIntosh and Schreiber, 1971	M Dev	—	X	—	X	—	?	—	—	—	—
Cestocrinus striatus Kirk, 1940	M Miss	X	—	—	X	—	?	—	—	—	—
Corynecrinus romingeri Kirk, 1934	M Dev	—	—	—	X	—	?	—	—	—	—
Tetrapleurocrinus eifelensis Wanner, 1942	M Dev	X	—	—	—	—	—	—	—	—	—
Porocrinidae											
Porocrinus conicus Billings, 1857; Kessling and Paul, 1968	L Ord	X	—	—	X	—	?	—	—	—	—
Carabocrinus radiatus Billings, 1857	L Ord	X	—	—	—	—	—	—	—	—	—
Palaeocrinus striatus Billings, 1859; Hudson, 1911	L Ord	X	—	—	—	—	—	—	—	—	—
Triboloporus cryptoplicatus Kesling and Paul, 1968	L Ord	X	—	—	X	—	—	—	—	—	—

Note.—All references prior to 1999 are fully cited in Webster (2003); references after 1999 are fully cited herein.

Abbreviations.—Arl, armlets; Bf, bifascial; CvPl, cover plates; Dev, Devonian; E, Early; L, Late; SCu, slightly cuneate; M, Middle; Miss, Mississippian; Ord, Ordovician; Penn, Pennsylvanian; Perm, Permian; Pin, Pinnules; R/A, ramules and armlets; Rm, Ramules; Rtlr, rectilinear; Sil, Silurian; Tf, trifascial; Uf, unifascial.

*Listing of the type
species of primitive
dendrocrinids.*

Primitive Dendrocrinida	Age	Uf	Bf	Mus	Rtlr	SCu	MCu	CvPl	Rm	Arl	Pin
Botryocrinidae											
Botryocrinus ramosissimus Angelin, 1878; Bather, 1893	M Sil	X	—	—	X	—	—	—	—	X	—
Costalocrinus dilatatus Schultz, 1867	M Dev	X	—	—	—	—	—	—	—	—	—
Gastrocrinus patulus (Müller, 1859)	E Dev	—	—	—	X	—	—	—	—	X	—
Gothocrinus gracilis Bather, 1893	M Sil	—	—	—	X	—	—	—	—	—	X
Imitatocrinus gracilor (Roemer, 1863)	E Dev	—	—	—	X	—	—	—	—	—	X
Jahnocrinus minutus Jaekel, 1918	M Dev	X	—	—	X	—	—	?	—	—	—
Pandoracrinus pinnulatus Jaekel, 1918	L Ord	X	—	—	X	—	—	—	—	—	X
Parabotryocrinus tschudovensis Yakovlev, 1941	L Dev	—	—	—	X	—	—	—	—	—	X
Pskovicrinus sofronizkyi Rozhnov and Arendt, 1984	L Dev	X	—	—	—	X	—	—	—	—	X
Rhadinocrinus rhenanus Jaekel, 1895	E Dev	—	—	—	X	—	—	—	—	X	—
Cupulocrinidae											
Cupulocrinus heterocostalis (Hall, 1850)	L Ord	—	—	—	X	—	—	X	—	—	—
Morenacrinus silvani Ausich et al., 2002	M Ord	—	—	—	X	—	—	?	—	—	—
Quinquecaudex glabellus Brower and Veinus, 1982	L Ord	X	—	—	X	—	—	?	—	—	—
Stewbrecrinus terryi Jell, 1999	E Dev	X	—	—	X	—	—	?	—	—	—
Dendrocrinidae											
Dendrocrinus longidactylus Hall, 1852.	L Sil	—	—	—	X	—	—	—	—	—	—
Brechmocrinus eos Ausich, Bolton and Cummings, 1998	M Ord	—	—	—	X	—	—	—	—	—	X

Primitive Dendrocrinida	Age	Uf	Bf	Mus	Rtlr	SCu	MCu	CvPl	Rm	Arl	Pin
Eckidocrinus interbrachiatus Jell and Theron, 1999	E Dev	X	—	—	X	—	—	—	X	—	—
Elpasocrinus radiatus Sprinkle and Wahlman, 1994	E Ord	—	—	—	X	—	—	X	—	—	—
Esthonocrinus laevior Jaekel, 1918	L Ord	—	—	—	X	—	—	?	—	—	—
Grenprisia billingsi (Springer, 1911)	L Ord	—	—	—	X	—	—	—	—	X	—
Ottawacrinus typus Billings, 1887	L Ord	—	—	—	X	—	—	X	—	—	—
Parisangulocrinus zeaeformis (Schultz, 1866)	E Dev	X	—	—	X	—	—	—	—	—	—
Mastigocrinidae											
Mastigocrinus loreus Bather, 1892	M Sil	X	—	—	X	—	—	—	—	—	—
Antihomocrinus tenuis (Bather, 1893)	M Sil	X	—	—	X	—	—	—	—	—	—
Atelestocrinus delicatus Wachsmuth and Springer, 1886	M Miss	—	—	—	X	—	—	—	X	—	—
Bathericrinus ramosus (Bather, 1891)	M Sil	—	—	—	X	—	—	—	X	—	—
Dictenocrinus decadactylus (Bather, 1891)	M Sil	—	X	—	X	—	—	—	—	—	X
Goniocrinus sculptilis Miller and Gurley, 1890	M Miss	—	—	—	X	—	—	—	X	—	—
?*Hebohenocrinus quasipatellus* Webster et al., 2004	L Miss	—	—	X	X	—	—	—	—	—	—
Lasiocrinus scoparius (Hall, 1859)	E Dev	—	—	—	X	—	—	—	—	—	—
Streptocrinus crotalurus (Angelin, 1878)	M Sil	X	—	—	X	—	—	—	—	—	—
Zygiosocrinus typicus Webster, 1997	E Miss	—	—	—	X	—	—	—	—	—	X
Merocrinidae											
Merocrinus typus Walcott, 1884	L Ord	—	—	—	X	—	—	X	—	—	—
Polycrinus ramulatus Jaekel, 1918; Prokop, 1984	L Ord	—	—	—	X	—	—	—	X	—	—
Praecupulocrinus conjugans (Billings, 1857); Brower, 1992	L Ord	—	X	—	X	—	—	—	—	—	—
Quintuplexacrinus oswegoensis (Meek and Worthen, 1868)	L Ord	X	—	—	X	—	—	?	—	—	—

Primitive Dendrocrinida	Age	Uf	Bf	Mus	Rtlr	SCu	MCu	CvPl	Rm	Arl	Pin
Metabolocrinidae											
Metabolocrinus rossicus Jaekel, 190	L Ord	—	—	—	—	X	—	—	—	—	X
Cyliocrinus rigidus (Angelin, 1878)	M Sil	—	—	—	—	X	—	—	—	—	X
Eopinnacrinus pinnulatus Brower and Veinus, 1982	L Ord	X	—	—	—	—	X	—	—	—	X
Ontariocrinidae											
Ontariocrinus deviatus Jaekel, 1918	L Ord	X	—	—	X	—	—	—	—	—	—
Plicodendrocrinidae											
Plicodendrocrinus casei (Meek, 1871); Brower, 1973	L Ord	X	—	—	X	—	—	?	—	—	—
Compagicrinus fenestratus Jobson and Paul, 1979	E Ord	—	—	—	X	—	—	X	—	—	—
Holmesocrinus enidae Jell, 1999	L Sil	—	—	—	X	—	—	?	—	—	—
Shintocrinus cometensis Jell, 1999	L Sil	—	—	—	X	—	—	—	—	—	—
Poteriocrinitidae											
Poteriocrinites crassus J. S. Miller, 1821	E Miss	—	X	—	X	—	—	—	—	—	—
Balearocrinus breimeri Bourrouilh and Termier, 1973	M Miss	X	—	—	X	—	—	—	—	—	?
Denariocrinus ferula Schmidt, 1942	M Dev	—	—	—	—	X	—	—	—	—	X
Rhabdocrinus scotocarbonarius (Wright, 1937)	E Miss	X	—	—	X	—	—	—	—	—	—
Springericrinus magniventrus (Springer, 1911); Kammer, 1984	M Miss	—	—	X	—	X	—	—	—	—	X
Thalamocrinidae											
Thalamocrinus ovatus Miller and Gurley, 1895	M Sil	X	—	—	—	—	—	—	—	—	—
Bactrocrinites fusiformis (Roemer, 1844)	M Dev	X	—	—	X	—	—	?	—	—	
Belanskicrinus westoni (Belanski, 1928)	M Dev	—	—	—	X	—	—	—	—	—	X
Cradeocrinus elongatus Goldring, 1923	L Dev	—	—	—	X	—	—	—	X	—	—
Eifelocrinus dohmi (Wanner, 1916)	L Dev	—	—	—	X	—	—	—	—	X	—
Follicrinus grebei (Follman, 1887)	L Dev	—	—	—	X	—	—	?	—	—	

Primitive Dendrocrinida	Age	Uf	Bf	Mus	Rtlr	SCu	MCu	CvPl	Rm	Arl	Pin
Illemocrinus amphiatus Eckert, 1987	L Ord	X	—	—	X	—	—	?	—	—	—
Kanabinocrinus thyaros Ausich, 1986	E Sil	—	—	—	—	X	—	?	—	—	—
Nuxocrinus crassus (Whiteaves, 1887); McIntosh, 1983	M Dev	—	X	—	X	—	—	—	—	X	—
Pagecrinus gracilis Kirk, 1929	E Dev	—	—	—	X	—	—	—	X	—	—
Pyrenocrinus bethanensis (Goldring, 1954); McIntosh, 1983	M Dev	—	X	—	X	—	—	—	—	X	—
Sacrinus gamkaensis Jell and Theron, 1999	E Dev	—	—	—	X	—	—	X	—	—	—
Sigambrocrinus laevis Schmidt, 1942	E Dev	—	—	—	X	—	—	X	—	—	—
Situlacrinus costatus Breimer, 1962	E Dev	—	X	—	X	—	—	—	—	—	—
Thenarocrinidae											
Thenarocrinus callipygus Bather, 1890	M Sil	—	—	—	X	—	—	?	—	—	—
Family Uncertain											
Aithriocrinus strahani Donovan and Veltkamp, 1993	L Ord	X	—	—	X	—	—	?	—	—	—
Chakomicrinus quinquepartitus Stukalina, 2000	M Sil	X	—	—	X	—	—	?	—	—	—
Kophinocrinus spiniferus Goldring, 1954	M Dev	X	—	—	X	—	—	—	—	—	—

Note.—All references prior to 1999 are fully cited in Webster (2003); references after 1999 are fully cited herein.

Abbreviations.—Arl, armlets; Bf, bifascial; CvPl, cover plates; Dev, Devonian; E, Early; L, Late; M, Middle; MCu, moderately cuneate; Miss, Mississippian; Ord, Ordovician; Penn, Pennsylvanian; Perm, Permian; Pin, Pinnules; Rm, Ramules; Rtlr, rectilinear; SCu, slightly cuneate; Sil, Silurian; Uf, unifascial.

Appendix 10.3

Listing of the type species of transitional dendrocrinids

Transitional Dendrocrinida, Glossocrinacea	Age	Bf	Mus	Rtlr	SCu	MCu	StCu	Rm	Arl	Pin
Amabilicrinidae										
Amabilicrinus iranensis Webster et al., 2003	E Miss	—	X	—	—	X	—	—	—	X
Bufalocrinus torus (Webster and Hafley, 1999)	L Dev	—	X	—	—	—	X	—	—	X
Cydrocrinus coxanus (Worthen, 1882)	M Miss	—	?	—	—	X	—	—	—	X
Gilmocrinus iowaensis Laudon, 1933	E Miss	—	X	—	—	—	X	—	—	X
Gulinocrinus bellus Chen, Wei, and Dai, 1997	E Miss	—	?	—	X	—	—	—	—	X
Hallocrinus ornatissimus Hall, 1843	L Dev	—	?	—	X	—	—	—	—	X
Corematocrinidae										
Corematocrinus plumosus Goldring, 1923	L Dev	—	?	—	X	—	—	—	—	X
Carcinocrinus stevensi Laudon, 1941	L Miss	—	?	—	—	X	—	—	—	X
Maragnocrinus portlandicus Whitfield, 1905	L Dev	—	?	—	X	—	—	—	—	X
Nudalocrinus jeffersonensis (Laudon and Severson, 1953)	E Miss	—	?	—	X	—	—	—	—	—
Glossocrinidae										
Glossocrinus naplensis Goldring, 1923	L Dev	—	—	X	—	—	—	—	—	X
Catactocrinus leptodactylus Goldring, 1923	L Dev	—	—	X	—	—	—	X	—	—
Charientocrinus ithacensis Goldring, 1923; McIntosh, 2001	L Dev	X	—	X	—	—	—	—	—	X
Gelasinocrinus revimentus Webster, et al., 2003	L Dev	—	X	X	—	—	—	—	—	X

Transitional Dendrocrinida, Glossocrinacea	Age	Bf	Mus	Rtlr	SCu	MCu	StCu	Rm	Arl	Pin
Liparocrinus batheri Goldring, 1923	L Dev	—	—	X	—	—	—	—	—	X
Rutkowskicrinidae										
Rutkowskicrinus patriciae McIntosh, 2001	M Dev	X	—	X	—	—	—	—	—	X
Decorocrinus arkonensis (Goldring, 1950); McIntosh, 2001	M Dev	X	—	X	—	—	—	—	—	X
Iteacrinus flagellus Goldring, 1923	L Dev	—	—	X	—	—	—	X	—	—
Nassoviocrinus pachydactylus (Sandberger and Sandberger, 1855)	E Dev	—	—	X	—	—	—	—	X	
Propoteriocrinus follmanni Schmidt, 1942	E Dev	?	—	X	—	—	—	—	—	X
Quantoxocrinus ussheri Webby, 1965	M Dev	—	—	X	—	—	—	—	—	X
?Schmidtocrinus winterfeldi (Schmidt, 1906)	E Dev	—	—	—	X	—	—	X	—	—

Note.—All references prior to 1999 are fully cited in Webster (2003); references after 1999 are fully cited herein.

Abbreviations.—Arl, armlets; Bf, bifascial; CvPl, cover plates; Dev, Devonian; E, Early; L, Late; M, Middle; MCu, moderately cuneate; Miss, Mississippian; Ord, Ordovician; Penn, Pennsylvanian; Perm, Permian; Pin, Pinnules; Rm, Ramules; Rtlr, rectilinear; SCu, slightly cuneate; Sil, Silurian; Uf, unifascial.

Appendix 10.4

Listing of the type species of advanced dendrocrinids.

Advanced Dendrocrinida	Age	Mus	Rtlr	SCu	MCu	StCu	FCBi	RCBi	WBi	Pin
Adinocrinidae										
Adinocrinus nodosus (Wachsmuth and Springer, 1885)	M Miss	X	—	—	X	—	—	—	—	X
Aesiocrinidae										
Aesiocrinus magnificus Miller and Gurley,1890	L Penn	X	—	X	—	—	—	—	—	X
Agassizocrinidae										
Agassizocrinus conicus Owen and Shumard, 1852	L Miss	X	—	X	—	—	—	—	—	X
Anartiocrinus lyoni Kirk, 1940	L Miss	X	X	—	—	—	—	—	—	X
Belashovicrinus gjelensis Arendt and Zubarev, 1993	L Penn	X	X	—	—	—	—	—	—	X
Epipetschoracrinus borealis Yakovlev in Y. and Ivanov 1956	E Perm	X	—	—	—	—	—	—	—	X
Paragassizocrinus tarri (Strimple, 1938).	L Penn	X	X	—	—	—	—	—	—	X
Petschoracrinus variabilis Yakovlev, 1928	E Perm	X	X	—	—	—	—	—	—	X
Polusocrinus avanti Strimple, 1951	L Penn	X	X	—	—	—	—	—	—	X
"Ampelocrinidae"										
Proampelocrinus himalayaensis Gupta and Webster, 1974	E Miss	X	—	X	—	—	—	—	—	X
Spheniscocrinus spinosus Wanner, 1937	E Perm	X	—	—	X	—	—	—	—	X
Ampullacrinidae										
Ampullacrinus marieae Webster et al., 2004	L Miss	X	—	X	—	—	—	—	—	X
Anobasicrinidae										
Anobasicrinus bulbosus Strimple, 1961	M Penn	X	X	—	—	—	—	—	—	X
Synphocrinus cornutus Trautschold, 1867	M Penn	X	X	—	—	—	—	—	—	X
Terpnocrinus ocoyaensis Strimple and Moore, 1971	L Penn	X	X	—	—	—	—	—	—	X

Advanced Dendrocrinida	Age	Mus	Rtlr	SCu	MCu	StCu	FCBi	RCBi	WBi	Pin
Aphelecrinidae										
Aphelecrinus elegans Kirk, 1944	M Miss	X	—	X	—	—	—	—	—	X
Apokryphocrinus wellsvillensis Webster, 1997	E Miss	X	—	X	—	—	—	—	—	X
Cosmetocrinus gracilis Kirk, 1941	M Miss	X	—	X	—	—	—	—	—	X
Paracosmetocrinus strakai Strimple, 1967	E Miss	X	—	X	—	—	—	—	—	X
Apographiocrinidae										
Apographiocrinus typicalis Moore and Plummer, 1940	L Penn	X	X	—	—	—	—	—	—	X
Paragraphiocrinus exornatus (Wanner, 1916).	M Perm	X	X	—	—	—	—	—	—	?
Arkacrinidae										
Arkacrinus dubius (Mather, 1915).	E Penn	X	—	—	—	—	—	—	—	—
Basleocrinidae										
Basleocrinus pocillum Wanner, 1916	M Perm	X	—	—	—	—	—	—	—	—
Laccocrinus scrobiculatus (Wanner, 1924)	M Perm	X	—	—	—	—	—	—	—	—
Blothrocrinidae										
Blothrocrinus jesupi (Whitfield, 1881)	M Miss	X	—	—	X	—	—	—	—	X
Culmicrinus regularis (Meyer, 1858)	M Miss	X	—	—	X	—	—	—	—	X
Elibatocrinus leptocalyx Moore, 1940	L Penn	X	—	X	—	—	—	—	—	X
Fifeocrinus tielensis (Wright, 1936)	M Miss	X	—	—	—	—	—	—	X	X
Moscovicrinus multiplex (Trautschold, 1867)	M Penn	X	—	X	—	—	—	—	—	X
Nebraskacrinus tourteloti Moore, 1939	E Perm	X	—	X	—	—	—	—	—	X
Scammatocrinus delicatus Burdick and Strimple, 1983	L Miss	X	—	—	—	X	—	—	—	X
Stinocrinus granulosus Kirk, 1941	M Miss	X	—	X	—	—	—	—	—	X
Bridgerocrinidae										
Bridgerocrinus fairyensis Laudon and Severson, 1953	E Miss	X	X	—	—	—	—	—	—	X
Derorhethocrinus elongatus Webster, Maples, et al., 2003	E Miss	X	—	X	—	—	—	—	—	X

Advanced Dendrocrinida	Age	Mus	Rtlr	SCu	MCu	StCu	FCBi	RCBi	WBi	Pin
?*Eireocrinus ornatus* Wright, 1951	E Miss	X	X	—	—	—	—	—	—	X
Gaelicrinus rostratus (Austin and Austin, 1843)	E Miss	X	—	X	—	—	—	—	—	X
Maevecrinus bothros Ausich and Sevastopulo, 2001	E Miss	X	—	X	—	—	—	—	—	X
Melbacrinus americanus Strimple, 1939	L Penn	X	?	—	—	—	—	—	—	X
Mixocrinus porosus Haude and Thomas, 1992	M Miss	X	X	—	—	—	—	—	—	X
Bursacrinidae										
Bursacrinus wachsmuthi Meek and Worthen, 1861	M Miss	X	—	X	—	—	—	—	—	X
Erincrinus austini (Wright, 1952)	E Miss	X	—	X	—	—	—	—	—	X
Lebetocrinus grandis Kirk, 1940	M Miss	X	—	—	X	—	—	—	—	X
Nactocrinus nitidus Kirk, 1947	M Miss	X	—	X	—	—	—	—	—	X
Cadocrinidae										
Cadocrinus variabilis (Wanner, 1916)	M Perm	X	—	—	—	—	—	—	—	—
Catacrinidae										
Arrectocrinus abruptus Moore and Plummer, 1940	L Penn	X	—	—	—	—	X	—	—	X
Cathetocrinus stullensis (Strimple, 1947)	L Penn	X	—	—	—	—	—	—	—	—
Delocrinus subhemisphericus Miller and Gurley, 1890	L Penn	X	—	—	—	—	X	—	—	X
Endelocrinus fayettensis (Worthen, 1873)	L Penn	X	—	—	—	—	X	—	—	X
Graffhamicrinus undulatus (Strimple, 1961)	M Penn	X	—	—	—	—	X	—	—	X
Lobalocrinus wolforum (Moore and Plummer, 1940	L Penn	X	—	—	—	—	X	—	—	X
Neoprotencrinus subplanus Moore and Plummer, 1940	M Penn	X	—	—	—	—	X	—	—	X
Palmerocrinus comptus Knapp, 1969	M Penn	X	—	—	—	—	X	—	—	X
Paraplasocrinus transitorius (Wanner, 1916)	M Perm	X	—	—	—	—	—	—	—	—
Pyndaxocrinus separatus (Strimple, 1949)	L Penn	X	—	—	—	—	—	—	—	—
Subarrectocrinus perexcavatus Moore and Plummer, 1940	L Penn	X	—	—	—	—	X	—	—	X

Advanced Dendrocrinida	Age	Mus	Rtlr	SCu	MCu	StCu	FCBi	RCBi	WBi	Pin
Cercidocrinidae										
Cercidocrinus bursaeformis (White, 1862)	M Miss	X	—	X	—	—	—	—	—	X
Ascetocrinus rusticellus (White, 1863)	M Miss	X	—	—	X	—	—	—	—	X
Coeliocrinus dilatatus (Hall, 1861)	M Miss	X	—	—	X	—	—	—	—	X
Clathrocrinidae										
Clathrocrinus clathratus Strimple and Moore, 1971	L Penn	X	—	X	—	—	—	—	—	X
Cromyocrinidae										
Cromyocrinus simplex Trautschold, 1867	M Penn	X	X	—	—	—	—	—	—	X
Aaglaocrinus expansus (Strimple, 1938)	L Penn	X	—	—	—	—	—	X	—	X
Aglaocrinus magnus (Strimple, 1949)	M Penn	X	—	—	—	—	—	X	—	X
Dicromyocrinus ornatus (Trautschold, 1879)	M Penn	X	—	—	X	—	—	—	—	X
Diphuicrinus croneisi Moore and Plummer, 1938	E Penn	X	—	X	—	—	—	—	—	X
Ethelocrinus magister (Miller and Gurley, 1890)	L Penn	X	—	—	—	—	X	—	—	X
Goleocrinus masonensis Strimple and Watkins, 1969	M Penn	X	—	—	X	—	—	—	—	X
Hemiindocrinus fredericksi Yakovlev, 1926	E Perm	X	—	—	—	—	—	—	—	—
Mantikosocrinus castus Strimple, 1951	L Miss	X	X	—	—	—	—	—	—	X
Mathericrinus grandis (Mather, 1915)	E Penn	X	—	—	—	—	—	—	—	X
Metacromyocrinus holdenvillensis Strimple, 1961	M Penn	X	—	—	—	—	—	X	—	X
Minilyacrinus williamburyensis Webster and Jell, 1992	E Perm	X	—	—	—	—	—	—	X	X
Moapacrinus rotundatus Lane and Webster, 1966	E Perm	X	X	—	—	—	—	—	—	X
Mooreocrinus geminatus (Trautschold, 1867)	M Penn	X	X	—	—	—	—	—	—	X
Paracromyocrinus vetulus (Lane, 1964)	E Penn	X	—	—	—	—	—	X	—	X
Parethelocrinus ellipticus Strimple, 1961	M Penn	X	—	—	—	—	X	—	—	X
Parulocrinus blairi (S. A. Miller and Gurley, 1893)	L Penn	X	—	—	—	—	—	X	—	X

Advanced Dendrocrinida	Age	Mus	Rtlr	SCu	MCu	StCu	FCBi	RCBi	WBi	Pin
Probletocrinus curtus Strimple and Moore, 1971	L Penn	X	—	—	—	—	—	X	—	X
Synarmocrinus brachiatus Lane, 1964	M Penn	X	—	X	—	—	—	—	—	X
Tarachiocrinus multiramus (Strimple, 1949)	M Penn	X	—	—	—	—	—	X	—	X
Tyrieocrinus laxus Wright, 1945	M Miss	X	—	—	—	—	—	—	—	—
Ulocrinus buttsi Miller and Gurley, 1890	L Penn	X	—	—	—	—	—	X	—	X
Ureocrinus bockschii (Geinitz, 1845)	M Miss	X	—	—	X	—	—	—	—	X
"Cymbiocrinidae"										
Adacrinus loeblichi (Moore, 1939)	L Penn	X	—	—	X	—	—	—	—	X
Kansacrinus cirriferous (Strimple, 1963)	E Perm	X	—	—	—	—	—	—	—	—
Lecobasicrinus kickapooensis Strimple and Watkins, 1969	M Penn	X	X	—	—	—	—	—	—	X
Paracymbiocrinus ormondi Burdick and Strimple, 1973	L Miss	X	X	—	—	—	—	—	—	X
Proallosocrinus glenisteri Moore and Strimple, 1973	E Penn	X	X	—	—	—	—	—	—	X
Sardinocrinus abruptus (Strimple, 1961)	M Penn	X	—	—	—	—	—	—	—	—
Decadocrinidae										
Decadocrinus scalaris Meek and Worthen, 1861	M Miss	X	—	—	X	—	—	—	—	X
Acylocrinus tumidus Kirk, 1947	M Miss	X	—	—	X	—	—	—	—	X
Aulocrinus agassizi Wachsmuth and Springer, 1897	M Miss	X	—	—	X	—	—	—	—	X
Eidosocrinus condaminensis Webster and Jell, 1999	E Perm	X	—	—	X	—	—	—	—	X
Glaukosocrinus parviusculus Moore and Plummer, 1940	M Penn	X	—	—	—	X	—	—	—	X
Grabauicrinus xinjiangensis (Lane, et al., 1997)	L Dev	X	—	—	X	—	—	—	—	X
Lanecrinus depressus (Meek and Worthen, 1870)	M Miss	X	—	—	X	—	—	—	—	X
Ramulocrinus nigelensis Laudon, Parks, and Spreng, 1952	E Miss	X	—	—	X	—	—	—	—	X

Advanced Dendrocrinida	Age	Mus	Rtlr	SCu	MCu	StCu	FCBi	RCBi	WBi	Pin
Trautscholdcrinus miloradowitschi Yakovlev, 1939	M Penn	X	—	—	X	—	—	—	—	X
Zostocrinus ornatus Kirk, 1948	M Dev	?	—	—	—	—	—	—	—	—
Erisocrinidae										
Erisocrinus typus Meek and Worthen, 1865	M Penn	X	—	—	—	—	X	—	—	X
Akiyoshicrinus isensis Hashimoto, 1995	M Penn	X	—	—	—	—	X	—	—	X
Eperisocrinus missouriensis Miller and Gurley, 1890	L Penn	X	—	—	—	—	—	—	—	—
Exaetocrinus argentinei (Strimple, 1949)	L Penn	X	—	—	—	—	—	—	—	—
Sinocrinus microgranulosus Tien, 1924	L Penn	X	—	—	—	—	X	—	—	?
Eupachycrinidae										
Eupachycrinus quatuordecembrachilis (Lyon, 1857)	L Miss	X	—	—	—	—	X	—	—	X
Bronaughocrinus figuratus Strimple, 1951	L Miss	X	—	—	—	—	X	—	—	X
Intermediacrinus asperatus (Worthen, 1882)	L Miss	X	—	—	—	—	—	—	X	X
Exocrinidae										
Exocrinus multirami Strimple, 1949	L Penn	X	X	—	—	—	—	—	—	X
Oxynocrinus spicata Strimple and Watkins, 1969	E Penn	X	X	—	—	—	—	—	—	X
Petalambicrinus craddocki Strimple, 1976	L Penn	X	—	—	—	—	—	—	—	—
Galateacrinidae										
Galateacrinus stevensi Moore, 1940 and Ivanov, 1956)	M Penn	X	—	X	—	—	—	—	—	X
Graphiocrinidae										
Graphiocrinus encrinoides de Koninck and Le Hon, 1854	E Miss	X	X	—	—	—	—	—	—	X
Contocrinus stantonensis (Strimple, 1939)	L Penn	X	—	X	—	—	—	—	—	X
Holcocrinus longicirrifer (Wachsmuth and Springer, 1890)	E Miss	X	—	—	—	X	—	—	—	X
Permiocrinus immaturus Wanner, 1949	M Perm	X	X	—	—	—	—	—	—	X

Advanced Dendrocrinida	Age	Mus	Rtlr	SCu	MCu	StCu	FCBi	RCBi	WBi	Pin
Tapinocrinus macurdai Webster, 1987	E Perm	X	—	X	—	—	—	—	—	X
Hydreionocrinidae										
Hydreionocrinus woodianus de Koninck, 1858	M Miss	X	—	—	—	—	—	—	X	X
Derbiocrinus diverus Wright, 1951	M Miss	X	—	—	—	—	—	—	—	—
Telikosocrinus caespes Strimple, 1951	L Miss	X		—	—	—	—	—	X	X
Indocrinidae										
Indocrinus elegans Wanner, 1916	M Perm	X	—	—	—	—	—	—	—	—
Contignatindocrinus contignatus (Wanner, 1931)	M Perm	X	—	—	—	—	—	—	—	—
Crinophagus permiensis Arendt, 1985	E Perm	X	—	—	—	—	—	—	—	—
Eoindocrinus praerimosus Arendt, 1981	E Perm	X	—	—	—	—	—	—	—	—
Proindocrinus piszowi (Yakovlev, 1926)	E Perm	—	X	—	—	—	—	—	—	—
Pumilindocrinus pumilius (Wanner, 1931)	M Perm	X	—	—	—	—	—	—	—	—
Rimosindocrinus rimosus Wanner, 1916	M Perm	X	—	—	—	—	—	—	—	—
Yakovlevicrinus subglobosus Arendt, 1981	E Perm	X	—	—	—	—	—	—	—	—
Laudonocrinidae										
Laudonocrinus subsinuatus (S. A. Miller and Gurley, 1894).	L Penn	X	—	—	—	—	—	—	—	—
Anchicrinus toddi Strimple and Watkins, 1969	M Penn	X	—	X	—	—	—	—	—	X
Athlocrinus placidus Moore and Plummer, 1940	L Penn	X	—	—	—	—	—	—	—	—
Bathronocrinus turioformis Strimple, 1962	M Penn	X	—	—	—	—	—	—	—	—
Paianocrinus durus Strimple, 1951	L Miss	X	—	X	—	—	—	—	—	X
Tetrabrachiocrinus fabianii Yakovlev, 1934	M Perm	—	—	—	—	—	—	—	—	—
Lophocrinidae										
Lophocrinus speciosus von Meyer, 1858	M Miss	?	—	X	—	—	—	—	—	X
Mollocrinidae										
Mollocrinus poculum Wanner, 1916	M Perm	—	—	—	—	—	—	—		

Advanced Dendrocrinida	Age	Mus	Rtlr	SCu	MCu	StCu	FCBi	RCBi	WBi	Pin
Hemimollocrinus uralensis Yakovlev, 1930	E Perm	X	—	—	—	—	—	—	—	—
Strongylocrinus molengraffi Wanner, 1916	M Perm	—	—	—	—	—	—	—	—	—
Pachylocrinidae										
Pachylocrinus aequalis (Hall, 1861)	M Miss	X	—	—	X	—	—	—	—	X
Depaocrinus ottowi Wanner, 1937	M Perm	X	—	—	—	—	—	—	—	—
Malaiocrinus sundaicus (Wanner, 1916)	M Perm	X	—	—	—	—	—	—	—	—
Nacocrinus elliotti Webster and Olson, 1998	M Penn	X	—	—	—	X	—	—	—	X
Plummericrinus mcguirei (Moore, 1939)	L Penn	X	—	—	—	X	—	—	—	X
Paradelocrinidae										
Paradelocrinus aequabilis Moore and Plummer, 1938	E Penn	X	—	—	—	—	—	—	—	—
Atokacrinus obscurus Knapp, 1969	M Penn	X	—	—	—	—	—	—	—	—
Lopadiocrinus granulatus Wanner, 1916	M Perm	X	—	—	—	—	—	—	—	—
Neocatacrinus protenus Moore and Plummer, 1940	L Penn	X	—	—	—	—	—	—	—	—
Sublobalocrinus iolaensis (Strimple, 1949)	L Penn	X	—	—	—	—	—	—	—	—
Parindocrinidae										
Parindocrinus oyensi Wanner, 1937	M Perm	X	—	—	—	—	—	—	—	—
Metaindocrinus cooperi Strimple, 1966	M Perm	X	—	—	—	—	—	—	—	—
Pelecocrinidae										
Pelecocrinus insignis Kirk, 1941	M Miss	X	—	—	—	X	—	—	—	X
Exoriocrinus lasallensis (Worthen 1875)	L Penn	X	—	—	—	X	—	—	—	X
Phanocrinidae										
Phanocrinus formosus (Worthen, 1873)	L Miss	X	—	X	—	—	—	—	—	X
Cryphiocrinus girtyi Kirk, 1929	L Miss	X	X	—	—	—	—	—	—	X
Hosieocrinus caledonicus (Wright, 1936)	M Miss	X	—	—	—	—	—	—	—	—
Idosocrinus bispinosus Wright, 1954	M Miss	X	X	—	—	—	—	—	—	X

Advanced Dendrocrinida	Age	Mus	Rtlr	SCu	MCu	StCu	FCBi	RCBi	WBi	Pin
Pentaramicrinus gracilis (Wetherby, 1880)	L Miss	X	—	—	—	—	—	—	—	X
Pirasocrinidae										
Pirasocrinus scotti Moore and Plummer, 1940	M Penn	X	X	—	—	—	—	—	—	X
Aatocrinus robustus (Beede, 1900)	L Penn	X	—	—	—	—	—	—	—	—
Affinocrinus concavus Knapp, 1969	M Penn	X	—	—	—	—	—	—	—	—
Dasciocrinus florialis (Yandell and Shumard, 1847)	L Miss	X	—	—	—	X	—	—	—	X
Eirmocrinus grossus Strimple and Watkins, 1969	M Penn	X	—	—	—	—	—	—	X	X
Exterocrinus pumilis (Moore and Plummer, 1938)	E Penn	X	—	—	—	—	—	—	—	—
Lasanocrinus daileyi (Strimple, 1940)	E Penn	X	—	—	—	—	—	—	—	—
Metaffinocrinus perundatus (Moore and Plummer, 1940)	M Penn	X	X	—	—	—	—	—	—	X
Metaperimestocrinus spiniferus Strimple, 1961	M Penn	X	—	X	—	—	—	—	—	X
Metutharocrinus cockei Moore and Strimple, 1973	E Penn	X	—	—	—	—	—	—		
Perimestocrinus nodulifer (S. A. Miller and Gurley, 1894).	L Penn	X	X	—	—	—	—	—	—	X
Platyfundocrinus typus Knapp, 1969	M Penn	X	—	—	—	—	—	—	—	—
Plaxocrinus crassidiscus (S. A. Miller and Gurley, l894)	L Penn	X	X	—	—	—	—	—	—	X
Polygonocrinus multiextensus Strimple, 1961	M Penn	X	—	—	—	—	—	X	—	X
Psilocrinus omphaloides (Moore and Plummer, 1940)	M Penn	X	—	—	—	—	—	—	—	—
Retusocrinus lobatus (Moore and Plummer, 1940)	M Penn	X	—	—	—	—	—	—	—	—
Schedexocrinus gibberellus Strimple, 1961	M Penn	X	—	—	—	—	—	X	—	X
Sciadiocrinus acanthophorus (Meek and Worthen, 1870)	L Penn	X	X	—	—	—	—	—	—	X
Separocrinus praevalens (Moore, 1939)	L Penn	X	—	—	—	—	—	—	—	—
Simocrinus modestus (Moore, 1939)	L Penn	X	—	—	—	—	—	—	—	—
Stenopecrinus planus (Strimple, 1952)	L Penn	X	X	—	—	—	—	—	—	X

Advanced Dendrocrinida	Age	Mus	Rtlr	SCu	MCu	StCu	FCBi	RCBi	WBi	Pin
Triceracrinus moorei Bramlette, 1943	E Perm	X	X	—	—	—	—	—	—	X
Utharocrinus pentanodus (Mather, 1915)	E Penn	X	—	—	—	—	—	—	—	—
Vertigocrinus subtilis (Moore, 1939)	L Penn	X	—	—	—	—	—	—	—	—
Zeusocrinus foveatus (Strimple, 1951)	L Miss	X	—	X	—	—	—	—	—	X
Protencrinidae										
Protencrinus moscoviensis Jaekel, 1918	M Penn	X	—	—	—	—	—	X	—	X
Rhenocrinidae										
Rhenocrinus ramosissimus Jaekel in Schmidt, 1906	E Dev	?	X	—	—	—	—	—	—	X
Scotiacrinidae										
Scotiacrinus tyriensis (Wright, 1937)	M Miss	X	—	X	—	—	—	—	—	X
Scytalocrinidae										
Scytalocrinus robustus (Hall, 1861)	M Miss	X	—	X	—	—	—	—	—	X
Anemetocrinus biserialis Wright, 1938	M Miss	X	—	—	—	—	—	X	—	X
Atrapocrinus mutatus Strimple, 1951	M Penn	X	—	X	—	—	—	—	X	
Bollandocrinus conicus (Phillips, 1836)	E Miss	X	—	X	—	—	—	—	—	X
Histocrinus grandis Wachsmuth and Springer, 1880	M Miss	X	—	—	X	—	—	—	—	X
Hydriocrinus pusillus Trautschold, 1867	M Penn	X	—	—	—	X	—	—	—	X
Hypselocrinus hoveyi (Worthen, 1875)	M Miss	X	—	—	X	—	—	—	—	X
Julieticrinus romeo Waters et al., 2003	L Dev	X	X	—	—	—	—	—	—	X
Lorocrinus zanguensis Webster et al., 2003	E Miss	X	—	X	—	—	—	—	—	X
Morrowcrinus fosteri Moore and Plummer, 1938	E Penn	X	—	X	—	—	—	—	—	X
Ophiurocrinus originarius (Trautschold, 1867)	M Penn	X	—	—	X	—	—	—	—	X
Parascytalocrinus validus (Wachsmuth and Springer, 1897)	M Miss	X	—	—	X	—	—	—	—	X
Pegocrinus bijugus (Trautschold, 1867)	M Penn	X	X	—	—	—	—	—	—	—

Advanced Dendrocrinida	Age	Mus	Rtlr	SCu	MCu	StCu	FCBi	RCBi	WBi	Pin
Phacelocrinus longidactylus (McChesney, 1860)	L Miss	X	—	—	X	—	—	—	—	X
Prininocrinus robustus Goldring, 1938	L Dev	X	X	—	—	—	—	—	—	X
Pulaskicrinus campanulus (Horowitz, 1965)	L Miss	X	—	X	—	—	—	—	—	X
Roemerocrinus gracilis Wanner, 1916	M Perm	X	—							—
Ulrichicrinus oklahoma Springer, 1926	E Penn	X	—	—	—	—	—	—	X	X
Wetherbyocrinus pulaskiensis (S. Miller and Gurley, 1896)	L Miss	X	—	—	—	—	—	—	—	—
Woodocrinus macrodactylus de Koninck, 1854	E Penn	X	X	—	—	—	—	—	—	X
Sellardsicrinidae										
Sellardsicrinus marrsae Moore and Plummer, 1940	M Penn	X	—	—	—	—	—	—	X	X
Sostronocrinidae										
Sostronocrinus superbus Strimple and McGinnis, 1969	E Miss	X	X	—	—	—	—	—	—	X
Amadeusicrinus subpentagonalis Lane, et al. 1997	L Dev	X	—	—	X	—	—	—	—	X
Haeretocrinus missouriensis Moore and Plummer, 1940	L Penn	X	—	X	—	—	—	—	—	X
Tundracrinus polaris Yakovlev, 1928	E Perm	X	X	—	—	—	—	—	—	X
Spaniocrinidae										
Spaniocrinus validus Wanner, 1924	M Perm	X	X	—	—	—	—	—	—	X
Missouricrinus admonitus Miller, 1891	E Miss	?	X	—	—	—	—	—	—	?
Parspaniocrinus beinerti Strimple, 1971	E Perm	X	X	—	—	—	—	—	—	?
Stuartwellercrinus turbinatus (Weller, 1909	E Perm	X	X	—	—	—	—	—	—	X
Stachyocrinidae										
Stachyocrinus zea Wanner, 1916	M Perm	X	X	—	—	—	—	—	—	X
Coenocrinus elegans Valette, 1934	M Perm	X	—	—	—	—	—	—		
Parastachyocrinus malaianus (Wanner, 1924)	M Perm	X	X	—	—	—	—	—	—	X

Advanced Dendrocrinida	Age	Mus	Rtlr	SCu	MCu	StCu	FCBi	RCBi	WBi	Pin
Staphylocrinidae										
Staphylocrinus bulgeri Burdick and Strimple, 1969	L Miss	X	—	X	—	—	—	—	—	X
Agnostocrinus typus Webster and Lane, 1967	E Perm	X	—	—	—	X	—	—	—	X
Abrotocrinus cymosus Miller and Gurley, 1890	M Miss	X	—	—	X	—	—	—	—	X
Borucrinus eirensis Wright, 1951	E Miss	X	—	—	X	—	—	—	—	X
Dinotocrinus compactus Kirk, 1941	L Miss	X	—	X	—	—	—	—	—	X
Exochocrinus tumulosus (S. A. Miller, 1892)	L Miss	X	—	—	—	—	—	—	—	—
Harmostocrinus porosus Strimple, 1975	L Miss	X	—	—	X	—	—	—	—	X
Hylodecrinus sculptus Kirk, 1941	M Miss	X	—	—	X	—	—	—	—	X
Microcaracrinus delicatus Strimple and Watkins, 1969	M Penn	X	—	—	X	—	—	—	—	X
Stellarocrinidae										
Stellarocrinus stillativus (White, 1880)	L Penn	X	—	—	—	—	X	—	—	X
Anechocrinus nalbiaensis Webster, 1990	E Perm	X	—	—	X	—	—	—	—	X
Brabeocrinus christinae Strimple and Moore, 1971	L Penn	?	—	—	—	—	—	—	X	X
Brychiocrinus texanus Moore and Plummer, 1940	M Penn	?	—	—	—	—	—	X	—	X
Celonocrinus expansus Lane and Webster, 1966	E Perm	X	—	—	—	—	X	—	—	X
Forthocrinus lepidus Wright, 1942	M Miss	X	—	—	—	—	—	—	—	—
Heliosocrinus aftonensis Strimple, 1951	L Miss	X	—	—	X	—	—	—	—	X
Pedinocrinus clavatus (Wright, 1937)	M Miss	X	—	—	X	—	—	—	—	X
Rhopocrinus spinosus Kirk, 1942	L Miss	X	—	—	X	—	—	—	—	X
Sundacrinidae										
Sundacrinus granulatus Wanner, 1916	M Perm	X	—	—	—	—	—	—	—	—
Paratimorocidaris problematicus Arendt, 1981	E Perm									
Timorocidaris sphaeracantha Wanner, 1920	M Perm	X	—	—	—	—	—	—	—	—

Advanced Dendrocrinida	Age	Mus	Rtlr	SCu	MCu	StCu	FCBi	RCBi	WBi	Pin
Trimerocrinus pumilus Wanner, 1916	M Perm	X	—	—	—	—	—	—	—	—
Texacrinidae										
Texacrinus gracilis Moore and Plummer, 1940	M Penn	X	—	X	—	—	—	—	—	X
Marathonocrinus bakeri Moore and Plummer, 1940	M Penn	X	X	—	—	—	—	—	—	X
Timorechinidae										
Timorechinus mirabilis Wanner, 1911	M Perm	X	X	—	—	—	—	—	—	X
Benthocrinus cryptobasalis Wanner, 1937	M Perm	X	—	—	—	—	—	—	—	—
Notiocrinus timoricus Wanner, 1924	M Perm	X	X	—	—	—	—	—	—	X
Parabursacrinus procerus (Wanner, 1916)	M Perm	X	X	—	—	—	—	—	—	X
Prolobocrinus permicus Wanner, 1937	M Perm	X	X	—	—	—	—	—	—	?
Zeacrinitidae										
Zeacrinites magnoliaeformis Troost, 1858	L Miss	—	X	—	—	—	—	—	—	X
Alcimocrinus girtyi (Springer, 1926)	E Penn	X	—	—	—	X	—	—	—	X
Bicidiocrinus wetherbyi (Wachsmuth and Springer, 1886)	L Miss	X	—	X	—	—	—	—	X	X
Eratocrinus elegans (Hall, 1858	M Miss	X	—	X	—	—	—	—	—	X
Lekocrinus divaricatus (Hall, 1860)	M Miss	X	—	X	—	—	—	—	—	X
Linocrinus wachsmuthi Kirk, 1938	L Miss	X	—	X	—	—	—	—	—	X
Neozeacrinus peramplus Wanner, 1937	M Perm	X	X	—	—	—	—	—	—	X
Parazeacrinites konincki (Bather, 1912)	M Miss	X	—	—	X	—	—	—	—	X
Sarocrinus nitidus Kirk, 1942	M Miss	X	X	X	—	—	—	—	—	X
Tholocrinus spinosus (Wood, 1909)	L Miss	X	—	X	—	—	—	—	—	X
Worthenocrinus paterus Kammer and Ausich, 1994	M Miss	X	X	—	—	—	—	—	—	X
Family uncertain										
Adiakritocrinus oviatti Webster, 1997	E Miss	X	X	—	—	—	—	—	—	X
Aenigmocrinus anomalus (Wetherby, 1889)	L Miss	X	—	X	—	—	—	—	—	X

Advanced Dendrocrinida	Age	Mus	Rtlr	SCu	MCu	StCu	FCBi	RCBi	WBi	Pin
Armenocrinus watersi Strimple and Horowitz, 1971	L Miss	X	X	—	—	—	—	—	—	X
Arroyocrinus popenoi Lane and Webster, 1966	E Perm	X	X	—	—	—	—	—	—	X
Aulodesocrinus parvus Wright, 1942	M Miss	X	—	—	—	—	—	—	—	—
Carlopsocrinus bullatus Wright, 1933	M Miss	?	X	—	—	—	—	—	—	X
Struveicrinites holleri Hauser, 1998	M Dev	?	X	—	—	—	—	—	—	X
Tarassocrinus synchlydus Webster and Hafley, 1999	L Dev	X	—	X	—	—	—	—	—	X

Note.—All references prior to 1999 are fully cited in Webster (2003); references after 1999 are fully cited herein.

Abbreviations.—Arl, armlets; Bf, bifascial; CvPl, cover plates; Dev, Devonian; E, Early; L, Late; M, Middle; MCu, moderately cuneate; Miss, Mississippian; Ord, Ordovician; Penn, Pennsylvanian; Perm, Permian; Pin, Pinnules; Rm, Ramules; Rtlr, rectilinear; SCu, slightly cuneate; Sil, Silurian; Uf, unifascial.

CONTRIBUTORS

William I. Ausich, School of Earth Sciences, The Ohio State University, Columbus, Ohio 43210

Tomasz K. Baumiller, Museum of Paleontology, 1109 Geddes Road, The University of Michigan, Ann Arbor, Michigan 48109-1079

Susan E. Bolyard, Geosciences Department, University of Arkansas, Fayetteville, Arkansas 72701

Carlton E. Brett, Department of Geology, University of Cincinnati, Cincinnati, Ohio 45221-0013

Yu-Ping Chin, School of Earth Sciences, The Ohio State University, Columbus, Ohio 43210

George H. Davis, Missouri Department of Transportation, Jefferson City Missouri 65211

Bradley L. Deline, Department of Geology, University of Cincinnati, Cincinnati, Ohio 45221-0013

Stephen K. Donovan, Department of Geology, Nationaal Natuurhistorisch Museum, Postbus 9517, NL-2300 RA, Leiden, Netherlands

Stephen Q. Dornbos, Department of Geosciences, University of Wisconsin–Milwaukee, Milwawkee, Wisconsin 53201-0413

Raymond L. Ethington, Geological Sciences Department, University of Missouri–Columbia, Columbia, Missouri 65211

Kevin R. Evans, Geography, Geology, and Planning, Missouri State University, Springfield, Missouri 65897

Fiona E. Fearnhead, School of Earth Sciences, Birkbeck College, University of London, Malet Street, Bloomsbury, London, WC1E 7HX, England

Forest J. Gahn, Department of Geology, Brigham Young University–Idaho, Rexburg, Idaho 83460-0510

Thomas E. Guensburg, Science Division, Rock Valley College, 3301 North Mulford Road, Rockford, Illinois 61114

Hans Hess, Naturhistorisches Museum, Augustinergasse 2, CH-4001, Basal, Switzerland

Hongfei Hou, Institute of Geological Sciences, Chinese Academy of Geological Sciences, Beijing, People's Republic of China

Thomas W. Kammer, Department of Geology and Geography, West Virginia University, Morgantown, West Virginia 26506-6300

N. Gary Lane (deceased), Department of Geological Sciences, Indiana University, Bloomington, Indiana 47405

David N. Lewis, Department of Palaeontology, The Natural History Museum, Cromwell Road, London, SW7 5BD, England

Zhouting Liao, Nanjing Institute of Geology and Paleontology, Nanjing, 210008, People's Republic of China

Lujun Liu, Nanjing Institute of Geology and Palaeontology, Nanjing, 210008, People's Republic of China

Christopher G. Maples, Desert Research Institute, 2215 Raggio Parkway, Reno, Nevada 89512

Sara A. Marcus, 3485 San Juan Drive, Reno, Nevada 89512

Patrick I. McLaughlin, Department of Geology, University of Cincinnati, Cincinnati, Ohio 45221-0013

Charles G. Messing, Nova Southeastern University Oceanographic Center, 8000 North Ocean Drive, Dania, Florida 33004

David L. Meyer, Department of Geology, University of Cincinnati, Cincinnati, Ohio 45221-0013

James F. Miller, Geography, Geology, and Planning, Missouri State University, Springfield, Missouri 65897

Clare V. Milsom, School of Biological and Earth Sciences, Liverpool John M. Moores University, Liverpool L3 3AF, England

James H. Nebelsick, University of Tübingen, Institute of Geology and Palaeontology, Sigwartstrasse 10, D-72076 Tübingen, Germany

Christina E. O'Malley, School of Earth Sciences, The Ohio State University, Columbus, Ohio 43210

Charles W. Rovey II, Geography, Geology, and Planning, Missouri State University, Springfield, Missouri 65897

Charles A. Sandberg, U.S. Geological Survey, Box 25046 MS 939, Federal Center, Denver, Colorado 80225

Hernán Santos, Department of Geology, University of Puerto Rico, Mayagüez Campus, P.O. Box 9017, Mayagüez, Puerto Rico 00681

Chris L. Schneider, Department of Geology, University of California, Davis, California 95616

George D. Sevastopulo, Department of Geology, Trinity College, Dublin 2, Ireland

James Sprinkle, Department of Geological Sciences, Jackson School of Geosciences, University of Texas, 1 University Station C1100, Austin, Texas 78712-0254

Colin D. Sumrall, Earth and Planetary Science, University of Tennessee, Knoxville, Tennessee 37996-1410

Thomas L. Thompson, Missouri Geological Survey, Rolla, Missouri 65402

Jorge Vélez-Juarbe, Department of Geology, University of Puerto Rico, Mayagüez Campus, P.O. Box 9017, Mayagüez, Puerto Rico 00681

Jinxing Wang, Institute of Geological Sciences, Chinese Academy of Geological Sciences, Beijing, People's Republic of China

Johnny A. Waters, Geology Department, Appalachian State University, Boone, North Carolina 28608

Andrew J. Webber, Department of Geology, Miami University Hamilton, Hamilton, Ohio 45011

Gary D. Webster, School of Earth and Environmental Sciences, Washington State University, Pullman, Washington 99164

Rosanne E. Widdison, 15 Manor Close, Notton, Wakefield, West Yorkshire, WF4 2NH, England

INDEX

Page numbers in bold indicate illustrations.

Milton Keynes UK
Ingram Content Group UK Ltd.
UKHW050145241223
434725UK00011B/56